Sustainable Engineering Forum 2020

Held at the 2020 AIChE Annual Meeting

Online
16 – 20 November 2020

ISBN: 978-1-7138-2316-2

Printed from e-media with permission by:

Curran Associates, Inc.
57 Morehouse Lane
Red Hook, NY 12571

CURRAN ASSOCIATES INC.
proceedings
.com

Some format issues inherent in the e-media version may also appear in this print version.

AIChE
120 Wall Street, FL 23
New York, NY 10005-4020

Phone: (800) 242-4363
Fax: (203) 775-5177

www.aiche.org

Additional copies of this publication are available from:

Curran Associates, Inc.
57 Morehouse Lane
Red Hook, NY 12571 USA
Phone: 845-758-0400
Fax: 845-758-2633
Email: curran@proceedings.com
Web: www.proceedings.com

TABLE OF CONTENTS

(19A) CATALYTIC REACTION PATHWAYS AND PROCESS SYNTHESIS FOR BIOMASS CONVERSION TO HIGH-VALUE CHEMICALS FOR ORGANIC DYE AND POLYMER APPLICATIONS 1
Hochan Chang, Ishan Bajaj, George W. Huber, Christos T. Maravelias, James A. Dumesic

(19B) SUPERACID POLYMER CATALYSTS FOR HYDROXYMETHYLFURFURAL PRODUCTION 2
Ibeh S. Omodolor, Sarah A. Walz, Subhash Kalidindi, Vimantha Bamunuarachchi, Ravikumar Gogar, Sridhar Viamajala, Dr. Maria R. Coleman, Ana Alba-Rubio

(19C) RATES OF LEVOGLUCOSANOL HYDROGENOLYSIS OVER BRØNSTED AND LEWIS ACID SITES ON PLATINUM SILICA-ALUMINA CATALYSTS SYNTHESIZED BY ATOMIC LAYER DEPOSITION 3
Siddarth Krishna, Lifeng Zhang, Ive Hermans, George W. Huber, Thomas F. Kuech, James A. Dumesic

(19D) TUNING PERFORMANCE OF KETONE PARTIAL OXIDATION OVER SUPPORTED VOX AND ITS UTILIZATION: SELECTIVE MALEIC ANHYDRIDE PRODUCTION FROM LEVULINIC ACID OXIDATIVE SCISSION 4
Ran Zhu, Anargyros Chatzidmitriou, Bowei Liu, Siwen Wang, Jesse Bond

(19E) SELECTIVE REDUCTION OF CARBOXYLIC ACIDS TO ALDEHYDES OVER PROMOTED MOO3 CATALYST 5
Laura A. Gomez, Mallikharjuna R. Komarneni, Jennifer A. Gutierrez, Steven Crossley

(19F) NOBLE METAL CATALYZED KETONE HYDROGENATION WITH MICROKINETIC MODELING 6
Xin Gao, Andreas Heyden, Omar Abdelrahman, Jesse Bond

(19H) BIFUNCTIONAL RU@N-DOPED MESOPOROUS CARBON CATALYSTS FOR CATALYTIC TRANSFER HYDROGENOLYSIS OF BIO-DERIVED POLYOLS 7
Wenxiang Zhang, Mingyu Zhao, Quanxing Zhang, Yurou Song, Jinyao Wang, Chaohe Yang, Xin Jin

(50A) A TUNEABLE PROCESS FOR THE CONVERSION OF LIGNIN TO JET FUEL OVER MOLYBDENUM CARBIDE CATALYSTS 8
Michael Stone, William P. Mounfield III, Ana R. C. Morais, Eric Anderson, Gregg T. Beckham, Yuriy Román-Leshkov

(50C) INVESTIGATING THE ROLE OF SOLID ACID SITES IN SURFACE REACTIONS OF A MODEL SUGAR 9
Sean Najmi, Andrew Medford, Carsten Sievers

(50D) MECHANISM OF POLYOL HYDRODEOXYGENATION ON MOLYBDENUM CARBIDE SUPPORTED COPPER CATALYST: A COMBINED DFT AND MICROKINETIC MODELING STUDY 10
Kyung-Eun You, Salai C. Ammal, Zhexi Lin, Weiming Wan, Jingguang G. Chen, Andreas Heyden

(50E) STRUCTURE-SENSITIVE CATALYTIC GLYCEROL OXIDATION ON LATE TRANSITION METALS 11
Geet Gupta, Luke Roling

(50G) SELECTIVE HYDRODEOXYGENATION OF 1,4-ANHYDROERYTHRITOL TO TETRAHYDROFURAN OVER MOOX-PD/TIO2 CATALYSTS 12
Sandra Albarracín-Suazo, Blake Macqueen, Jochen Lauterbach, Yomaira J. Pagan-Torres

(50H) LOW-TEMPERATURE HYDRODEOXYGENATION BY POLYOXOMETALATE-SUPPORTED PD1 SINGLE-ATOM CATALYSTS 13
Max J. Hülsey, Geng Sun, Philippe Sautet, Ning Yan

(112A) EMISSIONS CONTROL TECHNOLOGIES FOR RTI'S NON-AQUEOUS SOLVENT FOR CARBON CAPTURE 14
Paul Mobley, Jak Tanthana, Lucas Cody, Jacob Lee, David Barbee, Vijay Gupta, Roger Pope, Ryan Chartier, Jonathan Thornburg, Marty Lail

(112C) OPTIMIZATION AND TRANSPORT OF CO2 FROM REFINERIES IN SOUTHEAST TEXAS FOR OFFSHORE SUBSEA STORAGE 16
Adhish Madugula, Tracy Benson

(112D) DYNAMIC MODELING, OPTIMIZATION, AND CONTROL STUDIES OF A MOVING BED PROCESS FOR CO2 CAPTURE USING A MICRO-ENCAPSULATED SOLVENT 17
Goutham Kotamreddy, Ryan Hughes, Debangsu Bhattacharyya, Benjamin P. Omell, Michael S. Matuszewski

(112E) ENHANCED CO2 UPTAKE AND SELECTIVITY UPON INCORPORATION OF TASK-SPECIFIC IONIC LIQUIDS INTO HIGHLY POROUS MOF-177 AT POST-COMBUSTION CAPTURE CONDITION 19
Firuz Philip, Amr Henni

(148B) AN IMPROVED METHOD FOR PREDICTING AUTOIGNITION TEMPERATURES BASED ON FIRST PRINCIPLES 20
Mark E. Redd, Joseph M. Black, Glenn Seaton, Thomas A. Knotts Iv, Neil Giles, W. Vincent Wilding

(148C) WATER WASHING OF STARTUP CRUDE OIL 21
Paul M. Mathias

(148D) UNIVERSAL CORRELATION FOR THE BINARY INTERACTION PARAMETER OF HYDROGEN WITH PETROLEUM FRACTIONS IN REFINERY HYDROPROCESSING APPLICATIONS 22
Nevin Gerek Ince, Soham Kulkarni, David Bluck

(148E) DISTILLATION CURVE AND VAPOR PRESSURE PREDICTIONS OF ALCOHOL-GASOLINE MIXTURES USING AN ADVANCED EQUATION OF STATE 23
Joseph R. Vella, Bennett D. Marshall, Constantinos P. Bokis

(148F) DEVELOPMENT OF MEAN IONIC ACTIVITY COEFFICIENTS FOR MODIFICATION OF THE KRICHEVSKY – ILINSKEYA EQUATION FOR METHYLIMIDAZOLIM – TYPE IONIC LIQUIDS 25
Linh Doan, Tracy Benson

(201A) (INVITED TALK) MECHANISTIC INSIGHTS INTO DEOXYGENATION CHEMISTRY ON BIMETALLIC PHOSPHIDES 26
Varsha Jain, Woodrow Wilson, Neeraj Rai

(201B) ULTRAFAST HOT CARRIER INJECTION IN AU/GAN: THE ROLE OF BAND BENDING AND THE INTERFACE BAND STRUCTURE 27
Fan Zheng, Lin-Wang Wang

(201C) HALIDES INDUCE ANISOTROPIC GROWTH OF CU NANOCRYSTALS 28
Zihao Chen, Myung Jun Kim, Mutya A. Cruz, Kristen Fichthorn, Benjamin Wiley

(201D) FIRST-PRINCIPLES STUDY OF LITHIUM-AIR BATTERIES BASED ON TRI-MOLYBDENUM PHOSPHIDE (MO3P) NANOPARTICLES 29
Zhen Jiang, Arvin Kakekhani, Alireza Kondori, Mohammadreza Esmaeilirad, Mohammad Asadi, Andrew M. Rappe

(201E) FIRST-PRINCIPLES MODELING OF THERMAL DEGRADATION OF CO2-LOADED AQUEOUS AMINE SOLVENTS 30
Bohak Yoon, Gyeong S. Hwang

(201G) ASSESSMENT OF THE TWO-PHASE THERMODYNAMIC METHOD FOR CALCULATING ADSORPTION ISOTHERMS 32
Shubham Malviya, Peng Bai

(454A) SUSTAINABILITY AND RESILIENCE IN THE CHEMICAL PROCESS INDUSTRIES THROUGH PROCESS INTENSIFICATION AND MODULAR MANUFACTURING 33
Paul Yelvington

(454B) TECHNOLOGICAL OPPORTUNITIES IN SUSTAINABLE ENERGY AND WATER SYSTEMS 34
Ravi Prasher

(454C) PHOTOVOLTAIC AGLECTRIC FARMING FOR MEETING LOCAL FOOD, ENERGY,WATER AND CHEMICAL NEEDS 35
Rakesh Agrawal

(463A) A MACHINE LEARNING-BASED APPROACH FOR CHEMICAL PROCESS SUSTAINABILITY ANALYTICS 36
Majid Moradi Aliabadi, Yinlun Huang

(463B) ACCELERATING DISCOVERY OF HIGH-PERFORMANCE ELECTROCATALYSTS FOR AMMONIA OXIDATION VIA MACHINE LEARNING 37
Hemanth S. Pillai, Hongliang Xin

(463D) SAFETY CONSIDERATION IN CARBON DIOXIDE INTEGRATION NETWORKS IN INDUSTRIAL CLUSTERS 38
Nitin Roy, Dhabia Al-Mohannadi

(463F) PARALLEL PLANT: A SMART SYSTEM APPROACH FOR AMMONIA SYNTHESIS PLANT 39
Ming Zhu

(471A) SPATIALLY EXPLICIT OPTIMIZATION OF CELLULOSIC BIOFUEL SUPPLY CHAINS CONSIDERING LANDSCAPE DESIGN AND BIOMASS YIELD UNCERTAINTY 40
Eric O'Neill, Bruno Basso, Christos T. Maravelias

(471B) OPTIMAL DESIGN OF AN INTEGRATED THERMAL ENERGY CONVERSION SYSTEM FOR SUSTAINABLE ENERGY AND DESALINATED WATER PRODUCTION 41
Ilse María Hernández-Romero, Victor M. Zavala, Fabricio Nápoles-Rivera, Luis Fabian Fuentes-Cortes, Antonio Flores-Tlacuahuac

(471C) OPTIMIZATION OF A SPATIALLY-RESOLVED PETROCHEMICALS MANUFACTURING AND SUPPLY CHAIN NETWORK 42
Ioannis Giannikopoulos, Alkiviadis Skouteris, Michael Baldea, Thomas F. Edgar, Mark A. Stadtherr, David Allen

(471D) SPATIALLY-EXPLICIT TECHNO-ECOLOGICAL DESIGN FOR SUSTAINABLE MANUFACTURING APPLIED TO A POWER PLANT 44
Michael Charles, Bhavik Bakshi

(471E) ECONOMIC AND ENERGY IMPACTS OF WIND ENERGY EXPANSION IN THE US USING MULTI-REGIONAL INPUT OUTPUT (MRIO) TABLES. 45
Futu Faturay, Venkata Sai Gargeya Vunnava, Manfred Lenzen, Shweta Singh

(471F) CO-OPTIMIZED POWER AND H2 INFRASTRUCTURE PLANNING FOR MULTI-SECTOR DECARBONIZATION 46
Guannan He, Dharik Mallapragada, Abhishek Bose, Clara F. Heuberger, Emre Gençer

(471G) OPTIMAL SELECTION AND LOCATION OF NUTRIENT RECOVERY SYSTEMS CONSIDERING STANDALONE AND COORDINATED STRATEGIES 47
Edgar Martín-Hernández, Yicheng Hu, Mariano Martín, Victor M. Zavala, Gerardo Ruiz-Mercado

(471H) TECHNO-ECONOMIC ANALYSIS OF ELECTRICITY STORAGE TECHNOLOGIES FOR BEHIND-THE-METER APPLICATIONS 48
Jude O. Ejeh, Diarmid Roberts, Solomon F. Brown

(471I) DYNAMIC OPTIMIZATION OF LI-ION BATTERY SYSTEMS FOR GRID-CONNECTED HIGH CURRENT DENSITY OPERATION 50
Chungho Suh, Chrysanthos E. Gounaris, Debangsu Bhattacharyya

(471J) AMMONIA FOR GAS TURBINES FUELLING - EUROPEAN PROGRESS 52
Agustin Valera-Medina

(471K) OPTIMAL ELECTRIC VEHICLE CHARGING SCHEDULE: IMPACT OF USER TRAVEL BEHAVIOUR AND DISTRIBUTED ENERGY RESOURCE AVAILABILITY 53
Jude O. Ejeh, Donovan Aguilar-Dominguez, Alan Dunbar, Solomon F. Brown

(471L) OPTIMISATION OF FUEL SUPPLY CHAINS WITHIN PLANETARY BOUNDARIES: A CASE STUDY OF HYDROGEN FOR ROAD TRANSPORTATION IN THE UK 54
Michael Ehrenstein, Ángel Galán Martín, Victor Tulus, Gonzalo Guillén-Gosálbez

(506A) AN INTEGRATED MODEL FOR FOOD-ENERGY-WATER NEXUS TO STUDY GLOBAL SUSTAINABILITY 55
Apoorva Nisal, Neeraj Hanumante, Yogendra Shastri, Heriberto Cabezas, Roelof Boman, Urmila Diwekar

(506B) CONSUMPTION-BASED ACCOUNTING OF ENVIRONMENTAL IMPACTS ASSOCIATED WITH BEEF SUPPLY CHAINS IN THE UNITED STATES: IMPLICATIONS FOR FEW NEXUS 59
Anais Ostroski, Tomas Lagos, Oleg A Prokopyev, Vikas Khanna

(506C) A LIFE-CYCLE MODELING FRAMEWORK FOR DYNAMIC ENERGY ANDWATER FOOTPRINTS OF AGRICULTURE SYSTEMS 60
Kai Lan, Yuan Yao

(506D) UTILIZATION OF INSOLUBLE MG MINERALS AS AN ALTERNATIVE MG SOURCE FOR STRUVITE CRYSTALLIZATION FROM WASTEWATER 62
Manoj Silva, Jonas Baltrusaitis

(506E) ALL FOR ONE AND ONE FOR ALL RESEARCH: MORINGA OLEIFERA LAM. IN FOOD, HEALTH AND ENVIRONMENTAL APPLICATIONS 65
Andrea C. Landázuri, Andrés S. Lagos, Renata Paz, Andrea Álvarez, Alina L. Trávez, Arleth Gualle, Juan S. Villarreal, Javier Gándara, Andrés S. Lagos, Verónica Castañeda, Emilia Morales, Javier Garrido, María Gabriela Vernaza, Lucía Ramírez-Cárdenas, Danny Navarrete, Lorena Bejarano, Andrés Caicedo, Lourdes Orejuela

(570A) ADOPTION OF BIOFUELS FOR MARINE SHIPPING 71
Eric Tan, Darlene Steward, Christopher Kinchin, Kylee Harris

(570B) APPLICATION OF HYDROPROCESSING, FERMENTATION AND ANAEROBIC DIGESTION IN A PYROLYSIS REFINERY TO MAXIMIZE ECONOMIC VALUE 72
Arna Ganguly, Irene Mas Martin, Robert C. Brown, Mark Mba Wright

(570C) A SUPERSTRUCTURE OPTIMIZATION APPROACH FOR THE DESIGN OF ETHANOL UPGRADING PROCESSES TO DISTILLATE RANGE FUELS 73
Juan Manuel Restrepo-Florez, Christos T. Maravelias

(570D) OPTIMAL SOLVENT SELECTION IN CO-SOLVENT-BASED BIOMASS CONVERSION PROCESSES 74
Kaan Karacasulu, Alexander Echtermeyer, Christoph Kabatnik, Jan D. Scheffczyk, André Bardow, Jörn Viell, Alexander Mitsos

(570E) SEPARATION AND PURIFICATION OF 2,5-DIMETHYLFURAN: PROCESS DESIGN AND COMPARATIVE TEA AND SUSTAINABILITY EVALUATION OF ADVANCED ADSORPTION AND CONVENTIONAL DISTILLATION 75
Yadong Chiang, Hugo Bassas, Ryan P. Lively, Sankar Nair

(570G) MULTI-OBJECTIVE OPTIMISATION OF SUSTAINABLE BIO-AVIATION FUEL VALUE CHAINS: MODELLING THE PRODUCTION POTENTIAL OF DEVELOPED AND DEVELOPING COUNTRIES 76
Stephen S. Doliente, Sheila Samsatli

(579A) CIRCULAR ECONOMY SYSTEMS ENGINEERING: A CASE STUDY ON THE COFFEE SUPPLY CHAIN 78
Stefanos G. Baratsas, Styliani Avraamidou, Amanda Birnbaum, Leonardo Lombardini, Efstratios N. Pistikopoulos

(579B) THE INFLUENCE OF INCREASING POLYURETHANE RECYCLABILITY AND RENEWABLE CONTENT ON MATERIAL FLOWS 79
Chloe Liang, Troy R. Hawkins, Jennifer B. Dunn

(579C) SUPERCRITICAL CO2-INDUCED ALTERATION OF POLYMER-METAL MATRIX AND SELECTIVE EXTRACTION OF VALUABLE METALS FROM ELECTRONIC WASTES 80
Peng Peng, Alan West, Ah-Hyung Alissa Park

(579D) RECOVERY OPTIONS FOR IRON AND SULPHUR PRODUCED DURING NICKEL EXTRACTION FROM (BIO)LEACHING OF PYRRHOTITE TAILINGS 81
Joanne Gamage McEvoy, Yves Thibault, Suzanne Beauchemin

(579E) STRATEGIC MOLECULES IN LARGE REACTION NETWORKS FOR CIRCULAR CHEMICAL ECONOMY 82
Jana M. Weber, Pietro Liò, Alexei A. Lapkin

(579F) CO2 MASS PRODUCTION FROM THE AIR VERY CHEAPLY 84
David Judbarovski

(585A) SOLAR FOR INDUSTRIAL PROCESS HEAT: A TECHNICAL POTENTIAL ASSESSMENT FOR THE U.S MANUFACTURING SECTOR 86
Carrie Schoeneberger, Jingyi Zhang, Colin McMillan, Parthiv Kurup, William Xi, Robert Margolis, Eric Masanet

(585B) --TECHNO-ECONOMIC ANALYSIS OF SMALL HABER-BOSCH PROCESS 87
Bosong Lin, Theodore Wiesner, Mahdi Malmali

(585C) ANALYSIS OF GREEN AMMONIA PRODUCTION RESPONDING TO FLUCTUATING RENEWABLE ENERGY BY USING DYNAMIC SIMULATION 88
Yuki Hoshino, Yasushi Fujimura, Mototaka Kai, Sho Fujimoto, Keisuke Narita, Hiroyuki Isobe, Miho Kawamata

(585D) EFFECTS OF AMMONIA INJECTOR DESIGN ON CO-FIRING CHARACTERISTICS OF PULVERIZED COAL BURNER WITH HIGH AMMONIA CO-FIRING RATIO 89
Juwei Zhang, Takamasa Ito, Hiroki Ishii, Toshiyuki Suda

(585E) POWER PRODUCTION THROUGH AN AMMONIA BASED COMBINED CYCLE 90
Antonio Sánchez, Elena Castellano, Mariano Martín

(450A) KEYNOTE: THE REMARKABLE INFLUENCE OF ZR-DOPING ON THE THERMOCHEMICAL PROPERTIES OF CERIA 92
Sossina M. Haile

(450C) INVITED TALK: A THERMOCHEMICAL STUDY OF DOPED-SPINELS: A CANDIDATE CLASS OF REDOX INTERMEDIATES FOR ISOTHERMAL WATER SPLITTING 93
Kent J. Warren, Justin T. Tran, Alan W. Weimer

(504D) OXIDATION KINETICS OF HERCYNITE ALLOYS FOR SOLAR THERMOCHEMICAL FUEL PRODUCTION 94
Samantha L. Millican, Iryna Androshchuk, Justin T. Tran, Ryan Trottier, Alicia Bayon, Yahya Al-Salik, Hicham Idriss, Charles B. Musgrave, Alan W. Weimer

(450D) ASSESSING NONSTOICHIOMETRIC OXIDES IN SOLAR THERMOCHEMICAL FUEL PRODUCTION 95
Zhenyu Tian, Jiahui Lou, Xin Qian, Sossina M. Haile, Yong Hao

(442A) RESOURCES, SUSTAINABILITY, CIRCULAR ECONOMY, SUSTAINABLE DEVELOPMENT, AND? 96
Rafiqul Gani

(442B) TOWARD MULTISCALE CONSEQUENTIAL SUSTAINABLE PROCESS DESIGN: INCLUDING THE EFFECTS OF ECONOMY & RESOURCE CONSTRAINTS WITH APPLICATION TO GREEN UREA PRODUCTION IN A WATERSHED 97
Kyuha Lee

(504A) KEYNOTE: THERMODYNAMIC LIMITS FOR MATERIALS IN SOLAR THERMOCHEMICAL FUEL PRODUCTION 98
Ellen B. Stechel, Ivan Ermanoski

(504B) INVITED TALK: THE ROLE OF ENTROPY IN THE SUCCESS OF NONSTOICHIOMETRIC OXIDES FOR SOLAR THERMOCHEMICAL WATER AND CO2 SPLITTING 99
Thomas Cooper

(504C) THERMODYNAMIC GUIDING PRINCIPLES FOR DESIGNING EFFICIENT HIGH-TEMPERATURE SOLAR THERMOCHEMICAL REDOX MATERIALS FOR FUEL PRODUCTION 101
Sha Li, Vincent M. Wheeler, Apurv Kumar, Mahesh B. Venkataraman, Christopher L. Muhich, Yong Hao, Wojciech Lipinski

(450B) INVITED TALK: ENTROPY SOURCE ANALYSIS FOR FERRITES IN TWO-STEP THERMOCHEMICAL SPLITTING OF WATER AND CARBON DIOXIDE 102
Shang Zhai, Kipil Lim, Gopalakrishnan Sai Gautam, Jimmy Rojas, Joonhyun Nam, Kevin Stone, In-Ho Jung, Emily A. Carter, Michael Toney, William Chueh, Arun Majumdar

(111A) TRACKING THE ELEMENTARY KINETICS AND DYNAMIC EVOLUTION OF MOLECULAR STRUCTURES DURING BIOMASS PYROLYSIS 103
Ziwei Wang, Matthew Neurock

(111B) EVOLUTION OF CELLULOSE STRUCTURE THROUGHOUT BIOMASS DECONSTRUCTION IN GAMMA-VALEROLACTONE 104
Elise B. Gilcher, Theodore Walker, Nathaniel Kuch, Catherine F. M. Clewett, Brian G. Fox, Thatcher W. Root, James A. Dumesic

(111C) THEORETICAL ANALYSIS OF THE PERIODIC TRENDS IN THE ALDOL CONDENSATION OF ACETALDEHYDE OVER ISOLATED METAL SITES 105
Branden Leonhardt, Alexis T. Bell

(111D) ALDOL CONDENSATION OF CYCLOPENTANONE AND ACETONE ON MGO: A MECHANISM STUDY BY DFT 106
Yu Yan, Duong T. Ngo, Daniel E. Resasco, Bin Wang

(111E) COMPUTATIONAL ANALYSIS OF KINETICS OF KETONIZATION AND ALDOL CONDENSATION ON ISOLATED ZIRCONIUM SITES..... 107
Lance A. Bettinson, Shylesh Sankaranarayanapilla, Alexis T. Bell, Martin Head-Gordon

(111F) OXYGENATE REACTIONS OVER PDCU/PDAG CATALYST: ROLE OF SYNERGISTIC EFFECTS 108
Tanmayi Bathena, Truc Phung, Scott Ryan Svadlenak, Konstantinos Goulas

(111G) COMPUTATIONAL INSIGHTS INTO THE DIRECT ACYLATION OF 2-METHYLFURAN WITH ACETIC ACID OVER PHOSPHOTUNGSTIC ACID AND H-BEA ZEOLITE 109
Tso-Hsuan Chen, Stavros Caratzoulas, Omar Abdelrahman, Dionisios G. Vlachos

(111H) UNDERSTANDING SOLVENT AND PRETREATMENT EFFECTS ON THE SURFACE CHEMISTRY OF SMALL OXYGENATES ON MOLYBDENUM TRIOXIDE..... 110
Sean Najmi, Mathew Rasmussen, Chaoyi Chang, Giada Innocenti, Eli Stavitski, James Medlin, Andrew Medford, Carsten Sievers

(354A) N-NITROSODIMETHYLAMINE (NDMA) FORMATION DURING ADVANCED OXIDATION PROCESS (AOP) TREATMENT FOLLOWING REVERSE OSMOSIS FOR POTABLE WATER REUSE 111
Aleksandra Szczuka, Nan Huang, Zhong Zhang, William Mitch

(354B) ASSESSMENT OF ENHANCED ADVANCED OXIDATION PROCESSES FOR TREATMENT OF CHLORINATED SOLVENTS 112
Robert Peters, Yusuf G. Adewuyi

(354C) PLASMA-ASSISTED DEGRADATION OF PERFLUOROBUTANE SULFONATE (PFBS) 113
Faith Isowamwen, Rui Li, Thomas Holsen, Selma Mededovic Thagard

(354D) GRAPHENE QUANTUM DOTS BASED ZNO NANOMATERIALS FOR DEGRADATION OF POLLUTANT AND MICROBIAL DEACTIVATION IN WASTEWATER TREATMENT PLANT. 114
Charmaine S. Tshangana, Adolph A. Muleja

(354E) INCREASING SODIUM PERCARBONATE FENTON LIKE PROCESS BY GLUTAMIC ACID ASSISTED ZERO-VALENT IRON FOR THE DEGRADATION OF METHYLENE BLUE 115
Muhammad Danish, Ayyaz Ahmad, Usman Farooq

(354F) IONIZING RADIATION-INDUCED CATALYTIC DEGRADATION OF ANTIBIOTICS IN WATER MATRICES USING FE/C NANOMATERIALS DERIVED FROM IRON-BASED MOFS 116
Libing Chu, Dan Chen, Jianlong Wang

(600A) FROM BENCHTOP TO FIELD DEMONSTRATION: SCALING THE "CO2CONCRETE" TECHNOLOGY FOR THE MANUFACTURE OF CONCRETE PRODUCTS BY CARBON DIOXIDE MINERALIZATION 118
Gabriel Falzone, Iman Mehdipour, Gaurav Sant

(600B) IMPLEMENTATION OF ION EXCHANGE PROCESSES FOR CO2 MINERALIZATION USING PRODUCED WATER STREAMS 119
Steven Bustillos, Abdulaziz Alturki, Dale Prentice, Erika Callagon La Plante, Gaurav Sant, Dante Simonetti

(600C) ELECTROCHEMICAL CO2 CONVERSION TO VALUABLE CHEMICALS 120
Feng Jiao

(600D) ELECTROCHEMICAL CO2 UTILIZATION: SCALABLE SYSTEM OPERATION FOR FORMIC ACID PRODUCTION 121
Daniel Moreno, Ayokunle Omosebi, Keemia Abad, Jesse G. Thompson, Kunlei Liu

(600E) OXIDE-DERIVED TIN MICROSPHERES FOR EFFICIENT CO2 ELECTROREDUCTION TO FORMIC ACID 122
Thuy-Duong Nguyen-Phan, Douglas R. Kauffman

(600F) DESIGNING CATALYSTS AND REACTORS TOWARDS PRACTICAL CO2 REDUCTION 123
Haotian Wang

(600G) CARBON DIOXIDE UTILIZATION THROUGH PEM ELECTROLYSIS 124
Kendra Kuhl

(600H) SIMULTANEOUS CARBON UTILIZATION AND ALKANE CONVERSION USING SOLID OXIDE ELECTROLYTIC TECHNOLOGY 125
Samgopiraj Velraj, Ahmad Abu Hajer, Andrew Kasick, Damilola Daramola, Jason Trembly

(621A) EXAMINING ROUTES TO P-CYMENE THROUGH INTEGRATING BIOLOGICAL AND CHEMICAL CATALYSIS 126
Hsi-Hsin Lin, Xi Wang, Yan Cheng, Daniel Mendez-Perez, Taek Soon Lee, Brent H. Shanks

(621B) ASSESSING THE IMPACT OF ALLOSTERIC ENZYME REGULATIONS LIMITING ETHANOL TITER IN CLOSTRIDIUM THERMOCELLUM USING A CORE KINETIC MODEL 127
Charles Foster, Veda Sheersh Boorla, Tyler B. Jacobson, Ratul Chowdhury, Saratram Gopalakrishnan, Satyakam Dash, Daniel Olson, Daniel Amador-Noguez, Lee R. Lynd, Costas Maranas

(621C) TOWARDS THE POTENTIAL COUPLING OF MICROBIAL BIOSYNTHESIS AND HETEROGENEOUS CATALYSIS FOR THE PETROLEUM-FREE PRODUCTION OF ALKYLATED PHENOL COMPOUNDS 128
Adam Zuber, Zhenghong Li, Xiaonan Wang, Ashil Vekaria, Yingxi Lu, Haoran Zhang, George Tsilomelekis

(621D) OPTIMIZATION OF LIGNOCELLULOSE-BASED PRODUCTION OF BIOCONTROL AGENTS AS BIOREFINERY COPRODUCTS 129
Patricia J. Slininger, Nora L. Olsen, Andrew R. Schoepke, Maureen A. Shea-Andersh, Bruce S. Dien

(621E) TANDEM CHEMICAL AND BIOLOGICAL CATALYSIS: PRODUCTION OF 2-PYRONE-4,6-DICARBOXYLIC ACID VIA LIGNIN VALORIZATION 130
Canan Sener, Canan Sener, Miguel Perez, German Umana, Jason Coplien, Wayne Kontur, Steven D. Karlen, Daniel R. Noguera, Timothy J. Donohue, John Ralph

(621F) BIOLOGICAL ALCOHOL PRODUCTION FROM CO2 USING NOVEL ACETOGENS 131
Rahul Thunuguntla, Hasan Atiyeh, Raymond L. Huhnke, Ralph S. Tanner

(624A) TOWARDS SUSTAINABLE CIRCULAR ECONOMY: DESIGN FRAMEWORK AND APPLICATION TO GROCERY SACKS 132
Vyom Thakker, Bhavik Bakshi

(624B) IDENTIFYING STRATEGIES FOR SYSTEMS SCALE TRANSITION TO CIRCULAR ECONOMY VIA PHYSICAL SUPPLY USE TABLES DEVELOPED FROM BOTTOM-UP MODELS: A CASE STUDY ON ILLINOIS AGRO ECONOMY 134
Venkata Sai Gargeya Vunnava, Shweta Singh

(624C) SYSTEMS-LEVEL ANALYSIS OF GLOBAL PLASTIC TRADE: IMPLICATIONS FOR A CIRCULAR ECONOMY 135
Joseph Zappitelli, Eli Smith, Kevin Padgett, Melissa M. Bilec, Vikas Khanna

(624D) A GRAPH-ANALYTICAL APPROACH FOR THE DEVELOPMENT OF CIRCULAR ECONOMY NETWORKS 136
Theodore Chatzidimitriou, Thanassis Gentimis, Theodore Dalamagas, Antonis C. Kokossis

(624E) INTEGRATED FOOD FACILITIES: CIRCULAR ECONOMY APPLIED TO THE FOOD INDUSTRY 137
Manuel Taifouris, Arantza Criado, Marcos Corazza, Mariano Martín

(624F) TOWARDS SUSTAINABLE RESOURCE INTEGRATION NETWORKS IN INDUSTRIAL CLUSTERS 138
Dhabia Al-Mohannadi, Razan Ahmed, Patrick Linke

(643A) MULTI-OBJECTIVE OPTIMIZATION OF AN ELECTROCHEMICAL CO2 UTILIZATION PROCESS 139
Nishant V. Giridhar, Debangsu Bhattacharyya, Chrysanthos E. Gounaris

(643C) NOVEL CATALYTIC CHEMICAL LOOPING PROCESS TECHNOLOGY FOR UTILIZATION OF CO2 FOR ACRYLONITRILE PRODUCTION 141
Aravind V. Rayer, Qinghe Zheng, Jian-Ping Shen, Lindsey Chatterton, Tim Bellamy, Gary Howe, Kelly Amato, Vijay Gupta, Marty Lail

(643D) REFORMING OF METHANE WITH CARBON DIOXIDE OVER CERIUM OXIDE PROMOTED NICKEL NANOPARTICLES DEPOSITED ON 4-CHANNEL HOLLOW FIBERS BY ATOMIC LAYER DEPOSITION 142
Baitang Jin, Zeyu Shang, Shiguang Li, Xinhua Liang

(643E) CAPTURE AND CONVERSION OF CO2 INTO METHANE USING CATALYTIC SORBENT OF NANO3/MGO + RU/AL2O3 143
Sang Jae Park, Maxim P. Bukhovko, Christopher Jones

(643F) MOLTEN SALT FACILITATED OXIDATIVE DEHYDROGENATION OF ETHANE WITH IN-SITU CO2 CAPTURE AND UTILIZATION 144
Junchen Liu, Yunfei Gao, Fanxing Li

(643G) NATIONAL CARBON CAPTURE CENTER SUPPORTING DEVELOPMENT AND SCALE-UP OF CO2 UTILIZATION TECHNOLOGIES 145
Frank C. Morton

(643H) CATALYTIC CONVERSION OF CO2 INTO SOLAR FUELS VIA THERMOCHEMICAL CYCLE DRIVEN BY LA-BASED PEROVSKITES 146
Rahul Bhosale, Suliman Rashid

(671B) POLAR APROTIC SOLVENT AND ALKALI METAL ION EFFECTS ON THE CONVERSION OF C6 CARBOHYDRATES TO ALPHA-HYDROXY ACIDS USING LEWIS ACID BETA ZEOLITES 147
Isabel Hortal-Sánchez, Christian G. Rivera-Goyco, Nelson Cardona-Martínez

(671C) ELUCIDATING THE ROLES OF LATTICE AND MOLECULAR OXYGEN DURING THE OXIDATIVE SCISSION OF METHYL KETONES OVER SUPPORTED VANADIUM OXIDES 148
Siwen Wang, Ran Zhu, Bowei Liu, Jesse Bond

(671D) LATTICE DISTORTED MNCO OXIDE MATERIALS AS EFFICIENT CATALYSTS FOR TRANSFER HYDROGENATION OF LEVULINIC ACID USING FORMIC ACID AS H-DONOR 149
Jinyao Wang, Jiefeng Liu, Xiao Yu, Mengyuan Liu, Qi Xia, Wenxiang Zhang, Xin Jin, Chaohe Yang

(671E) SELECTIVE HYDROGENATION OF CROTONALDEHYDE OVER A PDSN CATALYST: A FIRST-PRINCIPLES STUDY 150
Yu Yan, Wenyu Huang, Bin Wang

(671F) CATALYTIC CONVERSION OF GLUCOSE TO TARTARIC ACID OVER SYNERGISTIC AUPT/TIO2 CATALYSTS 151
Mengyuan Liu, Tianqi Fang, Jinyao Wang, Guangyu Zhang, Xin Jin, Chaohe Yang

(693A) DESIGNING EFFICIENT SOLVENT RECOVERY PATHWAYS BY INTEGRATING ECONOMICS AND ENVIRONMENTAL METRICS 152
Emmanuel A. Aboagye, John D. Chea, Austin L. Lehr, Jake P. Stengel, Mariano Savelski, Stewart Slater, Kirti M. Yenkie

(693B) MODELING SPATIAL AND TEMPORAL EMISSIONS FOR ANIMAL FARMING USING MECHANISTIC MODELS 155
Venkata Sai Gargeya Vunnava, Martina Macaggi, Yonru Chen, Shweta Singh

(693C) CONFLICT RESOLUTION OF ECONOMIC AND ENVIRONMENTAL METRICS IN DAIRY WASTE MANAGEMENT 156
Yicheng Hu, Horacio Aguirre-Villegas, Rebecca Larson, Victor M. Zavala

(693D) OPTIMIZATION OF FUEL COMPOSITION FOR PLASTIC DERIVED FUEL OIL PRODUCED VIA TRASH TO TANK 158
Chandni Joshi, Jeffrey Seay

(693E) USE OF R2 STANDARD IN SUSTAINABLE MANAGEMENT OF EWASTE 159
Ashit Dalal

(693F) INFLUENCE OF CR(VI) AND CD(II) ON STRUVITE CRYSTALLIZATION IN SIMULATED WASTEWATER 160
John Paul Renzo Jucar, Bryan G. Alamani

(711A) ENERGY-EFFICIENT VIRUS REMOVAL WITH COTTON-BASED DIY WATER FILTERS 161
Laxmicharan Samineni, Sophie Derespino, Stephanie B. Velegol, Manish Kumar

(711B) LEVERAGING REVERSIBLE CHEMISTRY FOR GREEN AND EFFICIENT BATTERY RECYCLING 163
Zheng Chen

(711C) DEVELOPMENT OF A CLOSED-LOOP PROCESS FOR SEPARATION AND RECOVERY OF LI AND CO FROM LICOO2 164
Ankit Verma, David R. Corbin, Mark B. Shiflett

(711D) SPARGED BUT NOT STIRRED: ENZYME-CATALYZED DERACEMIZATION OF ALCOHOLS IN A BUBBLE COLUMN 165
Shelby Anderson, Bettina Bommarius, John M. Woodley, Andreas Bommarius

(711E) SELF-ASSEMBLED METAL/POLYMER HYBRID NANOPARTICLES AND THEIR USE AS NANOREACTORS TOWARDS CASCADE REACTIONS 166
Andrew Harrison, Michael Zeevi, Christina Tang

(711F) FLOW CHEMISTRY-ENABLED INVESTIGATIONS OF SWITCHABLE HYDROPHILICITY SOLVENTS 167
Suyong Han, Keshav Raghuvanshi, Milad Abolhasani

(711G) GREEN SOLVENTS FOR THE HYDROFORMYLATION OF LONG CHAIN ALKENES 169
Steffen Linke, Michael Jokiel, Kevin McBride, Kai Sundmacher

(504E) INVITED TALK: HIGH-TEMPERATURE SOLAR THERMOCHEMISTRY CO2 REDUCTION OVER CR2FEO4/ZRO2@ZRO2 OXYGEN EXCHANGE MATERIAL 172
Yong Shuai, Bachirou Guene Lougou, Jiupeng Zhao, Heping Tan

(504G) OPTIMAL POROSITY DISTRIBUTION OF NON-UNIFORM MACROPOROUS CERIA IN SOLAR THERMOCHEMICAL WATER SPLITTING 174
Xiaoyu Dai, Sophia Haussener

(504H) THERMALLY-DRIVEN OXYGEN PUMPING AND SWEEPING/SEPARATION 175
Ivan Ermanoski, Mai Xu, Shuguang Deng, Ellen B. Stechel

(504I) COMBINED HYDROGEN AND POWER PRODUCTION SYSTEMS USING THERMOCHEMICAL REDOX CYCLE 176
Wonjae Choi, Xiaoyu Wu, Aniket Patankar, Ahmed F. Ghoniem

(503A) ESTIMATING CHEMICAL RELEASES AND EVALUATING INDICATORS FOR MANUFACTURING PROCESSES: BIOREFINERY CASE STUDIES 178
Raymond Smith, Eric Tan, Gerardo Ruiz-Mercado

(559A) KEYNOTE: OPPORTUNITIES AND CHALLENGES FOR CST IN HIGH TEMPERATURE INDUSTRIAL PROCESSES 179
G. J. Nathan, Woei L. Saw, Bassam B. Dally, Mehdi Jafarian, Maziar Arjomandi, Peter J Ashman, Alfonso Chinnici, Zhao Tian, Philip Ingenhoven, Ahmad Seyfaee

(559B) INVITED TALK: GRID-LEVEL THERMOCHEMICAL ENERGY STORAGE USING THE MAGNESIUM-MANGANESE-OXIDE REDOX SYSTEM 180
Nima Rahmatian, Alessandro Bo, Kelvin Randhir, Keith King, James F. Klausner, Joerg Petrasch

(559C) THE THERMODYNAMICS OF MAGNESIUM MANGANATE REDUCTION AS APPLIED TO THERMOCHEMICAL ENERGY STORAGE 181
Jayni Hashimoto, Srashtasrita Das, Harsheen Rajput, Christopher L. Muhich

(559D) THERMAL REDUCTION OF IRON-MANGANESE OXIDE PARTICLES IN A HIGH-TEMPERATURE PACKED-BED SOLAR THERMOCHEMICAL REACTOR 182
Bo Wang, Lifeng Li, Florian Schäfer, Johannes J. Pottas, Apurv Kumar, Vincent M. Wheeler, Wojciech Lipinski

(658A) SUSTAINABILITY: LEGAL AND BUSINESS DRIVERS, AND RECENT TRENDS 183
Mary Ellen Ternes

(658B) SUSTAINABILITY: CONCEPTS, INNOVATION, AND ECONOMICS 184
Heriberto Cabezas, Csaba Deak

(658C) THE NEED FOR CONVERGENCE 185
Bhavik Bakshi

(658D) DO WE NEED NEW CLASSES OF METHODS, TOOLS AND STANDARDS? 186
Rafiqul Gani

(175A) ENABLING SELECTIVE CATALYTIC CONVERSION OF BIOMASS-DERIVED SATURATED FURANS TO C4-C5 DIENES 187
Gaurav Kumar, Limin Ren, Sha Li, Anatoliy Kuznetsov, Omar Abdelrahman, Michael Tsapatsis, Paul Dauenhauer

(175B) THEORETICAL INSIGHTS INTO THE SELECTIVE CONVERSION OF TETRAHYDROFURAN TO 1,3-BUTADIENE ON ZRO2 188
Sai Praneet Batchu, Yichen Ji, Stavros Caratzoulas, Raymond J. Gorte, Dionisios G. Vlachos

(175C) ACTIVE AND SELECTIVE ZRO2 BASED CATALYSTS FOR DIENE MONOMER PRODUCTION FROM BIOMASS. 189
Yichen Ji, Ajibola Lawal, Raymond J. Gorte, Omar Abdelrahman

(175D) THE CASCADE REACTION OF ETHANOL TO BUTADIENE OVER SILICA-SUPPORTED SILVER-ZIRCONIA CATALYSTS 190
Naomi Miyake, Gordon Brezicki, Robert J. Davis

(175E) ETHANOL CONVERSION TO BUTADIENE OVER ISOLATED ZINC AND YTTRIUM SITES IN DEALUMINATED BETA ZEOLITE 191
Liang Qi, Alexis T. Bell

(175G) KINETIC AND MECHANISM OF ACETONE CONVERSION TO ISOBUTENE OVER HF/SILICALITE-1 AND HF/SIO2 192
Yanfei Zhang, Alexis T. Bell

(175H) VAPOR PHASE ETHANOL SELECTIVE OXIDATION TO ACETALDEHYDE AND ACETIC ACID ON GOLD CATALYSTS 193
Priya D. Srinivasan, Hongda Zhu, Hashim A. Alzahrani, Juan J. Bravo-Suárez

(428A) DISSOCIATION, DISSOLUTION, AND DIFFUSION OF NITROGEN ON VXFEY AND VXCRY ALLOY MEMBRANES STUDIED BY FIRST PRINCIPLES 194
Peter C. Psarras, Ryther Anderson, Jennifer Wilcox, Diego Gomez Gualdron

(428B) IMPACT OF OXYGEN ON PHOTOCATALYTIC NITROGEN FIXATION 195
Marta Hatzell

(428D) MANIPULATING STRUCTURES AND CHEMICAL BONDING IN NITRIDES USING FIRST-PRINCIPLES METHODS FOR AMMONIA PRODUCTION 196
Bin Liu, Chaoran Huang, Nannan Shan, Peter H. Pfromm, Viktor Chikan

(513FM) ENHANCING POTENTIAL OF CATALYSTS FOR PLASMA NITROGEN OXIDATION 197
Hanyu Ma, William F. Schneider

(428F) THEORETICAL INSIGHT INTO THE SURFACE DYNAMICS OF LITHIUM-MEDIATED NITROGEN ACTIVATION 198
Thomas Ludwig, Jens Norskov

(428G) CYCLIC STABILIZATION METHOD FOR LITHIUM MEDIATED ELECTROCHEMICAL NITROGEN REDUCTION 199
Michael J. Statt, Suzanne Andersen, Vanessa J. Bukas, Sarah G. Shapel, Jakob B. Pedersen, Kevin Krempl, Mattia Saccoccio, Debasish Chakraborty, Jakob Kibsgaard, Peter C. K. Vesborg, Jens Norskov, Ib Chorkendorff

(428H) THEORETICAL INSIGHTS INTO THE POTENTIAL-DEPENDENT COMPETITION BETWEEN ELECTROCHEMICAL NITROGEN REDUCTION REACTION AND HYDROGEN EVOLUTION REACTION 200
Changhyeok Choi, Geun Ho Gu, Juhwan Noh, Yousung Jung

(481A) RENEWABLE NH3 AS ALTERNATIVE ENERGY FUEL FOR OCEAN EXPLORATION 201
Jian Liu, Yuyan Shao, Vassiliki-Alexandra Glezakou, Robert Cavagnaro, Li-Jung Kuo

(481B) MIXED-PHASE PEROVSKITE MATERIALS FOR H2 GENERATION AND HYBRID CHARGE STORAGE 202
Joseph Houck, Vinod S. Amar, Bharath Maddipudi, Anuradha Shende, Rajesh Shende

(481C) HYDROGEN AND AMMONIA FOR RENEWABLE ENERGY STORAGE: THE ECONOMICS OF LOCATION, SCALE, AND DEMAND TYPE 203
Matthew Palys, Prodromos Daoutidis

(481D) HYDROTHERMAL LIQUEFACTION OF BAGASSE FOR H2, BIO-OIL AND HIGH VALUE PRODUCTS 205
Vinod S. Amar, Anuradha Shende, Bharath Maddipudi, Rajesh Shende

(481E) CUO-CU2O REDOX THERMOCHEMICAL STORAGE FOR HIGH-TEMPERATURE SOLAR PROCESS HEAT 206
Marco Gigantino, Sebastian Sas Brunser, Aldo Steinfeld

(555A) DOWNSTREAM PROCESS WITH COUNTERCURRENT CHROMATOGRAPHY TO RECOVER BIOPRODUCTS FROM SUSTAINABLE FEEDSTOCKS 207
Hoon Choi, Patrick Saboe, Brenna Black, Dong Xueming, Stefan Haugen, Eric M. Karp

(555B) SEPARATION OF LIGNOCELLULOSIC SUGARS FROM PYROLYTIC BIO-OIL USING SIMULATED MOVING BED 208
Arpa Ghosh, John P. Stanford, Ryan G. Smith, Robert C. Brown

(555C) BLACK LIQUOR GASIFICATION CHEMICAL RECOVERY AND COMBINED CYCLE POWER AT PULP AND PAPER MILLS BIOREFINERY 209
Tapas K. Das

(555D) FROM BIOMASS TO POWER USING INTEGRATED COMPACT PYROLYSIS, COMBUSTOR, AND STIRLING ENGINE 210
Sampath Gunukula, Dat T. Tran

(555E) TECHNO-ECONOMIC ANALYSIS AND LIFE-CYCLE ASSESSMENT OF EMERGING TECHNOLOGIES FOR BIOPROCESSING SEPARATIONS 211
Lauren Valentino, Jennifer B. Dunn, Eric Tan, Charles J. Freeman, William L. Kubic Jr., Alex Rosenthal

(680A) BIO-BASED PYRROLIDONE FROM SUCCINIC ACID: PROCESS DESIGN AND LIFE CYCLE ASSESSMENT 212
Benedikt Winter, Moritz Haus, Regina Palkovits, André Bardow

(680B) GIS INTEGRATED OPTIMIZATION MODEL FOR LOCATION OF BIOGAS COMBUSTION POWER PLANTS UNDER CARBON OFFSETS IN WISCONSIN 213
Sharath Ankathi, David W. Watkins, David Shonnard

(680C) SUSTAINABILITY AND COST TRADEOFFS OF EPOXIDIZED SUCROSE SOYATE: A BIO-BASED THERMOSET RESIN 214
Shokoofeh Ghasemi, Mukund Sibi, Dean Webster, Ghasideh Pourhashem

(680D) CARBON AND ENERGY IMPLICATIONS OF FAST PYROLYSIS BIOREFINERIES WITH BLENDED FEEDSTOCKS AND DECENTRALIZED SUPPLY CHAIN DESIGN IN THE SOUTHEASTERN UNITED STATES 215
Kai Lan, Longwen Ou, Sunkyu Park, Stephen S. Kelley, Yuan Yao

(680E) INTEGRATING MACHINE LEARNING AND PROCESS SIMULATION TO ESTIMATE ENERGY AND GREENHOUSE GAS EMISSIONS OF ACTIVATED CARBON PRODUCTION 217
Mochen Liao, Yuan Yao

(680F) SPATIAL LIFE CYCLE ANALYSIS OF SOYBEAN-BASED BIODIESEL PRODUCTION IN INDIANA USING PROCESS MODELING 219
Venkata Sai Gargeya Vunnava, Shweta Singh

(685A) ADVANCED MANUFACTURED CARBONATE MATERIALS FOR CO2 UTILIZATION IN ALGAL BIOMASS PRODUCTION 220
Jennifer M. Knipe, Maira R. Ceron, Matthew A. Worthington, Mary Bao Tran-Gyamfi, Hannah M. Goldstein, Sarah Baker, Todd W. Lane

(685B) NH4OH LOOPING WITH MEMBRANE CO2 ABSORBER AND DISTRIBUTED STRIPPER FOR ENHANCED ALGAE GROWTH 222
Heather Nikolic, Kunlei Liu, Mark Crocker

(685C) ENHANCING MONOETHANOLAMINE REGENERATION AT LOW TEMPERATURES VIA IN-SITU PRECIPITATION OF SOLID CARBONATES........ 224
Anas Alaqeeli, Dante Simonetti

(685D) LIFE CYCLE ANALYSIS TOOLS FOR EVALUATING THE ENVIRONMENTAL FOOTPRINT OF CARBON UTILIZATION PROJECTS 225
Sheikh Moni, Gregory Cooney, Michele Mutchek, Timothy Skone, Evelyn Dale

(685E) FIRST OPERATION OF 5 KWCH4 LOCAL CARBON RECYCLING SYSTEM: PRODUCT GAS QUALITY OF CO2 METHANATION........ 226
Shogo Sayama, Kunitomi Seiichi, Kazuto Ogasawara, Seiji Yamamoto

(685F) INDUSTRIAL SCALE CO2-REFORMING FOR PRODUCTION OF CO-RICH SYNTHESIS GAS AT LOW STEAM TO CARBON RATIO........ 228
Marené Rautenbach, Peter M. Mortensen, Martin Østberg

(685G) EFFECTS OF CO2 METHANATION ON ECONOMIC FEASIBILITY AND GLOBAL CLIMATE CHANGE IN COAL-FIRED POWER PLANTS 231
Semie Kim, Young-Il Lim, Doyeon Lee, Myung Won Seo

(685H) CALCINATION-FREE PRODUCTION OF CALCIUM HYDROXIDE AT SUB-BOILING TEMPERATURE........ 232
Sara Vallejo Castaño, Erika Callagon La Plante, Sho Shimoda, Laurent Pilon, Gaurav Sant

(688A) "SUSTAINABILTY – THE NEED FOR SCALABLE SOLUTIONS & A FRAMEWORK TO MAKE DECISIONS" 233
Narayan Ramesh

(688B) THE IMPACT OF POLICY INSTRUMENTS ON DEPLOYMENT OF LOW-CARBON HEATING IN THE UK: A WHOLE SYSTEM OPTIMISATION STUDY 234
Jennifer Penman, Sheila Samsatli

(688C) THE ROLE OF CHEMICAL ENGINEERING IN SUSTAINABLE ENERGY POLICY 235
Quinta Nwanosike Warren

(688D) REVIEW OF US ENVIRONMENTAL PROTECTION AGENCY GENERAL EXPOSURE SCENARIO MODELING FOR EVALUATING CHEMICAL SAFETY 236
William Barrett, Raymond Smith, David E. Meyer, Gerardo Ruiz-Mercado, Michael A. Gonzalez

(688F) EFFECTIVELY COMMUNICATING SUSTAINABILITY AND ENGINEERING WITHIN FEDERAL POLICY 237
Ashley M. Pennington

(708A) MAKING AMINO ACIDS FROM BIOMASS VIA CHEMICAL CATALYSIS 238
Ning Yan

(708C) DEVELOPMENT OF SOLID BASE ACTIVATED CARBON CATALYSTS FOR KETONE CONDENSATION AND AVIATION FUEL SYNTHESIS........ 239
Seyedehsan Vasefi, James Kastner

(708E) A PATHWAY TOWARDS CARBON NEUTRAL FUELS: SYNTHESIS OF I-BUTANOL WITH RENEWABLE METHANOL 240
Johannes Häusler, Joachim Pasel, Friederike Woltmann, Ralf Peters, Detlef Stolten

(718A) HYBRIDIZATION POTENTIAL OF CARBON CAPTURE AND UTILIZATION: A SMOOTHER TRANSITION TOWARDS A DECARBONIZED ETHYLENE PRODUCTION 241
Iasonas Ioannou, Sebastiano C. D'Angelo, Antonio J. Martín, Javier Pérez-Ramírez, Gonzalo Guillén-Gosálbez

(718B) CLIMATE-RESILIENT PROCESS DESIGN USING THE FLEXIBILITY ANALYSIS APPROACH 242
Kyuha Lee, Sami Khanal, Bhavik Bakshi

(718C) TOWARDS INTEGRATION OF DESIGN AND OPERATION OF TECHNO-ECOLOGICAL SYNERGISTIC SYSTEMS 244
Utkarsh Shah, Joel Paulson, Bhavik Bakshi

(718D) MULTISCALE ANALYSIS FOR POWER-TO-GAS-TO-POWER FACILITIES. TECHNICAL AND SOCIAL ISSUES OF SUBSTITUTING BASE LOAD 246
Jorge Heras, Mariano Martín

(718E) INCORPORATING PROCESS SAFETY IN HEAT EXCHANGER NETWORK SYNTHESIS 248
Ahmed Harhara, M M Faruque Hasan

(718F) EFFLUENT REDUCTION IN PULP AND PAPER MANUFACTURING: WHITE WATER RECYCLE IN BLEACHING SECTION 249
Amod Parkhi, Selen Cremaschi, Zhihua Jiang

(718G) ON A SYSTEMATIC APPROACH TO SIMULTANEOUSLY INTEGRATE WATER AND ENERGY UTILITIES IN MULTI-PRODUCT BIOREFINERIES 250
Aikaterini D. Mountraki, Bouchra Benjelloun Mlayah, Antonis C. Kokossis

(720A) DEVELOPING PERFORMANCE-ADVANTAGED CHEMICALS UTILIZING BIOPRIVILEGED MOLECULES: ORGANIC CORROSION INHIBITORS 251
Jiajie Huo, William Bradley, Kyle Podolak, Bradley J Ryan, Luke Roling, George Kraus, Brent H. Shanks

(720B) ADVANCED ROBUST CATALYSTS FOR PRODUCING STABLE BIOMASS PYROLYSIS OILS FOR UPGRADING TO HYDROCARBON FUELS 252
Foster Agblevor, Huamin Wang, Daniel Santosa, Oleksandr Hietsoi, Sedat H. Beis

(720C) ONE-STEP CONVERSION OF BIODERIVED HYDROXY ACIDS TO A RENEWABLE ALKANEDIOL 253
Shylesh Pillai, Deepak Dugar, Karthikeyan K. Ramasamy

(720G) NITROGEN BEHAVIOR IN GLYCINE DECOMPOSITION UNDER HYDROTHERMAL CONDITION 254
Yukihiko Matsumura, Hiroki Yokoyama

(721A) HIGH-EFFICIENCY HYDROGEN PRODUCTION BY DECOUPLED WATER SPLITTING IN AN ELECTROCHEMICAL - CHEMICAL CYCLE 255
Avigail Landman, Hen Dotan, Gideon S. Grader, Avner Rothschild

(721B) EFFICIENCY LIMITS OF UNDERWATER SOLAR CELLS: AN EXCELLENT APPLICATION FOR WIDE-BAND-GAP SEMICONDUCTORS? 256
Jason A. Röhr, Jason Lipton, Jaemin Kong, Stephen A. Maclean, André D. Taylor

(721C) MICROVASCULAR CARBON-CARBON COMPOSITE FOR CONCENTRATED SOLAR POWER GAS RECEIVERS 257
Jose C. Cordeiro Jr., Jonathan Argot, Hema Ramsurn, Daniel W. Crunkleton, Todd Otanicar, Michael Keller

(721D) ON THE ECONOMICS OF A THERMAL MANAGEMENT SYSTEM FOR INDUSTRIAL-SCALE NATURAL GAS ADSORPTION STORAGE 258
Mohammed Alhashim, Jamal Alsmail, Lanre Oshinowo, Calogero Migliore

(721E) TRADING OFF THE COSTS AND BENEFITS OF ELECTRIC AND HYDROGEN TECHNOLOGIES IN DECARBONISING THE HEAT SECTOR: WHOLE-SYSTEM VALUE CHAIN OPTIMISATION 259
Jennifer Penman, Sheila Samsatli

(721G) SUBNATIONAL SUPPLY CHAIN ANALYSIS OF COBALT IN THE US TO GUIDE SUSTAINABLE MANUFACTURING OF BATTERIES FOR RENEWABLE ENERGY: A MULTI-REGIONAL INPUT-OUTPUT (MRIO) APPROACH. 260
Miriam Stevens, Shweta Singh

(721H) CAN COMBINED HEAT AND POWER (CHP) BE A FRIEND TO RENEWABLE ENERGY? 261
Hyeunguk(Henry) Ahn, William Miller, Paul Sheaffer, Vi Rapp

(739A) DISTRIBUTED MODULAR GREEN AMMONIA SYNTHESIS WITH ABSORPTION SEPARATION: UNDERSTANDING RATE LIMITATIONS, INSIGHTS TOWARD IMPROVEMENT 262
Deepak Ojha, Matthew Kale, Paul Dauenhauer, Edward L. Cussler, Alon V. McCormick

(739B) ALKALI & GREEN CATALYST ACTIVATED GEOPOLYMERS USING COAL AND WOOD FLY ASH 263
Arun Kumar, Yvonne Lin, Babu Cadiam Mohan, Wang Chi-Hwa

(739C) SYSTEMATIC STUDY OF CHARGE TRANSFER BETWEEN CATALYTIC AU NANOPARTICLES AND SUPPORT AND ITS IMPLICATION FOR CO OXIDATION 264
Haotian Yang, Jiajie Cen, Claron Ridge, Kyle Overdeep, Michael Lindsay, Xiao Tong, Alexander Orlov

(739D) TOXICITY STUDY OF POWDERED AND CAST INCINERATED SEWAGE SLUDGE ASH - CHEMICAL TREATMENT, LEACHING PATTERNS AND MARINE TOXICITY 265
Arun Kumar, Babu Cadiam Mohan, Teo Lay Ming Serena, Wang Chi-Hwa

(739E) DYNAMIC SIMULATION AND MODEL PREDICTIVE CONTROL OF POST-COMBUSTION CO2 CAPTURE PROCESS WITH ADVANCED FLASH STRIPPER 266
Howoun Jung, Jay H. Lee

(739F) EVALUATION OF AN INDUSTRIAL BIO-CATALYST FOR REDUCING HYDROGEN SULFIDE OF THE GAS STREAM FROM DOMESTIC WASTEWATER PLANT TREATMENT 267
José A. Pérez, Danitza Marín, Jonnathan Cárcamo

(559I) INVITED TALK: NUMERICAL AND EXPERIMENTAL INVESTIGATION OF AEROSOL PARTICLES REACTIONS DRIVEN BY CONCENTRATING SOLAR POWER 268
Konstantinos E. Kakosimos, Navaira Fathima, Agni E. Kalyva, Athanasios G. Konstandopoulos

(559E) REACTION SCREENING FOR THERMOCHEMICAL ENERGY STORAGE 269
Ishan Bajaj, Xinyue Peng, Christos T. Maravelias

(559F) THERMAL ENERGY STORAGE USING THE BOUDOUARD REACTION 271
Yu Miao, Alexandre Yokochi, Nick Auyeung, Annette Von Jouanne

(511G) OPPORTUNITIES TO ENHANCE SUSTAINABILITY OF COPPER PRODUCTION WITH THE HYS CYCLE 272
Ahmad Seyfaee, Mehdi Jafarian, G. J. Nathan

(0) SPONSORED TECHNOLOGY WORKSHOP - INSTRUMENTATION FOR ENERGY DEVICE RESEARCH; FROM MATERIAL STUDIES TO CELLS TO FULL STACKS UP TO 1000V - GAMRY INSTRUMENTS 274
Jerome Babauta, Andrew McCaskill, Jacob Ketter

(605C) LEGAL DRIVERS FOR WASTE PLASTIC LAW AND REGULATION 275
Mary Ellen Ternes

(605D) PUBLIC HEALTH PERSPECTIVES ASSOCIATED WITH WASTE PLASTICS 276
Lisa McCormick, Maryam Karimi, Rouzbeh Nazari, Robert Peters

(612A) KEYNOTE: SOLAR CHEMISTRY IN MEXICO AND THE NATIONAL LABORATORY OF SOLAR CONCENTRATION AND SOLAR CHEMISTRY 277
Claudio Estrada

(612B) INVITED TALK: REACHING BEYOND 1'500°C WITH THE SYNHELION ABSORBING GAS SOLAR RECEIVER: RESULTS OF EXPERIMENTAL CAMPAIGN 278
Gianluca Ambrosetti, Philipp Good, Philipp Furler, Lukas Geissbühler, David Rutz, Simon Ackermann

(612C) CFD MODELING AND PERFORMANCE EVALUATION OF THE SYNHELION ABSORBING GAS SOLAR RECEIVER FOR PROCESS HEAT AT 1500°C AND BEYOND 279
Simone A. Zavattoni, Davide Montorfano, Philipp Good, Gianluca Ambrosetti, Maurizio C. Barbato

(612D) OPTICAL ANALYSIS OF A NOVEL SOLAR THERMOCHEMICAL SYSTEM WITH A ROTATING TOWER REFLECTOR AND A RECEIVER–REACTOR ARRAY 283
Lifeng Li, Song Yang, Bo Wang, John Pye, Wojciech Lipinski

(511A) NUMERICAL STUDY ON A NOVEL SOLAR-DRIVEN INTERFACE EVAPORATION DEVICE COMBINED WITH THERMOELECTRIC MATERIAL 284
Jinhai Xu, Siwei Chen, Yuxuan Zhang, Bin Zhang, Peiyu Wang, Xinyi Li, Qiuwang Wang, Ting Ma

(511B) SOLAR PYROLYSIS PRODUCTION OF SELF-ACTIVATED BIO-DERIVED CARBONS FOR SUPERCAPACITORS 285
Diego R. Lobato-Peralta, Daniella E. Pacheco-Catalán, Patricia E. Altuzar-Coello, François Béguin, Alejandro Ayala-Cortés, Heidi I. Villafán-Vidales, Camilo A. Arancibia-Bulnes, Ana K. Cuentas-Gallegos

(511C) NON-ISOTHERMAL CONDITIONS IN COUNTERCURRENT THERMOCHEMICAL REACTORS FOR FUEL PRODUCTION 286
Alberto De La Calle, Ivan Ermanoski, Xiang Gao, Ellen B. Stechel

(511D) ACHIEVING HIGH TEMPERATURE AND ROBUSTNESS IN THERMOCHEMICAL REACTOR PROTOTYPES 287
Xiang Gao, Ivan Ermanoski, Alberto De La Calle, Ellen B. Stechel

(511E) NUMERICAL MODELLING OF CERIA UNDERGOING REDUCTION IN A PARTICLE–GAS COUNTER-FLOW: EFFECTS OF CHEMICAL KINETICS UNDER ISOTHERMAL CONDITIONS 288
Sha Li, Vincent M. Wheeler, Apurv Kumar, Wojciech Lipinski

(511F) POROUS SILICA-ENCAGED ULTRAFINE METAL NANOCATALYSTS FOR SOLAR THERMOCHEMICAL METHANE REFORMING 289
Qiming Sun, Xian Li, Wojciech Lipinski, Ning Yan, Wang Chi-Hwa

(511H) CORRECTION OF THE DRIFT IN SOLAR TRACKING SYSTEMS THROUGH CLOSED LOOP CONTROL 290
Claudio Estrada

(517A) PROCESS SIMULATION OF LIGNOCELLULOSIC MATERIAL PYROLYSIS TO PRODUCING HIGH ADDED-VALUE CHEMICALS 291
Ronaldo Santos, Bruna Giron

(517B) UPGRADING CHEMICALLY DEPOLYMERIZED LIGNIN INTO VALUE-ADDED PRODUCTS: MAXIMIZING FEEDSTOCK UTILIZATION 292
Canan Sener, Miguel Perez, German Umana, Jason Coplien, Wayne Kontur, Steven D. Karlen, Daniel R. Noguera, Timothy J. Donohue, John Ralph

(517C) OXIDATIVE STABILITY AND BIODEGRADABILITY OF BIO-BASED LUBRICANTS USING THE PETROOXY AND BIO-KINETIC METHODS 293
F. Murilo T. Luna, Célio L. Cavalcante Jr.

(517D) EFFECTS OF PREPROCESSING PARAMETERS ON MATERIAL ATTRIBUTES AND FLOW BEHAVIOR OF LOBLOLLY PINE 294
Tiasha Bhattacharjee, Jordan Klinger, Wencheng Jin, Yidong Xia

(517F) OPTIMAL ECONOMIC AND ENVIRONMENTAL DESIGN TO PRODUCE BIO-JET FUEL 295
Juan José Quiroz-Ramírez Sr., V. A. Suárez Toriello, Argel Gastelum-Arellan, Eduardo Sanchez-Ramirez, Juan Gabriel Segovia-Hernández

(517H) AUTONOMOUS MONITORING OF WELLBORE INTEGRITY APPLYING TIME REVERSE NONLINEAR ELASTIC WAVE SPECTROSCOPY (TR NEWS) AND FIBER OPTIC SENSING AND COMMUNICATION 296
Carly M. Donahue, J. W. Carey, Erin Dauson, Luke Frash, Liwei Hau, Paul Johnson, Lawrence Murdoch, Seiji Nakagawa, Chris Ren, Hai Xiao

(517I) OPTIMAL SELECTION AND LOCATION OF NITROGEN RECOVERY SYSTEMS FOR INTENSIVE PIG FARMING 297
Edgar Martín-Hernández, Mariano Martín

(517J) USED COOKING OIL SUPPLY CHAIN ANALYSIS, OPTIMIZATION AND INTEGRATION IN THE CITY OF BOGOTÁ, COLOMBIA. 299
Juan Sebastián Rodríguez Flórez, Alvaro Orjuela

(517K) ENGRO'S INNOVATIVE APPROACH FOR SUSTAINED PLANT OPERATION WITH 25+ YEARS LIFE OF REFORMER RADIANT TUBES 300
Kashif Jameel, Usman Asif

(517L) DESIGNING SUSTAINABLE SUPPLY CHAINS IN LIGHT OF PLANETARY BOUNDARIES: A CASE STUDY OF BIOETHANOL PRODUCTION FROM SUGARCANE IN ARGENTINA 301
Ángel Galán-Martín, Jonathan Wheeler Sr., Fernando Daniel Mele, Gonzalo Guillén-Gosálbez

(517M) STOACHASTIC TECHNO-ECONOMIC ANALYSIS FOR AN INTEGRATED STRATEGY TO COPRODUCE JET FUEL RANGE ALKENES AND PENTANEDIOLS 302
Jaewon Byun, Jeehoon Han, Hoyoung Park, Dongseong Kang, Oseok Kwon

(517N) RECYCLING OF CATHODE OF LITHIUM-ION BATTERIES: A UNIQUE APPLICATION OF OXALATE CHEMISTRY 303
Ankit Verma, David R. Corbin, Mark B. Shiflett

(517O) IMPROVE THE THERMAL PROCESSABILITY OF POLY(3-HYDROXYBUTYRATE-CO-3-HYDROXYVALERATE) (PHBV) BIOPLASTIC WITH NATURAL RUBBER 304
Xiaoying Zhao, Peter Dent, Kurt W. Koelling, Osvaldo H. Campanella, Yael Vodovotz

(463C) RISK-CENTRIC OPTIMAL SYNTHETIC-PATHWAY GENERATION SYSTEM USING MONTE CARLO TREE SEARCH AND OPEN DATA OF REACTIONS AND MATERIAL SAFETY 305
Nagyeong Lee, Joonsoo Jeong, Dongil Peter Shin

(626A) CHARACTERIZATION OF FACTORS AFFECTING THE GAS WELL DELIVERABILITY OF THE JIAOSHIBA SHALE GAS FIELD BY MULTIVARIATE STATISTICAL ANALYSIS 306
Yizhong Zhang, Maolin Zhang, Bin Ju, Fanhua Zeng, Haiyan Mei, Ping Guo

(626B) SOLUTIONS OF C6F13- AND C4F9-BASED POLYFLUOROACRYLATES IN CO2; SOLUBILITY, VISCOSITY, AND ADSORPTION ONTO CEMENT 328
Peter Lemaire III, Adel Alenzi, Peter Koronaios, Dustin Crandall

(626C) COMPETITIVE SORPTION OF CO2 IN SHALE NANOPORES FOR SEQUESTRATION AND ENHANCED GAS RECOVERY USING MOLECULAR DENSITY FUNCTIONAL THEORY 330
Jinlu Liu, Arjun Valiya Parambathu, Walter G. Chapman

(626D) SPATIAL AND TEMPORAL EVOLUTION OF NATURAL GAS AND NATURAL GAS LIQUID FEEDSTOCKS IN SHALE BASINS 331
Qining Chen, Jennifer B. Dunn, David Allen

(626E) A NEW GEOLOGY-ENGINEERING INTEGRATION SOLUTION FOR SHALE GAS EXPLORATION: A CASE STUDY FROM CHANGNING BLOCK IN CHINA 333
Jianfa Wu, Haoyong Huang, Cheng Chang, Yi Song, Shengxian Zhao

(626F) NEW TECHNOLOGY OF EOR OF ULTRA DEEP AND COMPLEX CONDENSATE GAS RESERVOIRS IN TARIM BASIN,CHINA 334
Jiang Tongwen, Xie Wei, Wang Yong

(626G) PRODUCTION CAPACITY EVALUATION METHOD OF CONDENSATE GAS RESERVOIR CONSIDERING THE INFLUENCE OF STRESS SENSITIVITY AND CONDENSATE OIL 335
Chunxiao Du, Fanhua Zeng, Jun Yang

(626H) NUMERICAL SIMULATION STUDY OF COMPLEX FRACTURE TYPE EXTRA HIGH CONDENSATE GAS RESERVOIR IN BOHAI SEA 342
Zhennan Gao, Yanchun Su, Zhigang Geng, Weilin Liu, Baogang Huang, Fanhua Zeng, Jun Yang

(679A) DEVELOPMENT OF A COST-EFFECTIVE AND MODULAR CLOSED LOOP PROCESS FOR METHANOL PRODUCTION IN WASTEWATER TREATMENT PLANTS 352
Pradeepkumar Sharma, Sameer Parvathikar, John R. Carpenter

(679B) DETERMINING MATERIALS SPECIFICATIONS BASED ON TECHNOECONOMIC AND LIFE CYCLE ASSESSMENT: A CASE ON OXYGEN CAREER MATERIAL FOR CHEMICAL LOOPING STEAM METHANE REFORMING 354
Yasuhiro Fukushima, Alexander Guzman-Urbina, Hajime Ohno, Kakeru Ouchi

(679C) MINIMIZATION OF CAPITAL AND OPERATING COSTS IN COOLING WATER DISTRIBUTION SYSTEMS 355
Ricardo F. F. Pontes, Beatriz L. B. C. Macedo

(679F) ECONOMIC EVALUATION OF WASTEWATER TREATMENT PLANT CONSTRUCTION IN SMALL COMMUNITIES 356
Nolan Kelly, Peipei Chen, Yunting Zhu, Jiaqi Zhu, Jennifer Escobar

(612I) INVITED TALK: UNDERSTANDING RADIATIVE TRANSPORT IN FLOWING PARTICLES VIA MONTE CARLO RAY TRACING AND DATA-DRIVEN TECHNIQUES 357
Bingjia Li, Zijie Chen, Rohini Bala Chandran

(612E) THROUGH-THE-GLASS THERMAL IMAGING OF SOLAR CAVITY RECEIVERS 358
Jushwin Mahal, Thomas Cooper

(612F) ROBUST AND SPECTRALLY SELECTIVE AEROGELS FOR NEXT-GENERATION CONCENTRATED SOLAR THERMAL CONVERSION 359
Zachary Berquist, Andrew J Gayle, Neil P Dasgupta, Andrej Lenert

(612G) MODELLING OF A SMALL SCALE SOLAR CENTRAL RECEIVER WITH PRESSURIZED WATER FOR HEAT COLLECTION 360
Ahmed Gamil, Peiwen Li

(659A) KEYNOTE: SOLAR FUEL PROCESSING: RESEARCH, TECHNOLOGY DEVELOPMENT, AND SCALING 375
Sophia Haussener

(659B) INVITED TALK: SOLAR WATER SPLITTING: A DECOUPLED PHOTOELECTROCHEMICAL BENCHTOP DEMONSTRATION SYSTEM 376
Avigail Landman, Rawan Halabi, Avner Rothschild, Hen Dotan, Gideon S. Grader

(659C) DYNAMIC PROCESS SIMULATION AND CONTROL OF A HIGH PERFORMANCE THERMALLY-INTEGRATED CONCENTRATOR PHOTOVOLTAIC AND ELECTROLYSIS CO-GENERATION DEVICE 377
Isaac Holmes-Gentle, Saurabh Tembhurne, Clemens Suter, Sophia Haussener

(659E) CO2 GASIFICATION OF SEWAGE SLUDGE, PLASTIC WASTE, AND MUNICIPAL SOLID WASTE IN AN INTERNALLY-CIRCULATED FLUIDIZED BED UNDER HIGH-FLUX SOLAR IRRADIATION 379
Xian Li, Qiming Sun, Wojciech Lipinski, Yanjun Dai, Wang Chi-Hwa

(60A) FINDING OUR WAY TO DRIVE SUSTAINABILITY FORWARD THROUGH A GLOBAL PANDEMIC 380
Debbie F Mielewski, Alper Kiziltas

(60B) CREATING NEW BIOPRODUCTS TARGETING "ZERO WASTE" PRODUCTION 381
William J. Orts, Lennard Torres, Delilah F. Wood, William Hart-Cooper, Zachariah McCaffrey, Artur P. Klamczynski, Gregory M. Glenn

(60C) CAN BIOPLASTICS ALLEVIATE PLASTIC WASTE ISSUES? 382
Ramani Narayan

(60D) CIRCULAR ECONOMY AND SUSTAINABLE MATERIALS 383
Amar K. Mohanty

(694A) LIGNIN-BASED CARBON FIBERS FROM HYBRID POPLAR AND CORN STOVER: RENEWABLE AQUEOUS SOLVENTS FOR LIGNIN FRACTIONATION AND SOLVATION 384
Graham W. Tindall, Sagar Kanhere, Villo E. Bécsy-Jakab, Amod Ogale, David Hodge, Mark Thies

(694B) PARTIAL SUBSTITUTION OF PHENOL IN PHENOL-FORMALDEHYDE (PF) RESIN BY PLANT BIOMASS FOR WOOD ADHESIVES 385
Archana Bansode, Mehul Barde, Osei Asafu-Adjaye, John Hinkle, Vivek Patil, Brian Via, Sushil Adhikari, Thomas Elder, Ramsis Farag, Maria Auad

(694C) HAIRY CELLULOSE NANOCRYSTALS: AN EMERGING CLASS OF NANOCELLULOSES FOR ADVANCED INDUSTRIAL APPLICATIONS 386
Amir Sheikhi

(694D) NOVEL DOUBLE-SHELL LIGNIN NANOCAPSULES ARE A STABLE VEHICLE FOR FUNGICIDE ENCAPSULATION AND RELEASE 387
Raisa Carmen Andeme Ela

(694E) THE EFFECT OF NATURAL FILLERS ON THE MARINE BIODEGRADATION BEHAVIOUR OF POLY(3-HYDROXYBUTYRATE-CO-3-HYDROXYVALERATE) (PHBV) 388
Kjeld Meereboer, Akhilesh K. Pal, Erick O. Cisneros-López, Manju Misra, Amar K. Mohanty

(703B) HYDROTHERMAL LIQUEFACTION FOR CO-VALORIZATION OF PLASTICS AND BIOPOLYMERS 389
Seshasayee Mahadevan Subramanya, Phillip E. Savage

(703C) POLYSTYRENE SULFONIC ACID AS A RECOVERABLE CATALYST FOR THE HYDROLYSIS OF PET 391
Hossein Abedsoltan, Dr. Maria R. Coleman, Ana Alba-Rubio

(703E) CONVERSION OF WASTE SOLIDS TO MONOMERS AND INTERMEDIATES 392
Gennaro Maffia, Anne M. Gaffney

(703F) VALORIZATION OF PLASTIC WASTE USING HOT PRESSURIZED WATER 393
Seshasayee Mahadevan Subramanya, Phillip E. Savage

(706A) IRON OXIDE SUPPORTED ON SILICALITE-1 AS A MULTIFUNCTIONAL MATERIAL FOR BIOMASS CHEMICAL LOOPING GASIFICATION AND SYNGAS UPGRADING 395
Xin Li, Lijun Wang, Bo Zhang, Afsaneh Khajeh, Abloghasem Shahbazi

(706B) EVALUATION OF THE IRON-BASED MATERIAL WITH AL2O3/ZRO2 ADDITION AS OXYGEN CARRIER APPLIED IN CHEMICAL LOOPING PROCESS 396
Chingti Kao, Juiyen Cheng, Chenghsien Shen, Houpeng Wan

(706C) FEASIBILITY ANALYSIS OF CHEMICAL LOOPING AMMONIA SYNTHESIS FROM STRANDED NATURAL GAS........ 407
Laron Burrows, George M. Bollas

(706D) OXYGEN VACANCY HEALING FOR DESORPTION OF STORED AMMONIA FROM A METAL/ZEOLITE NANOCOMPOSITE 409
James M. Crawford, Ryther Anderson, Diego Gomez Gualdron, Moises A. Carreon

(706E) A MOLTEN CARBONATE SHELL MODIFIED PEROVSKITE REDOX CATALYST FOR ANAEROBIC OXIDATIVE DEHYDROGENATION OF LIGHT ALKANES 410
Yunfei Gao, Xijun Wang, Junchen Liu, Fanxing Li

(706G) SOLAR DRIVEN IRON OXIDE BASED THERMOCHEMICAL METHANE REFORMING AND WATER SPLITTING FOR H2 AND SYNGAS PRODUCTION 412
Rahul Bhosale, Suliman Rashid

(706H) KINETIC STUDIES ON CHEMICAL LOOPING – OXIDATIVE DEHYDROGENATION OF ETHANE USING SURFACE MODIFIED MIXED OXIDES 413
Yuan Tian, Ryan B. Dudek, Leo Brody, Luke Neal, Phillip R. Westmoreland, Fanxing Li

(728B) A HIGH-THROUGHPUT METHOD FOR ANALYZING STRUVITE FORMATION AND MORPHOLOGY: EXPLORING THE EFFECT OF PEPTIDE ADDITIVES........ 414
Jacob Hostert, Olivia Kamlet, Zihang Su, Julie Renner

(728C) A TECHNO-ECONOMIC ANALYSIS OF GREENHOUSE WASTE AND WATER RECYCLING USING ANAEROBIC DIGESTION: A CASE STUDY IN FOOD-ENERGY-WATER NEXUS 415
Babu Joseph, José Luis Guzmán

(728D) ENGINEERING A PURPLE NON-SULFUR BACTERIUM TO EXPAND SYMBIOTIC NITROGEN FIXATION........ 417
Cheryl Immethun, Rajib Saha

(728E) A MICROALGAE-METHANOTROPH COCULTURE PLATFORM FOR FUELS AND CHEMICAL PRODUCTION FROM WASTEWATER........ 418
Nathan Roberts, Matthew Hilliard, Q. Peter He, Jin Wang

(728F) AUXIN DEGRADATION BY VARIOVORAX MAINTAINS STEREOTYPIC PLANT ROOT DEVELOPMENT WITHIN THE COMPLEX PLANT MICROBIOME........ 419
Jonathan M. Conway, Isai Salas-González, Omri M. Finkel, William G. Walton, Matthew R. Redinbo, Jeffery L. Dangl

(731A) CATALYTIC PYROLYSIS OF WASTE AUTOMOBILE TIRES TO LOW SULFUR FUELS........ 420
Foster Agblevor, Oleksandr Hietsoi, Sedat H. Beis, Hossein Jahromi

(731B) PROBABILISTIC ANALYSIS OF CONVERTING MARINE-BORNE PLASTICS INTO USABLE FUELS 421
Elizabeth Belden, Michael T. Timko, Nikolaos Kazantzis

(731C) STRUVITE RECOVERY FROM MANURE BY WAY OF HYDROTHERMAL CARBONIZATION 422
Saeed Vahed Qaramaleki, Charles Coronella

(731D) CONVERSION OF RECYCLED DRYWALL INTO NITROGEN EFFICIENT FERTILIZERS 423
Jonas Baltrusaitis

(731E) SYNTHESIS AND CHARACTERIZATION OF IMPROVED ENERGY MATERIALS FROM SPENT LITHIUM-ION BATTERY CATHODES 424
Ankit Verma, David R. Corbin, Mark B. Shiflett

(731F) UPGRADING BIOCHAR VIA CO-PYROLYSIS OF AGRICULTURAL BIOMASS AND SPENT TIRES/SULFUR WASTES FOR ENVIRONMENTAL REMEDIATION 425
Seokyoung Oh, Yongdeuk Seo, Junhwan Kim, Hyungwoo Lee, Dokyoum Kim

(731G) SYNTHESIS OF NITROGEN-CONTAINING CARBON DOTS AS FE (III) PROBE AFTER EXTRACTING HUMIC ACID FROM COMPOST OF SLUDGE AND LEAF 426
Yan Zhu, Yong Liu, Ying Bao, Chuang Xie

(731H) STRUCTURAL CHARACTERIZATION AND MOLECULAR REPRESENTATION OF REFINED UNBURNED CARBON RECOVERED FROM FLY ASH 428
Yash Jaiswal, Sunderlal Pal

(732A) ELECTROCHEMICAL CONTROL OF THE CO2 CAPTURE CAPACITY OF AMINE SORBENTS USING CU (II)/CU (I) REDOX CHEMISTRY 430
Jonathan Boualavong, Christopher Gorski

(732C) EPOXIDE-MODIFIED AMINE-IMPREGNATED CO2 ADSORBENT WITH LOWER REGENERATION ENERGY REQUIREMENT AND ENHANCED DURABILITY 431
Yao Ma, Joo-Youp Lee

(732D) INTEGRATED COMPUTATIONAL SCREEN AND DESIGN OF POLYMERS FOR POST-COMBUSTION CARBON CAPTURE 432
Wei Shi, Kevin P. Resnik, David Hopkinson

(732E) PROCESS MODELING AND TECHNO-ECONOMIC ANALYSIS OF A DIAMINE-APPENDED METAL–ORGANIC FRAMEWORK FOR CO2 CAPTURE 434
Ryan Hughes, Goutham Kotamreddy, Debangsu Bhattacharyya, Chinyere Ezeobinwune, Ana Monteiro, Stephanie Didas, Surya T. Parker, Jeffrey R. Long, Michael S. Matuszewski, Benjamin P. Omell

(732F) ECONOMIC OPTIMIZATION AND SENSITIVITY ANALYSIS OF IONIC LIQUID-BASED CARBON CAPTURE PLANTS 436
Kyeongjun Seo, Calvin Tsay, Bo Hong, Thomas F. Edgar, Mark A. Stadtherr, Michael Baldea

(732G) OPTIMAL TEMPERATURE PROFILES WITHIN A CCS ABSORBER 438
Thomas Moore, Jaisree K. Iyer, Joshuah K. Stolaroff

(732H) OPTIMIZATION OF THE AMINES-CO2 CAPTURE PROCESS BY A NONEQUILIBRIUM RATE-BASED MODELING APPROACH TO DETERMINE OPERATING POLICIES 441
Luis Angel López-Bautista, Antonio Flores-Tlacuahuac

(733A) PLANNING AND OPTIMIZATION OF A REGIONAL HYDROGEN SUPPLY CHAIN: AN ASEAN CASE STUDY 442
Xiaodong Hong, Vaishali B. Thaore, Iftekhar A. Karimi, Farooq Shamsuzzaman, Farooq Shamsuzzaman, Xiaonan Wang, Adam K. Usadi, Bryan R. Chapman

(733B) PREDICTING REACTION MECHANISM OF ELECTROCHEMICAL CO2 REDUCTION USING MACHINE LEARNING 443
Aditya Prajapati

(733D) REVEALING THE MECHANISM OF INTRIGUING PREDOMINANT UREA FORMATION FROM THERMAL DEGRADATION OF CO2-LOADED AQUEOUS ETHYLENEDIAMINE 444
Bohak Yoon, Gyeong S. Hwang

(733E) TECHNO-ECONOMIC AND SUSTAINABILITY ANALYSIS OF THE PRODUCTION OF ACETIC AND FROM METHANE AND CARBON DIOXIDE 446
Marco Avendano, Mahmoud El-Halwagi

(733F) AN OVERVIEW OF RECENT EFFORTS UNDER THE DEPARTMENT OF ENERGY'S CARBON STORAGE PROGRAM: CARBONSAFE 447
Mary Sullivan, Kanwal Mahajan, Darin Damiani, Tomas Mora

(737A) ANATOMICAL FRACTION SEGMENTATION IN BIOMASS BALES 472
Rahul Reddy Kancharla, William A. Smith, Dr. Elisa Barney Smith, Jordan Klinger

(737B) POPULATION BALANCE OF HAMMER MILLED LOBLOLLY PINE IN 'ONCE-THROUGH' AND 'FRACTIONAL MILLING' CONFIGURATIONS 480
Tiasha Bhattacharjee, Jordan Klinger, Neal Yancey, Vicki S. Thompson

(737C) ASSESSING THE EFFECT OF BIOMASS PROPERTIES ON BIOMASS TO BIOFUEL SYSTEM PERFORMANCE USING OVERALL OPERATING EFFECTIVENESS 481
Damon Hartley, David N. Thompson, L. Michael Griffel, Quang Nguyen, Roni Mohammad

(737D) NTH PLANT VS. NTH SUPPLY: FEEDSTOCK COSTS AND SUPPLY UNCERTAINTY AS FUNCTION OF NUMBER OF BIOREFINERIES 482
Matthew Langholtz, Damon Hartley, Nicole Samu, Daniela Jones, Craig Brandt, Michael Hilliard, Chad Hellwinckel, Erin Webb, Chris Daly, Mike Halbleib

(737E) LOCATING NTH-PLANTS FOR BLENDED FEESTOCK CONVERSION AND PREPROCESSING NATIONWIDE: BIOREFINERIES AND DEPOTS 486
Tasmin Hossain, Daniela Jones, Damon Hartley, L. Michael Griffel, Yingqian Lin, Pralhad Burli, David N. Thompson, Matthew Langholtz, Maggie Davis, Craig Brandt

(746A) PRODUCTION OF RARE EARTH ELEMENTS AND CRITICAL MINERALS FROM COAL-BASED RESOURCES 487
Mary Anne Alvin

(746B) ENABLING EARLY PLANNING AND DEVELOPMENT OF ECO INDUSTRIAL PARKS FOR PHOTOVOLTAIC CIRCULAR ECONOMY USING MULTI OBJECTIVE OPTIMIZATION TECHNIQUES 488
Nehika Mathur, John Sutherland, Shweta Singh

(746C) THE LIFE MATHER TOOL: CHEMICAL MONITORING OF DATA AND EVALUATION OF ALTERNATIVE MATERIALS ON CONSUMER PRODUCTS 489
Melina Psycha, Filopoimin Lykokanellos, Antonis C. Kokossis

(746D) RESILIENCE AND RISK MANAGEMENT FOR PANDEMICS IN THE CHEMICAL INDUSTRY 490
Joan Cordiner

(746E) DEVELOPMENT OF ENERGY EFFICIENT AND DEPOLLUTING CERAMIC ROOF TILES 491
Maria Kouroutzi, Antonis Stratidakis, Marianthi Kermenidou, Spyros Karakitsios, Dimosthenis Sarigiannis

(748A) DEEP EUTECTIC SOLVENT TREATMENT OF CORN STOVER HYDROLYSATE FOR THE PRODUCTION OF HIGH PURITY LIGNIN 492
Dylan Cronin, Xiao Zhang

(748C) CONVERSION OF LIGNIN INTO BIOPLASTIC BY RHODOPSEUDOMONAS PALUSTRIS CGA009 494
Brandi Brown, Cheryl Immethun, Mark Wilkins, Rajib Saha

(748D) OXIDATIVE CONVERSION OF LIGNIN TO VALUE ADDED PRODUCTS USING OXYGEN AND HYDROGEN PEROXIDE 495
Ajinkya More, Thomas Elder, Zhihua Jiang

(748E) LIGNIN-DERIVED COMPOUNDS FOR THE PRODUCTION OF POLYURETHANE PLASTICS AND FOAMS 496
Dylan Cronin, Xiao Zhang

(748F) SEPARATION AND PROCESSING OF COFFEE CHAFF BY DEEP EUTECTIC SOLVENTS 499
Joan G. Lynam, Narendra Kumar, Justin Stallings

(748G) STRUCTURAL INTEGRITY OF BIOMASS-DERIVED MATERIALS FOR HYDROGEN STORAGE AT HIGH PRESSURE AND CRYOGENIC TEMPERATURE 500
Nepu Saha, Al Ibtida Sultana, Toufiq Reza

(749A) LIFE CYCLE MODELING FOR EMERGING TECHNOLOGIES: A PARAMETRIC APPROACH TO GUIDE RESEARCH AND DEVELOPMENT FOR ADDITIVE MANUFACTURING 501
Yuan Yao, Runze Huang

(749B) INCORPORATING ECOSYSTEM SERVICES IN LIFE CYCLE ASSESSMENT: MANAGED POLLINATORS AS A CASE STUDY 502
Alex Jordan, Harland Patch, Christina Grozinger, Vikas Khanna

(749C) EXPLORING A COMPUTATIONAL FRAMEWORK FOR SPATIALLY-EXPLICIT ABSOLUTE SUSTAINABILITY ASSESSMENT BASED ON A MULTI-REGIONAL HYBRID APPROACH 503
Kyuha Lee, Soomin Chun, Jeffrey M. Bielicki, Bhavik Bakshi

(749D) PLASTICS: REDEFINING SINGLE USE 505
Rachael H. Rothman

(749E) LIFE CYCLE ASSESSMENT (LCA) OF IRON OXIDE NANOPARTICLES () SYNTHESIS BY CO-PRECIPITATION: TRADITIONAL AND NOVEL MICROFLUIDIC METHODS 506
Olga P. Fuentes, Miguel A. Casalins, Juan C Cruz, Johann F. Osma

(753A) EFFICIENT DESIGN OF A METAL-OXIDE MEDIATED RED-OX CYCLE TO PRODUCE HYDROGEN FROM AMMONIA 509
Mandar Kathe, L. S. Fan, Michael Petrecca

(753B) DESULFURIZATION OF JP-8 BY CU-Y ZEOLITE AT ELEVATED TEMPERATURES FOR REMOTE ELECTRICITY GENERATION HAS TWO DISTINCT STAGES: CHEMISORPTION AND SURFACE REACTION 510
Priscilla Dias Da Silva, Michael S. Wong, Kyriacos Zygourakis

(753C) MODULAR ELECTROCHEMICAL TREATMENT OF CONCENTRATED ANIMAL FEEDING OPERATIONS WASTE FOR SIMULTANEOUS NUTRIENT REDUCTION AND RECOVERY 513
Damilola Daramola, Jason Trembly

(753E) STATE-BY-STATE COMPARISON OF THE ECONOMIC, ENVIRONMENTAL, AND ENERGY IMPACTS OF MANUFACTURING FACILITIES INTEGRATING SOLAR PHOTOVOLTAIC OR COMBINED HEAT AND POWER SYSTEMS 514
Derek Machalek, Kody M. Powell

(753F) MODULAR PLATE COLD DBD PLASMA REACTORS FOR DRY METHANE REFORMING 516
Jonas Baltrusaitis, Howard Pearlman, Chien-Hua Chen

(754A) OVERVIEW OF POTENTIAL CARBON STORAGE COMPLEXES IN THE OFFSHORE UNITED STATES GULF OF MEXICO AND ATLANTIC 517
William O'Dowd, Kanwal Mahajan, Darin Damiani, Larry Myer

(754B) LABORATORY INVESTIGATIONS OF CO2-ENHANCED OIL RECOVERY AND ASSOCIATED STORAGE IN RESERVOIR AND SOURCE ROCKS FROM THE BAKKEN PETROLEUM SYSTEM 518
Steve Hawthorne, David Miller, Carol Grabanski, James A. Sorensen, Bethany Kurz, Steve Smith

(754C) OPTIMIZATION OF EOR IN HORIZONTAL FRACTURING THROUGH INJECTION OF RECOVERED CO2 FROM AMINE SWEETENING PROCESS 520
Rajib Mukherjee, Sepehr Arbabi

(754E) SMART: A NEW INITIATIVE TO TRANSFORM SUBSURFACE VISUALIZATION AND PREDICTION THROUGH MACHINE LEARNING 521
M. Kylee Underwood, Kanwal Mahajan, Darin Damiani, Grant Bromhal

(754F) IMPROVING CO2 ENHANCED OIL RECOVERY PERFORMANCE THROUGH DATA ANALYTICS AND NEXT-GENERATION CONTROLLABLE COMPLETIONS 522
Nicholas Bosshart, John A. Hamling, Nicholas Azzolina, James A. Sorensen, Santosh Patil

(754G) THERMODYNAMIC MODELLING OF FECO3 SOLUBILITY USING THE EXTENDED UNIQUAC MODEL 523
Randi Neerup, Isaac A. Løge, Diana Carolina Figueroa Murcia, Kaj Thomsen, Philip L. Fosbøl

(757A) CATALYST SCREENING FOR THE HYDROTHERMAL LIQUEFACTION OF FOOD WASTE 524
Bita Motavaf, Sofia H. Capece, Phillip E. Savage

(757B) BIFUNCTIONAL ACID/BASE CATALYSIS FOR IMPROVED BIOCRUDE RECOVERY FROM FOOD WASTE HYDROTHERMAL LIQUEFACTION 525
Heather O. Leclerc, Feng Cheng, Geoffrey Tompsett, Michael T. Timko, Andrew R Teixeira

(757D) SIMULTANEOUS BIOGAS UPGRADING AND HYDROGEN SULFIDE REMOVAL THROUGH ENHANCED ELECTROCHEMICALLY ASSISTED ANAEROBIC DIGESTION OF FOOD WASTE 526
Lingkan Ding, Bo Hu

(757E) APPLICATION OF BLACK SOLDIER FLY LARVAE TO CONVERT MUNICIPAL ORGANIC WASTE TO SURFACTANTS AND OTHER VALUE-ADDED CHEMICALS 527
Eleanor O'Doherty, James McGregor

(757F) ALGAE CULTIVATION IN A SPACE STATION-BASED BIOREFINERY 528
Remil Aguda, Sarah Simoneaux, Cody Stelly, Cory Orgeron, Brandon Leblanc, Riley Nguyen, Y Ho, William Holmes, Rafael Hernandez, Mark Zappi, Emmanuel Revellame

(695A) KEYNOTE: SOLAR HYDROCARBONS FROM AIR CAPTURED CARBON DIOXIDE AND WATER 529
Klaus S. Lackner

(659H) INVITED TALK: PLASMA-ASSISTED CO2RECYCLING: A TOOL FOR STORING SOLAR ELECTRICITY 530
Maria Mikhail, Stéphanie Ognier, Maria E. Gálvez

(659F) SOLAR-THERMAL AMMONIA PRODUCTION: A RENEWABLE, CARBON-NEUTRAL ROUTE TO AMMONIA 532
Andrea Ambrosini, Kevin Albrecht, H. Evan Bush, Alberto De La Calle, Ivan Ermanoski, Tyler Farr, Xiang Gao, Matthew W. Kury, Peter Loutzenhiser, Nhu "ty" Nguyen, Ellen B. Stechel

(659G) SUSTAINABLE THERMOCHEMICAL HIGH-PURITY NITROGEN PRODUCTION USING REDOX OXIDES 533
Lena Klaas, Dorottya Guban, Nicole Preisner, Bruno Lachmann, Stephan Petersen, Martin Roeb, Christian Sattler

(695B) INVITED TALK: FROM CARBON DIOXIDE AND WATER TO CARBON-NEUTRAL FUELS AND CHEMICALS 535
Martin Roeb, Andreas Rosenstiel, Stefan Brendelberger, Nathalie Monnerie, Thomas Fend, Christian Sattler

(695F) INVITED TALK: SCREENING AND CHARACTERIZATION OF REDUCTION/OXIDATION-ACTIVE REACTIONS OF MIXED IONIC-ELECTRONIC CONDUCTING MATERIALS FOR AIR SEPARATION 536
Nhu "ty" Nguyen, Tyler Farr, H. Evan Bush, Andrea Ambrosini, Peter Loutzenhiser

(695C) PERSPECTIVES OF SOLAR THERMOCHEMICAL FUELS FOR AVIATION 537
Christoph Falter, Valentin Batteiger, Andreas Sizmann

(695D) METHANOL FROM SUNLIGHT AND AIR USING A MODULAR SOLAR DISH REACTOR SYSTEM 541
Remo Schäppi, David Rutz, Fabian Dähler, Philipp Haueter, Philipp Furler, Aldo Steinfeld

(695E) INTEGRATED REACTOR/HEAT-EXCHANGER FOR DIRECT SYNTHESIS OF DIMETHYL-ETHER 543
Lucas Freiberg, Matthew Coblyn, Goran Jovanovic, Nick Auyeung

Author Index

596584 Catalytic Reaction Pathways and Process Synthesis for Biomass Conversion to High-Value Chemicals for Organic Dye and Polymer Applications

Abstract:

Decreasing prices of liquid fuels have attracted interest in the production of high-value chemicals from biomass. Biomass-derived 5-(hydroxymethyl) furfural (HMF) can be converted to high-value chemicals with versatile functional groups, and it generates new opportunities for practical applications in organic dye and polymer industries. We have developed an effective process to produce the HMF from glucose by combining catalytic reactions and Simulated-Moving-Bed (SMB) separation. Glucose feed is converted to fructose in acetone/water solvent by an enzymatic isomerization, and fructose is then converted to HMF by acid-catalyzed dehydration. For the modeling of SMB separation, a commercial chromatography resin (DOWEX 99) was examined to measure adsorption equilibrium parameters of glucose, fructose, and HMF. The estimated minimum selling price (MSP) of HMF from glucose was $1461/ton by techno-economic analysis (TEA). Additionally, as-synthesized HMF from sugar feed was reacted with acetone to produce an aldol-condensed product (HAH), where various functional groups can be inserted by using a base catalyst (NaOH) in acetone/water solvent. High purity HAH (>99% purity) was obtained by flowing water over the precipitated product. Finally, selective modifications of functional groups in HAH were investigated by catalytic etherification, reduction, and dimerization to demonstrate the possibility of applications as monomers in organic dyes and polyethers. TEA indicates that the process can produce HAH from fructose at a MSP of $1958/ton, showing its price potential to replace anthraquinone dye and bisphenol-A (BPA)-based polymer markets. The proposed processes with catalytic reaction routes can achieve effective control over functional groups in biomass and provide a facile approach to supply renewable high-value chemicals.

608006 Superacid Polymer Catalysts for Hydroxymethylfurfural Production

Abstract:

Hydroxymethylfurfural (HMF) is an important and versatile renewable platform chemical. HMF can be obtained from fructose, but its production from glucose is preferred due to its higher availability and lower cost. Glucose is first isomerized to fructose (Lewis acid), followed by subsequent dehydration of fructose to HMF (Brønsted acid). To overcome the challenges associated with the use of homogeneous and heterogeneous catalysts, we have developed novel polymer-based catalysts that join the advantages of both systems. We found that polystyrenesulfonic acid (PSSA), a catalyst containing Brønsted acid sites, was very active in the dehydration of fructose to HMF. PSSA is soluble in polar solvents; therefore, it is very active and cannot be deactivated through coking, with the additional advantage that it can be easily recovered from the reaction medium for further use. First, we explored the effect of the molecular weight on the catalytic activity and recoverability. Then, we added a second functionality (Lewis acidity) to PSSA by ion exchange with $AlCl_3$ to make it suitable for the production of HMF from glucose.[1] A series of soluble and reusable catalysts with different Brønsted: Lewis acid ratios was synthesized, characterized, and studied in the tandem catalytic conversion of glucose to HMF.

Figure 1 shows the glucose conversion and HMF yield obtained with PSSA and PSSA-$AlCl_3$ catalysts. The highest HMF yield was obtained with the PSSA-$AlCl_3$ 60:40 catalyst, in which 40 % of the Brønsted acid sites were replaced by $AlCl_3$. We found that further addition of $AlCl_3$ resulted in lower HMF yields, likely due to the reduction of the hydrophilicity (and solubility) of the polymer catalysts.

(1) A.C. Alba-Rubio, M.R. Coleman, S. Kalidindi, A.S. Joshi, I.S. Omodolor; Homogeneous and Reusable Superacid Polymer Catalysts Useful for the Synthesis of 5-Hydroxymethylfurfural from Glucose. WO 2019/089448 A1 (2019).

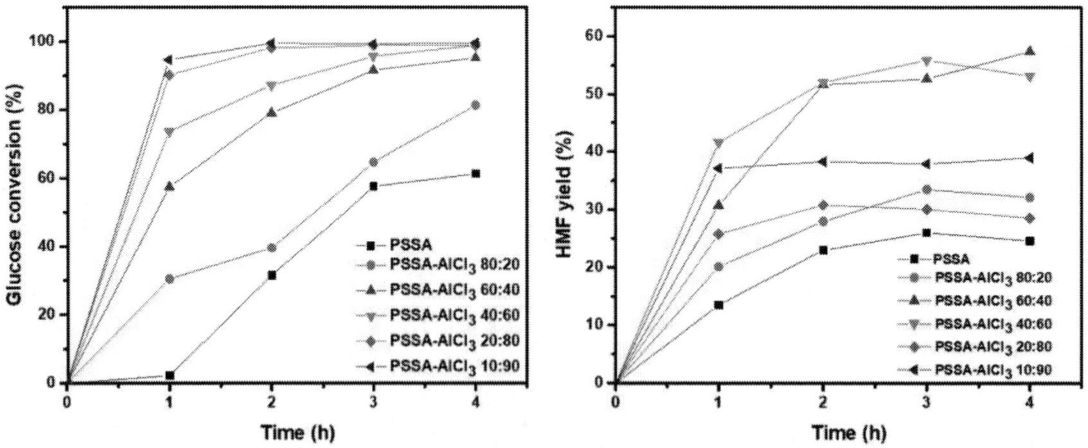

Figure 1. Glucose conversion and HMF yield with 200 kDa PSSA and PSSA-$AlCl_3$ catalysts.

595950 Rates of Levoglucosanol Hydrogenolysis over Brønsted and Lewis Acid Sites on Platinum Silica-Alumina Catalysts Synthesized By Atomic Layer Deposition

Abstract:

Silica-alumina materials (SiAls) are solid acid catalysts used in applications such hydrocarbon and biomass upgrading. While amorphous SiAls contain both Brønsted and Lewis acid sites, distinguishing their structures and reactivities remains a fundamental challenge [1]. Here, we employ Atomic Layer Deposition (ALD) to synthesize SiAls with the aim of understanding the reactivity of Brønsted and Lewis acid sites for the C-O bond hydrogenolysis of levoglucosanol (Lgol) to high-value chemicals [2]. Lgol is produced from biomass-derived levoglucosenone [3].

Bifunctional Pt/AlO_x/SiO_2 catalysts were synthesized using ALD to deposit sub-nanometer overcoats of AlO_x onto SiO_2, and kinetic measurements were made in the acid-limited regime. The ratio of Brønsted to Lewis acid sites, measured by NH_3-TPD and pyridine-FTIR, systematically decreases with Al_2O_3 loading, while the Al coordination, measured by solid state ^{27}Al NMR increases. Lewis acid sites on ALD-AlO_x/SiO_2 catalysts (potentially related to Al(V) species) have a 4-times higher turnover frequency (TOF) compared to Lewis acid sites on bulk γ-Al_2O_3. A simple model is used to describe the reactivity trends for ALD-AlO_x catalysts, assuming that is overall rate is the linear sum of independent contributions from Lewis and Brønsted acid sites. Based on this model, Brønsted acid sites on ALD-AlO_x/SiO_2 catalysts have a 6-times higher TOF than Lewis acid sites on these catalysts. With decreasing Al_2O_3 loading, the reactivity of these catalysts transitions from being entirely governed by Lewis acid sites, to largely governed by Brønsted acid sites. This work provides a new approach to synthesize amorphous SiAls with tunable Brønsted/Lewis acid site ratios, and reveals differences in the reactivity of Brønsted and Lewis acid sites on these materials.

References

[1] Hensen, E. J. M., et al. *J. Catal.* **2010,** *269* (1), 201-218.

[2] Krishna, S. H., et al. *ACS Catal.* **2018**, 3743-3753.

[3] Court, G. R., et al. *WO Patent* **2011**/000030 A1.

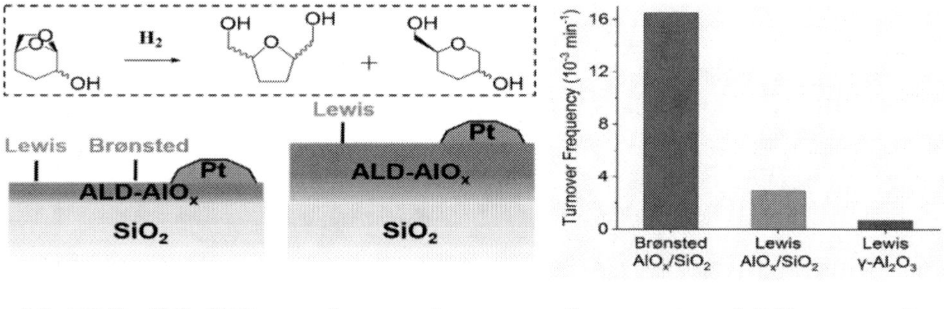

Schematic of Pt/ALD-AlO_x/SiO_2 catalysts and turnover frequencies of different acid site types for Lgol hydrogenolysis.

611196 Tuning Performance of Ketone Partial Oxidation over Supported VO_x and Its Utilization: Selective Maleic Anhydride Production from Levulinic Acid Oxidative Scission.

Abstract:

Aerobic partial oxidations are crucial for producing commodity and specialty chemicals. In many industrial applications, vanadium oxide catalysts are used to effect partial oxidations; however, the desired reactions often suffer poor selectivity losses to combustion. In this study, we consider ketone partial oxidation, namely oxidative scission, to produce carbonyl and carboxyl fragments. We consider the impact of support, solvent, vanadate structure, and ketone structure in dictating performance. Specifically, we examine the performance of various mono- and polyvanadate structures supported on SiO_2, γ-Al_2O_3, TiO_2, and CeO_2. Structures of each catalyst were characterized using Raman and/or DR-UV-Vis spectroscopy. Interestingly, we observe that while support and vanadium oxide structure have a significant impact on activity, oxidative scission selectivity is largely invariant. Further investigations with a range of methyl ketone probe molecules having different degrees of a-carbon substitution revealed further insights about the influence of ketone structure on oxidative scission selectivity. Finally, we turn our attention to suppressing combustion, and we observe that water is beneficial in this regard. Specifically, it enhances oxidative scission rates, suppresses combustion rates, and mitigates catalyst deactivation.

Toward potential applications, we study the oxidative scission of bio-based levulinic acid. We observe that oxidation of either affords maleic anhydride in excellent yields (60-70%), which contrasts expectations based on trends observed for monofunctional methyl ketones. An examination of primary, secondary, and tertiary reaction pathways during oxidative scission of levulinic acid reveals that its facile intramolecular dehydration leads to the formation of stable, cyclic angelicalactones, which quickly dehydrogenate to form protoanemonin—a polyfunctional molecule having both diene and lactone structures. Its unique structure suggests it may be a useful platform chemical in the synthesis of biopolymers; further, its formation is unique to this 5-carbon system, and its formation underlies the high selectivity to maleic anhydride observed in this system.

599925 Selective Reduction of Carboxylic Acids to Aldehydes over Promoted MoO_3 Catalyst

Abstract:

Selective activation of carboxylic acids to form alcohols and aldehydes with highly oxophilic catalysts could enable the formation of a variety of highly valuable renewable products. Selective reduction with highly oxophilic supports carries the advantage of activating the acid functional group without perturbing other functionalities. The use of traditional metal catalysts for this purpose generally favors C-O cleavage followed by rapid decarbonylation to yield alkenes. These alkenes are readily converted to less desirable alkanes in the presence of H_2 or often lead to coke and catalyst deactivation in the absence of co-fed H_2. This limits applicability to produce high value chemicals such as butadiene via these routes.

In this contribution, we report a new chemistry that has been explored with aldehydes and ketones, reverse mars van Krevelen chemistry over MoO_x, for the selective conversion of carboxylic acids. Our study focusses on the selective conversion of pentanoic acid (PA) to aldehyde over MoO_3 at different temperatures via hydrodeoxygenation (HDO) reaction. Here, we hypothesize that the selective C-O cleavage reactions are occurring over the defects on the sub stoichiometric oxide via reverse Mar-Vars Krevelen chemistry. Reducibility of MoO_3 is well known to be challenging at low temperatures, therefore in order to facilitate the reduction process, very low loadings of platinum (Pt) were incorporated on the catalyst (0.05%Pt/MoO_3). The net effect is the increase on the rate of conversion of PA at lower temperature over MoO_3 by hydrogen spillover to promote partial reduction of the MoO_x while minimizing side reactions that may occur on the metal directly. X-Ray photoelectron spectroscopy (XPS) was used to investigate effect of H_2 during the partial reduction on both MoO_3 and 0.05% Pt/MoO_3 catalysts. Kinetics analysis reveals that low loading of Pt on MoO_3 decrease the apparent activation barrier for acid conversion by over 33 kJ/mol.

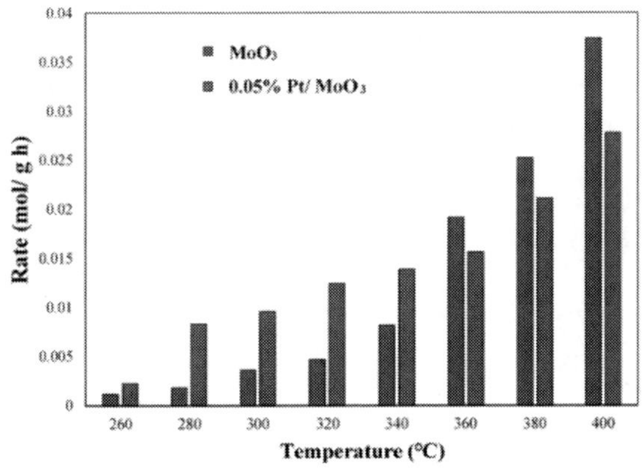

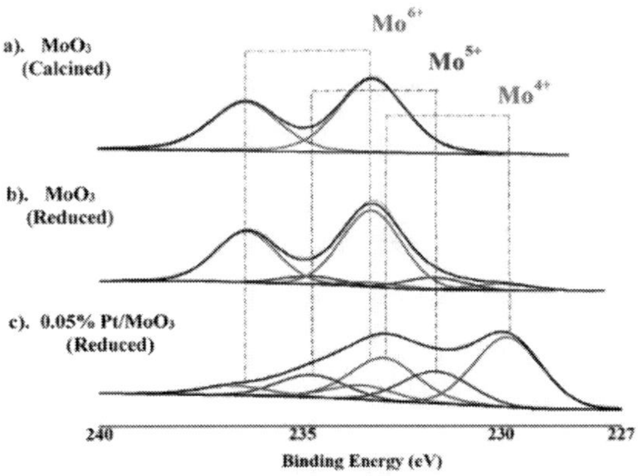

Fig.1 Rate conversion at different temperatures for MoO_3 and 0.05 % Pt/MoO_3. **Fig. 2** XPS analysis of MoO_3 calcined, reduced at in atmospheric flowing H_2 at 400°C for 1 hour, and 0.05wt% Pt/MoO_3 reduced under identical conditions.

594177 Noble Metal Catalyzed Ketone Hydrogenation with Microkinetic Modeling

Abstract:

Vapor-phase hydrogenation of acetone was studied over supported Pt catalysts in a packed bed reactor. To fundamentally understand both the mechanism and rate controlling steps during carbonyl hydrogenation, we performed a comprehensive macro- and micro-kinetic analysis of acetone hydrogenation. We report apparent reaction orders in acetone and hydrogen, activation energies, and observed kinetic isotope effects over a large range of temperatures and species partial pressures. Generally, ketone hydrogenations on metal surfaces follow a Horiuti-Polanyi mechanism, wherein hydrogen atoms add sequentially to the adsorbed ketone. Overall, our experimental observations are best described by assuming atomic hydrogen and surface-bound hydrocarbons do not compete for adsorption at the same active sites; this suggests that, under reaction conditions, there is a sub-set of surface sites that are accessible to hydrogen atoms, but not larger hydrocarbon fragments. We further conclude that two sequential surface reactions, respectively involving the formation of H-C and an H-O bonds, each exert partial rate control under most reaction conditions. The degree of rate control from each is ultimately determined by a combination of intrinsic kinetic parameters (e.g., elementary activation barriers) and coverages of reacting species under experimental conditions conditions. Despite our conclusion that H-X bond formation steps are rate controlling, we observe a very minor kinetic isotope effect upon switching between H_2 and D_2. We reconcile these observations by considering both kinetic and thermodynamic impacts of isotopic switching, as well as partial rate control from multiple elementary steps. Finally, we report an optimized set of elementary kinetic parameters that, when coupled with our microkinetic model, allow one to predict rates of acetone hydrogenation over a large range of temperatures and pressures.

References:

1. X. Gao, A. Heyden, O.A. Abdelrahman, J.Q. Bond, Journal of Catalysis, 374 (2019) 183-198.
2. O.A. Abdelrahman, A. Heyden, J.Q. Bond, Journal of Catalysis, 348 (2017) 59-74.

604844 Bifunctional Ru@N-Doped Mesoporous Carbon Catalysts for Catalytic Transfer Hydrogenolysis of Bio-Derived Polyols

Abstract:

The depleting fossil fuel reserves and increasing problems of environmental pollution have accelerated the development and utilization of renewable energy. Biomass resources, as the only widely distributed and accessible renewable organic carbon raw materials in the world, have attracted worldwide attention.[1, 2] In this context, hydrogenation of bio-derived polyols (sorbitol, glycerol, xylitol, etc.) is an important way to convert biomass into high value-added chemicals.[3]

Aqueous phase hydrogenation process is considered to be the most promising route for biomass conversion. But high temperatures (> 400 ^{o}C) and high H_2 pressure (> 4 MPa) make the side reactions very intense, leading to poor atomic utilization economics.[4] The operating conditions of the catalytic transfer hydrogenolysis process are much milder, using H_2 generated from water or renewable resources, so the energy consumption is small, the atomic utilization rate is high (~100%) and the environmental pollution is small.

According to our literature review, it is found that current studies on transfer hydrogenolysis of polyols involve alkaline species as promoters for simultaneous H_2 generation and hydrogenolysis reactions, which unfavorably cause corrosion and additional costs for product separation. In this work, we proposed a series of nitrogen doped mesoporous carbon (NMC) for efficient transfer hydrogenolysis of glycerol to propylene glycol. The unique interaction between Ru-nitrogen sites is key for anchoring of Ru particles on the inner and outer surfaces of the NMC. While metallic sites are active for H_2 production (C-H cleavage), H_2 transfer and hydrogenation (C-O cleavage) properties, the basic sites provided from NMC facilitate C-H and C-O bond cleavage during transfer hydrogenolysis of glycerol. The conversion of glycerol and the selectivity of 1,2-PDO are 45.1% and 46.7% at 200 ^{o}C , under N_2 atmosphere. Detailed characterization by XRD, XPS, TEM and BET will be conducted to reveal the structural features and correlated with catalytic performances.

599212 A Tuneable Process for the Conversion of Lignin to Jet Fuel over Molybdenum Carbide Catalysts

Abstract:

The aviation industry has a growing demand for sustainable aviation fuels (SAF) that can reduce greenhouse gas emissions while satisfying strict physical property requirements to meet safety and quality standards of jet fuel. Between 8% and 25% of such SAFs must be aromatic to satisfy requirements related to energy density, lubricity, material compatibility, and elastomeric swelling. One of the most promising feedstocks for the aromatic portion of SAFs is lignin, as it constitutes 15-30 wt% of all lignocellulosic biomass and is currently underutilized waste in the cellulosic-ethanol and pulp-and-paper-industries. The utilization of lignin in jet fuel hinges on effective methods for reducing oxygen content (lignin O/C ~0.3-0.4 vs jet fuel O/C<0.03), while limiting ring hydrogenation and maximizing yields of C9-C20 hydrocarbons. Molybdenum carbide (Mo_2C) has been shown to selectively cleave aryl-oxygen bonds in model compounds such as anisole and m-cresol, removing oxygen and preserving aromaticity. Herein, we test the efficacy of Mo_2C during hydrodeoxygenation of lignin oil produced via reductive catalytic fractionation (RCF), which is a stabilized mixture of oxygenated, lignin-derived monomers, dimers, trimers and tetramers. Flowing pure RCF lignin oil over a packed catalyst bed in a 3-phase trickle bed reactor enables the generation of steady-state partial-conversion data to understand and optimize catalyst performance while deoxygenating complex lignin streams. Through recycling products, we can obtain and characterize fully deoxygenated lignin oil, concluding that both monomer and dimer products are strong jet fuel candidates. Finally, by co-supporting small amounts of late transition metals, we can shift the product selectivity to tune the aromaticity, closing the loop to enable rational catalyst design for desired jet fuel properties.

597910 Investigating the Role of Solid Acid Sites in Surface Reactions of a Model Sugar

Abstract:

Sugars derived from biomass possess a high degree of functionality and offer promise as a potential feedstock for production of commodity chemicals. Erythrose is a cyclic C4 sugar with similar functionality to glucose, which makes it an attractive model feedstock to study. Heterogeneous catalysts, such as metal oxides, have been shown to convert biomass-derived sugars to range of chemicals via acid catalyzed reactions. This work analyzes the interactions of erythrose with a variety of metal oxides that possess different types of acid sites.

^{13}C MAS NMR spectroscopy with a direct polarization (DP) pulse sequence was used to identify surface species when the catalysts were impregnated with erythrose at room temperature. The following catalysts were tested due to varying degrees of Brønsted and Lewis acidity: γ-Al_2O_3, CeO_2, Nb_2O_5, SiO_2, SnO_2, TiO_2, and ZrO_2. CeO_2 and Nb_2O_5 promoted ring opening as evidenced by the appearance of two peaks in the carbonyl region of the spectrum. Nb_2O_5 also produced a significant amount of surface species containing aliphatic carbon suggesting further reaction to other species. SnO_2, TiO_2, and ZrO_2 showed the most drastic reactions when the temperature was increased from 25 °C to 50 °C. Both carbonyl and aliphatic carbon became significantly more abundant at elevated temperature. In addition, species on TiO_2 had the sharpest peaks suggesting it is weakly bound to the active sites due to the moderately acidic nature of TiO_2.

Contact time experiments were carried out to determine the mobility of specific functional groups within the surface species present at 25 °C and 50 °C. This was done by comparing the relaxation times of different moieties. Some groups were greatly affected by the temperature increase, while others were immune. This is a significant consideration because one can discriminate which moieties are preventing the molecules release into the bulk phase.

601141 Mechanism of Polyol Hydrodeoxygenation on Molybdenum Carbide Supported Copper Catalyst: A Combined DFT and Microkinetic Modeling Study

Abstract:

Producing valuable chemicals from renewable biomass has received consistent attention. For highly oxygenated biomass-derived chemicals, the selective hydrodeoxygenation without C/C cleavage is one of the desired reactions preventing loss of carbon as CO or CO_2. With the goal of designing non-precious metal catalysts, the Mo_2C supported Cu has been investigated for the hydrodeoxygenation of oxygenates using density functional theory and microkinetic modeling. Results are correlated to experimental surface science and reactor data.

Previously, we determined that a Cu modifier on a molybdenum carbide (Mo_2C) support reduces the oxophilicity of Mo_2C, such that the selectivity with regards to the number of C/O cleavages can be controlled with the Cu coverage. Here, we focused on a monolayer Cu/Mo_2C and a Cu(111) model to understand the support effect of Mo_2C on the reaction mechanism and on the active site performance under various reaction conditions. Our microkinetic analysis predicted the Cu/Mo_2C surface was more active for a single C/O dissociation of glycerol versus Cu(111) where C/O scission is slow and dehydrogenated products are primarily produced. The enhanced activity of the Cu/Mo_2C surface for the glycerol activation was attributed to electronic modification of Cu atoms by the Mo_2C support, which enhanced the adsorption strength of intermediates and reduced the activation barrier for C/O cleavage reactions.

Next, we examined the mechanism of the conversion of 1,2- and 1,3-propanediol on Cu/Mo_2C to understand the role of intramolecular hydrogen bonding on the selective C/O cleavage. We observed that one C/O cleavage can occur for both species which is consistent with our glycerol study. Furthermore, in agreement to experimental studies, 1,2-propanediol favors terminal C/O cleavage to produce acetone. Finally, we extended our C3 studies on Mo_2C supported Cu catalysts to larger linear polyols. Here, we extrapolate the energetics from our C3 studies to larger molecules via machine learning methods.

605828 Structure-Sensitive Catalytic Glycerol Oxidation on Late Transition Metals

Abstract:

Glycerol, a substantial byproduct of bio-diesel processing (about 10% w/w), has the potential to be selectively transformed to higher value chemicals and improve overall bio-diesel refining economics.[1] Various transition metals are suitable for catalyzing glycerol oxidation, but face a substantial selectivity challenge between various C_1, C_2, and C_3 products.[2] Further, despite recent developments in the shape-selective synthesis of metal nano particles, a detailed understanding of the impact of catalyst structure on the reaction selectivity is still limited. This suggests a substantial opportunity for selective glycerol oxidation if fundamental insights into the role of catalyst surface structure can be obtained and leveraged toward design.

In this presentation, we share density functional theory (DFT) calculations of glycerol oxidation mechanisms to yield glyceraldehyde and dihydroxyacetone on transition metal catalysts. Our results show that the preference for C-H vs. O-H bond breaking varies between (111) and (100) crystal facets, suggesting the possibility of manipulating product selectivity based on catalyst structure. Moreover, we show that a recently-developed energy scaling model is valid for molecules even as complex as glycerol, enabling prediction of structure-sensitive reaction energetics on nano particle models with improved computational efficiency.[3] We finally highlight the importance of considering the presence of adsorbed oxygen species in stabilizing glycerol adsorption and promoting the kinetics of dehydrogenation, as has been noted previously in the literature but not extensively studied for glycerol oxidation, and discuss implications for the design of catalysts targeting the selective formation of C_3 products.

1. S. Bagheri, N.M. Julkapli, and W.A. Yehye, *Renew. Sust. Energ. Rev.* (2015), 41, 113-127.
2. S. Carrettin, P. McMorn, P. Johnston, K. Griffin, C.J. Kiely, and G.J. Hutchings, *Phys. Chem. Chem. Phys.* (2003), 5, 1329-1336.
3. L. T. Roling and F. Abild-Pedersen, *ChemCatChem* (2018), 10, 1643-1650.

610406 Selective Hydrodeoxygenation of 1,4-Anhydroerythritol to Tetrahydrofuran over MoO_x-Pd/TiO_2 catalysts

Abstract:

The development of catalytic materials that can selectively cleave hydroxyl groups from oxygen-rich feedstocks is a challenge in the production of chemicals from lignocellulosic biomass. Recently, heterogeneous catalysts comprised of supported palladium and rhenium oxide species have been reported for the selective removal of vicinal hydroxyls in biomass-derived sugar alcohols, such as erythritol and 1,4-anhydroerythritol (1,4-AHERY).[1-3] In this presentation, we describe our findings on the development of heterogeneous catalysts comprised of molybdenum oxide and palladium supported on titanium oxide (MoO_x-Pd/TiO_2) for the hydrodeoxygenation (HDO) of 1,4-AHERY to tetrahydrofuran (THF) in high yields. The activity and stability of the catalysts are comparable to other materials reported for this chemistry. However, in contrast to other catalysts, it eliminates the use of expensive rhenium metal. The relationship between catalyst structure and performance of MoO_x-Pd/TiO_2 was studied by a combination of reaction kinetic and characterization studies. Reaction kinetic studies suggest there are possibly two distinct catalytic sites for hydrogen and 1,4-AHERY activation. Characterization studies, including X-ray diffraction (XRD), Raman spectroscopy, and X-ray photoelectron spectroscopy (XPS) suggest that isolated monomeric Mo^{6+} species supported on TiO_2 are reduced to catalytically active Mo^{4+} species by spillover hydrogen dissociated over Pd metal surfaces. Based on these results, we propose that the reaction pathway over MoO_x-Pd/TiO_2 involves the deoxydehydration of 1,4-AHERY to 2,5-dihydrofuran followed by hydrogenation of the C=C bond to produce THF. Furthermore, we demonstrate that MoO_x-Pd/TiO_2 shows remarkable catalytic stability upon multiple reaction cycles.

1. Ota, N., Tamura, M., Nakagawa, Y., Okumura, K. and Tomishige, K., *Angew. Chem. Int. Ed.*, **54**, 1897–1900 (2015).
2. Ota, N., Tamura, M., Nakagawa, Y., Okumura, K. and Tomishige, K., *ACS Catal.*, **6**, 3213–3226 (2016).
3. Xi, Y., Yang, W., Ammal, S.C., Lauterbach, L., Pagan-Torres, Y.J. and Heyden A., *Catal. Sci. Technol.*, **8**, 5750–5762 (2018).

608397 Low-Temperature Hydrodeoxygenation By Polyoxometalate-Supported Pd_1 Single-Atom Catalysts

Abstract:

Hydrodeoxygenation (HDO) is essential for the conversion of biomass resources into fuels and chemicals. Most catalysts suffer from low activity under mild conditions or low selectivity and stability under required harsh reaction conditions. Herein, we present a heterogeneous Pd_1/CsPTA catalyst that exhibits turnover-frequencies of up to 10,000 h^{-1} for the HDO of a variety of aromatic C=O and COH-containing compounds under mild reaction conditions (30 °C and 1 MPa H_2 pressure). Isotope labeling studies show that the HDO of carbonyl compounds follows a reduction, hydrogenolysis pathway in contrast to most previously reported acidic HDO catalysts. A combination of inelastic neutron scattering, infrared spectroscopy, kinetic studies and DFT calculations reveal the formation of hydride species from the heterolytic H_2 splitting explaining the catalytic activity significantly enhanced compared to all previously reported Pd particle catalysts. DFT calculations further confirm the heterolytic H_2 splitting with low energy barriers and suggest that water is actively participating in the catalytic cycle. The hydrogenolysis reaction is predicted to follow a two-step mechanism involving the initial protonation of the OH group preceded by the hydride transfer. Furthermore, we show that the substrate scope can be extended to the HDO of molecules derived from lignocellulosic biomass. The HDO of furfural to 2-methyltetrahydrofuran was achieved under conditions much milder than any reported so far.

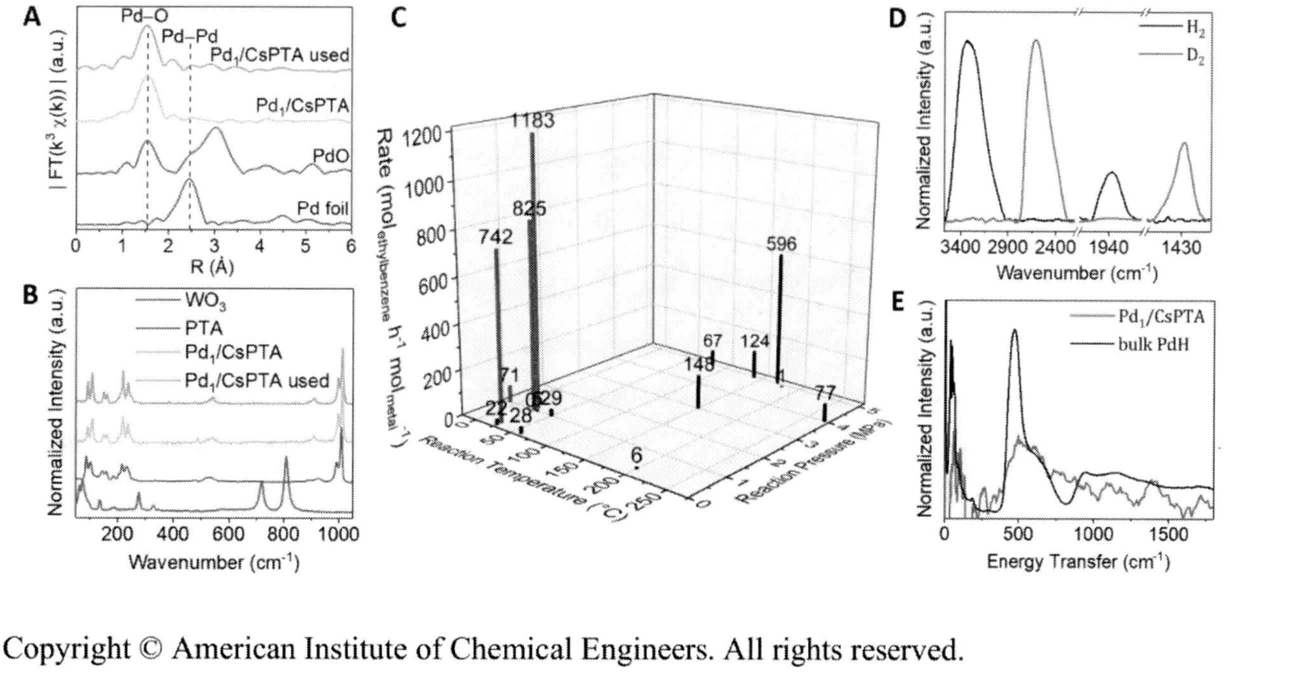

602024 Emissions Control Technologies for RTI's Non-Aqueous Solvent for Carbon Capture

Abstract:

Advanced water-lean solvents (WLS) for post-combustion CO_2 capture have been gaining interest due to their ability to reduce the parasitic penalty from energy needed for solvent regeneration. RTI International, with funding from the US Department of Energy, has been involved in development of its non-aqueous solvent (NAS), eCO$_2$Sol^{TM}. The solvent has demonstrated specific reboiler duties as low as 2.3 GJ/t-CO_2 at pilot plant scale CO_2 capture system at Tiller, Norway. Commercial implementation of these novel CO_2 capture technologies hinges on successful control of amine emissions. Vapor phase emissions can be controlled through use of water wash that scrub the amine components out of the treated gas as it exits the absorber column. While significant research work has been done to characterize and control aerosol-based emissions from aqueous solvents, similar information is not available for emissions from use of these novel water-lean solvents.

RTI has been conducting studies from fundamental and operational aspects to reduce the overall amine emissions from the WLS systems, specifically RTI's non-aqueous solvent eCO$_2$Sol^{TM}. The work has been carried out at 4-6 kW equivalent using the RTI's bench-scale gas absorption system (BsGAS). It consists of a CO_2 absorber column with intercoolers and water wash section, and a solvent regenerator column with a thermosiphon reboiler and inter-stage heaters. RTI's BsGAS was modified to add the aerosol generation system and monitoring equipment to determine the aerosol characteristics during the eCO$_2$Sol CO_2 capture process. The system is now capable of producing aerosols with the peak diameter and concentration of 50 mm and 1.2 x10^7 cm^{-3}, similar to those observed in the actual coal-fired power plant flue gas at the inlet of the CO_2 capture system.

RTI has conducted over 1,300 hours of parametric testing to evaluate the impact of the aerosols and operating conditions during the CO_2 capture with NAS on the overall amine emissions in the treated flue gas. A simulated flue gas, with 15% CO_2, 2.3-4.2% H_2O and balance air, was used in the study. SO_3 particulates were generated by reacting SO_2 with O_2 in air over a silica supported vanadium pentoxide (V_2O_5/SiO_2) catalyst at 450°C. The generated SO_3 particulates were injected into the simulated flue gas fed to the absorber. The SO_3 particulates mimic the presence of SO_X, fly ash and other particulates in the flue gas exiting the direct contact cooler and provides nucleation sites for growth of aerosols in the CO_2 capture process. A Scanning Mobility Particle Sizer and an Aerodynamic Particle Sizer were used to monitor the aerosol particles size and number distribution at various locations in the CO_2 capture system. The amine emissions at the absorber outlet and water wash outlet were monitored using a MKS MultiGas™ FTIR spectroscopy gas analyzer. Initial testing found that the CO_2 capture rate and water wash temperature were the two most dominant process factors on vapor emissions. Further testing fixed those two process variables to evaluate other variables including liquid/gas ratio (L/G, mass basis), intercooling profile (extent of intercooling at the top and bottom intercooler), temperature of the lean solvent to the absorber, the water wash L/G, temperature of the water wash return stream, and the inlet saturation temperature of the flue gas (which dictates both the extent of water in the flue gas, and the water wash temperature required to maintain water balance in the system). The CO_2 capture rate was maintained close to 90% capture rate.

Results from parametric testing suggested that the presence of the aerosols in the flue gas could increase the overall emissions by 10X compared to the baseline emissions from NAS's vapor pressure. The temperature difference between the temperature bulge seen in the absorber column and the water wash temperature impacts the particle growth, with the aerosol emissions increasing with the increase in the temperature difference. Most of the aerosols did not grow substantially in the system, and the particle concentrations remained nearly constant between the absorber inlet and wash outlet. Only a small portion of the particles were found to grow significantly. A second water-lean solvent was also tested and elicited insights into how to minimize vapor and aerosol emissions through solvent modification.

Results from parametric testing have established the emission baseline from water-lean solvent processes. The BsGAS has been modified to add in vapor and aerosol control technologies such as a water wash acidifier, an amine sorbent removal and regeneration system, high efficiency demisters, and a lime-coated filter which mimics a filter baghouse. Testing of the control technologies has shown their ability to significantly reduce the emissions from the water-lean solvents. Results from emission studies post-BsGAS modifications will be presented to show the effectiveness of the emission control strategies to minimize emissions from the CO_2 capture process using water-lean solvents.

606548 Optimization and Transport of CO_2 from Refineries in Southeast Texas for Offshore Subsea Storage

Abstract:

Behind fossil fuel-based power generation plants, petroleum Refineries are the second largest stationary source of CO_2emissions. Refineries often undergo process upgrades and plant expansions to maximize the crude throughput, thus maximizing profits. This in turn causes the amounts of CO_2 emissions from the refineries to vary annually, although predictably. For instance, process upgrades generally result in lower CO_2 emissions (~10%) based on total number of barrels of throughput. In addition, CO_2 emissions increase by 85 – 90% when the plant capacity is doubled. Due to these constant changes in the refineries, estimating the amount of carbon dioxide to predict the size of the compressor and the recompression station required for the purposes of Carbon Capture and Storage (CCS) becomes complicated.

In this study, the Petroleum Refineries present in the Beaumont and Port Arthur, Texas area were analyzed for CO_2output using data available from EPA's Facility Level Information on GreenHouse Gases Tool (FLIGHT). Refinery Capacity and actual production was determined from quarterly and annual reports from the respective refineries. The petroleum refineries considered for this study include Valero Refinery (Port Arthur, TX), Total Petrochemicals & Refining USA, Inc. (Port Arthur, TX), ExxonMobil Beaumont Refinery (Beaumont, TX), and, Motiva Enterprises, LLC (Port Arthur, TX).

Among the six different types of CO_2 emission sources found within refineries, Stationary Fuel Combustion Sources, Catalytic Cracking and Reforming Units, and Sulfur Recovery Units accounted for about 80 – 85% of the total carbon dioxide emissions. The CO_2 emissions produced from these units was found to have the highest correlation with crude throughput. These findings were used to size compressors and a design CO_2 gathering network for delivery of $CO_2$from these facilities to offshore, subsea locations in the Gulf of Mexico.

607589 Dynamic Modeling, Optimization, and Control Studies of a Moving Bed Process for CO_2 Capture Using a Micro-Encapsulated Solvent

Abstract:

A number of potential solvents for CO_2 capture exhibit undesirable characteristics like high viscosity, solids formation thus making it difficult to use them in conventional towers [1]. Micro encapsulation of carbon sorbents (MECS) can help in overcoming these issues seen in highly viscous and phase change materials. Micro-encapsulation of solvents is a technology proposed [2] for post combustion carbon capture where the solvent of interest is encapsulated inside a polymer. Microencapsulation of carbon sorbents (MECS) involves fabrication of microcapsules containing a solvent with CO_2 capture capabilities. Even though capsules are manufactured with a known solvent concentration, water is known to permeate some capsule materials. H_2O gain or loss due to diffusion into or out of the microcapsules can significantly affect the solvent concentration and is one of the key operational challenges for reliable and efficient operation of MECS technology. While there are a few studies [2-3] that have investigated different reactor configurations for MECS, no studies have been conducted with the intent to understand and mitigate the issue of time-varying water concentration. Extracting solvent from microcapsules to measure the concentration may require collecting huge number of microcapsules from the reactor. The ambient conditions can further change the water content and may not represent the true concentration value in the reactor. Therefore one must rely on other measurable process variables to get an estimate of capsule concentration.

The present work is focused on dynamic modeling, optimization, and control of a moving bed reactor for MECS, where controlling the desired solvent concentration is one of the key control objectives. Moving bed reactors have a strong potential for solid-sorbent based CO_2 capture [4-5]. The counter-current contact between gas and capsules can help to achieve larger CO_2 driving forces in the absorber and regenerator and so is examined in this work. Also, the heat recovery from the capsules leaving the regenerator can help in pre-heating the capsules flowing from absorber to regenerator. Overall, this can reduce the regeneration energy requirements of the process. A 1-D multi-scale moving bed absorber and regenerator using a pre-existing microcapsule model [4] is developed in Aspen Custom Modeler (ACM) for MECS using Na_2CO_3 as the encapsulated solvent. The capsule level model used in the moving bed reactor is validated with experimental data in our previous study [4]. A steady-state optimization of the moving bed configuration that minimizes the Equivalent Annual Operating Cost (EAOC) by considering key operating and design variables is performed. Traditional and model-based advanced controllers are developed for controlling the capture percentage, and solvent concentration within the capsule in the face of input disturbances. An ARMAX type model based on information about measured variables from the rigorous process model will be exploited to estimate solvent concentration and to provide satisfactory control performance.

References

1. Stolaroff, J. K. et al., Microencapsulation of advanced solvents for carbon capture. Faraday Discussions. DOI: 10.1039/c6fd00049e (2016).
2. Vericella, J. J. et al. "Encapsulated liquid sorbents for carbon dioxide capture", Nature Communications, 6:6124, 1-7, 2015.

3. Hornbostel, et al, Packed and fluidized bed absorber modeling for carbon capture with micro-encapsulated sodium carbonate solution, Applied Energy 1192-1204, 2019.
4. Kotamreddy, et al., Process Modeling and Techno-Economic Analysis of a CO_2 Capture Process Using Fixed Bed Reactors with a Microencapsulated Solvent, Energy & Fuels 2019 33 (8), 7534-7549 DOI: 10.1021/acs.energyfuels.9b01255.
5. Knaebel, K.S., Temperature swing adsorption system. 2009, Google Patents.
6. Mondino, Giorgia et al. "Effect of Gas Recycling on the Performance of a Moving Bed Temperature-Swing (MBTSA) Process for CO_2 Capture in a Coal Fired Power Plant Context." Energies, 10(6), 745. 2017.

608069 Enhanced CO_2 Uptake and Selectivity upon Incorporation of Task-Specific Ionic Liquids into Highly Porous MOF-177 at Post-Combustion Capture Condition

Abstract:

Recently, Metal-organic-frameworks (MOFs) have gained tremendous attraction for application in CO_2 capture and storage due to an exceptional surface area, pore volume and for their tunability. Although some of the MOFs have shown excellent CO_2 uptake at high pressure, lower uptake at post combustion condition (partial pressure of CO_2 is 0.15 bar) is a major hindrance for widespread application in CO_2 capture operations. Researchers have deployed various strategies to enhance CO_2 uptake such as grafting amines functionalities covalently to the ligands, impregnating amines or ionic liquids (ILs). Post-synthesis modification of pores of solid sorbents by impregnation of amino functionalized ionic liquid also known as task-specific ionic liquid (TSILs) to enhance CO_2 capture capacity has shown promising results.

In this study, 1-ethyl-3-methylimidazolium [Emim] cations with Glycine [Gly] and Alanine [Ala] as reactive Amino Acid (AA) anions forming [Emim][Gly] and [Emim][Ala] as TSILs were immobilized into the pores of MOF-177 through the incipient wetness technique. TSIL@MOF-177 composites sorbents were prepared for three different loadings (10, 20 and 30) wt.% of TSIL and the developed composites were characterized and investigated from the engineering perspective in terms of CO_2 capture capacity, selectivity, and enthalpy of adsorption. CO_2 adsorption isotherms were measured at three different temperatures (303, 313 and 323 K) in the pressure range varying from (0.1 to 10) bar. In addition, N_2 adsorption isotherms were also measured for all composites and the pristine MOF-177 from which CO_2/N_2 ideal selectivities were calculated. In addition, experimental adsorption isotherms were modelled using the Freundlich model and the isosteric heats of adsorption were calculated and compared to those of the pristine MOF-177.

Impregnation of TSILs into the highly porous metal organic frameworks (MOF-177) had a remarkable impact on CO_2 capture and selectivity. The highest CO_2 uptake achieved by [Emim][Gly]@MOF-177 composite was 0.45 mmol/g-solid and 0.42 mmol/g-solid at 0.2 bar, 303 K for [Emim][Ala]@MOF-177 composite which were 3 and 2.8 times higher than the pristine MOF-177 at the same conditions, respectively. It was also observed that the optimum loading of TSIL was 20 wt.% as further increase to a loading to 30 wt.% did not results in any incremental CO_2 uptake. CO_2 selectivity over N_2 was also enhanced from 5 for the pristine MOF-177 to 13 and 11 for [Emim][Gly] and [Emim][Ala], respectively, at 0.2 bar and 313 K. This significant improvement of both CO_2 uptake and selectivity can be attributed to the chemisorption interaction between CO_2 and $-NH_2$ functional group present in the amino acid anion in TSIL which was also reflected in the increase in the isosteric heat of adsorption (Q_{st}) values from the value of the pristine MOF. Significantly high CO_2 uptake and CO_2/N_2 selectivity at the low pressure region (<1.0 bar) makes these functionalized MOFs promising candidates for the post-combustion CO_2 capture operation.

606173 An Improved Method for Predicting Autoignition Temperatures Based on First Principles

Abstract:

Methods for predicting autoignition temperatures (AIT) have been historically inaccurate and are rarely based on the underlying physical phenomena leading to observed AIT. Previous attempts at predicting AIT have been developed based on Quantitative Structure-Property Relationships (QSPR) and additive group contribution methods. This work presents an improved method for predicting AIT based on the method originally developed by the late Dr. William H. Seaton. The method of Seaton is described in detail including its use of first principles to predict AIT. Improvements on the method of Seaton and underlying principles are presented and discussed.

611821 Water Washing of Startup Crude Oil

Abstract:

During startup of cold sub-sea wells, methanol is injected at the wellhead to prevent the formation of hydrates in the pipeline/riser system. Methanol injection continues until the fluid arrival temperature exceeds hydrate-formation temperature at the operating pressure, and the typical duration is a few days. Methanol contamination is problematic, and hence the methanol specification of the crude oil is usually less than 20 ppmw. Water or brine wash is often used to reduce the methanol content of the startup crude oil. This paper studies the design of these systems using process simulation. The enabling need is for a reliable and accurate thermodynamic model. The model must describe the three-phase system containing the aqueous and oil liquid phases and the vapor phase, and the key needed feature is description of the distribution coefficient of methanol between the oil and aqueous phases. A model has been developed based upon the ElectrolyteNRTL model in Aspen Plus, and the presentation describes the key data sources and the parameter development. The model enables design of an optimum process flowsheet, and a two-stage countercurrent has been determined to be the optimum option.

607019 Universal Correlation for the Binary Interaction Parameter of Hydrogen with Petroleum Fractions in Refinery Hydroprocessing Applications

Abstract:

Hydrotreating and hydrocracking applications in oil refineries use large quantities of hydrogen. Hydrotreating removes hetero-atoms of sulfur, nitrogen and oxygen and can saturate aromatic ring structures, producing more desirable transportation fuels, such as ultra-low sulfur diesel. Hydrocracking is a more severe reaction, which breaks carbon-carbon bonds and improves the ratio of hydrogen to carbon in the fluid, again, making more suitable transportation fuels. Hydrogen is expensive and is either made on-site, perhaps through steam reforming of methane, or is bought from outside as a utility.

To design new hydroprocessing facilities and to improve the performance of existing plant, it is essential to be able to predict the solubility of hydrogen in typical hydrocarbon fluids. The hydrogen has to be separated from the product streams in high and low pressure separators and is then recycled to the reactor feed, where it is combined with hydrogen make-up.

Although Detailed Hydrocarbon Analysis (DHA) enables us to define the composition of reactor feedstocks in terms of a larger slate of defined chemical components, it is still common practice to represent the heavier portion of the feed in terms of undefined components, or petro fractions, characterized by narrow cut boiling points and densities from a petroleum assay.

Cubic equations of state (EoS), such as Soave-Redlich-Kwong and Peng-Robinson are the workhorse thermodynamic methods for calculating vapor liquid equilibrium (VLE) in refinery processes. From the study of the VLE of binary systems of hydrogen as a solute in a defined hydrocarbon solvent, it is evident that a significant correction to the basic Van der Waals mixing rule in the EoS is required if we are to predict accurately the composition of hydrogen in the liquid phase. Moreover, we can show that this correction, through a binary interaction parameter (k_{ij}) needs to be temperature-dependent. While we can correlate k_{ij} for hydrogen binaries with defined components through regression of experimentally measured VLE data, the problem is how we can predict the interaction between hydrogen and any petro fraction? We observe that increasingly large values of k_{ij} are required as the molecular weight of the hydrocarbon increases and any petro fractions are likely to be in the higher molecular weight range. Ignoring the interaction parameter between hydrogen and a petro fraction will result in significant inaccuracy in any prediction of solubility.

In this study, we will show how the wealth of binary VLE data for hydrogen with defined hydrocarbon solvents can be reduced through regression and conversion to Henry's Law constants that correlate with the reduced temperature and critical pressure of the solvent [1,2]. We propose a new universal correlation for these Henry's Law constants that will allow the subsequent prediction for k_{ij} between hydrogen and a petro fraction, based only on knowledge of the critical temperature, critical pressure and acentric factor of the petro fraction that can be calculated from well-established estimation methods. The validity of this approach will be shown by comparing the predicted solubility of hydrogen in a well-analyzed diesel stream with a base case using methods currently described in the Technical Data Book of the American Petroleum Institute [3].

[1] Trinh TKH, de Hemptinne JC, Lugo R, Ferrando N, Passarello JP, "Hydrogen Solubility in Hydrocarbon and Oxygenated Organic Compounds." J. Chem. Eng. Data 2016,61,119-34

[2] Fernández-Prini R, Alvarez JL, Harvey AH, "Henry's Constants and Vapor–Liquid Distribution Constants for Gaseous Solutes in and at High Temperatures "J. Phys. Chem. Ref. Data, 3012, 32, 903-912

[3] API Tech Data Book 10, 2019

600821 Distillation Curve and Vapor Pressure Predictions of Alcohol-Gasoline Mixtures Using an Advanced Equation of State

Abstract:

Due to federal environmental regulations, such as the Energy Policy Act of 2005 in the United States, ethanol is frequently blended into gasoline in an effort to reduce tailpipe emissions. Moreover, there is a shift towards increasing the use of more sustainable energy resources such as biofuels, which includes ethanol, as transportation fuels. A fundamental understanding of how blending alcohol into a gasoline stream affects the properties of said stream is of critical importance. Specifically, the change in the distillation curve and vapor pressure is essential from a manufacturing and engine performance perspective. In the plant setting, there exists empirical correlations to predict these changes. However, these are typically very limited in scope as a correlation is only applicable to a specific plant gasoline mixed with a specific alcohol in a certain blend ratio.

Process simulators of petroleum process rely on equation of state (EoS) predictions of the phase behavior of petroleum pseudo-components defined by normal boiling point T_b, specific gravity *SG,* and *MW*. For gasoline blending these pseudo-components are typically generated from D86 curves. Hence, any method to be widely used in the prediction of alcohol / gasoline blending in refining processes should be accurate for petroleum pseudo-components.

In this work, we present a fundamental methodology for alcohol / gasoline blend property prediction using the polar PC-SAFT EoS[1] combined with the recently developed petroleum parameterization methodology, TAPPS (formerly referred to as EMPETRO)[2-6]. TAPPS is a highly accurate methodology used to determine hydrocarbon PC-SAFT parameters for petroleum species defined solely by T_b, *SG,* and *MW*. In addition to highly accurate predictions for pure component vapor pressures and densities, TAPPS accounts for polar features of petroleum such as the fact that aromatic species receive hydrogen bonds. This makes it well suited to describe alcohol / gasoline blending.

The approach presented here employs the base gasoline D86, average gasoline SG, and dry vapor pressure equivalent (DVPE) to generate gasoline pseudo-components. Using these pseudo-components, it is demonstrated the PC-SAFT / TAPPS approach allows for the quantitative prediction of the D86 and DVPE of alcohol / gasoline mixtures. We validate the approach by reproducing both literature[7,8] and ExxonMobil refinery data[9] for various blends. This work demonstrates the strength of PC-SAFT / TAPPS in predicting the physical properties of non-traditional blends of petroleum streams and lays the foundation for the development of a predictive blending tool that will be valuable when considering emerging biofuel markets.

References

[1] J. Gross; G. Sadowski, *Ind. Eng. Chem. Res.*, **40** (2001) *1244.*

[2] B. D. Marshall; C. P. Bokis, *Fluid Ph. Equilibria*, **489** (2019) *83.*

[3] B. D. Marshall; C. P. Bokis, *Fluid Ph. Equilibria*, **478** (2018) *34.*

[4] B. D. Marshall, *Fluid Ph. Equilibria*, **493** (2019) *153.*

[5] B. D. Marshall, *Fluid Ph. Equilibria*, **497** (2019) *79.*

[6] B. D. Marshall, *Fluid Ph. Equilibria*, **507** (2020) *112420.*

[7] V. F. Andersen; J. E. Andersen; T. J. Wallington; S. A. Mueller; O. J. Nielsen, *Energy Fuels,* **24** (2010) *2683.*

[8] V. F. Andersen; J. E. Andersen; T. J. Wallington; S. A. Mueller; O. J. Nielsen, *Energy Fuels,* **24** (2010) *3647.*

[9] J. R. Vella; B. D. Marshall, *Ind. Eng. Chem. Res.* (2020) *accepted.*

606540 Development of Mean Ionic Activity Coefficients for Modification of the Krichevsky – Ilinskeya Equation for Methylimidazolim – Type Ionic Liquids

Abstract:

Removal of H_2S from oil and gas streams has been a priority, oftentimes a necessity, for upstream crude oil/gas facilities for decades. However, the advances in production, namely horizontal drilling and hydraulic fracking, have produced oils and natural gases that contain higher concentrations of H_2S than previously seen. Crude oils containing > 0.5 % sulfur are considered sour and must undergo separation and/or conversion processes to lower the H_2S content prior to entering the midstream pipelines and railcars. H_2S scavengers can be categorized into two types: non-regenerative and regenerative. Non-regenerative scavengers, such as triazines, use an irreversible reaction to convert H_2S forming sulfur-containing hydrocarbons. With regenerative scavengers, such as alkanolamines, absorb H_2S using reversible absorption with stripping recovery processes.

Ionic Liquids (ILs) are a new family of absorbent materials that are being sought to replace amine-type reversible absorbents due to lower production of volatile organic carbons (VOCs) and reduced costs of the regenerative stripping phase. The ILs that have been chosen for this work were 1-butyl-3-methylimidazolium hexafluorophosphate, 1-butyl-3-methylimidazolium chloride, and 1-butyl-3-methylimida-zolium tetrafluoroborate. In this work, the mean ionic activity coefficient and the adjustable parameters for the Pitzer Ion Interaction Model have been determined. An infinite dilution procedure has been employed where a concentrated mixture of H_2S in IL was stripped using helium gas. The decrease in H_2S concentration (measured through gas chromatography) over time was used to determine the pertinent thermodynamic parameters.

In addition, the Henry's constant and partial molar volume of H_2S in IL were determined. The Krichevsky – Kasarnovsky equation has been shown to be applicable over a wide range of temperatures, pressures, and solute/solvent systems. However, in some cases (i.e. where T is near T_c) the molar volume ( ) is not constant, and there are strong deviations between the measured solubility and the calculated Henry's constants. From preliminary estimates, the Krichevsky – Ilinskeya equation (a modification of the Krichevsky – Kasarnovsky equation) does not model the phenomenon well. Ultimately, this work seeks to establish a new thermodynamic model that replaces the 2-suffix Margules component of the Krichevsky – Ilinskeya equation with the mean ionic activity coefficient found in Pitzer Ion Interaction model. The experimental results and model development will be presented.

610534 (Invited Talk) Mechanistic Insights into Deoxygenation Chemistry on Bimetallic Phosphides

Abstract:

Bimetallic catalysts are crystalline solids that consist of two different metals that can facilitate specific chemical reaction. The cooperative interactions between the two metals often result in higher stability, and functionality when compared to their monometallic counterparts. Adding non-metallic elements, such as phosphorous, allows for charge distribution between the metal and nonmetal atoms, which improves on these properties. Although bimetallic catalysts have been used in traditional chemical reactions, their application to biomass conversion is largely unexplored. In this presentation, I will share our work on deoxygenation of model aromatic compounds over different composition contained bimetallic catalysts ($M_XMo_{(2-X)}P$; M = Fe, Ru, Ni; X = 0.8, 1.0, 1.2) [3-5]. We explain our results in terms of binding energies, energetics along chemical transformation, and Bader charges.

608166 Ultrafast Hot Carrier Injection in Au/GaN: The Role of Band Bending and the Interface Band Structure

Abstract:

Plasmon photochemistry can potentially play a significant role in photocatalysis. To realize this potential, it is critical to enhance the plasmon excited hot carrier transfer and collection. However, the lack of atomistic understanding of the carrier transfer across the interface, makes it challenging to design more efficient system. In this work, we develop and apply the non-adiabatic molecular dynamics simulation to study hot carrier dynamics in the system of Au nanocluster on top of GaN surface. This method by including the electron-phonon interaction allows for a direct *ab initio* simulation of excited carrier dynamics in a large-scale system and for a long period of time. By setting up the initial excited hole in Au, the carrier transfer from Au to GaN is found to be on a sub-pico second time scale. After the hole has cooled down to the band edge of GaN, we find some of the charges can return back to Au. By applying different external potentials to mimic the Schottky-barrier band bending, the returning charge can be reduced effectively. Finally, with the understanding of the carrier transfer's pathway, we suggest that a ZnO layer between GaN and Au can effectively block the "cold" carrier from returning back to Au but still allow the hot carrier to transfer.

*This work is funded by the Joint Center for Artificial Photosynthesis, supported through DOE under Award number DE-SC0004993. NERSC and Oak Ridge Leadership Computing Facility provide computational support.

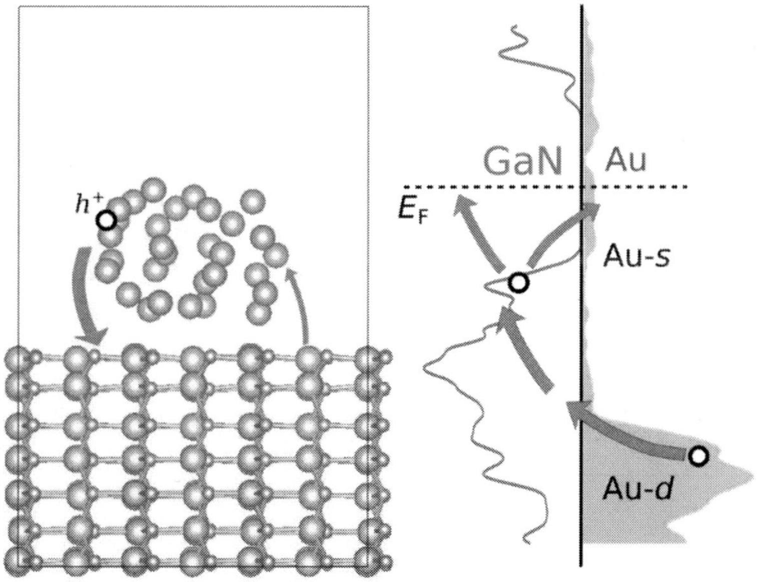

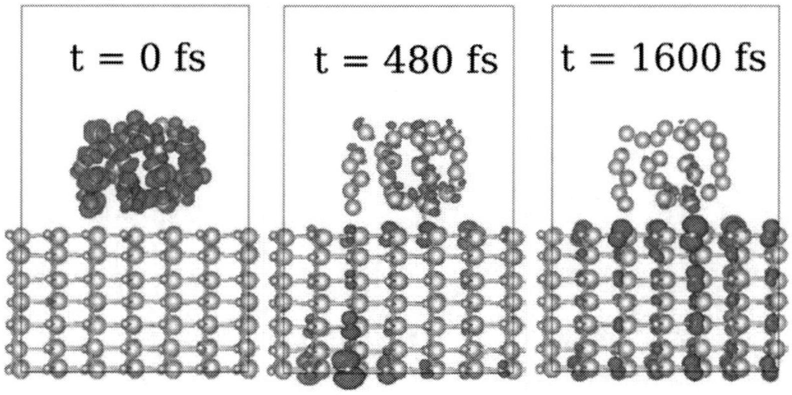

598436 Halides Induce Anisotropic Growth of Cu Nanocrystals

Abstract:

Properties of metal nanocrystals are generally controlled by their morphologies. Though Ag nanocrystals are used in commercial applications, the properties of Cu are comparable so that this earth-abundant material has drawn increasing attention. There have been many accounts of experimental, solution-phase syntheses of various Cu nanocrystals, including nanowires and nanoplates – however, the mechanisms of such anisotropic growth are still not fully understood.

It is commonly believed that capping agents, such as hexadecylamine (HDA) can induce anisotropic growth by selective binding to certain crystal facets. However, the emergence of round Cu nanoparticles synthesized in the presence of HDA does not seem to support this. Only when halide is present as an additive during nanocrystal synthesis are distinctive shapes observed. When Cl^- is the additive, nanowires form, but I^- leads to the formation of nanoplates. To elucidate the roles of halides in anisotropic Cu nanocrystal growth, we use density functional theory (DFT) to study the co-adsorption of halides and HDA on Cu surfaces with an analysis based on *ad initio* thermodynamics.

Both Cl^- and I^- are found to displace a chemisorbed HDA self-assembled monolayer (SAM) on Cu(100) and Cu(111). In the case of Cl^-, HDA SAMs are disrupted at lower Cl coverages on Cu(111) than on Cu(100). This can result in full protection for Cu(100) from Cu ion addition and a lack of protection for Cu(111) and is consistent with the experimentally observed formation of nanowires with long {100} side facets. *Ab initio* thermodynamics calculations show that, under the conditions for which nanowires are predicted and observed, introducing a trace amount of I^- can displace both Cl^- and HDA on both Cu surfaces, which creates a strong thermodynamic driving force for nanoplates with large {111} facets. The predicted evolution of Cu nanocrystal shape under different synthetic conditions agrees well with experimental results.

606718 First-Principles Study of Lithium-Air Batteries Based on Tri-Molybdenum Phosphide (Mo_3P) Nanoparticles

Abstract:

Lithium-air batteries are promising competitors to replace traditional lithium ion batteries due to the extremely high theoretical energy density. Despite extensive experimental research for decades, it is still a challenge to find a stable cathode material for both discharge (ORR) and charge (OER) reactions with a low overpotential. Here, we employ density-functional-theory (DFT) calculations to examine the mechanism, thermodynamic overpotential (η) and cyclability of a novel Mo_3P-based Lithium-air battery. We also compare and rationalize our findings with direct comparison with the recent experimental results of our collaborators. In this type of battery, Li_2O_2 is found as the only discharge product up to 1000 cycles due to the two-electron transfer reaction between Li^+ and O_2. Our DFT results show that the Mo-terminated Mo_3P (110) surface will be spontaneously oxidized in the oxygen-rich environment, thus forming a kinetically stable MoO layer to bring about the initial Li^+/e^- transfer from aprotic electrolyte to cathode. Furthermore, we find that the initial Li^+/e^- transfer reaction is the potential determining step (PDS) during ORR with a quite low DFT-based overpotential ($\eta = 0.06$ V), in agreement with our experimental detection ($\eta = 0.08$ V). In OER, the DFT-based overpotential was determined as 0.33V coupled with the desorption reaction of LiO_2 on the MoO layer covered catalyst, which is in agreement with experimental value ($\eta = 0.27$ V) but larger than that ($\eta = 0.12$ V)[1] on the kink sites of the crystal majority Li_2O_2 (0001) facet.[2,3] Therefore, we reveal that the MoO layer terminated Mo_3P cathode gives rise to a low-overpotential discharge reaction. However, the starting of charge process happens on the surface of agglomerated Li_2O_2 discharge product, while Mo_3P cathode becomes more and more important after most of the Li_2O_2 dissolved. Our results help shed fundamental insight on the role of ultrathin oxide monolayers on phosphide nanoparticles, and how this can potentially lead to a new nanomaterials design dimension.

1. Hummelshøj, J. S., Luntz, A. C. & Nørskov, J. K. Theoretical evidence for low kinetic overpotentials in Li-O_2 *Journal of Chemical Physics* **138**, 34703 (2013).

2. Radin, M. D., Tian, F. & Siegel, D. J. Electronic structure of Li_2O_2 (0001) surfaces. *Journal of Materials Science* **47**, 7564–7570 (2012).

3. Radin, M. D., Rodriguez, J. F., Tian, F. & Siegel, D. J. Lithium peroxide surfaces are metallic, while lithium oxide surfaces are not. *Journal of the American Chemical Society* **134**, 1093–1103 (2012).

610798 First-Principles Modeling of Thermal Degradation of CO_2-Loaded Aqueous Amine Solvents

Abstract:

Chemical absorption using an aqueous amine-based solvent appears to be the most promising near-term solution for post-combustion capture of CO_2 from flue gas.[1,2] Monoethanolamine (MEA) has been widely studied and is considered the benchmark solvent. Although it exhibits several advantages including a relatively high absorption rate, high solubility, low volatility, and low viscosity, widespread commercial implementation of MEA-based CO_2 capture systems remains hindered by costs associated with problems arising from solvent degradation and corrosion, in addition to the high parasitic energy consumption during regeneration.[1–8] Makeup of MEA loss from the solvent contributes about (~10 %) to the total cost of CO_2 capture, including the energy requirements for the reclamation process.[9–11] Moreover, some products of degradation may be corrosive and hazardous if released to the environment.[12,13]

Thermal degradation has typically been defined as degradation occurring due to elevated temperatures in the stripper, and tends to increase with CO_2 loading.[14,15] Thermal degradation of primary amines including MEA is thought to be initiated by cyclization of carbamates, which may result in the formation of 2-oxazolidinone (OZD) that is a minor product and proposed intermediate.[16] The OZD may undergo further reaction with amines to form diamines, imidazolidinones and urea, which are typically reported to be major degradation products.[14–16] Thermal degradation of CO_2-loaded aqueous amines could be mitigated by reducing the stripper operating temperature. However, lowering the temperature also typically increases the overall energy cost required for solvent regeneration.[17,18] The molecular mechanisms underlying thermal degradation of aqueous amines remain unclear, while an improved mechanistic understanding may provide valuable hints on how to improve the performance of existing solvents and in the rational design and synthesis of more degradation-resistant solvent materials.

In this talk, we will present our recent computational studies on the molecular mechanisms of thermal degradation of aqueous MEA in the CO_2 capture process. We have evaluated key elementary reactions using various computational methods. Static quantum mechanical calculations with an implicit solvent model were used to assess the relative stabilities of possible intermediates and products, and *ab initio* molecular dynamics (AIMD) simulations combined with metadynamics sampling were employed to predict free-energy barriers for key reaction steps, with particular focus on the effect of direct interaction between reaction intermediates and surrounding water molecules. Our work highlights that the thermal degradation mechanisms are largely governed by the competition of intramolecular and intermolecular interactions, besides steric constraints, which is a strong function of the arrangement and dynamics of water surrounding the intermediates.

References

[1] G. T. Rochelle, *Science*, 2009, **325**, 1652–4.

[2] J. D. Figueroa, T. Fout, S. Plasynski, H. McIlvried and R. D. Srivastava, *Int. J. Greenh. Gas*

Control, 2008, **2**, 9–20.

[3] U.S. Department of Energy, *Carbon Capture Technology Program Plan*, 2013.

[4] D. M. D'Alessandro, B. Smit and J. R. Long, *Angew. Chem. Int. Ed. Engl.*, 2010, **49**, 6058–82.

[5] B. R. Strazisar, R. R. Anderson and C. M. White, *Energy and Fuels*, 2003, **17**, 1034–1039.

[6] M. R. M. Abu-Zahra, J. P. M. Niederer, P. H. M. Feron and G. F. Versteeg, *Int. J. Greenh. Gas Control*, 2007, **1**, 135–142.

[7] M. R. M. Abu-Zahra, L. H. J. Schneiders, J. P. M. Niederer, P. H. M. Feron and G. F. Versteeg, *Int. J. Greenh. Gas Control*, 2007, **1**, 37–46.

[8] H. M. Stowe and G. S. Hwang, *Ind. Eng. Chem. Res.*, 2017, **56**, 6887–6899.

[9] A. B. Rao and E. S. Rubin, *Environ. Sci. Technol.*, 2002, **36**, 4467–4475.

[10] A. J. Reynolds, T. V. Verheyen, S. B. Adeloju, E. Meuleman and P. Feron, *Environ. Sci. Technol.*, 2012, **46**, 3643–3654.

[11] G. S. Goff and G. T. Rochelle, *Ind. Eng. Chem. Res.*, 2006, **45**, 2513–2521.

[12] G. S. Goff, 2005, 283.

[13] S. Chi and G. Rochelle, *Ind. Eng. Chem. Res.*, 2002, **41**, 4178–4186.

[14] J. Davis and G. Rochelle, *Energy Procedia*, 2009, **1**, 327–333.

[15] C. Gouedard, D. Picq, F. Launay and P. L. Carrette, *Int. J. Greenh. Gas Control*, 2012, **10**, 244–270.

[16] S. J. Vevelstad, I. Eide-Haugmo, E. F. Da Silva and H. F. Svendsen, *Energy Procedia*, 2011, **4**, 1608–1615.

[17] B. A. Oyenekan and G. T. Rochelle, *AIChE J.*, 2007, **53**, 3144–3154.

[18] G. T. Rochelle, *Curr. Opin. Chem. Eng.*, 2012, **1**, 183–190.

610484 Assessment of the Two-Phase Thermodynamic Method for Calculating Adsorption Isotherms

Abstract:

Adsorption isotherms are fundamental quantities of interest for understanding separation and catalysis phenomena. They are typically computed by grand-canonical or Gibbs ensemble Monte Carlo methods. These methods rely on potentially challenging particle insertion/deletion moves and may present a particular challenge for simulations using a first-principles potential energy surface. As an attractive alternative, the two-phase thermodynamic (2PT) model[1] allows one to calculate the free energy of a system from vanilla canonical molecular dynamics simulations. The 2PT method decomposes the total density of states into a gas-like component and a solid-like component. The former accounts for all fluidic effects and is treated as a hard-sphere gas. The solid-like component represents the non-zero, higher frequency modes and is handled as harmonic oscillators, which allows for straightforward quantum corrections by using the appropriate statistics. The 2PT method has been demonstrated to be capable of calculating the vapor-liquid coexisting properties and has recently been applied to study the free energy and stability of molecules adsorbed in carbon nanotubes[2] and zeolites[3]. In this talk, we will describe the development of the 2PT method for calculating adsorption isotherms and present the results for a united-atom methane model and the TIP4P water. We will discuss important caveats with the direct application of the original 2PT method and the modifications to the method that are necessary to obtain a correct description of adsorption equilibria.

(1) Lin, S.-T.; Blanco, M.; Goddard, W. A. The Two-Phase Model for Calculating Thermodynamic Properties of Liquids from Molecular Dynamics: Validation for the Phase Diagram of Lennard-Jones Fluids. *J. Chem. Phys.* **2003**, *119*, 11792–11805.

(2) Kumar, H.; Mukherjee, B.; Lin, S.-T.; Dasgupta, C.; Sood, A. K.; Maiti, P. K. Thermodynamics of Water Entry in Hydrophobic Channels of Carbon Nanotubes. *J. Chem. Phys.* **2011**, *134*, 124105.

(3) Bukowski, B. C.; Bates, J. S.; Gounder, R.; Greeley, J. Defect-Mediated Ordering of Condensed Water Structures in Microporous Zeolites. *Angew. Chem. Int. Ed.* **2019**, *58*, 16422–16426.

613566 Sustainability and Resilience in the Chemical Process Industries through Process Intensification and Modular Manufacturing

Abstract:

The manufacturing sector accounts for a quarter of the total U.S. energy use, and the process industries (chemicals, petroleum refining, pulp and paper, iron and steel) are the largest energy consumers in the sector. Likewise, these industries significantly impact other dimensions of sustainability including water use, climate change, clean energy, waste management, and human health. Process intensification (PI) is an often radical rethinking of chemical processes to remove bottlenecks, improving productivity and safety while reducing waste, energy consumption, production cost, and carbon footprint. PI is often a shift from the conventional "unit operations" paradigm in chemical processing to a more integrative paradigm. Modular manufacturing, a complementary concept, is an approach to building chemical plants that favors assembly of pre-fabricated, standardized modules rather than conventional "stick-built" construction. Modular manufacturing also provides opportunities for distributed chemical manufacturing that utilizes stranded resources such as waste biomass feedstocks or renewable wind and solar power. Similarly, distributed manufacturing has the potential to increase supply chain resilience by enabling flexibility in how production assets are deployed. Modular manufacturing enables a shift from the bigger-is-better paradigm in chemical production to a small, modular paradigm where economies of scale are replaced with economies of mass production. Both PI and modular manufacturing are tools that can be used to improve the sustainability and resilience of the chemical process industries, which are under pressure to change given recent shifts in supply and demand, the rise of inexpensive renewable power, and an increased need to reduce operational footprints. Our work uses research and education to promote better process technologies and to define appropriate metrics to capture the impact of these technologies on productivity and sustainability.

606158 Technological Opportunities in Sustainable Energy and Water Systems

Abstract:

The disproportionate use of fossil fuels over the last 250 years has resulted in the accumulation of ~2000 $GtCO_2$ in our atmosphere and ~1.2°C global average temperature rise compared to preindustrial levels, which has led to significant weather extremes and other consequences.. There is headroom for an additional 1400 $GtCO_2$ of total emissions in the future if we are to maintain the global average temperature rise below 2°C . With today's global emissions of ~40 $GtCO_2$/year, we have about 20-30 years left before the world exceeds this headroom. Therefore there is a critical need to develop sustainable energy systems.

Global water demand is projected to increase by 55% over the next three decades owing to population growth, industrialization and climate change. Anthorpogenic warming is also leading to severe droughts in various parts of the world. Desalination technologies can meet this demand by extracting clean water from saline sources such as ocean or contaminated sources such as produced waters, which are a byproduct of oil and gas extraction.

In this talk the speaker will provide an overview of various science and technology challenges which if solved can dramatically enable the development of sustainable clean energy and water systems.

602667 Photovoltaic Aglectric Farming for Meeting Local Food, Energy,Water and Chemical Needs

Abstract:

A solar economy is a promising sustainable alternative to the current fossil fuel economy. While solar energy is abundant, recent studies have shown that for many densely populated countries, a solar economy could be severely land-constrained by the highly diffuse nature of sunlight. To relax land constraints, we propose the concept of "aglectric" farming, where solar irradiation on agricultural land will be sustainably shared for food and energy coproduction. Adjusting the intensity, spectral distribution, and duration of shading allows innovative photovoltaic (PV) systems to achieve significant power generation without potentially diminishing agricultural output. The feasibility has been proved through shadow modeling. Local photons have a potential to satisfy local needs of food, energy, water and chemicals. The proposed PV aglectric farms have potential to prepare humanity for a sustainable solar economy that supports the "Full Earth" of over 10 billion people.

607663 A Machine Learning-Based Approach for Chemical Process Sustainability Analytics

Abstract:

Sustainability has become a focal point of future development manufacturing sectors, as it aims at continuous improvement of economic, environmental and social sustainability performance of manufacturing systems. In the process of developing strategies and solutions for goal achievement, the time-variant sustainability performance needs to be monitored, and the collected system input-output data needs to be analyzed. For a sophisticated manufacturing system, the amount and types of process data and beyond could be huge. The sustainability status of a chemical system is highly dependent on process parameters that need to be optimized for improving process sustainability performance. Data-driven modeling can build a mapping from process parameters to sustainability indicators. This is a very challenging problem because of high dimensionality of the input space (i.e., large number of process parameters) and many uncertain internal and external parameters.

In this paper, we propose a machine learning approach for chemical processes sustainability analytics. LASSO, a machine learning technique, can be used to build a regression model between process parameters and sustainability indicators. Such a technique can perform both parameter selection and regularization in order to enhance the prediction accuracy and interpretability of the statistical model. Using this technique, the most important process parameters with the highest effect on sustainability of the system can be specified. The trained model then can be used to generate the trade-off among sustainability objectives. This type of information is very valuable for process engineers to better understand the effect of process parameters on sustainability performance of the system. A case study on the methanol synthesis process is illustrated to show the effectiveness of the proposed approach.

---

* All correspondence should be addressed to Prof. Yinlun Huang (Phone: 313-577-3771; Fax: 313-577-3810; E-mail: yhuang@wayne.edu).

604820 Accelerating Discovery of High-Performance Electrocatalysts for Ammonia Oxidation Via Machine Learning

Abstract:

Electrooxidation of ammonia to dinitrogen plays a crucial role within the nitrogen transformation cycle. While platinum (Pt) based catalysts, specifically terminated with (100) facets, have shown promising activity. They still suffer from a large overpotential (~0.5 V) and surface deactivation[1]. Various alloys such as PtIr, PtRh and NiCu have offered an approach to tackle these issues. However due to vast combination of metals, the material space consisting of bimetallic alloys can quickly become overwhelming to navigate through. Additionally finding good electrocatalysts which not only have a lower overpotential but are also resistant to surface poisoning can be very difficult in this material space. Building upon our previous work[2], machine learning models are built in order to screen through the bimetallic alloy material space.

Our recent work showed that the optimization of *NHx and *OH adsorption energies is key to find thermodynamically efficient electrocatalysts. Additionally it was also shown that *N adatoms could be a precursor to surface poisoning. Using both these insights we have developed machine learning models to search for materials with optimal *NHx and *OH adsorption energies. This allows us to suggest potentially efficient electrocatalysts for ammonia electrooxidation[3].

(1) Katsounaros, I.; Figueiredo, M. C.; Calle-Vallejo, F.; Li, H.; Gewirth, A. A.; Markovic, N. M.; Koper, M. T. M. On the Mechanism of the Electrochemical Conversion of Ammonia to Dinitrogen on Pt(1 0 0) in Alkaline Environment. J. Catal. 2018, 359, 82–91.
(2) 1. H. S. Pillai, H. Xin, New Insights into Electrochemical Ammonia Oxidation on Pt(100) from First Principles. Ind. Eng. Chem. Res. 58, 10819–10828 (2019).

(3) Y. Li, X. Li, H. S. Pillai, J. Lattimer, N. Mohd Adli, S. Karakalos, M. Chen, L. Guo, H. Xu, J. Yang, D. Su, H. Xin, G. Wu, Ternary PtIrNi Catalysts for Efficient Electrochemical Ammonia Oxidation. ACS Catal. 10, 3945–3957 (2020).

604866 Safety Consideration in Carbon Dioxide Integration Networks in Industrial Clusters

Abstract:

The need to reduce carbon dioxide emissions can come at a high cost. Carbon integration (Al-Mohannadi and Linke, 2016) highlights opportunities to reduce costs associated with carbon dioxide mitigation through a systematic analysis of Carbon Capture Utilization and Storage (CCUS) networks in industrial clusters. It analyzes CO_2 point sources treatment, transportation, compression and utilization in CO_2 converting sinks to minimize costs while adhering to given emission targets. There are several process hazards in carbon integration and utilization which make the network vulnerable to process safety incidents and failure. Process safety engineering is the study of hazards and risks in the chemical process industry and deals with the prevention and mitigation of the risks. The concept of process safety has grown exponentially during the last two decades. Multiple methods exist to assess hazards of products, transportation and process. However, there is a gap in evaluating CO_2 utilization options and network. This work proposes an optimization-based framework of minimizing the process hazards as well as maximizing the profit in carbon integration. This multi-objective approach will not only reduce the hazards but also increase the reliability of the network against process safety incidents.

References

Al-Mohannadi, D.M., Linke, P. , 2016, On the systematic carbon integration of industrial parks for climate footprint reduction, Journal of Cleaner Production, 112, 4053-4064

592237 Parallel Plant: A Smart System Approach for Ammonia Synthesis Plant

Abstract:

Smart plant contains high intensive information and massive operational parameters, not only extensive chemical reactions. With the development of computation capability and future 5G, the ACP theory (i.e., artificial systems, computational experiments, and parallel computing) will play a more important role in modeling and control of complex systems like ammonia synthesis plant. The necessity of making accurate predictions of production conversion rate out of a large number of operational parameters has become a crucial problem in smart plant. Previous attempts have been made to seek reactive rate based on empirical equations in ammonia synthesis. However, there are still uncertainties about parallel plant chemical reaction mechanism using a large number of operational parameters. This work proposes a novel deep learning prediction approach that utilizes long short-term memory and gated recurrent unit in conjunction with ACP theory. The proposed approach is tested and validated by real-world dataset, and the results outperformed traditional prediction models.

604819 Spatially Explicit Optimization of Cellulosic Biofuel Supply Chains Considering Landscape Design and Biomass Yield Uncertainty

Abstract:

Planting dedicated bioenergy crops on lands less suitable for food production has the potential to lower the competition between food and fuel. Additionally, fuels made from second generation cellulosic feedstocks provide an environmentally attractive sustainable alternative to fossil fuels and first-generation grain-based biofuels. Because cellulosic feedstocks have yet to be planted in large quantities and biorefineries have yet to be constructed, there is an opportunity to use mathematical programming techniques to better understand the major environmental and economic drivers, and ultimately optimize the biofuel supply chain network design (SCND) from field to product.

When considering supply chains covering large study areas, there is considerable variation in local weather and soil quality which leads to uncertainty in biomass yields that, if not taken into account, can lead to sub-optimal facility siting and biomass establishment decisions. The varying soil quality provides an opportunity to optimize the landscape design (i.e. deciding where in the landscape to plant which bioenergy crop and how to manage that land) along with supply chain network design decisions simultaneously while exploring their tradeoffs. The optimization of landscape design and SCND together has the potential to protect against uncertainty and exploit the spatially explicit nature of land for biomass to achieve better solutions both economically and environmentally.

While the area of biofuel supply chain optimization is relatively mature, a flexible high resolution spatially explicit model that considers feedstock yield uncertainty, environmental outputs, and landscape design simultaneously has yet to be proposed. We present an integrated two-stage stochastic mixed integer linear model that, given (i) the spatially explicit available land, (ii) biomass yield potential at each site, (iii) potential biorefinery and preprocessing depot locations, (iv) potential preprocessing and conversion technologies, (v) cost parameters and environmental parameters finds the optimal (i) biorefinery and depot locations, (ii) technology, (iii) capacity, (iv) transportation, production, and inventory planning, (v) crop establishment locations, (vi) and land management; the combination of which minimizes the total annualized cost of the supply chain.

We integrate the yield and environmental outputs of SALUS biogeochemical crop model simulations over multiple years of weather data with the spatially explicit supply chain network design model and present a case study in southern Michigan, USA. Our results show that fertilization closer to the biorefinery leads to a decrease in costs and greenhouse gas emissions related to transportation. Additionally, the two-stage stochastic formulation leads to solutions which perform more favorably in expectation with respect to the uncertain biomass yields. By adjusting the cost of carbon (\$/MgCO2e), decision makers can determine how much to value reductions in GHG emissions with respect to SC costs.

Model complexity is sensitive to the number of uncertainty scenarios, the size of the study area, and the number of supply chain nodes considered. As model complexity increases, approximation methods are introduced to maintain computational tractability and solve larger problems. Because large, realistic, problems are difficult to solve with an extensive form of the two-stage stochastic model, we present and compare solution strategies and decomposition methods to solve detailed models more efficiently.

606172 Optimal Design of an Integrated Thermal Energy Conversion System for Sustainable Energy and Desalinated Water Production

Abstract:

The aim of this paper is to present a rigorous model for the optimal design of the open-cycle Ocean Thermal Energy Conversion system. The formulation allows the design and analysis of the system as energy and desalinated water distribution market. The operation of the system is subject to the maximum energy and water demand required by a given complex in a given geographical location. A supply and demand analysis of these products is carried out which considers a government incentive to strengthen the development and competitiveness of this system compared to other types of renewable energy. This economic incentive consists of carbon savings per kWh of energy and is also compared to oil-based electricity generation. Considering this approach, a cost analysis based on emissions generated is proposed to manage the operation policy of the system. The solution to the problem is to maximize social welfare by proposing a balance and a pricing policy based on supply and demand for the products derived from the operation of this thermal power cycle.

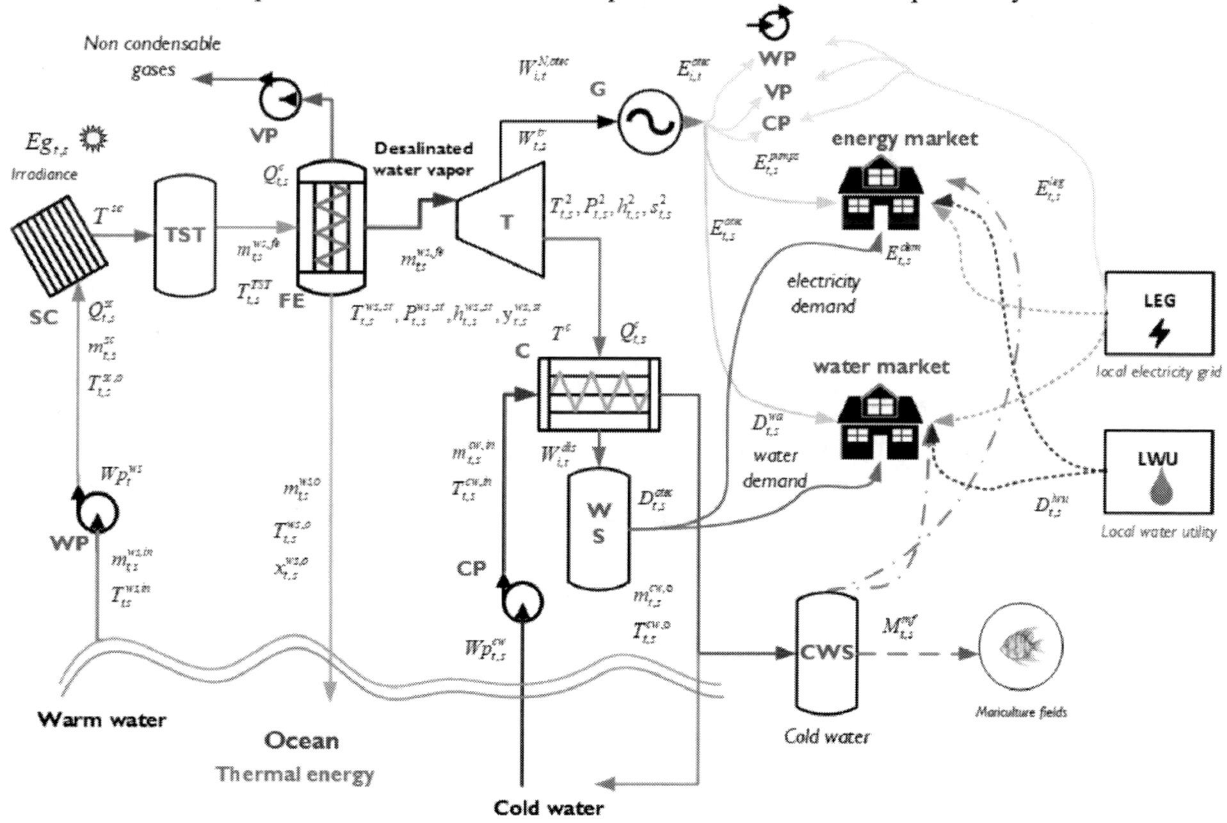

607762 Optimization of a Spatially-Resolved Petrochemicals Manufacturing and Supply Chain Network

Abstract:

The increase in oil and gas production from shale formations has resulted in significant increases in the production of Natural Gas Liquids (NGLs), including predominantly ethane, propane, *n*-butane and isobutane, with smaller amounts of pentane and less volatile hydrocarbons. The abundance of these hydrocarbons is an opportunity to boost chemical manufacturing because NGLs can be used as feedstocks for chemicals and fuel production [1]. The availability of NGLs has driven investments in chemical manufacturing and supply chains, in the form of new plants, expansions of existing plants, improvements in technologies, as well as the development of new technologies, aimed at harnessing the newly available resources. However, if this sort of infrastructure is not available, wasteful and environmentally unsound practices such as gas venting and flaring may result.

In this work, we develop an optimized, spatially-resolved network of petrochemical manufacturing technologies and supply chain transportation resources in order to best take advantage of the supply of NGLs, with a focus on determining an optimal product slate and the optimal placement of new manufacturing facilities based on new technology. As a starting point, we use a framework developed by DeRosa et al. [2], which involves a geographically detailed network model of the U.S. petrochemicals industry, with supply, demand, plant production capacity and trade data from the ICIS Supply and Demand Database [3]. In our work, we are adapting this model for use in specific regions, and incorporate the capability to expand the network by adding new manufacturing nodes in unspecified locations (to be identified via economic optimization).

The new model is formulated as a mixed-integer optimization problem with the objective of minimizing the cost of the manufacturing and supply chain network through choices of old and new technologies used and placement of new manufacturing facilities. Optimal placement of the new plants will contribute to lowering transportation costs, as well as establishing a more sustainable supply chain, reducing the distances required for transport of raw materials, intermediates and products. Various scenarios involving the introduction of new technologies [e.g, 4] are studied, focusing on the Marcellus/Utica region in the Northeast U.S. Sensitivity analyses are performed to study the impact of different supply, demand and capacity constraints. It is expected that this work will define targets for the development of more cost-effective and sustainable manufacturing and supply chain networks to take advantage of newly available light hydrocarbon resources, as well as provide guidance for the development of new technologies that utilize these feedstocks.

References

[1] U.S. Energy Information Administration (EIA), "What are natural gas liquids and how are they used?" [Online]. Available: http://www.eia.gov/todayinenergy/detail.cfm?id=5930. [Accessed: 02-Feb-2020].

[2] S. E. DeRosa, Y. Kimura, M. A. Stadtherr, G. McGaughey, E. McDonald-Buller, and D. T. Allen, "Network Modeling of the U.S. Petrochemical Industry under Raw Material and Hurricane Harvey Disruptions," *Ind. Eng. Chem. Res.*, 58, 12801–12815, 2019.

[3] ICIS Supply and Demand Database, 2017; http://www.icis.com/explore/services/analytics/supply-

demand-data/supply-and-demand-window [Accessed: 28 April 2020]

[4] T. Ridha, Y. Li, E. Gençer, J. J. Siirola, J. T. Miller, F. H. Ribeiro, and R. Agrawal, "Valorization of Shale Gas Condensate to Liquid Hydrocarbons through Catalytic Dehydrogenation and Oligomerization," *Processes*, 6, 139, 2018.

Acknowledgement

This work is supported in part by the National Science Foundation under Cooperative Agreement No. EEC-1647722 (CISTAR – NSF Engineering Research Center for Innovative and Strategic Transformation of Alkane Resources, http://cistar.us). Any opinions, findings and conclusions or recommendations expressed in this material are those of the author(s) and do not necessarily reflect the views of the National Science Foundation.

609436 Spatially-Explicit Techno-Ecological Design for Sustainable Manufacturing Applied to a Power Plant

Abstract:

Far too often, fields of study are kept separated and can result in many unintentional impacts, specifically environmental concerns. These impacts affect ecosystems that provide the services on which we rely for survival and well-being. Ironically, in search for technology to improve mankind's livelihood, we risk the health and well-being provided by the natural world. Therefore, to make better, holistic decisions, there is a need for bridging engineering and ecology to design systems that include both technology and ecology. Along with decreasing negative environmental impacts, including ecosystems in sustainable design can lead to innovative solutions which utilize the functions and co-benefits of nature as we search for sustainable solutions amid increasing population and demand for these services.

Previous work in techno-ecological process design has lacked inclusion of spatial heterogeneity of ecosystem service supply and demand. It is important to understand how, where, and when mass and energy flow across components of techno-ecological systems to inform design opportunities and impacts. This work will focus on gaseous pollutants and air quality regulation. To understand the atmospheric dispersion of the flue gas contaminants and the capacity of local ecosystems to uptake those chemicals, through dry deposition, we incorporate a detailed geophysical, meteorological, and dispersion model into the design framework. This research will **introduce a spatially-explicit techno-ecological design framework** and present an application of the framework to a coal-fired power station in Eastern Ohio.

In the case study, we focus on the SO_2, NO_2, and PM_{10} emissions of the power station and the adsorption of these chemicals onto the leaf surfaces of nearby land cover. Our framework will compare the pollution removal between technological equipment and land-use change in the region, optimizing minimum-cost solutions that meet the pollution removal requirements. To show the applicability and practicality of the framework, we will demonstrate a few different scenarios including: models at two different scales of land area, multiple technological scales of the power station, a deforestation (or natural disaster) simulation, inclusion of carbon sequestration, and different ecological budgets. The results yield both long-term operating conditions of the technology and maps or animations of the land changes over time. Further, we will discuss critical variables in determining each solution and conduct a sensitivity analysis on the price of land afforestation.

Initial results suggest that land management can provide limited air pollution removal of toxins like SO_2, NO_2, and $PM_{10,}$ however, the land management can still be economically-competitive options on a "per ton" basis. Including the valuation of other ecosystem services, like carbon sequestration, can heavily favor land management projects for industries. Bringing a balance of both technological and ecological solutions can bridge the gap between the amount of pollutant emitted from a manufacturing site and the uptake of local ecosystems. This introduces an absolute sustainability metric rather than a relative one, comparing emissions to ecological capacity. However, it is also recognized that only focusing on air quality regulation can shift environmental impacts onto other ecosystem services. We will discuss initial insights into land use competition and inclusion of other ecosystem services into this spatially-explicit design framework. Exploring spatially-explicit design options for techno-ecological systems enables smarter industrial site design and is one step closer towards bridging ecological knowledge with engineering practice.

611371 Economic and Energy Impacts of Wind Energy Expansion in the US Using Multi-Regional Input Output (MRIO) Tables.

Abstract:

Renewable energy sources help reduce the environmental load of the energy sector. However, when new renewable energy sources, such as wind energy infrastructure is deployed in a region, it can have economic and energy consumption impacts both in the region being deployed and the regions which interact with the deployed region due to supply chain impacts. It is important to quantify these multi regional impacts associated with wind energy expansion to understand what are the regional economic sectors that are being impacted and how their energy consumption pattern changes. In this study, we use an economic multi-regional input output (MRIO) table developed for the US to assess the multi-regional economic and energy impacts of wind energy expansion in the US. The US MRIO table was developed using the Industrial Ecology Virtual Lab project that uses non-survey approaches to build MRIO tables with the help of different national and regional level statistics followed by constrained optimization. In this study it was assumed that 500 megawatts capacity wind energy farms were installed in the already top ten wind energy producing states. The developed US MRIO table, coupled with input output analysis, was then used to find out how the deployment of wind energy in these ten states affect the remaining 42 regions in the US. The local impact data of installing wind turbines in each of the ten states was obtained from the National Renewable Energy Lab's Jobs and Economic Development Impact (JEDI) model which was used as the feed data to perform input output analysis on the developed US MRIO table.

The changes in economic throughputs of different economic sectors in each of the 52 regions was quantified. The total economic impact was found to be 26 billion dollars of which 3 billion dollars was associated with the states where no new wind energy capacity was installed. Installation of new energy production capacity also results in change in energy consumption by different economic sectors across US. Using the US-MRIO model and the energy intensity of manufacturing sectors, the energy consumption increase due to addition of wind farms was found to be about 6952 trillion of btu for the total change in economic throughput. Primary metal manufacturing and Machinery manufacturing sectors stood out amongst other manufacturing sectors with considerable change in energy consumption with an increase of 3074 trillion of btu and 1537 trillions of btu. The change in energy consumption was found at a four census (North, South, Midwest and West regions of the US) regional level.

600321 Co-Optimized Power and H_2 Infrastructure Planning for Multi-Sector Decarbonization

Abstract:

The past decade has seen significant progress in CO_2 emission reduction in the power sector, while progress in other sectors has remained sluggish. With cost reductions in fuel cell and electrolysis technologies and power sector decarbonization, Hydrogen (H_2) is becoming an increasingly appealing option for reducing emissions from other end-use sectors, like heavy-duty transport, heating and industry, where electrification may be challenging. Meeting spatially and temporally varying H_2 demand requires significant infrastructure investments in H_2 generation, storage and transport. To date, most studies on hydrogen infrastructure planning have focused on H_2 use in transportation and on evaluating the trade-offs within the H_2 supply chain, without consideration to the dynamics of its interactions with the electricity sector. The latter could include: a) H_2-based energy storage to manage wind, solar, and demand variability at multiple time-scales, b) flexible electrolytic H_2 production coordinated with electricity adequacy and c) transporting H_2 or derived carriers instead of electricity to balance spatial variations in energy supply and demand. Understanding the implications of these interactions on the overall cost competitiveness of H_2 use requires developing scalable decision-support frameworks with adequate representation of temporal and spatial variability in cross-sectoral interactions.

We have developed a framework for power and H_2 infrastructure planning that determines the least-cost mix of electricity and H_2 generation, storage, and transmission infrastructures to meet power and H_2 demands subject to a variety of operational and policy constraints. Notably, the model includes hourly representation of power and H_2 system operation, that is made computationally tractable using judicious approximations and offline time-domain reduction strategies. The developed optimization model can represent a wide range of power and H_2 technology options, including renewables, carbon capture and storage (CCS) applied to power and H_2 generation, and truck (gaseous and liquid) and pipelines for H_2 transportation.

We apply the model to study future electricity and H_2 infrastructure needs for the U.S. Northeast region under various carbon policy and H_2 demand scenarios. The study reveals several interesting insights. First, CCS is deployed for H_2 production via steam methane reforming at lower CO_2 prices (< \$100/tonne-CO_2) than for power generation (~ \$200/tonne-CO_2). Second, for scenarios with relatively small H_2 demand, the operational synergies between the power and H_2 sectors, especially at times when electricity prices are low or high, reduces the cost of power sector decarbonization compared to the case without H_2 demand. These findings indicate the importance for joint planning of electricity and hydrogen infrastructure for cost-effective energy system decarbonization.

600703 Optimal Selection and Location of Nutrient Recovery Systems Considering Standalone and Coordinated Strategies

Abstract:

Human farming and agricultural activities have altered nutrient cycles; this has been the result of the technification and intensification of production processes, which decreases the mobility and availability of nitrogen and phosphorus. Among the anthropogenic activities, the agricultural sector is the main source of phosphorus emissions [1]. Livestock farming generates substantial amounts of organic waste, being linked with an in-excess nutrient concentration in soils and water [2]. The implementation of sustainable nutrient management processes aids the recycling of non-renewable resources essential to supporting life and mitigating nutrient pollution of waterbodies and soil degradation [3].

In this work, the selection, implementation, and performance evaluation of different state-of-the-art nutrient recovery technologies at livestock facilities are assessed from both standalone and logistics network perspectives and are coupled with electricity production through anaerobic digestion of manure [4]. On the one hand, following a standalone approach, the individual facilities are evaluated considering their characteristics regarding size, type, as well as its location to reflect local environmental and nutrient pollution conditions due to legacy and new inputs of nutrients. In addition, the location and choice of nutrient recovery systems are evaluated using a logistics network framework where the potential coordination between facilities for nutrient recovery is included in the analysis. Allowing scenarios in which there is coordination is beneficial due to the economies of scale by transporting organic waste to centralized facilities for energy and nutrient recovery. The evaluation of both approaches allows studying the economic feasibility of operating nutrient recovery facilities, determining if the concentrated animal feeding operations can operate nutrient recovery processes in a standalone mode or, conversely, they should be centralized in large management facilities, even if this accounts for transportation costs of high-water content livestock waste. Besides, the effect of different policy incentives for installing nutrient and energy recovery technologies, including RECs (Renewable Energy Credits) and hypothetical nutrient credits, are evaluated.

The developed framework is applied to a series of real case studies of cattle facilities in the U.S. by considering the location and size of livestock farms. Therefore, this determines the optimal distribution of nutrient and energy recovery processes. Finally, this framework develops heuristic rules that can be used as guidelines by the stakeholders involved in the decision-making process for designing and implementing coordinated nutrient management systems.

References

[1] Dzombak, D. A. (2011). Nutrient Control in Large-Scale U.S. Watersheds. The Bridge, 41 (4), 13–22.

[2] Sampat, A., Martın, E., Martın, M., & Zavala, V. (2017). Optimization formulations for multi-product supply chain networks. Comput. Chem. Eng., 104 , 296-310.

[3] Cervantes, Francisco J. Environmental Technologies to Treat Nitrogen Pollution. IWA Publishing, 2009.

[4] Martín-Hernandez, E., Sampat, A., Zavala, V., & Martın, M. (2018). Optimal integrated facility for waste processing. Chem. Eng. Res. Des., 131 , 160-182.

602435 Techno-Economic Analysis of Electricity Storage Technologies for behind-the-Meter Applications

Abstract:

The electric power grid is one of the most sophisticated, complex and vital inventions of modern times (Erenoğlu et al., 2019). Consisting of infrastructure for generating, transmitting and distributing electricity to hundreds of millions of end-users around the globe, it has become an indispensable amenity in present day. However, owing to unprecedented energy demands attributed to population growth, combined with growing concerns of climate change, a number of challenges have arisen (Bartolucci et al., 2019; Erenoğlu et al., 2019). Some of these include the need for increased energy generation to meet future demand, reduction in greenhouse gas emissions, increased power reliability and flexibility.

Proposed solutions to address these challenges can be categorised in three: to decarbonise, decentralise and digitise energy. This entails, although it is not limited to, a reduction in fossil-based fuel use towards an increased share of cleaner and renewable energy sources (RES), and development of decentralised electrical power systems. RES such as solar, wind, hydro, biomass, have been introduced to a greater extent in the past decade in most regions of the world (REN21, 2019) both in large generating facilities and behind-the-meter (BTM) applications. These changes further introduce challenges of their own, particularly issues relating to power quality and reliability owing to the intermittent nature of renewable energy sources.

Electrical energy storage (EES) technologies have been proven to mitigate some of these problems. With EES, excess electrical energy generated can be stored for use at a different period – with higher loads or during blackouts. Whether they are used in BTM applications – small businesses, households, microgrids, etc. – or in the utility grids, EES devices have shown to improve power reliability, quality and flexibility (Ma et al., 2018). As the use of EES has grown from more traditional roles of just storing excess energy for use during periods of high demand, to more diverse roles such as load levelling, energy arbitrage, ancillary services, etc., it has become more important, and complicated, to determine the optimal choice and size of EES technologies, especially in BTM applications.

This work presents a techno-economic optimisation model for BTM EES technologies. Building on existing works (Fisher et al., 2019; Roberts & Brown, 2019), the proposed model aims to obtain the optimal size of a range of EES assets in light of their ability to access certain revenue streams. Storage technologies explored in this work are restricted to battery energy storage systems – Li-ion, flow, lead-acid batteries – over a range of sizes, capacities, degradation levels and/or power ratings. Revenue streams comprise a reduction in demand charges, energy costs, long term asset costs and/or participation in wholesale and ancillary services markets. Using available load profiles from a portfolio of commercial building types, the proposed model assesses the optimal choice and size of storage technologies that maximises the total financial benefits within different markets.

References

Bartolucci, L., Cordiner, S., Mulone, V., & Rossi, J. L. (2019). Hybrid renewable energy systems for household ancillary services. *International Journal of Electrical Power & Energy Systems*, *107*, 282–297. https://doi.org/10.1016/j.ijepes.2018.11.021

Erenoğlu, A. K., Erdinç, O., & Taşcıkaraoğlu, A. (2019). History of electricity. In *Pathways to a Smarter Power System* (pp. 1–27). Elsevier. https://doi.org/10.1016/B978-0-08-102592-5.00001-6

Fisher, M., Apt, J., & Whitacre, J. F. (2019). Can flow batteries scale in the behind-the-meter commercial and industrial market? A techno-economic comparison of storage technologies in California. *Journal of Power Sources, 420*, 1–8. https://doi.org/10.1016/j.jpowsour.2019.02.051

Ma, T., Shen, L., & Li, M. (2018). Electrical energy storage for buildings. In *Handbook of Energy Systems in Green Buildings* (pp. 1079–1107). Springer Berlin Heidelberg. https://doi.org/10.1007/978-3-662-49120-1_44

REN21. (2019). *Renewables 2019 Global Status Report.* https://www.ren21.net/wp-content/uploads/2019/05/gsr_2019_full_report_en.pdf

Roberts, D., & Brown, S. (2019). Flow batteries for energy management: Novel algebraic modelling approaches to properly assess their value. *Journal of Energy Storage, 26*(September), 100977. https://doi.org/10.1016/j.est.2019.100977

603291 Dynamic Optimization of Li-Ion Battery Systems for Grid-Connected High Current Density Operation

Abstract:

Title: Dynamic Optimization of Li-Ion Battery Systems for Grid-Connected High Current Density Operation

Authors: Chungho Suh, Chrysanthos E. Gounaris, Debangsu Bhattacharyya

Traditionally, energy generation has relied heavily on fossil fuels. However, in order to achieve the goals set out by the Paris Agreement, reliance on fossil fuels must be dramatically diminished by increasing renewable energy penetration. One of the greatest challenges with the renewable energy sources is due to their intermittency. This not only causes an increased cycling operation of the power plants leading to efficiency loss, higher emissions, and an adverse effect on plant health, but also prevents higher penetration of renewables to the grid [1]. To promote higher penetration of renewables to the grid, development of efficient energy storage technologies is absolutely critical. While many candidate energy storage technologies are being investigated, grid-level Li-ion batteries have great promise as a storage technology due to their wide-ranging power ratings, high cycle efficiency, and fast response times. While dynamics and charging/discharging characteristics of Li-ion batteries have been investigated earlier for applications in electric cars, dynamic optimization of these batteries for grid-level applications is yet to be thoroughly investigated [2][6]. Grid-level operation has some distinct characteristics compared to electric car applications including extremely fast yet uncertain charging/discharging requirements, very high current density operation, and resource constraints (i.e. if excess energy is not available, the battery may not be desired even if the state of charge is low). In addition, it is desired to maximize the roundtrip efficiency [5].

As a first step to the model-based dynamic optimization problem, a 1-D dynamic model of a Li-ion battery stack is developed [3][4]. Governing equations for the conservation of both Li-ion species and charge are considered. These equations are applied to both the electrolyte and solid phases. Polynomial approximation and volume-averaging are used to eliminate the radial dependence of the solid-phase concentration, resulting in a system of partial differential equations (PDEs). Coordinate transformation, spatial discretization and orthogonal collocation are used for developing a system of equations that are used for optimization [3].

For grid-connected systems, battery packs can be charged when there is excess energy available while discharging them when there is discrepancy between the demand and supply [5]. Given the current cost of the Li-ion batteries, it is unlikely that in the near-future, battery packs that can provide grid-level power for long duration will be economically viable. They are likely to be deployed for addressing discrepancy between the demand and supply that are high but of shorter duration [7]. Such scenarios have considerable uncertainty in the magnitude of discrepancy, its duration, and time of the day/night. Thus it is desired to solve a multi-scenario nonlinear programming (NLP) problem. Furthermore, high current density operation can cause overheating and fire. High current density operation can also significantly reduce the battery efficiency while improving the grid resiliency and reducing the ramp rate for the fossil-fired generators. Therefore it is desired to solve a multi-objective dynamic optimization problem. If the rating of the battery pack is increased, operating current density at any instant of time can be reduced for the same level of discrepancy. While a lower current density operation can improve the cell efficiency, it also increases the capital cost. Therefore rating of the battery pack is a critical decision variable. In addition to rating, cell dimensional variables and its charging/discharging profiles are also optimized. Overall, a multi-scenario, multi-objective dynamic optimization problem is solved for optimal rating, design and operation of the grid-connected Li-ion batteries. For tractability of this large scale NLP, sparsity in the structure of the system of equations is exploited.

References

[1] N. Kumar, S. Paterson, K. Coleman, C. Lee, D. Agan, and S. Lefton, "Power plant cycling measure - Evaluating historical cycling to model future grid operations," in *2013 IEEE Power & Energy Society General Meeting*, Vancouver, BC, 2013, pp. 1–5, doi: 10.1109/PESMG.2013.6672754.

[2] P. M. Attia *et al.*, "Closed-loop optimization of fast-charging protocols for batteries with machine learning," *Nature*, vol. 578, no. 7795, pp. 397–402, Feb. 2020, doi: 10.1038/s41586-020-1994-5.

[3] P. W. C. Northrop, V. Ramadesigan, S. De, and V. R. Subramanian, "Coordinate transformation, orthogonal collocation, model reformulation and simulation of electrochemical-thermal behavior of lithium-ion battery stacks," *J. Electrochem. Soc.*, vol. 158, no. 12, p. A1461, 2011, doi: 10.1149/2.058112jes.

[4] V. R. Subramanian, V. Boovaragavan, V. Ramadesigan, and M. Arabandi, "Mathematical model reformulation for lithium-ion battery simulations: galvanostatic boundary conditions," *J. Electrochem. Soc.*, vol. 156, no. 4, p. A260, 2009, doi: 10.1149/1.3065083.

[5] X. Luo, J. Wang, M. Dooner, and J. Clarke, "Overview of current development in electrical energy storage technologies and the application potential in power system operation," *Applied Energy*, vol. 137, pp. 511–536, Jan. 2015, doi: 10.1016/j.apenergy.2014.09.081.

[6] N. Xue, W. Du, T. A. Greszler, W. Shyy, and J. R. R. A. Martins, "Design of a lithium-ion battery pack for PHEV using a hybrid optimization method," *Applied Energy*, vol. 115, pp. 591–602, Feb. 2014, doi: 10.1016/j.apenergy.2013.10.044.

[7] A. Nottrott, J. Kleissl, and B. Washom, "Energy dispatch schedule optimization and cost benefit analysis for grid-connected, photovoltaic-battery storage systems," *Renewable Energy*, vol. 55, pp. 230–240, Jul. 2013, doi: 10.1016/j.renene.2012.12.036.

592768 Ammonia for Gas Turbines Fuelling - European Progress

Abstract:

A hydrogen economy has been the focus of researchers and developers over the decades. However, the complexity of moving and storing hydrogen has always been a major obstacle to deploy the concept. Therefore, other materials can be employed to improve handling whilst reducing cost over long distances and long periods. Ammonia, a highly hydrogenated molecule, can be used to store and distribute hydrogen easily, as the molecule has been employed for more than 120 years for fertilizer purposes. Being a carbon-free chemical, ammonia (NH3) has the potential to support a hydrogen transition thus decarbonising transport, power and industries.

However, the complexity of using ammonia for power generation lays on the appropriate use of the chemical to reach high power outputs combined with currently low efficiencies that bring up overall costs. This complex scenario is also linked to the production of combustion profiles that tend to be highly polluting (with high NOx emissions and slipped unburned ammonia). There is still no technology capable of using ammonia whilst producing both low emissions and high efficiencies in large power generation devices, thus efficiently enabling the recovery of hydrogen and reconversion of stranded, green energy that can be fed back to the grid. Tackling these problems can resolve one of the most important barriers in the use of such a molecule and storage of renewable energies. Many countries have recently engaged in ambitious programs to resolve these issues, aiming for large power units to run on ammonia by 2030. Thus, European counterparts need also to engage in these technological advancements to fully unlock a hydrogen, cost-effective economy.

Therefore, the current presentation provides an insight of British led attempts to support the transition towards a zero-carbon economy via ammonia. Ammonia has recently been recognised for its potential to fuel gas turbines. Works around the subject have focused on the use of pure ammonia or its various blends to enable the production of extra power using stranded renewable energy converted into this hydrogen carrier. Therefore, interest across Europe has followed current developments from Japan in order to ensure that European technologies are capable of using the chemical without incurring into large amounts of emissions, operational problems or flame instabilities. Projects go from small power units to large power and propulsion applications establishing new set of guidelines to employ ammonia blends for the production of zero-carbon power.

Results demonstrate that ammonia can support the transition towards a zero-carbon economy, producing negligible emissions due to the self-consuming effect of unwanted nitrogen oxides. Moreover, doping and improved injection strategies have shown enhanced stability, with new trials that have been conceived to determine the impact of these aggressive atmospheres towards hot combustor components. Numerical simulations and fundamental studies, also employed to define new insights into the chemistry of these blends, have evolved to ensure that complex designing tools (i.e. DNS/LES CFD) can be afforded for more accurate designs. All these works go hand-by-hand to demonstrate the potential of ammonia to supplement the long-time storage capability of hydrogenated fuels in gas turbines. Therefore, apart from all these crucial points, this presentation will also highlight efforts that Europe has mustered to remain a key player around the subject.

602433 Optimal Electric Vehicle Charging Schedule: Impact of User Travel Behaviour and Distributed Energy Resource Availability

Abstract:

The role that greenhouse gas (GHG) emissions play especially as it relates to the potential impacts and associated risks of climate change have been acknowledged in most parts of the world. In light of this, up to eleven countries (the United Kingdom inclusive) have set a net-zero emission target by 2050 of GHG emissions (Committee on Climate Change, 2019). In the United Kingdom, transport has been identified as the largest contributor to GHG. To this end, one of the scenarios outlined towards achieving the net-zero goal involves 'extensive electrification, particularly of transportation and heating' (Committee on Climate Change, 2019; Küfeoğlu & Khah Kok Hong, 2020), which directly involves a transition to electric-powered surface transport vehicles.

In the UK, there has been a growing adoption of Electric Vehicles (EVs). This is attributed to benefits from capital subsidies, lower fuel and vehicle taxation as well as the increasing cost-competitiveness of EVs compared with traditional internal combustion engine-type vehicles. This growth has its impact on the utility grid. Unregulated connections of EVs can result in a substantial increase in aggregate demand, impact power quality, or cause an outright destabilization of the grid (Ahmadian et al., 2020; Xiong et al., 2017). There are also benefits to the grid. EVs act as additional energy storage devices providing energy through vehicle-to-home (V2H) and/or vehicle-to-grid (V2G) services. This allows for peak load shaving, reduction in household energy costs and backup power supply during outages. What then becomes important is the need to provide an optimal scheduling strategy to ensure that these benefits are leveraged whilst minimising the negative impacts to the grid.

This work thus proposes an optimal EV charging schedule for households considering the availability of EVs, stationary battery energy storage systems and/or solar power generation. Given a number of EV types and properties providing V2H services, as well as a number of different classes of EV user travel behaviour, the proposed optimisation model obtains the day-to-day charging schedule. The schedule will ensure minimal costs based on available electricity tariffs and household energy demand data. Using real-world cases, the most cost-effective electricity tariff can be selected, with its associated charging schedule, for a specific household and/or EV user type based on the aforementioned considerations.

References

Ahmadian, A., Mohammadi-ivatloo, B., & Elkamel, A. (2020). *Electric vehicles in energy systems* (A. Ahmadian, B. Mohammadi-ivatloo, & A. Elkamel (eds.)). Springer International Publishing. pp. 183. https://doi.org/10.1007/978-3-030-34448-1

Committee on Climate Change. (2019). *Net Zero: The UK's contribution to stopping global warming* (Issue May). https://www.theccc.org.uk/publication/net-zero-the-uks-contribution-to-stopping-global-warming/%0Awww.theccc.org.uk/publications

Küfeoğlu, S., & Khah Kok Hong, D. (2020). Emissions performance of electric vehicles: A case study from the United Kingdom. *Applied Energy*, *260*, 114241. https://doi.org/10.1016/j.apenergy.2019.114241

Xiong, Y., Chu, C., Gadh, R., & Wang, B. (2017). Distributed optimal vehicle grid integration strategy with user behavior prediction. In A. Ahmadian, B. Mohammadi-ivatloo, & A. Elkamel (Eds.), *2017 IEEE Power & Energy Society General Meeting* (Vols. 2018-Janua, pp. 1–5). IEEE. https://doi.org/10.1109/PESGM.2017.8274327

599100 Optimisation of Fuel Supply Chains within Planetary Boundaries: A Case Study of Hydrogen for Road Transportation in the UK

Abstract:

Current studies in sustainable transportation often neglect environmental impacts beyond climate change or consider them using metrics which are difficult to interpret from a world-wide sustainability viewpoint. We apply here the concept of Planetary Boundaries (PBs), which define a global environmental "safe operating space", to address the optimal design of sustainable fuel supply chains (SCs) focusing on hydrogen for vehicle use. Applying the PBs concept in a mixed-integer linear programming model, SC configurations are identified which satisfy a given transportation demand while minimising transgression of PBs, hence reducing the risk of exceeding the carrying capacity of the Earth system. We apply this methodology to a UK hydrogen case study, finding that while current transgression levels could be significantly reduced by moving from fossil fuels to hydrogen, it would yet be insufficient to operate fully within all PBs concurrently. The SC design with the lowest transgression would produce and compress hydrogen using nuclear power, while avoiding most transportation needs due to a decentralised structure. This work opens new avenues for the incorporation of PBs in the optimisation and assessment of energy systems and highlights the potential of emerging technologies to significantly improve the status quo in environmental terms. Nevertheless, further work will be required to find solutions operating within all PBs, and ultimately resulting in an economy fully consistent with the concept of absolute environmental sustainability.

601104 An Integrated Model for Food-Energy-Water Nexus to Study Global Sustainability

Abstract:

At its core, the concept of sustainability is about ensuring that the conditions necessary for civilized human existence on Earth are maintained over the long term, or for as long as possible [1-6]. Successful implementation of sustainability ideas in food-energy-water (FEW) systems requires a basic understanding of the often nonlinear and non-intuitive relationships among different dimensions of the network [7], particularly the system-wide implications of human actions [8-11]. This basic understanding further includes a sense of the time scale of possible future events and the limits of what is and is not likely to be possible. In this work, we present an integrated model that has different components like food-web ecosystem dynamics, microeconomics components, including energy producers and industries, and various socio-techno-economic policy dimensions.

Mathematical models featuring the critical components of a real ecosystem can aid in the formal study of sustainability. For instance, ecologists have used mathematical models to determine the stability of a community based on the size and connectivity of the food web in the face of fluctuations of the system . Lotka-Volterra models used to describe the predator-prey interactions represent one classical theory used to study the dynamics of these biological systems. Also, there have been ecological models that have incorporated varying degrees of socio-economic features. Acutt and Mason [12] and Heal [13] proposed valuing the various environmental services in an ecosystem in their decision-making models. Some of the other works which included primarily simplified economic features in their ecological models are the ones developed by [15-17]. The first such comprehensive model is the model from the Club of Rome [9, 10]. This model has components like the human population, natural resources, pollution, industrial production, and agricultural land use. The results of this model established growth patterns and limits for the first time. Models accounting for processes such as resource extraction, waste assimilation, recycling, and pollution in an integrated ecological, economic framework have also been proposed in the literature [18]. In recent years, united global metamodel of the biosphere (GUMBO) [11] was developed to include dynamic feedbacks among human technology, economic production and welfare, and ecosystem goods and services. Recently, the model predictions were compared with data from 2000-2018 [12]. Whitmore et al. [19] have proposed a model that integrated an economy under imperfect competition with a widely used twelve-cell ecological model [3]. This twelve-cell ecological model developed by Cabezas et al. [3] mimics a global ecosystem with a very rudimentary social system that regulates the flows of mass according to its criteria and is the first model we have developed. The intermediate model includes the industrial sector and the rudimentary economic system [3]. The third more comprehensive and recent model is an integrated model, including ecology, economics, social system, technology, and time [20]. The food-web is modeled by Lotka–Volterra type expressions, whereas the economy is represented by a price-setting model wherein firms and the human households attempt to maximize their well-being. In this work, we combine the food-web model with a microeconomics model similar to Kotecha et al. [20]. Although our global model is conceptually similar to the World and GUMBO model and includes additional information like land use, water capacity, and pollution, our focus is on food-energy-water nexus, unlike existing global sustainability models. The microeconomic model is augmented by a reinforced learning (RL) based model based on time series data.

A schematic of the food-web model we have developed is shown in Figure 1. The model consists of 15 compartments and represents a simplified ecological food web set in a macroeconomic framework along with a rudimentary legal system. The model consists of three primary producers (P1, P2, and P3), three herbivores (H1, H2, and H3), two carnivores (C1 and C2) along with human households (HH). The Energy source (ES) represents a finite nonrenewable energy source. The Inaccessible Resource Pool (IRP) results from waste flows from the ES and IS that are not biologically accessible to the ecosystem. Primary producers P2 and P3 feed on IRP and make this mass available to the rest of the ecosystem but at a slow rate symbolizing bacterial actions, while other ecological processes absorb other wastes such as CO_2. As a stock of pollutants, the IRP degrades the ecological fund. All biological compartments recycle mass back to RP through death. The Energy Producer (EP) is an industrial compartment that uses labor to transform the energy source into a usable form of energy. This energy is supplied to the human households (HH) and the Industrial sector (IS). The Energy Producer is also capable of producing energy using P1, representing the production of energy using biomass. The Industrial Sector (IS) produces products valuable to HH using P1 and RP. The use of the IS products does not increase the mass of the human households, but it does raise the mass of IRP. The biological compartments of the systems can be aggregated as shown in the figure into domesticated species with economic value and wild species with no economic value. A legal system assigns property rights to domesticated biological species, the product of the industrial sector, and the non-renewable energy source.

The model construction and validation are carried out using global-scale data for stocks and flows of food, energy, and water, which can be used to parameterize this model. We have also collected data (e.g., pollution, land utilization, etc.) from the World3 and GUMBO model as these models are validated for past data. We will use past data to generate the parameters of this lumped compartmental model. The last ten years of data for key variables like the human population, GDP growth, global warming trends, and land use are used to validate the model.

The model will then use to conduct different scenario analyses forecasting the sustainability of the planet for the next century or more. In particular, we will look at the following aspects:

- Impact/feasibility of achieving US sustainable development goals in the domain of food, energy, and water, and identify interlinkages that may restrict achieving the targets.
- Identify optimal pathways for ensuring sustainability of food, energy, and water subsystems.
- Study the effect of biomass based energy on FEW sustainability.

The results will provide policy guidelines and recommendations for ensuring long term sustainability.

Bibliography

1. World Commission on Environment and Developement, 1987; Our Common Future, Oxford University Press, Oxford, United 383 pp.
2. Cabezas and Fath B. 2002, Towards a theory of sustainable systems, Fluid Phase Equilibria, 194-197, 3-14.

3. Cabezas , C. W. Pawlowski, A. L. Mayer, and N. Theresa, 2002; Sustainable Systems Theory: Ecological and Other Aspects, 15-th International Congress of Chemical and Process Engineering, Praha, Czech Republic.
4. Brian Fath, Heriberto Cabezas and Christopher W. Pawlowski, 2003; Regime Changes in Ecological Systems: An Information Theory Approach. Journal of Theoretical Biology., 222, 517-530.
5. Shastri and U. Diwekar, 2006. Sustainable ecosystem management using optimal control theory: Part 1 (de- terministic systems), Journal of Theoretical Biology, 241, 506521
6. Shastri, , Diwekar, U, Cabezas, H., Williamson, J. 2008. Is sustainability achievable? exploring

the limits of sustainability with model systems. Environ. Sci. Technol. 42, 6710-6716.
7. Liu , Slotine J., and Barabasi A.,Controllability of complex networks, nature, 473, 167-173.
8. Doshi, R. Diwekar, U, Benavides, P. Yenkie, K. Cabezas, H. (2015), "Maximizing sustainability of ecosystem model through socio-Economic policies derived from multivariable optimal control theory", Journal Clean Tech Environ Policy, 17,
9. Meadows, D. H., Meadows, D. L., Randers, J., & Behrens, W. W. (1972). The limits to growth. Jorgen Randers, New York,.
10. Meadows D. H., Limits to Growth: The 30-year update (2014), Jorgen Randers, New York.
11. Roelof Boumans, Robert Costanza, Joshua Farley, Matthew A. Wilson, Rosimeiry Portela, Jan Rotmans, Ferdinando Villa, and Monica Grasso, 2002, Economic and Ecological Perspectives Modeling the dynamics of the integrated earth system and the value of global ecosystem services using the GUMBO model, Ecological Economics 41, 529–560
12. Boumans R., 2019, Revising GUMBO Model, 2019 Global Symposium on Sustainability, Rome, Italy, Oct 1-3.
13. Acutt M, Mason 1998. Environmental valuation, economic policy and sustainability: Recent advances and environmental economics. Cheltenham, UK: Edward Elgar.
14. Ludwig D, Carpenter S, Brock 2003. Optimal phosphorous loading for a potentially eutrophic lake. Ecological Applications 13: 1135–1152.
15. Brock W, Xepapadeas 2002. Optimal ecosystem management when species compete for limiting resources. Journal of Environmental Economics and Management 33: 189–220.
16. Costanza R, Voinov A, Boumans R, Maxwell T, Villa F, Wainger L, Voinov 2002. Integrated ecological economic modeling of the Patuxent River watershed, Maryland. Ecological Monographs 72: 203–231.
17. Carpenter S, Brock W, Hanson 1999. Ecological and social dynamics in simple models of ecosystem management. Conservation ecology 3.
18. van den Bergh 1996. Ecological economics and sustainable development, chapter A multisectoral growth model with materials flows and economic-ecological interactions. Cheltenham, UK: Edward Elgar, 147–172.
19. Whitmore, ; Pawlowski, C.; Cabezas, H. 2006. Integration of an economy under imperfect competition with a twelve-cell ecological model. Technical Report EPA/600/R-06/046, 2006.
20. Kotecha P., U. Diwekar, and H. Cabezas (2013), "Model Based Approach to Study the Impact of Biofuels on the Sustainability of an Integrated System", Clean Technologies and Environmental Policy, 15, 21.

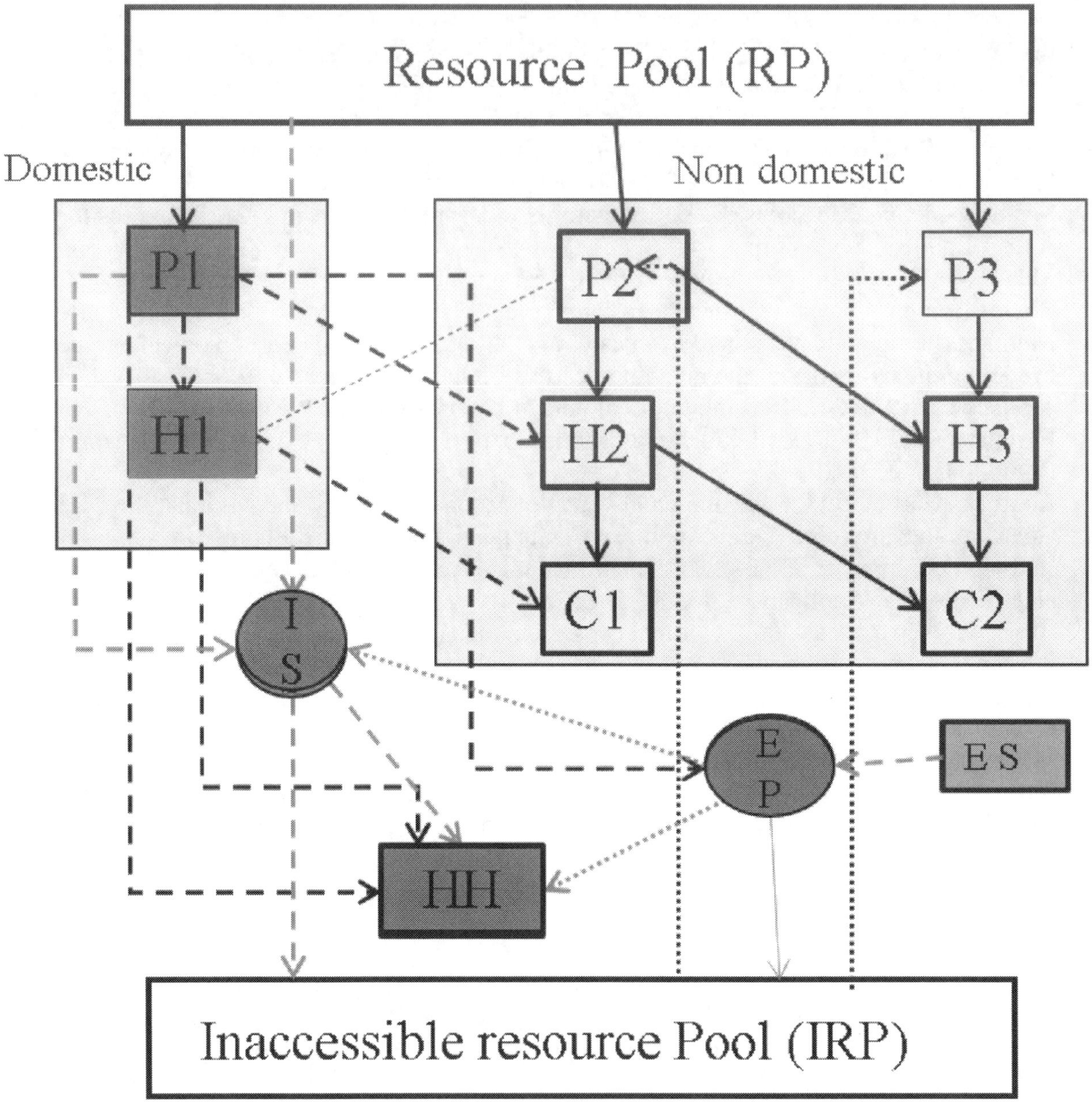

Resource Pool (RP)
Domestic
Non domestic
P1
H1
P2
H2
C1
P3
H3
C2
I
S
E
P
E S
HH
Inaccessible resource Pool (IRP)

600971 Consumption-Based Accounting of Environmental Impacts Associated with Beef Supply Chains in the United States: Implications for FEW Nexus

Abstract:

With increasing population, food supply chains have evolved to be robust and often associated with great distances between the food production and consumption locations. The consumption of food commodities, and specifically of animal-based products, is associated with environmental impacts including greenhouse gas (GHG) emissions and vast quantities of embodied resources, such as energy and water. A crucial step for enhancing the environmental sustainability of intertwined food-energy-water (FEW) systems is quantification of both production and consumption-based impacts of the food system. Such an understanding can aid in devising interventions for improving the environmental sustainability of FEW systems that are effective from both production and consumption perspective. While there exists a rich body of work on production-based impacts of diverse food commodities, consumption-based accounting remains challenging. In addition, most studies focus on impacts at state or national levels, producing results that are limited in their ability to guide decision-making and identify improvement opportunities. We address these challenges by presenting a systems-based framework to quantify the consumption-based impacts of beef in the United States.

Using publicly available data, we present a network model of beef production-consumption for the US at the county-level. We utilize network analysis to understand the structure and robustness of the beef consumption network, and to highlight counties critical for the beef supply chain. The model is used to attribute the production locations to beef consumption elsewhere in the network. By spatially modeling the movement of beef products, we identify environmental impacts that are specific to certain consumption locations and supply chains. The environmental aspects of beef consumption are elucidated by translating the beef production-consumption network into networks of embodied virtual water, energy, and GHG emissions. Evaluating the patterns of consumption and production reveals synergistic opportunities for impact reduction and sustainable management. Preliminary results show that approximately 50 counties are responsible for half of beef cattle production in the US. Further, the counties consuming the most meat usually do not produce any, and almost 75% of the states consume more meat than they produce in terms of equivalent cattle. Analysis indicates that the network of live cattle trade is neither dense nor robust. The implications of consumption-based accounting for the sustainable management of FEW systems will be discussed.

607310 A Life-Cycle Modeling Framework for Dynamic Energy Andwater Footprints of Agriculture Systems

Abstract:

Given the rapid increase in food demand, it is critical to develop effective policy and management strategies to enhance the sustainability of agricultural activities. System evaluations for different strategies and their effects on agricultural productivity and environmental impacts are needed for decision making and strategy implementation. Quantifying the energy and water footprints of those strategies are particularly important as such information will be helpful for synergetic management and optimization of food-energy-water (FEW) nexus. Previous studies have used Life Cycle Assessment (LCA) and Techno-Economic Analysis (TEA) to evaluate the environmental impacts and economic gains of different farming activities and strategic scenarios.[1–3] However, most of the previous models are static without considering geographic and temporal dynamics andindividual variabilities of farm decisions, both of which could have large impacts on the productivity of agriculture systems and environmental footprints.

To fill the gap, this study developed a modeling framework integrating LCA, TEA, and agent-based modeling (ABM).[4] ABM is a practical simulation tool to model dynamic systems that consist of autonomous agents. In this study, each farm was modeled as an autonomous agent who could make decisions on crop selection and fertilizer usage according to interactions with other farms and different attributes. These attributed included farm size, profitability, environmental awareness, soil quality, and forecasting accuracy. More attributes could be added in the future given the flexibility of the modeling framework. A stochastic approach was adapted to model the parameter uncertainties related to farmers' decision making. TEA was built upon dynamic simulation models of crop cultivation costs, crop selling prices, and crop yields. TEA, LCA, and ABM were linked to allow dynamic simulations of yields, profitability of plantation activities and life-cycle environmental impacts including primary energy consumption and water footprints. A case study of 1,000 farms in a 30-year period in North Carolina was conducted for demonstration. Different scenarios were developed to investigate the impacts of farms' attributes, socio-economic factors, and intervention strategies on farmers' decision-making and the environmental impacts of overall agriculture systems.

The results of the case study indicated that information exchanges among farmers, environmental awareness of farmers, access to environmental footprint information, and farm size are the key factors driving the system-level environmental impacts. Enhancing the environmental awareness of farmers has the potential to significantly reduce life-cycle primary energy consumption by 10–29% and water footprint by 8–35%. Switching to more environmentally friendly options generally reduced profits compared to the baseline (farms without environmental awareness). However, enhancing farmers' environmental awareness can mitigate such profit decrease due to the significant cost benefits of reduced chemical usage. In general, farms with higher environmental awareness and more information exchange with other farmers have less profit loss than those with lower environmental awareness with limited information sharing with others. Another finding of the case study is that providing farms with timely feedback on the environmental impacts of farmer's past decisions could affect farmers' choices in the following years and as a result leads to more reductions in energy and water consumption.

The results of this study can provide a variety of stakeholders (e.g., policymakers, nonprofits, agriculture companies) with better understanding of FEW interactions and useful information for effectively managing the large-scale agriculture systems . Although this study mainly used the data in North Carolina for 1,000 farms, this modeling framework can be applied to other regions and different sizes of agricultural systems. In addition, the framework can be integrated with other agricultural, energy, water, and socio-economic models to assist the understanding of complex FEW systems with different geographic variations, climate conditions, crop species, and system boundaries.

Refenrences

(1) Kim, S.; Dale, B. E. Life Cycle Assessment of Various Cropping Systems Utilized for Producing Biofuels: Bioethanol and Biodiesel. *Biomass and Bioenergy* **2005**, *29* (6), 426–439.

(2) Roy, P.; Nei, D.; Orikasa, T.; Xu, Q.; Okadome, H.; Nakamura, N.; Shiina, T. A Review of Life Cycle Assessment (LCA) on Some Food Products. *Journal of Food Engineering* **2009**, *90* (1), 1–10.

(3) Chavas, J. P.; Shi, G. An Economic Analysis of Risk, Management, and Agricultural Technology. *Journal of Agricultural and Resource Economics* **2015**, *40* (1), 63–79.

(4) Lan, K.; Yao, Y. Integrating Life Cycle Assessment and Agent-Based Modeling: A Dynamic Modeling Framework for Sustainable Agricultural Systems. *Journal of Cleaner Production* **2019**, *238*, 117853.

602807 Utilization of Insoluble Mg Minerals As an Alternative Mg Source for Struvite Crystallization from Wastewater

Abstract:

Introduction

The significant discharge of N and P nutrients from agricultural systems into wastewater has posed the challenge of sustainable treatment in order to mitigate significant environmental consequences. Nutrient laden wastewater can be treated as a feedstock and the NH_4^+ and PO_4^{3-} nutrients can be captured and recycled back into agricultural systems via struvite crystallization. Struvite ($MgNH_4PO_4.6H_2O$) is a slow-release fertilizer that can be produced from wastewater by adding an Mg source. While soluble Mg sources such as $MgCl_2$ have been studied in previous literature reports, owing to cost of precursor production it is more sustainable and cost-effective to use naturally abundant insoluble Mg minerals.[1] Furthermore, detailed characterization of products as well as reaction intermediate studies using *in situ* spectroscopic methods are required in order to fully elucidate mechanistic insights and to asses product quality. Our studies have demonstrated that insoluble Mg minerals such as bulk periclase (MgO)[1], magnesite ($MgCO_3$)[2], and dolomite ($CaMg(CO_3)_2$)[3] all result in struvite formation from simulated wastewater.

Materials and Methods

All bulk reagents were purchases from Millipore-Sigma. *In situ* Raman spectroscopy has been utilized to elucidate mechanistic insights of phosphate adsorption on the mineral surface. Aqueous PO_4^{3-} concentration was monitored using ion chromatography for kinetic studies. *Ex situ* Raman spectroscopy and powder X-ray diffraction were utilized to characterize all precursors and products. Simulated wastewater in all studies utilized 100 ppm NH_4^+ and 500 ppm PO_4^{3-} to reflect reported agricultural wastewater. The total reaction time was 2 hours.

Results and Discussion

Figure 1(a) shows the nutrient removal efficiency for the optimal loading of each Mg source used in the struvite crystallization studies. Mg^{2+}:NH_4^+:PO_4^{3-} ratio of 1:1:1 or slightly higher Mg2+ loading was found to be preferable MgO and $MgCO_3$ both exhibit ~80% NH_4^+ removal and over 90% PO43- removal, while producing pure struvite. Dolomite was less effective with 37% NH_4^+ removal and a resulted in a mixture of struvite and amorphous hydroxyapatite. MgO exhibited a 2-regime behavior for struvite formation; a kinetic regime for low MgO loadings which resulted in struvite formation, and a thermodynamic regime where high MgO loadings resulted in NH_4^+ adsorption and desorption (shown in Figure 1(b)) with the final product being $Mg_3(PO_4)_2.22H_2O$. Dolomite displayed a similar behavior while all $MgCO_3$ loadings resulted in struvite formation as shown in Figure 1(c). *Ex situ* Raman analysis of the intermediate reaction product after 5 minutes for MgO showed that the reaction intermediate is in the form of magnesium hydroxycarbonate. Furthermore, kinetic fitting of MgO concentration data showed the best fit for pseudo-second-order, indicating that a surface reaction is rate limiting for struvite formation. $MgCO_3$ kinetics followed pseudo-first-order. The change in order is hypothesized to be due to the fact that $MgCO_3$ carbonate surface is closer to the reaction intermediate observed in the MgO study. This further demonstrates that the oxide surface transformation to carbonate is of significance for phosphate adsorption. Figure 1(d) shows normalized Raman band intensity ratios for *in situ* monitoring of the calcined dolomite surface during struvite formation. This study demonstrates clearly that the MgO surface sites undergo significant transformation prior to struvite formation. In phase 1 of the reaction the oxide surface is transformed to a carbonate-rich surface by dissolved CO_2 reacting with the surface hydroxyl groups. These carbonate groups are replaced by HPO_4^{2-} and $H_2PO_4^-$ group chains as shown by the increasing normalized Raman band intensity ratio. Finally, the phosphate chains transform to PO_4^{3-} groups that result in the struvite formation. Further work will include extending this time-resolved Raman study methodology to other Mg minerals to identify the surface reaction mechanism.

Figure 1. (a) Removal of NH_4^+ and PO_4^{3-} for each Mg source. (b) NH_4^+ concentration profile for each loading of MgO (c) pXRD patterns for reaction products for each loading of $MgCO_3$ (d) Normalized band intensity variation for struvite formation reaction on the calcined dolomite surface.

Significance

Modern agricultural practices, as well as overfertilization due to the increasing demand for food production has led to significant anthropogenic impact on the nitrogen and phosphorus cycles.[4] Loss of ammonium and phosphate from agricultural systems in wastewater has led to significant environmental consequences such as eutrophication, which has been demonstrated to restrict access to freshwater resources, adversely affect aquatic ecosystems, and emit significant quantities of greenhouse gases. Utilization of low-cost and abundant natural Mg minerals have been demonstrated to be sustainable Mg sources for the struvite crystallization process, which allow for nutrient recycle and reuse.

References

(1) Kiani, D.; Sheng, Y.; Lu, B.; Barauskas, D.; Honer, K.; Jiang, Z.; Baltrusaitis, J. Transient Struvite Formation during Stoichiometric (1:1) NH4+ and PO43– Adsorption/Reaction on Magnesium Oxide (MgO) Particles. *ACS Sustain. Chem. Eng.* **2018**, *7*, 1545–1556, DOI: 10.1021/acssuschemeng.8b05318.

(2) Lu, B.; Kiani, D.; Taifan, W.; Barauskas, D.; Honer, K.; Zhang, L.; Baltrusaitis, J. Spatially Resolved Product Speciation during Struvite Synthesis from Magnesite (MgCO 3) Particles in Ammonium (NH 4 +) and Phosphate (PO 4 3–) Aqueous Solutions. *J. Phys. Chem. C* **2019**, *123*, acs.jpcc.8b12252, DOI: 10.1021/acs.jpcc.8b12252.

(3) Kiani, D.; Silva, M.; Sheng, Y.; Baltrusaitis, J. Experimental Insights into the Genesis and Growth of Struvite Particles on Low-Solubility Dolomite Mineral Surfaces. *J. Phys. Chem. C* **2019**, DOI: 10.1021/acs.jpcc.9b05292.

(4) Smol, M. The importance of sustainable phosphorus management in the circular economy (CE) model: the Polish case study. *Journal of Material Cycles and Waste Management*. Springer Tokyo March 15, 2019, pp 227–238.

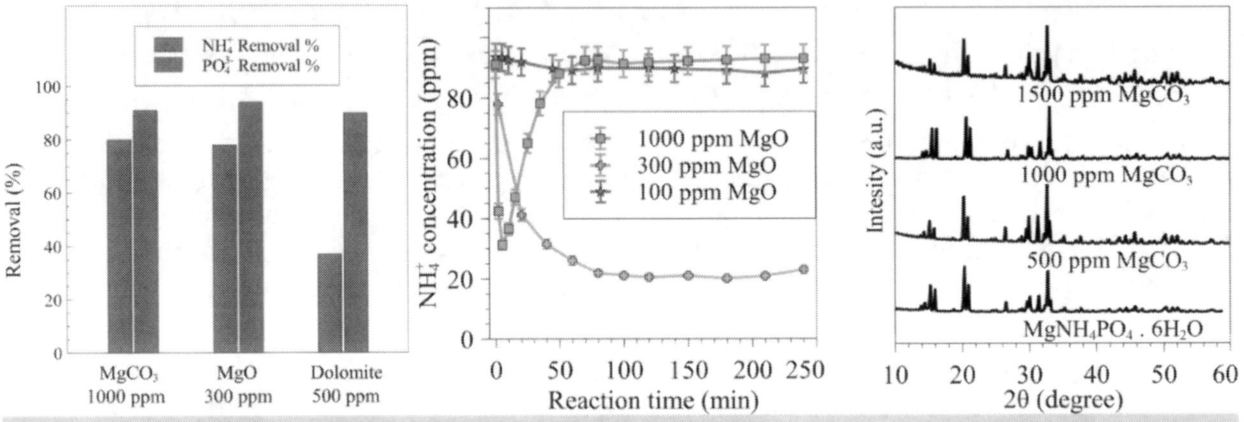
NH4+ Removal %
PO43- Removal %
Removal (%)
MgCO3 1000 ppm
MgO 300 ppm
Dolomite 500 ppm
NH4+ concentration (ppm)
1000 ppm MgO
300 ppm MgO
100 ppm MgO
Reaction time (min)
Intesity (a.u.)
1500 ppm MgCO3
1000 ppm MgCO3
500 ppm MgCO3
MgNH4PO4 . 6H2O
2θ (degree)

(a) (b) (c)

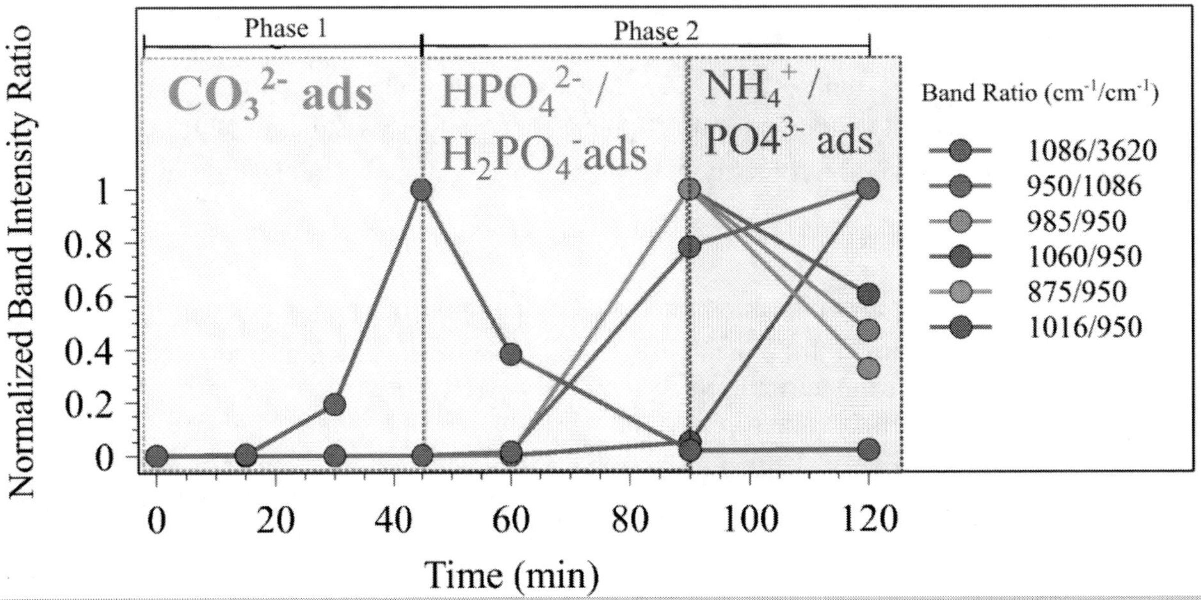
Phase 1
Phase 2
CO32- ads
HPO42- /
H2PO4-ads
NH4+ /
PO43- ads
Band Ratio (cm-1/cm-1)
1086/3620
950/1086
985/950
1060/950
875/950
1016/950
Normalized Band Intensity Ratio
Time (min)

(d)

ALL FOR ONE AND ONE FOR ALL RESEARCH: *MORINGA OLEIFERA* LAM. IN FOOD, HEALTH AND ENVIRONMENTAL APPLICATIONS

Andrea C. Landázuri[1,2,3], Renata Paz[1], Andrea Álvarez[4], Alina Trávez[1], Arleth Gualle[1], Juan S. Villarreal[1], Javier Gándara[1], Andrés Lagos[1], Verónica Castañeda[3,5] Emilia Morales[3,5], Javier Garrido[4], Gabriela Vernaza[4], Lucía Ramírez[4], Danny Navarrete[6], Lorena Bejarano[7], Andrés Caicedo[1,3,8,9], Lourdes Orejuela[1,2,3]*

[1]Universidad San Francisco de Quito USFQ, Colegio de Ciencias e Ingenierías; Engineering, Applied Sciences & Simulation Group (GICAS), Departamento de Ingeniería Química, Diego de Robles y Vía Interoceánica, P.O. Box 17-0901, Quito, Ecuador.

[2]Universidad San Francisco de Quito USFQ, Instituto de Investigaciones Biológicas y Ambientales BIÓSFERA, Diego de Robles y Vía Interoceánica, P.O. Box 17-0901, Quito, Ecuador.

[3]Universidad San Francisco de Quito USFQ, Instituto de Investigaciones en Biomedicina, 17912841, Ecuador.

[4]Universidad San Francisco de Quito USFQ, Colegio de Ciencias e Ingenierías, Departamento de Ingeniería de Alimentos, Diego de Robles y Vía Interoceánica, P.O. Box 17-0901, Quito, Ecuador.

[5]Universidad San Francisco de Quito USFQ, Colegio de Ciencias Biológicas y Ambientales, Departamento de Ingeniería en Procesos Biotecnológicos, Diego de Robles y Vía Interoceánica, P.O. Box 17-0901, Quito, Ecuador.

[6]Universidad San Francisco de Quito USFQ, Colegio de Ciencias e Ingenierías, Departamento de Ingeniería Industrial, Diego de Robles y Vía Interoceánica, P.O. Box 17-0901, Quito, Ecuador.

[7]Universidad San Francisco de Quito USFQ, Colegio de Ciencias e Ingenierías, Departamento de Ingeniería Mecánica, Diego de Robles y Vía Interoceánica, P.O. Box 17-0901, Quito, Ecuador.

[8]Universidad San Francisco de Quito USFQ, Colegio de Ciencias de la Salud, Escuela de Medicina, Quito, 1712841, Ecuador.

[9]Sistemas Médicos, SIME, Universidad San Francisco de Quito USFQ, Quito,17912841, Ecuador.

*Corresponding author: alandazuri@usfq.edu.ec

Abstract

The *Moringa oleifera* Lam. (MO) plant has characteristics that makes it suitable for various uses related to environmental remediation, development of cosmetics, pharmaceutical and food products, among others. This project contemplates and put together several areas of science, including: engineering, biomedicine with several objectives framed within a circular economy, bioeconomy and sustainability approach. The intention is to *i.* promote the use of MO available in Ecuador, as a biomaterial with rheological contributions; *ii.* extract bioactive chemical compounds; *iii.* evaluate its antioxidant, antimicrobial and cytotoxicity activities; *iv.* take advantage of the waste generated during moringa oil extraction processes for use in water recovery processes; and, *v.* develop food, cosmetic, nutraceutical, therapeutic and environmentally applicable products.

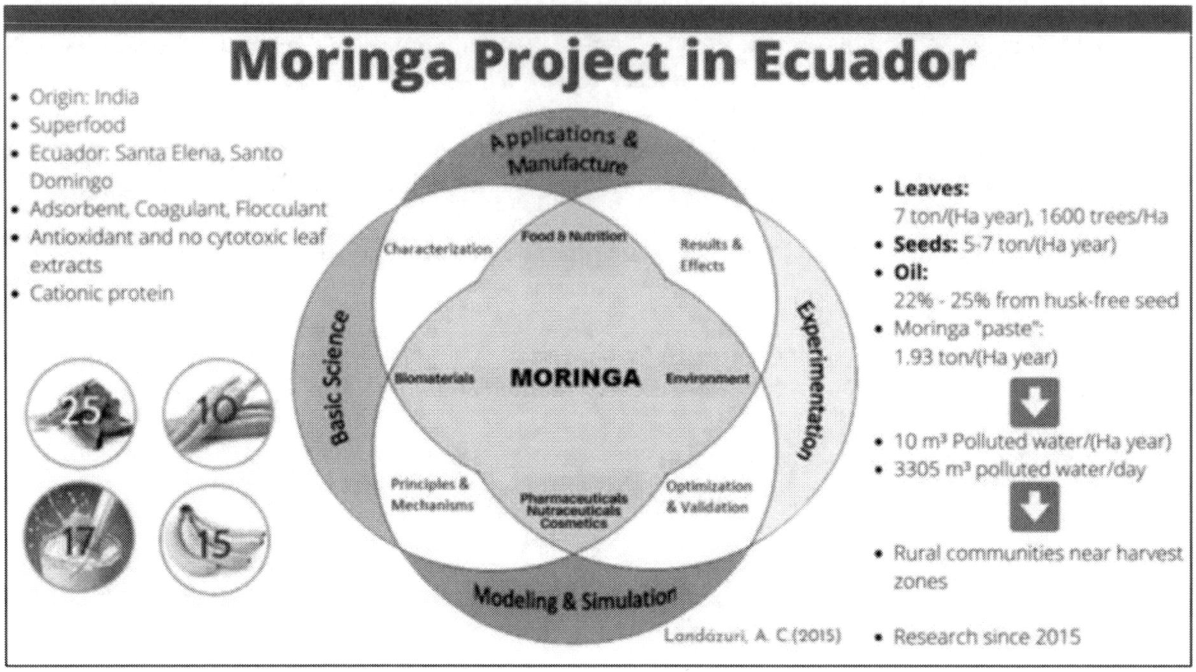

Research findings

Moringa biorefinery and nutritional aspects

The *Moringa oleifera* Lam. (MO) species has characteristics that makes it suitable for various uses related to environmental remediation, development of cosmetics, nutraceuticals, pharmaceutical and food products, among others. MO is a native species of India, it is considered a superfood product and it was introduced in Ecuador approximately 12 years ago. It is mainly harvested in The Santa Elena and Santo Domingo provinces. Our research group has demonstrated the adsorbent, coagulant, flocculant, antioxidant and non-cytotoxic capabilities of Ecuadorian moringa (under different dosages and treatments). For example, in terms of water treatment, from the pressed moringa cotyledons, also known as moringa paste or powder, we could be able to treat 3300 m^3 of polluted water per day.

The moringa project in Ecuador contemplates and put together several areas of science, including: engineering and biomedicine with several objectives framed within a circular economy, bioeconomy and sustainability approach. Since 2015, the efforts have been centered towards second generation biorefinery supported by four main pillars (**Figure 1**): Circular bioeconomy, Zero waste technology, Environment, and Sustainable Development and Manufacturing [1]. From biobased materials and residues, bioactive materials can be recovered, other by-products can be reused and recycled. For this, creativity and innovation are key aspects. During these new green processes, environmental and sustainable awareness increase, promoting economic growth.

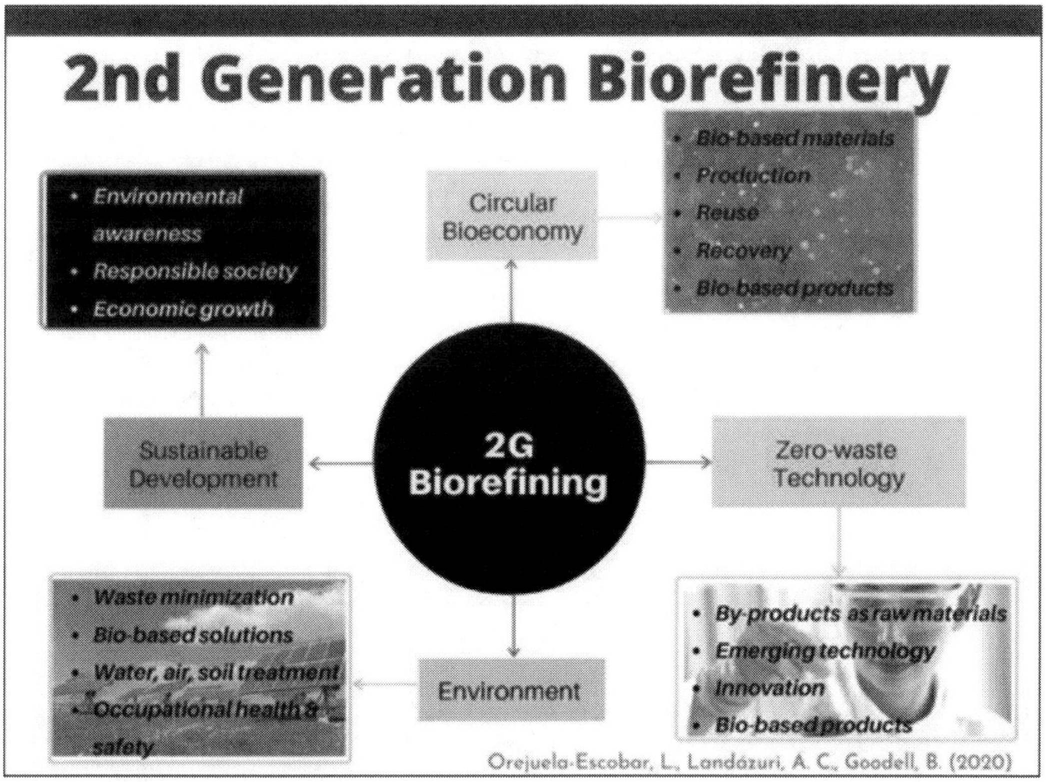

Figure 1. Pilars of Second Generation Biorefining

In Ecuador, MO is mostly used in nutraceuticals and as a dietary supplement [2]. MO has high nutritional content and provides rheological fluid characteristics. The protein content of Ecuadorian MO powder from its leaf is 29.46 g/100 g and from cotyledon (oil free, and provided as waste, that is, residual lignocellulosic biomass) is 48.28 g/100 g. These values exceed those present in foods of plant origin such as soybeans, quinoa and beans; and, also surpassing the values in foods such as beef and poultry [3]. The content of total carbohydrates in the powder of dry moringa leaves is lower than that of products such as quinoa, beans, corn, rice and wheat [4]; while crude fiber content is comparable to that found in foods such as oats, bread, whole wheat pasta and cooked lentils [5].

Development of a cereal-based food product and rheology

Due to these characteristics, a novel cereal-based food product was developed in which wheat flour was partially substituted in 0%, 1%, 3%, 5%, 7%, and 9%. The addition of the same cotyledon powder was found to significantly ($p < 0.05$) increase the protein content for all formulations, and iron content for the 7% and 9% formulations; Furthermore, sodium and calcium content changed significantly, while fat content remained constant. In parallel, the viscosity behavior of these formulations was evaluated as a function of temperature, concentration and shear stresses. The behavior was found to be pseudoplastic and thixotropic [6].

Metals and emerging contaminants removal from water

The leaves, seeds and pods are characterized by having adsorbing, coagulating, flocculating and filtering properties [7]–[10]. In environmental science, this project promotes research on the improvement of the removal of metals and emerging organic compounds, coagulation-flocculation processes, and filtration of pollutants [2], [9], [10]. Among recent results, around 90% of lead present in artificial water was adsorbed by seeds, while cadmium was removed by 60%. In addition, the adsorbent capacity during the removal of nickel, copper, and chromium was evidenced with removal percentages of 50%, 40%, and 25%, respectively [2]. On the issue of emerging contaminants, such as caffeine, a protein extract (with NaCl) from the cotyledon managed to remove caffeine present in water. When the initial concentration of caffeine was 5 ppm, the Fenton process reached a removal of 67% in 5 min, while the removal with 5 mL of MO protein was 61%, reached in 60 min. Although the removal time is longer, the process described with MO seed extract is technically feasible to remove organic emerging pollutants, such as caffeine.

Regarding coagulation-flocculation processes, experimental design was used for optimum particle size range selection. COD, turbidity and *E. Coli* levels were reduced below legislation requirements [9]. When coagulation-flocculation process was applied as a sole process after the recollection of crude water, the statistical results from a Randomized Complete Block Design (RCBD) over the sources concluded that water quality was improved by 41.90–51.94%, 55.32–71.10%, and 74.42–87.73%, for chemical oxygen demand (COD), turbidity and *E. Coli*, respectively. In addition, the use of moringa powder waste and complete moringa seeds also proved suitable for COD, iron, and turbidity removal, where moringa powder was significantly apt for turbidity removal, and moringa treatments with husk, were significantly apt for iron reduction. The process did not require the addition of chemicals.

(MO) seeds, pods and leaves have coagulant-flocculant and adsorptive properties, and it is a worth candidate for water treatment that can partially substitute commonly used adsorbents such as granular activated carbon, sand and zeolite. Lead (Pb) and iron (Fe) adsorption using borosilicate columns packed with MO seed husks for water purification purposes was investigated. A model (Bulk Balance Filtration – BBFM or Bulk Model for short) for contaminant adsorption was derived and applied to iron and lead data [10]. Columns of 1.1 cm-diameter were packed at different heights. On average, removals greater between 91-98.5 % were found. Isotherm modeling results, FTIR and SEM-EDS analyses support a physisorption process [11].

Antioxidant activity and cytotoxicity

As for MO leaves, they have a powerful mix of direct and indirect antioxidants that could be responsible for several beneficial health effects [12]. The antioxidant capacity was determined to be greater under the influence of methanolic solvent than under aqueous by a factor of 17 times. However, in the cytotoxic evaluation by means of MTT assay of the lyophilized extracts, the aqueous extract showed less cytotoxicity in comparison to the methanolic, as cells resulted to be more susceptible to concentrations ranging from 0.05 to 5 mg/L. These results concluded that moringa leaf extracts are a natural source of antioxidants and could be incorporated into formulations in food and health industries [13].

Skin-care products

Another example of product development is the encapsulation of skin-care products with MO and other lignocellulosic biomass extracts. Some advantages are that waste is minimized, transportation packing becomes lighter and easier, and the versatility of this method of conservation is its main strength.

Conclusions

GICAS Lab at Universidad San Francisco de Quito (USFQ) promotes green science and engineering within a zero-waste, circular bioeconomy and sustainable approach. Through multi-, inter-, and trans-disciplines, we can make a change. The development of simple, cheap, environmental solutions for different applications is an opportunity for communities that grow and harvest moringa, as well as for industries which can take advantage of moringa by-products for several subsequent processing.

MO cultivated in Ecuador and its waste shows the potential to: remove contaminants from water, increase nutritional power, modify the rheology of products, act as antimicrobial, and incorporate its bioactive compounds from this sustainable biomaterial in innovative food, pharmaceutical and environmental products, within green engineering concepts.

Acknowledgments

We are grateful to our undergraduate students, that actively supported our lab initiatives. More than 20 undergrads are now GICAS members. We thank Ecuamoringa, the College of Sciences and Engineering, Instituto Biósfera, Instituto de Investigaciones en Biomedicina, and Institute of Materials Research from Universidad San Francisco de Quito. Our projects were funded by the Chancellor, Collaboration and Poli Grants.

References

[1] L. Orejuela-Escobar, A. C. Landázuri, B. Goodell, "Second Generation Biorefining in Ecuador: Circular Bioeconomy, Zero Waste Technology, Environment and Sustainable Development – The Nexus," Accepted for publication in *J. Biores. and Bioprod.*, 2020.

[2] A. C. Landázuri, J. Cahuasqui, and A. Lagos, "Metal adsorption in aqueous media using Moringa oleifera Lam. seeds produced in Ecuador as an alternative method for water treatment," *Av. en Ciencias e Ing. USFQ*, vol. 11, no. 2, pp. 190–205, 2019, doi: http://dx.doi.org/10.18272/aci.v11i2.

[3] F. Marangoni *et al.*, "Role of poultry meat in a balanced diet aimed at maintaining health and wellbeing: An Italian consensus document," *Food Nutr. Res.*, vol. 59, pp. 1–11, 2015, doi: 10.3402/fnr.v59.27606.

[4] M. J. Kozioł, "Chemical composition and nutritional evaluation of quinoa (Chenopodium quinoa Willd.)," *J. Food Compos. Anal.*, vol. 5, no. 1, pp. 35–68, Mar. 1992, doi: 10.1016/0889-1575(92)90006-6.

[5] I. Elmadfa and A. Meyer, *Tabla de contenido de Fibra de los alimentos*. Hispano Europea, 2015.

[6] R. Paz, A.C. Landázuri, M.G. Vernaza, "Development of a cereal-based product using residual Moringa oleifera Lam. seed powder biomass and pseudo-plastic behavior of the dough mixtures", *Nutrition & Food Science*, Vol. ahead-of-print No. ahead-of-print, 2020, https://doi.org/10.1108/NFS-05-2020-0161

[7] G. S. Madrona, M. R. S. Scapim, L. A. C. Tonon, M. H. M. Reis, C. M. Paraiso, and R. Bergamasco, "Use of Moringa oleifera in a combined coagulation-filtration process for water treatment," *Chem. Eng. Trans.*, vol. 57, no. 2011, 2017, doi: 10.3303/CET1757200.

[8] M. C. and S. Neogi, "A natural coagulant protein from Moringa oleifera: isolation, characterization, and potential use for water treatment," *Mater. Res. Express*, vol. 4, no. 10, p. 105502, 2017.

[9] A. C. Landázuri *et al.*, "Experimental evaluation of crushed Moringa oleifera Lam. seeds and powder waste during coagulation-flocculation processes," *J. Environ. Chem. Eng.*, vol. 6, no. 4, pp. 5443–5451, Aug. 2018, doi: 10.1016/J.JECE.2018.08.021.

[10] A. C. Landázuri, J. S. Villarreal, J. C. Andrade, I. Sornoza, and A. S. Lagos, "Bulk balance filtration model (BBFM) for lead and iron physisorption through Moringa oleifera Lam. seed husks," *J. Environ. Chem. Eng.*, vol. 7, no. 5, p. 103302, 2019, doi: 10.1016/j.jece.2019.103302.

[11] J. S. Villarreal, J. Gándara, D. Navarrete, L. Bejarano, A. C. Landázuri, "Lead (Pb^{2+}) adsorption by means of pristine and prewashed residual Moringa oleifera Lam. seed husk biomass for water treatment applications," Under review in *Int. J. Sustain. Eng.,* 2020.

[12] T. B. Tumer, P. Rojas-Silva, A. Poulev, I. Raskin, and C. Waterman, "Direct and indirect antioxidant activity of polyphenol- and isothiocyanate-enriched fractions from moringa oleifera," *J. Agric. Food Chem.*, vol. 63, no. 5, pp. 1505–1513, 2015, doi: 10.1021/jf505014n.

[13] A.C. Landázuri, A. Gualle, V. Castañeda, E. Morales, A. Caicedo, L. Orejuela-Escobar. "Moringa oleifera Lam. leaf powder antioxidant activity and cytotoxicity in human primary fibroblasts." *Nat. Prod. Res., 2020. https://doi.org/10.1080/14786419.2020.1837804*

611175 Adoption of Biofuels for Marine Shipping

Abstract:

Biomass-derived fuels can undoubtedly play an essential role in a future of marine fuel that is more renewable and offers potential synergistic benefits when blended with petroleum fuels. Biomass-derived marine fuel would reduce sulfur content, as well as improve overall emission profiles and ultimately environmental sustainability (including decarbonization) for marine transport. For example, biofuel derived from biomass such as forest residues is expected to deliver possibly up to 90% of life cycle (or well-to-propeller) greenhouse gas reduction versus fossil equivalents, and virtually eliminate SOx emissions – potentially without any requirement for engine modifications. For the benefits of biofuels to be realized and to be impactful, a significant fraction of marine fossil fuel will need to be displaced with biofuel. The direct replacement of marine fossil fuel with a biofuel necessitates that the production volume of biofuel will be high enough to meet the demand. Furthermore, the potential of biofuel adoption also depends on the future biofuel price. Hence, the long-term price and availability of marine biofuel are two critical factors for establishing the feasibility of marine biofuel adoption for operating vessels. This study will shed light on the prospect of biofuel to be adopted for the maritime shipping sector.

602904 Application of Hydroprocessing, Fermentation and Anaerobic Digestion in a Pyrolysis Refinery to Maximize Economic Value

- Download/View Extended Abstract

Abstract:

This study investigates the economic and environmental benefits of integrating hydroprocessing, fermentation and anaerobic digestion into a pyrolysis refinery. Two scenarios were developed for upgrading and/or utilizing the primary products of pyrolysis (bio-oil, gas and char). The first (hydroprocessing) scenario hydroprocesses whole bio-oil into gasoline and diesel. The second (fractionation) scenario fractionates the bio-oil into sugars for fermentation to cellulosic ethanol and phenolic oil for esterification into phenyl esters (py diesel). Both scenarios use the gaseous product of pyrolysis for process heat in the plant and employ biochar to enhance anaerobic digestion of manure for production of methane for power generation.

The fast pyrolysis plant processes 2000 ton/day of corn stover while the anaerobic digester employs 430 ton/day of manure to generate power. The hydroprocessing scenario produces gasoline at a minimum fuel-selling price (MFSP) of \$2.77 per gallons of gasoline while the fractionation scenario produces ethanol for \$2.25 per gallon (\$3.38 per GGE). Sensitivity analysis indicates that the MFSP for both scenarios is highly sensitive to the fixed capital cost. Fixed capital costs for the hydroprocessing and fractionation scenarios were estimated to be \$643 million and \$285 million, respectively. Fuel production rates for the hydroprocessing and fractionation scenarios are 60.5 and 7.7 million GGE per year, respectively. Life cycle greenhouse gas emissions were calculated as -7.6 and –108.9 gm $CO_{2,\ eq}$ per MJ for the hydroprocessing and fractionation scenarios, respectively. LCA emissions are sensitive to by-product credits derived from biochar sequestration, phenolic oil utilization, and power generation. This study shows that both systems produce transportation fuels at competitive market prices with an additional reduction in atmospheric CO_2 levels in comparison to fossil fuel sources.

601722 A Superstructure Optimization Approach for the Design of Ethanol Upgrading Processes to Distillate Range Fuels

Abstract:

The use of ethanol as a sustainable fuel alternative to fossil fuels although well established is not exempt of challenges. First, ethanol can only be blended in fuels at levels below 10%. With an increasing offer of ethanol, coupled with the potential to have lignocellulosic ethanol production in the near future, one question that arises is how can we can find strategies to upgrade ethanol to other molecules such that this 10% constraint can be relaxed. A second important challenge stems from the fact that ethanol is mainly used to replace gasoline. Since the production of ethanol is likely to increase, it would be beneficial to find strategies to upgrade ethanol into the middle distillate range components which can also be used for diesel. In response to these challenges a growing body of literature dealing with the catalytic upgrading of ethanol to other fuel molecules has been produced in the last 30 years. Numerous chemistries have been explored (e.g. Guerbet coupling, dehydration, condensation etc.), and a large number of catalysts has been developed. The corpus of this literature is so vast that it is not trivial to answer the question of which strategy is optimal for the production of a particular fuel.

Accordingly, the goal of this work is to use process systems engineering strategies, and specifically of superstructure based optimization techniques, to address some of the important questions in this area. Specifically, we present a hierarchical superstructure optimization approach with the aim of finding optimal pathways for ethanol upgrading into fuels. Importantly, we formulate the problem such that three decision levels are considered simultaneously. In the first level we decide the species whose chemical transformation is allowed in a particular fuel production strategy (e.g. ethanol and ethylene). In the second level, we decide the specific chemistries that will be used to transform each of the species selected in the first level (e.g. ethanol dehydration, and ethylene oligomerization). Finally, in the third level, we select a specific process to realize the chemistry selected in the second level; this process is characterized by a specific catalyst and separation strategy. The contributions of this work are twofold: methodologically we introduce a new superstructure optimization strategy that can be used for the design of fuel upgrading strategies; from an application perspective, we present a complete and comprehensive analysis of the different strategies used for ethanol upgrading and provide insights into the major cost drivers for different ethanol upgrading strategies.

602074 Optimal Solvent Selection in Co-Solvent-Based Biomass Conversion Processes

Abstract:

The replacement of fossil-based products by renewables will require advanced processes with high economic and ecological performance. To that end, a promising option is to use gamma-valerolactone (GVL), a biomass-derived solvent, to solubilize all fractions of lignocellulosic biomass. This approach was successfully used to produce soluble carbohydrates, levulinic acid, and furfural [1,2,3]. To separate GVL effectively, a co-solvent can be used to switch the miscibility of the system by lowering the temperature inside the process. Thereby, the carbohydrates are immediately separated in an aqueous stream, while the GVL is recycled back to the reactor [4]. The resulting aqueous stream is then further deprived of GVL in a separation section, including an extraction and distillation column [5]. A key to unlock the potential of such a process lies in the right (co-)solvent choice. Today, the solvent selection is still mainly conducted experimentally and decoupled from process design.

In this work, we combine computational and experimental methods to increase the molecular set of co-solvent candidates and simultaneously decrease the manual effort to conduct experiments, compared to conventional solvent selection. A quantum mechanical approach [6] to predict physical properties allows us to investigate a broad range of molecules. Integration of reduced-order process models [7,8], embedded in an optimization, enables evaluation of each solvent candidate on the minimal process operating costs. Also, environmental, health, and safety (EHS) [9] are evaluated. The suggested approach allows us to identify the co-solvent, leading to the best process performance from over 4 600 molecules. We test the best system experimentally for the co-solvent hydrolysis reaction and confirm the predicted phase compositions through IR spectroscopy. The results show toluene and ethylbenzene as promising co-solvents. Ethylbenzene shows a lower experimental extraction efficiency than toluene and would not be a leading candidate when the choice is solely based on experimental evaluation. We show that ethylbenzene is even better than the conventionally selected toluene with operating costs savings of 15% and a lower EHS score. The presented approach allows the rapid in silico assessment of a large number of co-solvents on the process level and therefore balances the trade-offs leading to optimal process performance.

References:

[1] J. S. Luterbacher, J. M. Rand, D. M. Alonso, J. Han, J. T. Youngquist, C. T. Maravelias, B. F. Pfleger and J. A. Dumesic, Nonenzymatic Sugar Production from Biomass Using Biomass-Derived γ-Valerolactone, Science, 2014, 343, 277–280.
[2] S. G. Wettstein, D. M. Alonso, Y. Chong and J. A. Dumesic, Production of levulinic acid and gamma-valerolactone (GVL) from cellulose using GVL as a solvent in biphasic systems, Energy Environ. Sci., 2012, 5, 8199–8203.
[3] D. M. Alonso, S. G. Wettstein and J. A. Dumesic, Gamma-valerolactone, a sustainable platform molecule derived from lignocellulosic biomass, Green Chem., 2013, 15, 584–595.
[4] A. H. Motagamwala, W. Won, C. T. Maravelias and J. A. Dumesic, An engineered solvent system for sugar production from lignocellulosic biomass using biomass derived gamma-valerolactone, Green Chem., 2016, 18, 5756–5763.
[5] W. Won, A. H. Motagamwala, J. A. Dumesic and C. T. Maravelias, A co-solvent hydrolysis strategy for the production of biofuels: process synthesis and technoeconomic analysis, React. Chem. Eng., 2017, 2, 397–405.
[6] A. Klamt, Conductor-like Screening Model for Real Solvents: A New Approach to the Quantitative Calculation of Solvation Phenomena, J. Phys. Chem., 1995, 99, 2224–223.
[7] J. Bausa, R. v. Watzdorf and W. Marquardt, Shortcut methods for nonideal multicomponent distillation: I. Simple columns, AIChE J., 1998, 44, 2181– 2198.
[8] C. Redepenning, S. Recker and W. Marquardt, Pinch-based shortcut method for the conceptual design of isothermal extraction columns, AIChE J., 2017, 63, 1236– 1245.
[9] D. Prat, A. Wells, J. Hayler, H. Sneddon, C. R. McElroy, S. Abou-Shehada and P. J. Dunn, CHEM21 selection guide of classical- and less classical-solvents, Green Chem., 2016, 18, 288–296.

609199 Separation and Purification of 2,5-Dimethylfuran: Process Design and Comparative TEA and Sustainability Evaluation of Advanced Adsorption and Conventional Distillation

Abstract:

The furanic chemical 2,5-dimethylfuran (DMF) is an attractive biobased fuel and feedstock. High-yield biphasic processes for DMF synthesis from sugars employ higher alcohols such as *n*-butanol (BuOH) as solvents. This leads to a product containing < 8 wt% DMF and byproducts in BuOH solvent along with water. DMF process economics is strongly influenced by separation costs for DMF purification and recycle of BuOH and water. We previously demonstrated efficient adsorptive separation of DMF with a stable ZIF-8 metal-organic framework adsorbent [1]. Here we present a detailed, comparative technoeconomic analysis and sustainability evaluation of adsorptive versus conventional distillation separation processes to produce 98 wt% pure DMF at 100 metric tons/day. The novel adsorptive process integrates a simulated moving bed (SMB) unit with water removal and desorbent recovery systems, whereas the distillation process relies on multiple columns operating at different pressures to handle the strongly non-ideal multicomponent thermodynamics. Rigorous process modeling is conducted by a combination of the Aspen Plus flowsheet simulation package and our in-house SMB modeling and optimization package written in AMPL, with realistic multicomponent adsorption and vapor-liquid equilibrium models parametrized by experimental data. This is followed by detailed calculations and sensitivity analysis of bare-module, utility, and material costs. The net present values (NPVs) of the process alternatives show a clear long-term economic advantage of the adsorptive process, mainly due to large reduction in utility costs. When translated to a large potential DMF biofuel market size similar to that of bioethanol, the SMB-based adsorptive process would save >$8B/yr globally in separation costs. Furthermore, calculation of energy and CO_2 sustainability indices shows very clear advantages of the adsorption process.

References:

[1] Y. Chiang, W. Liang, S. Yang, C. R. Bond, W. You, R. P. Lively, S. Nair, Separation and Purification of Furans from n-Butanol by Zeolitic Imidazole Frameworks: Multicomponent Adsorption Behavior and Simulated Moving Bed Process Design, *ACS Sustainable Chemistry & Engineering* 7 (19), 16560-16568 (2019)

606407 Multi-Objective Optimisation of Sustainable Bio-Aviation Fuel Value Chains: Modelling the Production Potential of Developed and Developing Countries

Abstract:

Increasing global aviation activity is projected to contribute to about 3.1 billion tonnes of greenhouse gas emissions by 2050 [1]. Bio-aviation fuel, a blend of biomass-derived synthetic paraffinic kerosene and conventional jet fuel, is considered the best decarbonisation measure to limit the emissions to half of the 2005 baseline [2, 3]. High costs and limited availability of feedstocks pose the main barriers for large-scale production of bio-aviation fuel [4, 5]. Biomass already contributes a large share in the renewable energy supply of developed countries but available feedstocks are inadequate. They will need more plantations but stricter measures to conserve biodiversity and environment could hamper feedstock accessibility [6, 7]. Biomass in many developing countries of the tropics is generally more diverse and high-yielding, which could potentially supplement the relatively low feedstock supply in developed countries [8]. However, there are several barriers to the efficient use of biomass in many countries, such as lack of economic incentives, inhibitive government policies, concerns regarding deforestation and biodiversity loss issues [9]. The lack of developed and optimised biomass value chains is also a major hurdle in its implementation [10]. Process system engineering tools and techniques have been developed for the planning and design of bio-energy provision in order to support complex decision-making [11]. There are several bio-aviation fuel supply chain models in the literature [12] but these studies focus on bio-aviation fuel provision specific for the country or region of study. A comparative study on the bio-aviation fuel potential and prospective study of bio-aviation fuel trade between two or more countries is currently a research gap. In this study, the Value Web Model [13, 14], a spatio-temporal multi-objective optimisation model, based on mixed integer linear programming, is developed for bio-aviation fuel value chains in the UK and the Philippines. The feedstocks considered are Miscanthus and wheat straw for the UK, and Jatropha, rice straw and rice husk for the Philippines. The production pathways included are Fischer-Tropsch and alcohol-to-jet to convert the lignocellulosic feedstocks in both countries, as well as hydroprocessed esters and fatty acids to process Jatropha oil in the Philippines. These models can determine the optimal planning, design and operation of the value chains based on a variety of objectives such as maximising profit, maximising fuel production, minimising greenhouse gas emissions and minimising water consumption. Scenarios are developed for promising bio-aviation fuel value chains that can efficiently and sustainably satisfy domestic demands in both countries and meet prospective international demands. In this presentation, key insights on the implementation of bio-aviation fuel value chains in a developed and developing country and their potential collaboration through trade between countries will be reported.

Keywords: bio-aviation fuel; developing countries; developed countries; feedstock; production pathways; bio-aviation fuel value chains; biomass trade; multi-objective optimisation; mixed-integer programming.

Corresponding author: Dr Sheila Samsatli. Email: s.m.c.samsatli@bath.ac.uk

References:

1. Staples, M.D., et al., *Aviation CO_2 emissions reductions from the use of alternative jet fuels.*

Energy Policy, 2018. **114**: p. 342-354.
2. Yang, J., et al., *An overview on performance characteristics of bio-jet fuels*. Fuel, 2019. **237**: p. 916 - 936.
3. Schmidt, P., et al., *Power-to-Liquids as Renewable Fuel Option for Aviation: A Review*. Chemie Ingenieur Technik, 2018. **90**(1-2): p. 127 - 140.
4. US DOE, *Alternative aviation fuels: Overview of challenges, opportunities, and next steps*. 2017, Office of Energy Efficiency and Renewable Energy (EERE), US Department of Energy (US DOE). p. 1 - 80.
5. de Jong, S., et al., *The feasibility of short-term production strategies for renewable jet fuels - a comprehensive techno-economic comparison*. Biofuels, Bioproducts and Biorefining, 2015. **9**(6): p. 778 - 800.
6. IEA Bioenergy, *The availability of biomass resources for energy: Summary and conclusions – ExCo58 workshop*. 2008, IEA Bioenergy.
7. Sihvonen, J. and S. Eräjää, *How much sustainable biomass does Europe have in 2030?* 2016, European Federation for Transport and Environment AISBL. p. 1 - 7.
8. Oosterveer, P. and A.P.J. Mol, *Biofuels, trade and sustainability: a review of perspectives for developing countries*. Biofuels, Bioproducts and Biorefining, 2010. **4**(1): p. 66 - 76.
9. Elauria, J.C., et al., *Assessment of sustainable energy potential of non-plantation biomass resources in the Philippines*. Biomass and Bioenergy, 2005. **29**(3): p. 191-198.
10. Raychaudhuri, A. and S.K. Ghosh, *Biomass Supply Chain in Asian and European Countries*. Procedia Environmental Sciences, 2016. **35**: p. 914-924.
11. Tapia, J.F.D., et al., *Design of biomass value chains that are synergistic with the food–energy–water nexus: Strategies and opportunities*. Food and Bioproducts Processing, 2019. **116**: p. 170 - 185.
12. Doliente, S.S., et al., *Bio-aviation fuel: A comprehensive review and analysis of the supply chain components*. Under review in Frontiers in Energy Research.
13. Samsatli, S. and N.J. Samsatli, *A multi-objective MILP model for the design and operation of future integrated multi-vector energy networks capturing detailed spatio-temporal dependencies*. Applied Energy, 2018. **220**: p. 893-920.
14. Samsatli, S. and N.J. Samsatli, *The role of renewable hydrogen and inter-seasonal storage in decarbonising heat – Comprehensive optimisation of future renewable energy value chains*. Applied Energy, 2019. **233-234**: p. 854-893.

607287 Circular Economy Systems Engineering: A Case Study on the Coffee Supply Chain

Abstract:

The rising populations, the continuous effort for improvement in the standards of living, and the economic advancement under the current models of production and manufacturing lead to the depletion of natural resources and environmental degradation. Circular Economy (CE) appears as one of the most promising ways to reduce these impacts by looking beyond the current linear take-make-dispose models [1]. CE aims to solve resource, waste, and emission challenges confronting society by creating a production-to-consumption total supply chain that is restorative, regenerative, and environmentally benign. This transition is not trivial since it requires a holistic approach to quantitatively consider the multi-scale, multi-faceted and interconnected CE supply chains, identify opportunities for beneficial improvement, systematically explore interactions and trade-offs as well as assist the quantitative assessment and decision making [2].

Process Systems Engineering (PSE) could play a pivotal role towards this transition. As such, we present our novel framework along with an indicative case study [3]. First, we introduce CE assessment metrics for evaluating different aspects of CE at various levels. This is vital for effective decision making, and benchmarking, but it needs to be conducted holistically. Having established the key assessment metrics, then superstructure optimization is used to holistically evaluate alternative pathways for the transition from a linear to a circular economy. A techno-economic feasibility analysis of these pathways at a large scale through modeling and optimization will be also presented. Finally, a superstructure representation of the supply chain of coffee is presented, that involves alternative pathways for coffee harvesting and processing, waste utilization, product distribution and new/alternative products introduction. This analysis, in conjunction with feedback from experimentalists lead to the identification of the optimum CE supply chain for coffee.

References:

1. Ellen MacArthur Foundation, Towards the Circular Economy. Vol 1. 2013.
2. Avraamidou, S., Baratsas, S.G., Tian, Y., Pistikopoulos, E.N., 2020. Circular economy - a challenge and an opportunity for process systems engineering. Comput. Chem. Eng. 133, 106629. doi: 10.1016/j.compchemeng.2019.106629 .
3. A. Figueroa, T. Homann, H. M. Rawel. Coffee Production Wastes: Potentials and Perspectives, Austin Food Science, 1(3) (2016) 1014.

603683 The Influence of Increasing Polyurethane Recyclability and Renewable Content on Material Flows

Abstract:

Today, polyurethanes (PU) are effectively non-recyclable and are made from non-renewable, fossil-fuel derived resources. With technology to increase PU recyclability and its renewable content under development, this *status quo* could be dramatically altered, causing shifts in the PU supply chain, markets for precursors, and use patterns of downstream products. To understand the sustainability implications of these shifts, we first use material flow analysis (MFA) to characterize today's PU supply chain starting with raw materials. These include isocyanate, a compound that poses key sustainability challenges. The MFA also includes information about the relative uses of PU in a variety of products in building and construction, furniture and bedding, packaging, and electronics. Finally, it evaluates end-of-life options for PU, which are dominated by disposal. We explore how increasing PU recyclability and renewability could change MFA and influence its overall sustainability.

607702 Supercritical CO2-Induced Alteration of Polymer-Metal Matrix and Selective Extraction of Valuable Metals from Electronic Wastes

Abstract:

Supercritical CO2-induced alteration of polymer-metal matrix and selective extraction of valuable metals from electronic wastes

Peng Peng, Alan West, and Ah-Hyung (Alissa) Park

Waste electrical and electronic equipment (WEEE), one of the fastest growing waste segments, has become one of the most difficult challenges faced by humanity. Despite increasing efforts in recycling and reuse, a large fraction of WEEEs, such as waste printed circuit boards (PCBs), are being disposed into the environment. Even worse, the amount of WEEEs being transferred/exchanged between states and countries (particularly to developing countries) are causing significant social and economic complications.

The current PCB recycling procedures include physical separation, followed by metal extraction via pyro/hydro metallurgical pathways. The intensive physical separation of PCBs introduces environmental concerns including toxic gas emissions, and energy deficiency. Pyrometallurgy is also environmentally hazardous due to the formation of toxic gas, slag and other industrial wastes. Hydrometallurgy involves a series of acid or caustic leaches of e-waste followed by metal recovery techniques. Although hydrometallurgy is more controllable, solvents that are economical and less corrosive (e.g. sulfuric acid) cannot be used to extract gold, which is the most valuable metal within most PCBs. The use of concentrated, or highly reactive acids to extract gold is costly and generates gaseous and liquid pollutants.

In this research, we aim to develop a greener, and more selective way to extract both precious (e.g., gold) and common metals (e.g., copper) in waste PCBs utilizing supercritical CO_2 ($scCO_2$). The use of supercritical CO_2 has resulted in the reduction of the use of concentrated acids. Unlike cyanide or aqua regia leaching, gold in this process is recovered as solids rather than dissolved ions to eliminate additional separation steps, thus making the overall PCB treatment technology more intensified and sustainable. Furthermore, the proposed process can inherently separate metals copper and nickel at different stages of the treatment.

The mechanisms of how $scCO_2$ interacts with various components in the PCBs are studied and it has been found that the CO_2 interaction with the polymeric portion of PCBs has significantly influenced the subsequent leaching behaviors of metals. Once leached, a wide range of separation options including precipitation and electrowinning are considered to selectively recover different metals.

595134 Recovery Options for Iron and Sulphur Produced during Nickel Extraction from (bio)Leaching of Pyrrhotite Tailings

Abstract:

The development and application of a circular economy model to tailings management in the mining industry has the potential to address issues of mineral resource shortages and environmental pollution, while minimizing waste and generating economic profits[1]. For example, the valorization of the 100 million metric tons of pyrrhotite tailings contained in the Sudbury area, having an estimated average grade of ~0.8 wt% nickel[2], may be economically viable and would eliminate remediation required to prevent acid mine drainage. While this approach is technically feasible using (bio)leaching strategies, pyrrhotite ($Fe_{1-x}S$) oxidation results in the generation of large volumes of residual iron and sulphur in the form of goethite (a-FeOOH) and jarosite ($KFe_3(SO_4)_2(OH)_6$). Consequently, iron and sulphur management options will have a major impact on the economics and overall environmental footprint. In this context, the objective of this study was to investigate the low-temperature (<500°C) transformation of goethite and jarosite to an inverse spinel (e.g. magnetite, maghemite) amenable to magnetic separation using reducing gas produced from biomass, and fixation of the released sulphur in a recoverable phase. Magnetization of hematite (a-Fe_2O_3) was also considered as it represents an important product of goethite and jarosite decomposition.

Based upon the microstructural characterization of the products obtained from experiments performed on crystalline hematite, a solid-state mechanism for magnetic spinel formation, initiated at surface defects, was identified, where the low porosity of the reacted layer results in slow kinetics. On the other hand, goethite was readily converted to magnetite, as it involves an intermediate transition to a highly porous nanocrystalline reactive hematite phase allowing effective percolation of the reducing gases throughout the grains. The transformation of jarosite led to a phase assemblage consisting of magnetic spinel and arcanite (K_2SO_4), although persistence of minor residual hematite occurred. However, for a jarosite (75 wt%) / goethite (25 wt%) mixture, which represents a more realistic ferric residue of pyrrhotite (bio)leaching, complete transformation to magnetic spinel was observed, suggesting that the early transition of goethite to spinel promotes magnetite nucleation during K-jarosite decomposition.

Considering that about 55% of the total sulphur was lost to the gas phase during the transformation of the jarosite/goethite mixture, addition of potassium in the form of KOH and/or K_2CO_3 to the iron-bearing load was investigated as a means to trap the volatile sulphur and to access the minor amount locked in the magnetized product. This strategy was found to be very effective, especially in the case of KOH, where more than 90% of the sulphur could be fixed in the form of arcanite. As this phase is water soluble, it can easily be recovered through rinsing of the product. Furthermore, exchange with $CaCl_2$ led to stabilisation of sulphur as high-purity gypsum ($CaSO_4{\cdot}2H_2O$), suitable for use in the construction industry, where the potassium remaining in solution could be precipitated as sylvite (KCl) for use as fertilizer.

[1] Kinnuen P.H.-M. & Kaksonen, A.H., "Towards circular economy in mining: Opportunities and bottlenecks for tailings valorization", *J. Clean. Prod.*, **228**, pp. 153-160 (2019).

[2] Peek E., ***et al.***, "Nickeliferous pyrrhotite – Waste or resource?", *Miner. Eng.*, **24**, pp. 625-637 (2011).

602152 Strategic Molecules in Large Reaction Networks for Circular Chemical Economy

Abstract:

Chemical supply chains are currently based on linear valorisation methods: we extract raw materials – we process a variety of intermediates – we use products – and we dispose of waste. The depletion of fossil resources,[1] the generation of large amounts of waste,[2] and overconsumption or other societal implications [3] are only a few of the many disadvantages of linear economic models. A key concept of the future sustainable society is the transition to a circular industrial structure, requiring the reintegration of waste process streams as feedstock.

One of the fundamental challenges for building circular process structures is the question where to integrate the often highly functionalised waste as a novel feedstock. Many works focus on specific direct chemical transformation, e.g. the valorisation of crude-glycerol,[4] the hydration reactions on crude sulphate turpentine (CST) components,[5] or the conversion of lignocellulosic materials.[6] There is also a trend within the literature which considers wider applicability of newly introduced streams and functionalities, e.g. by investigating possible bio-based building blocks for industries,[7] or by highlighting the molecular geometries (privileged scaffolds) that lead to a wider range of potential end products.[8]

Due to the increase in available data in chemistry/chemical engineering, we can now intensively study the connectedness of chemical reactions and identify strategic integration points, i.e. *strategic molecules*, which are most relevant for overall chemical supply chains.[9] Large sets of known chemical reactions have been investigated as the Network of Organic Chemistry by representing molecules as nodes and reactions as edges connecting the nodes.[10] Inter alia, the networks historic evolution,[11] the statistics of the network,[12] and its usefulness for reaction development[13] and retrosynthesis[14] have recently been recognised. Thus, the data-driven networks are promising to also contribute to the understanding of integration points for novel feedstocks.

Herein, we introduce the concept of strategic molecules and present a method to identify those in chemical reaction networks. In particular, we mine Reaxys®[1] database for network assembly and describe molecules through a portfolio of graph theoretical features. We mined a network of more than 500,000 molecules connected through 955,905 edges and used graph centrality measures, e.g. betweenness centrality and pagerank, to describe the location of each molecule in the network. We then selected the strategic molecules based on an isolation forest detection algorithm with regard to the graph centrality values. The results reconcile a variety of chemical structures, main intermediates from the chemical industries as well as parts of the aforementioned bio-based building blocks and privileged scaffolds. The algorithm detects relevant chemical structures without any prior knowledge about the structures. In a case study on the use of CST we illustrate its potential for the valorisation of the waste stream. This indicates a generic solution of finding key intermediates in chemical supply chains, i.e. strategic molecules, which we hypothesise will play a major role for the transition towards circular economy.

Acknowledgment: J. M. Weber would like to thank the Department of Chemical Engineering for funding in the form of a PhD studentship and Philipp-M. Jacobs for his support and expertise. We gratefully acknowledge collaboration with RELX Intellectual Properties SA and their technical support, which enabled us to mine Reaxys. Reaxys data were made accessible to our research project via the Elsevier R&D Collaboration Network. This work was funded, in part, by the EPSRC project "Terpene-based manufacturing for sustainable chemical feedstocks" EP/K014889. This project was funded, in part, by the National Research Foundation (NRF), Prime Minister's Office, Singapore under its Campus for Research Excellence and Technological Enterprise (CREATE) program as a part of the Cambridge Centre for Advanced Research and Education in Singapore Ltd (CARES).

References:

1. Kätelhön, A., Meys, R., Deutz, S., Suh, S. & Bardow, A. Climate change mitigation potential of carbon capture and utilization in the chemical industry. *Proc. Natl. Acad. Sci. U. S. A.* **166**, 11187–11194 (2019).
2. Guo, Z., Yan, N. & Lapkin, A. A. Towards circular economy: integration of bio-waste into chemical supply chain. *Curr. Opin. Chem. Eng.* **26**, 148–156 (2019).
3. Clark, J. H., Farmer, T. J., Herrero-Davila, L. & Sherwood, J. Circular economy design considerations for research and process development in the chemical sciences. *Green Chem.* **18**, 3914–3934 (2016).
4. Yang, F., Hanna, M. A. & Sun, R. Value-added uses for crude glycerol - a byproduct of biodiesel production. *Biotechnol. Biofuels* **5**, 1–10 (2012).
5. Pakdel, H., Sarron, S. & Roy, C. A-Terpineol From Hydration of Crude Sulfate Turpentine Oil. *J. Agric. Food Chem.* **49**, 4337–4341 (2001).
6. Arevalo-Gallegos, A., Ahmad, Z., Asgher, M., Parra-Saldivar, R. & Iqbal, H. M. N. Lignocellulose: A sustainable material to produce value-added products with a zero waste approach—A review. *Int. J. Biol. Macromol.* **99**, 308–318 (2017).
7. Werpy, T. & Petersen, G. *Top Value Added Chemicals From Biomass: volume I -- results of screening for potential candidates from sugar and synthesis gas*. (2004).
8. Schneider, P. & Schneider, G. Privileged Structures Revisited. *Angew. Chemie - Int. Ed.* **56**, 7971–7974 (2017).
9. Weber, J. M., Lió, P. & Lapkin, A. A. Identification of strategic molecules for future circular supply chains using large reaction networks. *React. Chem. Eng.* **4**, 1969–1981 (2019).
10. Bishop, K. J. M., Klajn, R. & Grzybowski, B. A. The core and most useful molecules in organic chemistry. *Angew. Chemie - Int. Ed.* **45**, 5348–5354 (2006).
11. Llanos, E. J., Leal, W., Luu, D. H., Jost, J., Stadler, P. F., & Restrepo, G. Exploration of the chemical space and its three historical regimes. *Proc. Natl. Acad. Sci.* **116**, 12660-12665 (2019).
12. Jacob, P. M. & Lapkin, A. Statistics of the network of organic chemistry. *React. Chem. Eng.* **3**, 102–118 (2018).
13. Jacob, P. M., Yamin, P., Perez-Storey, C., Hopgood, M. & Lapkin, A. A. Towards automation of chemical process route selection based on data mining. *Green Chem.* **19**, 140–152 (2017).
14. Szymkuć, S., Gajewska, E. P., Klucznik, T., Molga, K., Dittwald, P., Startek, M., Bajczyk, M. & Grzybowski, B. A. Computer-Assisted Synthetic Planning: The End of the Beginning. *Angew. Chemie - Int. Ed.* **55**, 5904–5937 (2016).

594986 CO_2 Mass Production from the Air Very Cheaply

Abstract:

$Ca(OH)_2$ solution produced from CaO and water in separate small tanks, gradually intakes in corresponding ponds opened to the air, and capturing its diffused carbon dioxide by the said $Ca(OH)_2$ in form of $CaCO_3$ sediment, gradually extracted from the said ponds and thermally decomposed in CaO recycled and in CO_2 as a product of USD 15.0/ton or much cheaper in quantity up to 320 ton annually per each m2 of the ponds' area.

CO_2 concentration is constant of 0.6 gram/m3 at the solution surface under the open air and diffused through the water at D (diffusion coefficient) = 0.185 m2/s

The system productivity is determined by $Ca(OH)_2$ intake in the pond = Q2(g/s),

and by CO_2(aq) diffusion through the pond being

Q1(g/s) = 0.185 m2/s * (0.6 g/m3 /H m) * S m2 = (0.11 / H) * S

Here H (m) is depth below which CO_2 is completely captured, and S (m2) is its surface area.

Q2 (g/s) fully react with Q1 = (M=44/M=74) * Q2 = (0.11/ H) * S, so

Q2 = 0.11 * 74/44 * V / (H^2) = 0.187 * V / H^2, or

Q2 / V = 0.187 / H^2.

Here V (m3) = S * H

Q2 / V = $Ca(OH)_2$ intake to the pond (g/s) per m3 of V, and can be less or equal to saturated concentration of $Ca(OH)_2$ can be unchanging. Let it be equal 1600 g/m3/s, saturated.

So H^2 = (0.187/1600), so H = 0.01075 m.

The pond depth must be bigger than 2 * H, e.g. to be 1 m or less.

Q1 = 0.11 / H = 10.2 g/s / m2. CO_2 is diffused from the air to the pond almost instantly, because diffusivity of CO_2 in air is great D = 16.0 mm2/s, and practically instantly captured in thin layer by much more concentrated $Ca(OH)_2$ especially, if pH >=10.0

Now we are ready to evaluate CO_2 production:

Q1 = 10.0 g/s CO_2/m2 of the pond, or 320 ton/m2/year. It is tremendously bigger, than any other rational alternatives from atmospheric air, and total worldwide 100.0 km2 ponds would be enough to provide the artificial oil production in quantity of all mankind demand.

Further we must extract the said carbon dioxide by thermal decomposition $CaCO_3$ sediment in quantity 1.0 mol. CO_2 per 1.0 mol. $CaCO_3$, and energy for it being 180.0 kJ/mol., is a lion share of expenditures, while the others are relatively negligible.

Really,

$CaCO_3 = CaO + CO_2 - 1207 kJ + 635 + 393 = - 180 kJ$

1.0 mol. CO_2 needs 180 kJ/3600 sec = 0.05 kWh, or 1140.0 kWh/ton CO_2

Note: Referring to my archive of some decades of work, being constantly upgraded and updated, I can declare that solar and wind energy can be cheaper USD 0.01//kWh, even several times cheaper (as heat as electricity), even if at engineering level of past ages.

So CO_2 cost is 1140 kWh * $ 0.01/kWh = USD 11.4/ton + 20 % for others = USD 13.7/ton.

So 15.0 USD/ton CO_2 production was my estimation with good reserve.

595645 Solar for Industrial Process Heat: A Technical Potential Assessment for the U.S Manufacturing Sector

Abstract:

Renewable energy integration in energy-consuming sectors is essential for addressing global sustainability challenges, notably those linked to air pollution, resource conservation, and climate change. In recent years, despite short periods of stagnant growth, energy-related global carbon dioxide emissions remain at their highest levels, making decarbonization pathways even more urgent and necessary at a large scale. Transitioning to clean energy technologies in the industrial sector is critical since it has the highest energy consumption among the major energy-consuming sectors worldwide. Due to its diversity and complexity, the industrial sector has seen slower technology adoption than other sectors, yet decarbonizing industry remains an essential step toward achieving low-carbon production systems.

In the U.S., the vast majority of energy use in the industrial sector is associated with industrial process heat (IPH). In total, IPH amounts to 8% of U.S. primary energy use and relies predominantly on the combustion of fossil fuels, such as natural gas, coal, and oil, to meet process heat energy demand.

With the emergence of increasingly cost-effective and efficient solar technologies, solar for industrial process heat (SIPH) is gaining renewed attention as an alternative method for industrial heat production. A broad range of solar thermal and electric heating technologies can be used for process heating, and our work provides analyses that assess these technologies to determine the technical and economic potential of SIPH in U.S. manufacturing industries.

This national analysis, which has been conducted by researchers at the National Renewable Energy Laboratory and Northwestern University, specifically evaluates:

- the thermal energy demand of IPH that can be provided by a collection of solar technologies, for all manufacturing industries and regions in the U.S., at an hourly basis; and associated fuel savings
- methods for developing and applying a cost comparison framework for conventional IPH and SIPH systems with industry-relevant metrics, such as levelized cost of heat and payback period
- persisting data gaps in industrial energy usage and barriers to SIPH adoption

The presentation will provide a critical synthesis of our findings, with a focus on the technical potential analysis, where process-level thermodynamic models were integrated with national energy use and technical equipment data to estimate achievable fossil fuel savings. To this end, we developed a model that evaluates thermal energy demand and solar energy supply on a spatial temporal scale, accounting for the effect of variable demand and the addition of thermal energy storage.

We will also discuss identified barriers, both technical and economic, that must be overcome to promote future adoption, and suggest a research agenda for addressing those barriers moving forward.

612347 Techno-Economic Analysis of Small Haber-Bosch Process

Abstract:

Techno-Economic Analysis of Small Haber-Bosch Process

In this presentation, the techno-economic analysis of a 20,000 metric tons (MT) green ammonia production facility will be presented. This facility is 30 times smaller than a large-scale conventional process, producing renewable ammonia from totally renewable resources: hydrogen from the water electrolysis and nitrogen from air cryogenic distillation. Two different forms of the Haber-Bosch (HB) process are investigated: high pressure reaction-condensation (RXN-CON) and low-pressure reaction-absorption (RXN-ABS).[1–3] Process simulation was implemented using ASPEN Plus to model the ammonia synthesis loop, including the converter, heat exchangers, and separation units. The results obtained were then used to estimate the total capital and operating costs associated with high-pressure RXN-CON and low-pressure RXN-ABS. The performance analysis of HB process indicates that ammonia production cost is significantly affected by the operating pressure. The high-pressure processing enhances the single-pass conversion and synthesis loop efficiency but consumes more energy, whereas the low-pressure processing is more favorable for saving capital and operating costs. By replacing condensation with absorption in a 20,000 metric tons per year process, capital costs and annual operating costs are reduced by $1.69 million and $4.72 million, respectively, and the levelized cost of ammonia (LCOA) decreases when RXN-ABS configuration is utilized. Our techno-economic analysis results was finally compared with the other technologies.[4]

(1) Reese, M.; Marquart, C.; Malmali, M.; Wagner, K.; Buchanan, E.; McCormick, A.; Cussler, E. L. Performance of a Small-Scale Haber Process. *Ind. Eng. Chem. Res.* **2016**, *55* (13), 3742–3750. https://doi.org/10.1021/acs.iecr.5b04909.

(2) Malmali, M.; Reese, M.; McCormick, A. V.; Cussler, E. L. Converting Wind Energy to Ammonia at Lower Pressure. *ACS Sustain. Chem. Eng.* **2018**, *6* (1), 827–834. https://doi.org/10.1021/acssuschemeng.7b03159.

(3) Malmali, M.; Wei, Y.; McCormick, A.; Cussler, E. L. Ammonia Synthesis at Reduced Pressure via Reactive Separation. *Ind. Eng. Chem. Res.* **2016**, *55* (33), 8922–8932. https://doi.org/10.1021/acs.iecr.6b01880.

(4) Rouwenhorst, K. H. R.; der Ham, A. G. J. Van; Mul, G.; Kersten, S. R. A. Islanded Ammonia Power Systems: Technology Review & Conceptual Process Design. *Renew. Sustain. Energy Rev.* **2019**, *114*, 109339. https://doi.org/https://doi.org/10.1016/j.rser.2019.109339.

600587 Analysis of Green Ammonia Production Responding to Fluctuating Renewable Energy By Using Dynamic Simulation

Abstract:

Ammonia is a promising hydrogen carrier to transport green hydrogen from overseas to Japan at lower cost and resulting in lower lifecycle CO_2 emissions. Low-carbon ammonia will be produced by fossil fuel reforming with CCS or CO_2 -EOR at the early stage of the introduction of ammonia fuel to the market. Green ammonia, which is synthesized from green hydrogen and nitrogen, is the ultimate goal, but there are some issues facing its commercialization.

One of big issues is the low capacity factor which is caused by the fluctuation of solar irradiation or wind speed and makes ammonia production costs expensive. Hence, at the 2019 AIChE Annual Meeting's Ammonia Energy Conference, we introduced the results of optimization studies of the capacity and operation of units that comprise the ammonia process using photovoltaics as a renewable power source. Moreover, we also introduced the test results of the fluctuating operation in the ammonia synthesis demonstration plant. From the results of this study, it was put forward that, in order to absorb the fluctuations of solar or wind energy, the optimization of battery and hydrogen storage tank capacities and the ammonia plant load changing operation are effective for the ammonia production cost reduction. Regarding system capacities optimization, based on each equipment unit cost and specification, the optimal configuration can be determined by case-study calculation.

In this presentation, we would like to explain the results of the dynamic simulation of ammonia synthesis imitating the fluctuating renewable power source combined with a buffering system for verifying the influence of ammonia plant load changing operation. The influence, which is caused by the dynamic operation corresponding to renewable energy source fluctuation, against ammonia plant stability has to be simulated. As an example of non-steady state operation, there is a risk of overheating vis-a-vis the design temperature or oscillation of the reactor temperature caused by increasing the feed rate of hydrogen rapidly. Moreover, there is a probability of surge in the recycle compressor without actuating the anti-surge system when the feed rate suddenly decreases.

The dynamic model is a convenience to analyze the process condition and can show the plant stability without actual operation in a real plant. The objective of this study is to establish the design concept for the ammonia synthesis loop and identify the margin to instability of the ammonia reactor and relevant equipment. By a combination this study and overall system configuration optimization, we will establish the Green Ammonia Production system concept and design basis for the commercial Green Ammonia plant using renewable energy.

601962 Effects of Ammonia Injector Design on Co-Firing Characteristics of Pulverized Coal Burner with High Ammonia Co-Firing Ratio

Abstract:

Recently, the co-firing ammonia with coal in boiler has been paid more and more attention. A CFD simulation has been undertaken to investigate the co-firing characteristics of ammonia with coal in an experimental facility (1 MW_{th}), in which a pulverized coal burner was mounted. One case for coal combustion and three for ammonia co-firing were simulated. The three cases for ammonia co-firing had different ammonia injectors inserted into the center of burner, but the same co-firing ratio of 60 cal.%. The different ammonia injectors were designed corresponding to different flow distributions of ammonia into axial and radial directions. According to the simulation results, the following conclusions could be drawn. Firstly, basically, the cases of ammonia co-firing have higher NO concentration, higher unburnt carbon content in fly ash at exit, lower CO concentration at exit and less heat amount absorbed by walls, compared with that in the case of coal combustion. Additionally, as the increase of ammonia amount into radial direction, the NO concentration at exit is decreased. This could be mainly attributed to the de-NOx effect of ammonia inside the flame and inhabitation of ammonia after-burning. In the case with the most ammonia amount into radial direction, a lower NO concentration at exit than that in the case of coal combustion can be obtained. However, as the increase of ammonia amount into radial direction, the unburnt carbon content in fly ash and CO concentration at exit are increased. This could be mainly attributed to the competition of ammonia with char in O_2 consumption. These conclusions will be very helpful in the design of a high-performance burner for co-firing of ammonia with coal even under the conditions of quite high co-firing ratio.

599368 Power Production through an Ammonia Based Combined Cycle

Abstract:

The share of renewable energies in the power generation system has been increasing during the last years due to the new concerns in sustainability. Around 26% of the total power generation come from renewables worldwide. However, the share of renewables increases up to more than 60% in countries such as Denmark, New Zeeland or Norway. Hydropower, solar and wind are the three main renewable resources in the world. The increase in the use of intermittent energies, mainly wind and solar, require the implementation of energy storage to be able to meet the power demand regardless of the availability of wind or solar energy. Different ways to storage power have been proposed (Gür, 2018). The different storage alternatives present different energy densities, storage range or discharge time. Therefore, a combination of the different alternatives could be necessary to meet all the requirements of the power grid. Pumped-hydro could be one of the most implemented nowadays, however, a limited capacity expansion is expected. A wide range of batteries is being used for storing at small scale and the grid integration is being studied. The use of chemicals is attracting attention to storage energy in a long-term horizon and the possibility of transporting the chemicals to distribute the energy in different areas. However, further studies about the integration of the chemical with the renewables and their use to produce power must be conducted.

Different chemicals have been proposed as energy carriers. Hydrogen is one of the most extended due to its high energy density. However, its storage is difficult and expensive. Therefore, alternative chemicals have been proposed as hydrogen resources. Methanol or dimethyl ether (DME) has been used combining carbon dioxide and hydrogen. Ammonia is also attracting attention due to the possibility of producing it using hydrogen and nitrogen. Consequently, ammonia is a carbon free energy carrier. There are two main steps to use ammonia as an energy carrier: production using renewable power and its transformation into power. In previous works, the production of ammonia using renewables energies and raw materials have been studied using a process design perspective (Sánchez & Martín, 2018). Different alternatives have been proposed to generate power using ammonia. Three are the main ones (Valera-Medina et al., 2018): fuel cell using ammonia, ammonia cracking to produce hydrogen and, then, a hydrogen fuel cell and, finally, the ammonia combustion.

In this work, the ammonia transformation into power have been studied. The evaluation of the combustion way has been carried out. According to previous experimental works, the combustion of only ammonia is difficult and different mixture of ammonia with methane or hydrogen have been proposed. Therefore, a mixture of ammonia and hydrogen has been used in this work (Otomo et al., 2018). The hydrogen is produced from ammonia through cracking. Then, the mixture of hydrogen and ammonia is introduced in the combustion chamber to produce energy. A combined cycle is set up to obtain the power. The first stage is the gas turbine. Then, the Rankine cycle is introduced. Three turbines which different pressure are used in this cycle. The outlet gases are treated to meet the legislation about emissions (mainly nitrogen oxides) and to recover the valued products in the outlet stream. An equation based approach have been used to optimize the operating conditions of this process to produce power from renewable ammonia (temperature, pressure, ratios...).

References:

Gür, T.M., 2018. Review of electrical energy storage technologies, materials and systems: challenges and prospects for large-scale grid storage. Energy & Environmental Science, 11, 2696.

Sánchez, A., Martín, M., 2018. Scale up and scale down issues of renewable ammonia plants: Towards modular design. Sustainable Production and Consumption, 16, 176-192.

Valera-Medina, A., Xiao, H., Owen-Jones, M., David, W.I.F., Bowen, P.J., 2018. Ammonia for power. Progress in Energy and Combustion Science, 69, 63-102.

Otomo, J., Koshi, M., Mitsumori, T., Yamada, K., 2018. Chemical kinetic modeling of ammonia oxidation with improved reaction mechanism for ammonia/air and ammonia/hydrogen/air combustion. International Journal of Hydrogen Energy, 43(5), 3004-3014.

609572 Keynote: The Remarkable Influence of Zr-Doping on the Thermochemical Properties of Ceria

Abstract:

Ceria is a well-known material for the production of hydrogen by thermochemical cycling. While ceria offers several advantageous properties such as rapid surface reaction rate, high bulk oxygen diffusivity, and excellent stability over hundreds of redox cycles, it suffers from the requirement of reduction at relatively high temperatures (preferably as high as 1500 °C) in order to generate non-negligible quantities of hydrogen fuel per cycle. To address this challenge, several laboratories have evaluated zirconia-substituted ceria as an alternative to neat ceria. Introduction of zirconia has the surprising consequence of increasing the oxygen non-stoichiometry of the host oxide, despite the fixed 4+ valence of Zr. This behavior is reflected in a decrease in both the enthalpy and the entropy of reduction of $(Ce_{1-x}Zr_x)O_{2-\delta}$ with increasing x, as determined by thermogravimetric analysis of bulk, porous samples. The thermodynamic properties suggest enhanced efficiency for solar-to-fuel conversion as well as the possibility of cycling at overall reduced temperatures without loss in hydrogen production rates. The kinetic properties, however, appear to be negatively impacted by introduction of Zr. Using angle-resolved X-ray Absorption Near Edge Spectroscopy (XANES), we quantify under technologically relevant conditions the Ce^{3+} concentration in the surface (2-3 nm), as well as bulk regions, of ceria-zirconia films. We find that the surface is in all cases more reduced than the bulk, however, with increasing Zr, the differential between the two regions diminishes. In parallel, using electrical conductivity relaxation methods, we find that both the surface reaction rate and the bulk chemical diffusion of oxygen decreases in the presence of Zr. These observations are tentatively explained in terms of the preferred lower coordination of the small Zr^{4+} ion relative to the larger Ce^{4+} and Ce^{3+} ions and likely trapping of oxygen vacancies in the vicinity of Zr. Despite these effects, zirconia substitution remains attractive because of the overwhelming importance of thermodynamic properties.

610123 Invited Talk: A Thermochemical Study of Doped-Spinels: A Candidate Class of Redox Intermediates for Isothermal Water Splitting

Abstract:

Technologies such as solar thermochemical hydrogen (STCH) production, which leverage the oxygen-exchange capacity of metal oxides to split water, are thermodynamically attractive, as the entire solar spectrum can be exploited to supply process heat. Here, a metal oxide is subjected to cyclic reduction-oxidation reactions that facilitate either oxygen or hydrogen production, depending on the operating conditions. Chemical thermodynamics suggest that the forward direction of each redox half-reaction is favorable in different temperature and/or oxygen partial pressure regions, and as a result, in practice, large irreversibilities generally arise from sensible heating between redox regimes. To address this issue, recent research has primarily focused on: (1) investigating novel perovskite materials with tunable thermodynamic properties that enable greater reduction extents at more moderate temperatures; and (2) integrating auxiliary techniques, such as mechanical or thermochemical pumping, that reduce the operating oxygen partial pressure (and thus temperature) in the endothermic reduction step. However, until highly effective and reliable solid-solid heat recovery methods are devised, inefficiencies inherent to temperature-swing operation will continue to negatively impact performance. Alternatively, at the expense of lower hydrogen productivity, the water-splitting (oxidation) step can be operated at temperatures where reduction is favorable, thereby minimizing thermal stresses and avoiding complexities associated with solid-solid heat recuperation. Herein, motivated by the immediate practicality of isothermal operation, the behavior of low-cost iron-based aluminate spinels (e.g., hercynite) was characterized under conditions relevant to isothermal STCH. Using thermogravimetry, equilibrated changes in oxygen content were quantified at different oxygen partial pressures and temperatures to extract thermodynamic properties, namely partial molar enthalpy and entropy, and thus enable performance predictions. Results indicate that this class of redox intermediates is capable of higher hydrogen productivities under isothermal operation than previously established benchmark materials (e.g., ceria and lanthanum manganite-based perovskites) evaluated under temperature-swing. The favorable thermodynamic properties of spinels at conditions better suited for industrialization present a step forward towards sustainable hydrogen production with solar thermal energy.

602660 Oxidation Kinetics of Hercynite Alloys for Solar Thermochemical Fuel Production

Abstract:

Hydrogen (H_2) is a clean energy source that can replace fossil fuels in the transportation and power sectors, two of the largest contributors of anthropogenic greenhouse gas emissions. Globally, more than 95% of H_2 is produced from fossil fuels, resulting in carbon dioxide CO_2 emissions and a significant carbon footprint. To address this, researchers have sought to produce fuels, such as H_2 and carbon monoxide (CO), via solar thermochemical fuel (STCF) production using solar energy to drive the endothermic gas splitting reaction via a reduction/oxidation cycle. The development of an economically viable solar thermochemical fuel production process relies largely on identifying redox active materials with optimized thermodynamic and kinetic properties. Hercynite ($FeAl_2O_4$) has been demonstrated as a viable redox-active material for this process with higher H_2 productivities than its cobalt-iron aluminate counterpart ($Co_xFe_{1-x}Al_2O_4$). However, a qualitative tradeoff between the favorable thermodynamic properties of hercynite and its slower kinetic properties was observed when cobalt is introduced.

In this work, we evaluate four spinel aluminate materials with varying cobalt contents from 0 to 40% ($FeAl_2O_4$, $Co_{0.05}Fe_{0.95}Al_2O_4$, $Co_{0.25}Fe_{0.75}Al_2O_4$, and $Co_{0.40}Fe_{0.60}Al_2O_4$) in order to further understand the role of cobalt in these materials and to quantify its effect on the thermodynamic and kinetic properties for CO_2 reduction. A solid state kinetic analysis was performed on each sample to model its oxidation kinetics in CO_2 splitting experiments at temperatures ranging from 1200°C to 1350°C using a thermogravimetric analyzer (TGA). An F1 model representing first-order reaction kinetics was found to most accurately represent the experimental data for all materials evaluated. The computed rate constants, activation energies, and pre-exponential factors all increase with increasing cobalt content. However, lower productivities are seen with increasing cobalt content. High temperature in-situ XPS study was utilized, for the first time on these oxides, to characterize their surfaces and indicated the presence of metallic states of the reduced cobalt-iron alloys, which are not present in hercynite. These species provide a new site for the CO_2 reduction reaction and enhance its rate through an increased pre-exponential factor.

604959 Assessing Nonstoichiometric Oxides in Solar Thermochemical Fuel Production

Abstract:

The high temperatures at which two-step solar thermochemical fuel production proceeds (e.g. 1000 to 1500 °C) can render both surface and bulk kinetics of porous nonstoichiometric infinitely fast relative to gas sweep rates. In such case, the material operates under quasi-equilibrium conditions, and the macroscopically observed oxygen evolution and hydrogen production profiles are limited by the thermodynamic characteristics of the oxide. Recognition of this behavior enables the development of material-specific cycling strategies that maximize the process efficiency taking into account factors such as the energy for sweep gas and solid state heating and for mechanical pumping. Building on a previously established and experimentally validated model for predicting gas evolution profiles in the quasi-equilibrium regime [T. C. Davenport, M. Kemei, M. J. Ignatowich, and S. M. Haile, *Int. J. Hydr. Energy* **42**, 16932-16945 (2017)], we develop here a computational approach for predicting cycles that maximize solar-to-fuel efficiency. The optimization is carried out using as inputs the experimentally measured enthalpy and entropy of reduction of known and fully characterized nonstoichiometric oxides. The optimized cycles are defined in terms of the temperature, the duration time, and the sweep gas flow rate of each half cycle. Significantly, despite a large energy penalty of heating and cooling the oxide, for most materials considered, the overall efficiency is highest when the temperature for the water splitting half-cycle is relatively low. In such case, the thermodynamic driving force for the hydrogen evolution reaction is large, hastening the pace of the reaction. Achieving the predicted efficiencies, however, may require surface engineering to avoid limitations due to slow surface reaction kinetics at reaction temperatures below ~1000 °C. Most importantly, this approach serves as a framework for assessing the efficacy of candidate thermochemical materials on an optimized rather than *ad hoc* basis. That is, for each candidate, the maximum efficiency and optimal conditions, within some range of constraints, and can be determined, rather than comparing materials at arbitrary cycling conditions which may inherently favor one material over another.

614693 Resources, Sustainability, Circular Economy, Sustainable Development, and?

Abstract:

There is no doubt that resources, sustainability, circular economy and related topics continue to gain increasing importance and have a major role to play in tackling the challenges facing planet earth. However, after the publication of numerous scientific papers in journals on sustainability and related topics as well as other scientific journals, what is the status now? The presentation will highlight the activities within chemical and biochemical engineering related to sustainability and associated topics in the immediate past, the present and what should be in future.

614694 Toward Multiscale Consequential Sustainable Process Design: Including the Effects of Economy & Resource Constraints with Application to Green Urea Production in a Watershed

610470 Keynote: Thermodynamic Limits for Materials in Solar Thermochemical Fuel Production

Abstract:

Solar-thermochemical (STC) carbon dioxide and water splitting is a pathway for the direct thermal production of solar fuels—an alternative to direct thermolysis. Currently, the most widely researched implementation is the two-step redox active metal oxide cycle. Reducing the two-step cycle to practice is challenging, and there is a substantial ongoing effort to improve reactor designs, and to discover and synthesize functional materials that can exceed the STC redox capacity potential at milder operating conditions than ceria, the current state of the art redox active metal oxide. We have analyzed the underlying general thermodynamic boundaries that lead to materials performance tradeoffs between reduction enthalpy, productivity, and yield. These boundaries apply to all nonstoichiometric oxides without significant phase changes during redox cycles. We quantify the tradeoffs between the desirable decrease of the reaction enthalpy of reduction and penalties such as decreasing yields and increasing temperature difference between the two steps (reduction and re-oxidation), and show that operating parameters pose firm limitations on cycle performance and on the desirable materials properties space.

611427 Invited Talk: The Role of Entropy in the Success of Nonstoichiometric Oxides for Solar Thermochemical Water and CO_2 Splitting

Abstract:

Owing to their high theoretical efficiency, solar thermochemical water and CO_2 splitting cycles provide a promising pathway to store solar energy in a stable chemical bond. Because direct thermolysis of H_2O and CO_2 proceeds at prohibitively high temperatures, the two-step cycle using a metal oxide intermediate has been extensively investigated as a practical lower-temperature process. In Step 1 of the redox cycle, a metal oxide MO_{ox} is reduced to MO_{red} at high temperature T_{red} using concentrated solar energy to drive the reaction endotherm. In Step 2, the reduced oxide is cooled to T_{ox} and H_2O and/or CO_2 are introduced. The now oxygen-hungry MO_{red} is capable of stripping an oxygen from H_2O or CO_2 to produce H_2 or CO fuel, while recovering the initial MO_{ox} such that the cycle can repeat. The efficiency of the process is highly dependent on the MO_{ox}/MO_{red} pair used to facilitate the reaction. To date, the most successful demonstrations have utilized non-stoichiometric oxides, for example $CeO_{2-\delta}$, a special class of materials capable of continuously transitioning from MO_{ox} to MO_{red} without changing their crystallographic phase. The success of non-stoichiometric oxides is typically attributed to their ability to maintain their crystallographic phase, allowing them to be cycled many times without destabilizing. This is a tremendous *practical* advantage.

In this work, we will present an alternative explanation of the success of non-stoichiometric oxides from a *performance* perspective. In particular, we first illuminate the importance of entropy as a key driver in the efficiency of the two-step redox cycle. The entropy change ΔS_{rxn} of the $MO_{ox} \rightarrow MO_{red} + O_2$ reaction is a gauge of the effect of the temperature swing between Step 1 and Step 2: the higher ΔS_{rxn}, the lower the $\Delta T = T_{red} - T_{ox}$ needed to thermodynamically switch between Step 1 and Step 2. Because sensible heating between T_{red} and T_{ox} constitutes the major energy loss in the cycle, minimizing ΔT is key to maximizing efficiency. In this light, maximizing ΔS_{rxn} can be seen as a first step towards finding an optimal redox material.

Stoichiometric most metal/oxide pairs, for example $SnO_2 \rightarrow Sn + O_2$, suffer from an inherent flaw: the reduced metal is typically more ordered than its oxide counterpart. The thermodynamic significance of this is paramount. The reaction entropy is $\Delta S_{rxn} = S_{MOred} + S_{O2} - S_{MOox} = S_{ss} + S_{O2}$, where $S_{ss} = S_{MOred} - S_{MOox}$ is the entropy change in the solid state and S_{O2} is the entropy associated with the liberated oxygen. In this reaction ΔS_{rxn} will be a positive quantity, and here we would like to maximize it. Now S_{O2} is fixed by the temperature and partial pressure of the system. Holding those operating parameters constant, the only way we can increase ΔS_{rxn} is by making S_{ss} positive and as large as possible. However, recall that for a typical stoichiometric pair the reduced metal is *more* ordered (lower entropy) than its oxide. Therefore, for most stoichiometric pairs S_{ss} is *negative*, and so contributes negatively to ΔS_{rxn}.

For a non-stoichiometric pair, for example $CeO_2 \rightarrow CeO_{1.8} + 0.1O_2$, the above analysis procedure remains the same, but the result can be starkly different. Oxygen liberation in a non-stoichiometric oxide is concomitant with oxygen vacancy formation in the crystal. Therefore, reduced non-stoichiometric oxides are *less* ordered than their oxidized counterpart, which can be attributed to the configurational entropy associated with vacancy formation in the lattice. Because of this, S_{ss} can be *positive* for non-stoichiometric pairs, thereby giving them the unique capability of boosting to ΔS_{rxn} above that of S_{O2}.

In this work, the above phenomenon is investigated from an empirical and statistical thermodynamic approach. In particular, we investigate the potential to maximize ΔS_{rxn} through the appropriate choice of non-stoichiometric material, and nonstoichiometry, δ, endpoints for both the reduction and oxidation step. We then explore how this unique property of non-stoichiometric can be exploited to maximize the theoretical conversion efficiency of two-step thermochemical water and CO_2 splitting. Ultimately, we hope that this works will shed some light on the role of entropy in redox cycles based on non-stoichiometric oxides, and may pave a path to new materials discovery, guided by a principle of maximizing the solid-state entropy change.

611795 Thermodynamic Guiding Principles for Designing Efficient High-Temperature Solar Thermochemical Redox Materials for Fuel Production

Abstract:

The advancement of solar thermochemical fuel production technology critically relies on improvements in active redox materials and solar reactor designs. Previous studies typically predicted the solar-to-fuel efficiency of candidate active materials by adopting a simplified equilibrium approach, which could overpredict the efficiency. In this study, we describe a general thermodynamic framework to predict upper efficiency limits for both state-of-the-art and hypothetical materials in order to guide material design. The framework is based on conservation of mass and energy as well as Gibbs' criterion and allows for efficiency optimization by simultaneous consideration of material properties and reactor system conditions. A global map of solar-to-fuel efficiencies versus change in reaction entropy and enthalpy is presented for both existing and hypothetical materials revealing performance trade-offs due to competing factors: thermodynamic favorability, thermal losses, sweep gas demand, and oxidizer preheating. A hypothetical material design resulting in peak efficiency can be identified for a set of reactor operating conditions.

601514 Invited Talk: Entropy Source Analysis for Ferrites in Two-Step Thermochemical Splitting of Water and Carbon Dioxide

Abstract:

Greenhouse gases are being emitted at a rate of ~40 Gt(gigaton)-CO_2-equivalent/year. To keep the global average temperature rise to less than 2°C above that of the pre-industrial age, zero or even negative emissions need to be achieved within a couple of decades. Gt-scale carbon-free hydrogen (H_2) with cost below $2/kg has the potential to decarbonize industrial heat and feedstocks, electricity generation, and transportation, and it is also a potential energy carrier for intermittent renewables. Water splitting can produce carbon-free H_2 through electrolysis, photo(electro)chemistry and thermochemistry. The current chemical and energy industry relies almost exclusively on thermochemical processes because they are the only ones reaching Gt-scale with economies of scale.

Two-step thermochemical splitting of water and CO_2 has shown promise for large-scale operation due to its relative simplicity. The main challenges for reducing cost in the two-step thermochemical cycle include very high thermal reduction temperatures, low H_2O-to-H_2 (and CO_2-to-CO) conversion and limited oxygen-exchange reaction capacity. The thermodynamics of metal oxides for these reactions holds the key to achieving the needed performance. This research will present a deeper thermodynamic understanding of solid-solid metal oxide phase transitions, which forms the foundation for controlling these reactions.

For a metal oxide to spontaneously conduct both thermal reduction and water splitting reactions, increasing its h_O and s_O (partial molar enthalpy and entropy of oxygen, respectively) is necessary for lowering temperatures and improving H_2O-to-H_2 (and CO_2-to-CO) conversion. While state-of-the-art metal oxides for these reactions have h_O values spanning a large range, s_O values usually fall in a narrow range and their manipulation has remained poorly understood and difficult. Therefore, analyzing sources of s_O is critical for directing materials design.

Recently, iron-poor (Fe-poor) ferrites ($Fe_yM_{1-y}O_x$ where $y < 2/3$ and M is another one or more metals) were found to split CO_2 with unusually high yield compared to traditional Fe-rich ferrites, where Fe is the redox active element. The exceptional performance is achieved through tuning solid-solid phase transitions by changing the Fe:M ratio; specifically, Fe-poor ferrites show advantageous thermodynamic properties compared to the traditional Fe-rich ones. However, the fundamental sources of solid-state entropy change have not yet been analyzed. Here we have used $Fe_{0.45}Co_{0.55}O_{x1}$ and $Fe_{2/3}Co_{1/3}O_{x2}$ as model materials to quantify the entropic driving forces of the oxygen exchange reactions. Cation configurational entropy is determined by resonant x-ray diffraction measurements and CALPHAD-FactSage thermodynamic simulations. Lattice vibrational entropy is quantified by first-principles calculations using Hubbard-corrected density functional theory. Both sources are found to be important as components of entropic driving forces and the analysis explains the advantageous performance of Fe-poor ferrites compared to traditional Fe-rich ones.

608313 Tracking the Elementary Kinetics and Dynamic Evolution of Molecular Structures during Biomass Pyrolysis

Abstract:

Fast pyrolysis offers one of the most accessible ways to convert biomass into fuels and chemicals. The complexity of the feedstock and the dynamic changes in condensed phase chemical environments make it difficult to elucidate elementary kinetics and mechanisms and control the selectivity to products. These effects also manifest in macroscopic phenomena such as measurable differences in kinetics and changes in product distribution with changes in reaction temperature, feedstock composition, catalytic metal ion compositions and even dimensions of the feedstock. These complexities are ultimately dictated by molecular transformations controlled by the molecular structure and local chemical environments. Understanding the molecular processes in biomass pyrolysis is crucial for large scale and economical biomass conversion facilities.

Herein we present an ab initio-based kinetic Monte Carlo and Molecular Dynamics (kMC+MD) simulation that can model the kinetics of the fast pyrolysis and thermal reaction networks for biomass and other complex feedstocks. This approach tracks detailed atomic structural information including 3D coordinates of atoms and connectivity of chemical bonds in the feedstock molecules. It uses Stochastic Simulation Algorithm (SSA) to carry out elementary reaction steps and uses classical MD simulations to follow the dynamics of the feedstock and the reaction environment as reactions proceed. The elementary step kinetics for the kMC simulations are established from first-principle calculations.

Atomic-structural information is retained throughout the simulation, thus allowing the simulations to follow molecular transformations, the local molecular environment and the changes that result as a function of reaction conditions. As such, the simulations capture the unique kinetic manifestations that can result from the local structure of these feedstocks otherwise lost in composition-based deterministic models. This KMC+MD simulation approach is used herein to examine the pyrolysis reaction pathways of cellulose including levoglucosan and furan formation with some success in predicting kinetics and product distribution.

608094 Evolution of Cellulose Structure throughout Biomass Deconstruction in Gamma-Valerolactone

Abstract:

Biomass recalcitrance during deconstruction remains a key inhibitor to successful implementation of affordable biomass processing technologies. A clear connection between the cell wall structure and biomass deconstruction is necessary to understand how lignocellulosic material is broken down, and to identify defining features of residual cellulose. In this work, we have employed solid-state ^{13}C cross-polarization magic angle spinning (CP/MAS) nuclear magnetic resonance (NMR) spectroscopy to track domain changes in cellulose microfibrils throughout various chemical and enzymatic hydrolysis treatments. We also demonstrate the capability of an NMR observable metric to predict enzymatic hydrolysis sugar yields of pretreated cellulose, regardless of pretreatment co-solvents or biomass feed. P39 poplar was pretreated in GVL-water co-solvents with mild sulfuric acid concentrations at increasing temperatures and subsequently hydrolyzed with an engineered cellulase. Residual solids from each step were characterized using ^{13}C CP/MAS NMR and spectra were fit with subpeaks corresponding to different cellulose microenvironments at the C-4 carbon center. The changes of the C-4 carbon subpeaks were tracked throughout each treatment to understand the physical changes occurring within different cellulose microfibril domains. We show that the xylan-cellulose and inaccessible fibril surface resonances decrease significantly upon hydrolysis with mild acidic GVL-water co-solvent pretreatment. Enzymatic hydrolysis lead to further depletion of the inaccessible fibril surface peak. This behavior can be interpreted as an opening of bound microfibril surfaces previously inaccessible to the solvent. The cleaving of xylan-cellulose linkages and opening of inaccessible fibril surfaces by acid co-solvent pretreatment primes the cellulose for enzymatic attack. This technique can monitor the evolution of structural changes to the cell wall and allows for comparison between different cellulose deconstruction methods. It can guide larger questions of recalcitrance and give insight to improvement of subsequent steps for specific deconstruction methods based on residual solids structures.

601384 Theoretical Analysis of the Periodic Trends in the Aldol Condensation of Acetaldehyde over Isolated Metal Sites

Abstract:

Aldol condensation of aldehydes and ketones occurs during the heterogeneously catalyzed conversion of bioethanol to commodity chemicals and fuels. This reaction can be catalyzed by isolated metal cations grafted onto a high surface area support. Our interest is in understanding the effects of metal cation identity on the mechanism and the kinetics of aldol condensation. We have investigated this problem using electronic structure calculations (DFT) and statistical mechanics to compute Gibbs free energy profiles for aldol condensation of acetaldehyde over isolated metal cations grafted onto the surface of amorphous silica. As a model of the active center, we use a silsesquioxane cube with one corner replaced by the metal of interest. We consider Zr (IV), and less well known metal ion species for this reaction, such as Ta(V) and Sn(IV) for comparison across three groups on the periodic table. Figure 1 displays the Gibbs free energy landscape for acetaldehyde aldol condensation over isolated Zr, Ta, and Sn sites on silica. We predict enolization (state $C^{\ddagger}$) to be rate-limiting for the Ta(V), but C-C bond formation (state $D^{\ddagger}$) to be rate-limiting for Zr(IV) and Sn(IV) cations. The similarity in the Gibbs free energies for both transition states (enolization and C-C bond formation) for the Zr site suggests that both steps influence the observed kinetics and apparent activation barrier. We have also considered other metals in groups 4,5, and 14 and identified how the rate-controlling steps vary with the identity of the metal cation. The influence of the Lewis acid/base properties of the active center on the strength of acetaldehyde adsorption and the activation barrier for its enolization will also be discussed.

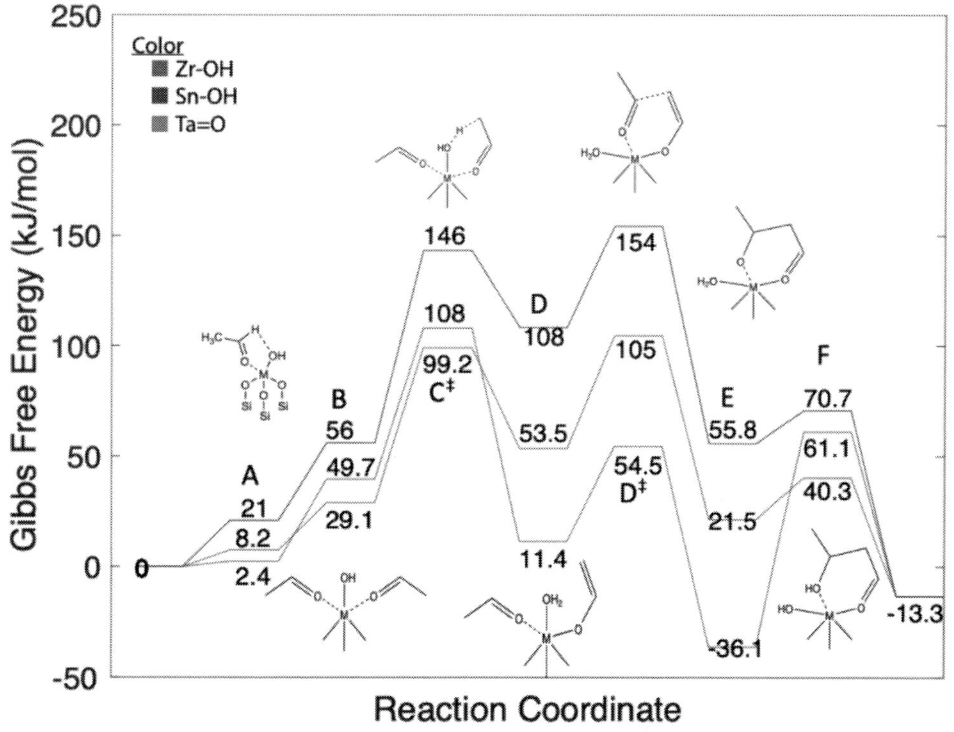

Figure 1: Gibbs free energy profile for the aldol condensation of acetaldehyde over isolated Ta(V), Zr(IV), and Sn(IV) cations

600043 Aldol Condensation of Cyclopentanone and Acetone on MgO: A Mechanism Study By DFT

Abstract:

Aldol condensation is an effective reaction to elongate the carbon chain for biomass conversion. Here we report a mechanistic study of the self- and cross-aldol condensation of cyclopentanone [C] and acetone [A] using density functional theory (DFT) calculations, though which we provide an explanation for the experimental finding of aldol condensation catalyzed by MgO, a solid base catalyst. Experimental results suggest that the [C]-activated products play a dominant role, as indicated by the initial rates with various A-to-C feed ratios. When more acetone is added to the system, the TOF of [C]-activated products remains considerably higher than that of [A]-activated ones. Calculations of the activation barriers show that α-H abstraction is the rate-determining step, and that the selectivity of aldol condensation (self vs cross) is controlled by the sequential C-C bond forming step. The selectivity to [C]C over [C]A is observed in experiment, while calculations show that formation of the [C]A product is more kinetically favorable. Both adsorption measurement and calculations suggest that cyclopentanone competes strongly for adsorption on MgO(100) and dominates the catalytic surface. The higher coverage of cyclopentanone than acetone on MgO may thus lead to higher selectivity toward the self-aldol condensation products. This conclusion is further supported by results from zeolite-catalyzed reactions.

609114 Computational Analysis of Kinetics of Ketonization and Aldol Condensation on Isolated Zirconium Sites

Abstract:

Two chemical reactions of interest for upgrading bio-oils to more valuable chemicals and fuels are ketonic decarboxylation (ketonization) and aldol condensation. Experimental work in our group has found that isolated Zr-hydroxy sites grafted onto silica are highly selective catalysts for these reactions. However, the mechanisms by which these reactions occur are not fully understood. Using DFT and statistical mechanics, we calculated the free energy profiles for both reactions. This involved the calculation of enthalpies and entropies for adsorption and reactions. We then used the calculated free energy profile to determine the turnover frequency for each reaction, the apparent activation energy, and the reaction order in reactant partial pressure and then compare these with the values determined from experiments. We also examined the degree of rate control for individual elementary steps as a function of reactant partial pressures. This analysis allowed us to identify limitations in both our computational and experimental methods. By testing the sensitivity of the predicted experimental quantities to potential errors in the assumptions and approximations we used to determine adsorption entropies, we were able to quantify the extent to which the calculated entropy of adsorption may be in error. We found that for our proposed mechanisms, we can replicate experimentally observed activation energies, rate-limiting steps, and zero order rate constants for both reactions within reasonable error. We speculate adsorbates retain significant rotational entropy unaccounted for by harmonic approximations, suggesting that our methods could be improved by explicit treatment of molecular surface rotations.

609667 Oxygenate Reactions over PdCu/PdAg Catalyst: Role of Synergistic Effects

Abstract:

Catalytic processes that involve C-C bond forming and chain elongation reactions such as Guerbet and aldol condensation are essential for upgrading biomass-derived oxygenates to fuel-range hydrocarbons. In this context, bimetallic catalysis has garnered extensive attention due to their tunable properties that result in selective C-H and C-O bond activation. In this study, we investigate and compare the activity of Pd_xCu_y/SiO_2 and Pd_xAg_y/SiO_2 catalysts in tandem hydrogenation-aldol reaction. We measure the rates of decarbonylation and hydrogenation of butyraldehyde, the reactive intermediate for Guerbet condensation, and correlate them with the properties of the catalysts via a range of characterization efforts. The selectivity of Pd_xCu_y/SiO_2 catalysts towards C-H bond formation increases with increasing Cu fraction, pointing to the geometric effect of Cu on the selectivity. In contrast, the turnover frequency peaks at a 3:1 Cu:Pd ratio, pointing to the promotional effect of Pd on Cu. Temperature-programmed reduction experiments indicate that alloying takes place, while chemisorption experiments show that surface Pd is replaced by Cu. Consistent with this, EXAFS data for Pd_xCu_y/SiO_2 shows a segregation of Cu to the surface. The charge transfer from Cu to Pd is observed in Pd LIII XANES data; this modifies the electronic properties of the metals, thereby improving the chemical reactivity of metallic surfaces. On the other hand, in Pd_xAg_y/SiO_2, Ag atoms significantly segregate to the surface, but do not show any charge transfer with Pd, negating any promotional effect. Hence PdCu is more reactive than PdAg. We use the implications of these findings to provide valuable design principles for oxygenate catalysis.

601057 Computational Insights into the Direct Acylation of 2-Methylfuran with Acetic Acid over Phosphotungstic Acid and H-BEA Zeolite

Abstract:

It has recently been demonstrated that the Friedel-Crafts acylation of furanic compounds is a low-carbon-footprint synthesis of renewable oleo-furan sulfonate (OFS) surfactants;[1] a class of surfactants with enhanced properties.[1] Current industrial acylation processes that use carboxylic chloride or anhydrides as acylating agents have several drawbacks: formation of corrosive side products and significant amounts of carbon waste requiring additional separation/recycling processes. In contrast, carboxylic acid acylating agents only generate water as a by-product and are 100% carbon atom efficient. Thereby, executing the direct acylation with carboxylic acids is of significant interest for developing renewable surfactants and green processes in general.[2]

A major challenge to realizing this in practice is accelerating the acyl formation, which entails dehydration of an acid. In this paper, we perform electronic structure calculations and microkinetically analyze an extensive reaction network to gain insights into the Brønsted acid catalyzed acylation of 2-methylfuran with acetic acid over phosphotungstic acid (HPW), a superacid, and compare with the catalytic activity of H-BEA zeolite. The reaction entails two steps: formation of the acyl group, either as acylium ion or stabilized acyloxy surface species; and formation of the Wheland intermediate by electrophilic aromatic addition (EAS).

Among the major predictions from our models are: *(a)* In HPW, the dominant pathway is through the acylium intermediate, which is very unstable and extremely reactive and as result the EAS requires no activation. *(b)* In H-BEA, the preferred pathway is via the stable acyloxy intermediate and the rate-limiting step is the EAS. *(c)* We predict higher turnover frequencies in HPW than in H-BEA even though HPW under low-coverage conditions. We compare with experimental kinetics.

References:

(1) Park, D. S.; Joseph, K. E.; Koehle, M.; *et al. ACS Cent. Sci.* **2016**, *2* (11), 820–824.

(2) Gumidyala, A.; Wang, B.; Crossley, S. *Science Advances* **2016**, *2* (9), e1601072.

597921 *Understanding Solvent and Pretreatment Effects on the Surface Chemistry of Small Oxygenates on Molybdenum Trioxide*

Abstract:

Sugars derived from biomass possess a high degree of functionality and offer promise as a potential feedstock for production of commodity chemicals. Understanding the pathway sugars take towards their desired end point is key because it allows for a more controlled chemical route. Heterogeneous catalysts are able to promote different routes based on their acidity and other physical parameters. Identifying descriptors for sugar interactions with a catalyst is key, but the role of solvents in these systems must not be neglected. Pretreatments can also affect the catalyst surface and must be considered as well. This work focuses on retro aldol condensation (RAC) and similar reactions by considering the reaction with simple oxygenates over different states of molybdenum trioxide (MoO_3).

This study highlights the role of ethanol on the surface of fully oxidized and partially reduced MoO_3 containing defect sites created through partial reduction. By combining diffuse reflectance infrared spectroscopy (DRIFTS) with temperature programmed desorption (TPD), surface species were monitored along with changes to the catalyst. X-ray absorption spectroscopy (XAS) measurements determined how the oxidation state was affected by the temperature and chemical environment during pretreatments. Ethanol, acetaldehyde and crotonaldehdye were introduced over the three different pretreated catalysts, and DRIFTS showed how these molecules bind to the surface. TPD analysis showed that reducing pretreatments significantly increased the yields of dehydrogenation and coupling products. Theoretical calculations gave vibrational frequencies and showed the catalyst with a vacancy defect adsorbed oxygenates more easily. These results collectively show that alcoholic species can interact with the fully oxidized surface, whereas aldehydes prefer surfaces containing defect sites. The oxidation state was also shown to be heavily dependent on which solvents were used and what pretreatment was done prior to reaction.

595426 N-Nitrosodimethylamine (NDMA) Formation during Advanced Oxidation Process (AOP) Treatment Following Reverse Osmosis for Potable Water Reuse

Abstract:

A 0.7 ng/L concentration of N-nitrosodimethylamine (NDMA) in drinking water is associated with a 10^(-6) lifetime excess cancer risk. NDMA is a particularly challenging compound for utilities purifying municipal wastewater effluents, where sewage contains higher concentrations of NDMA compared to conventional water sources, and NDMA can form from precursors when chloramines are applied upstream of microfiltration and reverse osmosis (RO) within advanced treatment. In order to reduce NDMA concentrations to below the 10 ng/L California Notification Level, UV-based advanced oxidation processes (AOP), reliant on radical production, such as UV/hydrogen peroxide (UV/H2O2), and, more recently, UV/free chlorine (UV/HOCl) are applied to RO permeates. While UV treatment effectively degrades NDMA, the potential for simultaneous NDMA formation during UV-based AOP treatment to partially offset this degradation has not been evaluated.

The objective of this laboratory-scale study is to evaluate the potential for NDMA formation during UV/hydrogen peroxide (UV/H2O2) and UV/free chlorine (UV/HOCl) AOP treatment. Interest in UV/HOCl processes has recently increased, as radical production from HOCl photolysis can be higher than from H2O2 photolysis resulting in more effective contaminant degradation. However, previous research suggests that the potential for greater NDMA formation during UV/HOCl AOP treatment. In this study, we demonstrate that dark reactions associated with residual chloramines in RO permeates increase NDMA formation during UV treatment. Addition of HOCl to RO permeates for UV/HOCl AOP treatment results in rapid NDMA formation from dark reactions with chloramine and UV-mediated breakpoint chlorination reactions. While these NDMA formation pathways and NDMA photodegradation occur simultaneously, our study shows that NDMA concentrations after UV/HOCl treatment can be double those post UV/H2O2 treatment. A mechanism for formation will be proposed. Given our results, in order to meet NDMA California Notification Levels with UV/HOCl treatment, higher UV fluence is required, and must be considered alongside higher radical production prior to implementation.

610968 Assessment of Enhanced Advanced Oxidation Processes for Treatment of Chlorinated Solvents

Abstract:

This paper addresses the process interactions resulting from combinative advanced oxidation processes, involving Fenton's reagent, ultraviolet light, titanium dioxide, ozone (O_3), hydrogen peroxide (H_2O_2), sonication or acoustic cavitation, among others. A comprehensive review of the technical literature has been performed, and the data ascertained have been critically analyzed and evaluated to determine whether the oxidative effects are additive, synergistic, or antagonistic. Predictions based on the individual unit operations are made and compared against the behavior of the combined unit operations. The data reported in this paper focus primarily on treatment of petroleum hydrocarbons (e.g., benzene, toluene, ethylbenzene, and xylene) and chlorinated solvents. The kinetic data included in this presentation are obtained both from the technical literature and from research investigations of the authors. The results of this study are assessed and discussed in this paper. The first-order kinetic rate constants are compared with other researchers.

595922 Plasma-Assisted Degradation of Perfluorobutane Sulfonate (PFBS)

Abstract:

This study investigates plasma-assisted degradation of perfluorobutane sulfonate (PFBS), a compound very few advanced oxidative and reductive technologies are able to degrade. PFBS, a member of a group of per- and polyfluoroalkyl substances (PFAS), is used as a replacement of perfluorooctane sulfonate (PFOS) in many consumer products such as nonstick pans, wire coating and firefighting foams. Recent studies have shown that PFBS is as persistent in the environment and bio-accumulative in people as its long-chain counterparts and human exposure to it can lead to serious health effects such as cancer, liver damage and immune system and thyroid changes. Here, the degradation of 10 mg/L of PFBS was investigated using a point-ring plasma reacor electrode configuration in which plasma is formed in argon gas and is contacting the surface of water containing dissolved PFBS. Experiments were conducted with and without cetyltrimethylammonium bromide (CTAB), a surfactant, whose purpose was to interact with and concentrate PFBS at the plasma-liquid interface, the region of chemical reactivity. No removal of PFBS was observed without the surfactant. With CTAB, ~99% of PFBS was removed from the bulk liquid and concentrated at the interface. At the interface, 67% of the 99% concentrate was degraded which resulted in 43% defluorination and 42% desulfonation. The dosage of the surfactant also played a role in the degradation. With step-wise CTAB dosing, defluorination rate increased markedly from 43% to 67%, and sulfate production from 42% to 68%. Experiments conducted with different surfactant types revealed that only cationic surfactants such as CTAB affect PFBS degradation, mainly through electrostatic attraction. Electrical discharge plasma appears to be a viable technology for the degradation of non-surfactants, in particular short-chain PFAS.

597753 Graphene Quantum Dots Based ZnO Nanomaterials for Degradation of Pollutant and Microbial Deactivation in Wastewater Treatment Plant.

Abstract:

Advanced oxidation processes (AOPs) offer enhanced contaminants destruction and low by-product and secondary pollution problems compared to conventional treatments. Photocatalysis is the mostly used form of AOPs. The photocatalyst can inactivate microorganisms and degrade organic materials by producing reactive oxygen species (ROS) upon photon irradiation. ZnO is an effective photocatalyst. It is characterized by a large bandgap energy, cost-effectiveness, biocompatibility and a high electron mobility. However, the efficiency of ZnO is reduced due to electron-hole (e^-/h^+) pairs rate recombination and inability to harvest a large part of the sunlight. In this work, different shapes of ZnO nanomaterials nanoflakes (ZnO-NFs) and zinc oxide nanorods (ZnO-NRs) were covalently linked to graphene oxide quantum dots (GQDs). They were linked to reduce the bandgap and decrease the e^-/h^+ pairs rate of recombination of the photocatalysts. The photocatalytic and antimicrobial behaviour of the graphene quantum dots based ZnO nanomaterials (ZnO-NFs@GQDs and ZnO-NRs@GQDs) were investigated. Several microscopic, spectroscopic and thermal analytic techniques were employed to confirm the successful preparation of the nanomaterials. Optical analysis showed that the calculated bandgap decreased from 2.98 eV to 2.61 eV and from 3.00 eV to 2.79 eV for the linked ZnO-NFs@GQDs and ZnO-NRs@GQDs, respectively. Data obtained from photolumiscence (PL) showed the same trend with a larger decrease in PL obtained for ZnO-NFs@GQDs indicating a large decrease in the e^-/h^+ pairs rate of recombination, thus a better photocatalyst. The photocatalytic experimental results revealed that the dyes were absorbed at 573 nm. A noticeable red-shifting of the dyes peak was evident with increase in time. The maximum absorption peak of the dyes at 573 nm decreased significantly with time for ZnONFs@GQDs than for ZnO-NRs@GQDs. The reduction in absorbance in the visible region indicated the decolourization through deprotonation (cleavage of the N–N bond) and desulfonation (ring opening). However, ZnO-NRs@GQDs showed greater efficiency in inhibiting bacterial growth for *E. coli* and *S. aureus* strains under the experimental conditions. The antimicrobial efficiency of the nanocomposites was in the order of ZnO-NFs < ZnO-NFs@GQDs < ZnO-NRs < ZnO-NRs@GQDs with greater efficiency being against *S. aureus*. The findings in this study imply that graphene quantum dots based ZnO nanomaterials can be used for the removal of dye pollutants and bacterial deactivation in wastewater treatment plant.

608957 Increasing Sodium Percarbonate Fenton like Process By Glutamic Acid Assisted Zero-Valent Iron for the Degradation of Methylene Blue

Abstract:

Zero valent iron (ZVI) catalyzed sodium percarbonate (SPC) Fenton like process assisted by glutamic acid (GA) for the degradation of methylene blue (MB) was evaluated in this study. Almost complete degradation of MB was observed within 40 min under optimum conditions when the molar ratios of ZVI/GA/SPC were 1/3/4 and the mineralization efficiency achieved was 42% within 3.5 hours. The detection of reactive oxygen species (ROSs) elucidated the significant generation of hydroxyl radicals and their key role in the degradation of MB. The experimental results and density functional calculations were found in good agreement thus, proving the more generation of ROSs in the ZVI/GA/SPC system for the efficient degradation of MB. The major intermediate products were identified and pathway mechanism was proposed based on mass spectrum analysis. Matrix effect of water constituents were examined to elucidate the efficiency of ZVI/GA/SPC system. In conclusion, actual wastewater results demonstrated significant potential of ZVI/GA/SPC system for the future aspirations.

Keywords

Sodium percarbonate, Fenton like process, Reactive oxygen species, Glutamic acid, DFT simulation.

594973 Ionizing Radiation-Induced Catalytic Degradation of Antibiotics in Water Matrices Using Fe/C Nanomaterials Derived from Iron-Based MOFs

Abstract:

The serious threat of antimicrobial resistance to human health has aroused the development of potential technologies to abate antibiotics from aqueous solutions. High-energy ionizing radiation, using electron beam or gamma rays, is considered to be more economic and efficient on a large-scale application than other AOPs such as UV radiation. H_2O molecules absorb radiation energy and water radiolysis occurs to generate the species involving the powerful oxidant ·OH and reductant e_{aq}^- and H·. ionizing irradiation has the advantages of good penetration in water matrices, insensitivity to suspended particles, no residuals produced and operation at room temperature. The public safety concern about radiation and high cost of investment and operation are the major factors to limit its application.

Recently, the iron-based metal organic frameworks (MOFs) have increasingly aroused great research attention in catalyzing the Fenton-like reaction and photocatalytic reaction to degrade organic pollutants in aqueous solution. MOFs are the hybrid porous crystalline materials constructed of the central metal /clusters coordinated to organic ligands, which have the large surface area and porosity, tunable topology and flexible structure. The Fe-based MOFs-derived Fe/C nanomaterials (DMOFs) are fabricated by using Fe-based MOFs as precursors through thermal pyrolysis under controlled temperatures and inert atmosphere. DMOFs remained the high specific surface area and porosity of MOFs and exhibited high catalytic activity and stability.

In present work, we proposed a novel Fe/C nanomaterial fabricated using Fe-based MOFs as precursors through thermal pyrolysis to catalyze gamma irradiation-induced degradation of antibiotics, cephalosporin C (CEP-C) and sulfamethazine (SMT) in aqueous solution. The adsorption kinetics of the two antibiotics on DMOFs were firstly investigated. The performances of gamma/DMOFs system was evaluated in terms of the antibiotics removal and mineralization extent in comparison to that using gamma irradiation alone. The degradation intermediates of antibiotics were identified and the physicochemical properties of DMOFs before and after irradiation was assessed. This study will provide a new insight about an efficient technique to decompose antibiotics in water matrices by catalytic ionizing radiation.

Results showed that DMOFs have the regular octahedrons structure of MOFs with high porosity. It consists of element C, Fe and O and Fe^0 with a fraction of Fe_3O_4 and Fe_2O_3 were identified. DMOFs addition was more efficient to enhance the degradation of antibiotics having a higher adsorption capacity like SMT. The degradation rate of CEP-C and SMT raised by 1.3 times and 1.8 times, and the TOC reduction at 1.0 kGy reached 42% and 51%, respectively by gamma/DMOFs treatment, compared to TOC removal of 20.2% (CEP-C) and 4.5% (SMT) by gamma irradiation alone. This was attributed to the enhanced generation of ·OH catalyzed by DMOFs during gamma irradiation. Overall, the degradation rate of both CEP-C and SMT was higher at acidic conditions and declined at alkaline pH of 9.0. It is noted that the adsorption capacity of both antibiotics declined with increasing the solution pH, which might have negative effect on antibiotic degradation. Regarding to the iron leached, it was much higher (up to 0.9-1.2 mg/L) at pH 4.0, while it maintained at low levels (0-0.2

mg/L) when the solution pH was higher than 5.2. Taken into account of the higher antibiotic decomposition rate and the minor iron leaching, pH value of 5.7 ± 0.5 was suggested for practical application.

The antibiotic molecules were destroyed, leading to the formation of intermediates of the N-S bond cleavage for SMT, C-N bond cleavage and β-lactam ring and dihydrothiazine ring opening for CEP-C. DMOFs exhibited good stability and recyclability as exposed to gamma irradiation. The crystal structure, functional groups and magnetism of DMOFs changed slightly after gamma irradiation, which made it possible to be recycled and reused. DMOFs were promising to enhance the degradation of antibiotics during gamma irradiation.

593211 From Benchtop to Field Demonstration: Scaling the "CO_2concrete" Technology for the Manufacture of Concrete Products By Carbon Dioxide Mineralization

Abstract:

Carbon capture and utilization processes that exploit "CO_2 mineralization" reactions to produce concrete products offer a transformative platform for gigaton-scale CO_2 utilization, globally, due to the vast market for concrete (> 20 billion tons, ~ $1 trillion annually). The reaction of portlandite ($Ca(OH)_2$), an archetypal alkaline solid, with carbon dioxide (CO_2) is one of such reactions that can be exploited to produce cementation agents with a low embodied-carbon intensity. This process is the core of the CO_2Concrete technology, which is supported by detailed investigations into the effects of reaction temperature, relative humidity (*RH*), and CO_2 concentration on the carbonation of portlandite in the form of finely divided particulates and monolithic compacts, and of the influences of pore saturation and CO_2 diffusivity on the carbonation kinetics and strength evolution of portlandite-enriched composites ("mortars").

Special focus is paid to uncover the factors affecting the extent of reactant (i.e., $Ca(OH)_2$) conversion in relation to the process conditions, moisture state (i.e., the presence of molecular or condensed water), and the size and surface area-to-volume ratio (SA/V, mm^{-1}) of monoliths. Notably, the carbonation of portlandite is not limited by surface passivation of reactants with products; rather, reaction progress is limited by the mobility of adsorbed water. Scaling from particulates to monolithic components, the imposition of a porous microstructure (through which CO_2 must diffuse prior to reaction) induces secondary conversion limits related to blocking of pore networks by condensed water. Further, the carbonation kinetics of monoliths are strongly linked to the rates of moisture transport and vaporization/condensation. Reducing saturation increases the gas diffusivity, and carbonation kinetics, so long as saturation exceeds a critical value ($S_{w,c} \approx 0.10$); independent of microstructural attributes.

The outcomes of these bench-scale investigations are foundational to the scaleup of the CO_2Concrete process for operation in a field demonstration setting i.e., in a system using flue gas from actual coal-fired power plants to produce construction-ready prefabricated concrete products. The design of the CO_2Concrete process to produce masonry products from coal flue gas is described, considering aspects of site operations, materials handling, reaction kinetics, and integration into the flue gas source (host site). Built on a rigorous foundation of experimental and modeling investigations, the CO_2Concrete technology is scaling up rapidly, with the aim of realizing large-scale, economical, and efficient utilization of CO_2 emissions.

602462 Implementation of Ion Exchange Processes for CO_2 Mineralization Using Produced Water Streams

Abstract:

Large coal-fired power plants (>500 MW) account for 46% of global CO_2 emissions, and long-term management of this CO_2 to is urgently needed mitigate global temperature increases. Sequestration of CO_2 within stable mineral carbonates (e.g., $CaCO_3$) represents an attractive emission reduction strategy because it offers a leakage-free alternative to geological storage of CO_2 in an environmentally friendly form. This work describes a mineralization process in which divalent cations are sourced from various waste streams (e.g., produced water and brackish water) and alkalinity is induced via regenerable ion-exchange materials. In our process, aqueous streams with pH > 8 are produced by contacting fresh water with various ion-exchange materials (e.g., Na form zeolites or ion exchange resins). These streams are mixed with produced water containing high concentrations (~0.1 M) of Ca^{2+} leading to the precipitation of solid calcium carbonate (PCC). This process has the advantages of using regenerable solids in a simple and continuous process to increase the pH of water by ion exchange instead of relying on the consumption of costly and unsustainable sources of alkalinity (e.g., sodium hydroxide).

The pH shift in our process is effected by the exchange of protons for Na^+ ions. Thus, the Na^+-H^+ exchange isotherms and adsorption rates of 13X and 4A zeolites and TP-207 and TP-260 resins in CO_2-saturated water (pCO_2 = 1 atm, pH 4) were studied via batch equilibrium experiments. Adsorption isotherms obey the Langmuir model, with the resins achieving higher capacities than zeolites: 2.4 mmol H^+/g for TP-207, 2.2 mmol H^+/g for TP-260, 1.8 mmol H^+/g for 4A and 1.6 mmol H^+/g for 13X. Consequently, ion exchange resins were capable of producing aqueous solutions with of pH = 11.3 and 10.9 for TP-207 and TP-260 respectively, compared to a pH = 10.2 and 9.8 for solutions contacted with 4A and 13X respectively. Multi-component mixtures of Ca^{2+} and Mg^{2+} cations in CO_2-saturated water were used to probe competitive ion exchange. The presence of divalent cations in solution significantly inhibited H^+ uptake because of their larger field strength, reducing capacities to as low as 1.2 mmol H^+/g for resins and 0.8 mmol H^+/g for zeolites. These results indicate the alkalinity inducing reaction for this process must occur in the absence of divalent cations to achieve the desired pH for CO_2 mineralization.

The extent of calcite precipitation was evaluated experimentally by mixing the alkaline CO_3^{2-}-rich water solution that is obtained from the ion-exchange column, with a simulated liquid waste stream solution (1.4 M NaCl, 0.1 M $CaCl_2$, 0.056 M $MgCl_2$, 0.01 M $CaSO_4$, 0.0001 M KCl, 0.00044 $FeCl_2$). Experimental calcite yields were 26 mmol/L for cation exchange resins and 8 mmol/L for zeolitic materials, with the formation of goethite (an iron-hydroxide phase, FeOOH) as the primary contaminant phase (99% calcite, 1% goethite). Yields calculated via Gibbs Energy minimization were 26.5 mmol/L and 8.8 mmol/L for the resin and zeolite systems, respectively, indicating that the experimental process was able to achieve thermodynamic maximum production of calcite.

The results from these studies indicate that ion exchange processes can be used as an alternative to the addition of stoichiometric bases to induce alkalinity for the precipitation of $CaCO_3$. The high calcium carbonate yields (1 ton CO_2/day utilizing a produced water flow rate = 6.2 L/min and a total bed volume = 0.02 m^3) obtained for the materials examined and the successful operation at standard temperature and pressure conditions support their potential for industrial implementation.

592779 Electrochemical CO_2 Conversion to Valuable Chemicals

Abstract:

The electrochemical conversion of carbon dioxide to value-added chemical products has been heavily explored as a promising strategy for carbon utilization. However, the direct synthesis of multi-carbon (C_{2+}) products suffers from undesired side reactions and relatively low selectivity. Electrochemically converting CO_2 to single-carbon products is much more effective and being commercially deployed. Recent studies have shown that CO can be electrochemically transformed further to C_{2+} at high reaction rates, high C_{2+} selectivity and inherently improved electrolyte stability, raising the prospect of a two-step pathway to transform CO_2. In this talk, we will discuss the progress towards high-rate CO conversion alongside mechanistic insights and device designs which can improve performance even further. A techno-economic analysis of the two-step conversion process and cradle-to-gate life cycle assessment shows the economic feasibility and improved environmental impact of a high-volume commercial process generating acetic acid and ethylene compared to the current state-of-the-art.

600965 Electrochemical CO_2 Utilization: Scalable System Operation for Formic Acid Production

Abstract:

In recent years, research into the electrochemical reduction of CO_2 has grown exponentially. Formic acid (FA) has become a particular value product from CO_2 due to its versatility [1], optimal atom economy [2], and thermodynamic favorability [3]. FA can be produced electrochemically via a direct two-electron transfer process involving CO_2 with a proton source, requiring less energy input and fewer reaction steps than the conventional Kemira process. However, product selectivity, electrode stability, and low Faradaic efficiency due to increased resistance under constant current, remain key issues for CO_2 reduction to FA [3].

University of Kentucky's Center for Applied Energy Research (UK CAER) is currently investigating reactor designs to address the challenges associated with electrochemical CO_2 conversion to FA [4]. The current reactor system employs: (1) an organic-based charge carrier which shuttles charge directly to the catalyst to enable CO_2 reduction to FA; (2) novel electrode materials to mitigate large voltages and improve conductivity; and (3) a flow system which not only allows for the volumetric scale-up of both charge carrier and catalyst, but also decouples the charger carrier re-energized and FA production processes to protect the catalyst's stability due to overpotential. Through careful adjustment of electrode, charge carrier, and electrolyte selection, cell resistances have decreased by over 15%, facilitating conduction necessary for effective CO_2 reduction to FA. The system uses a highly specific engineered catalyst, which has produced over 100 mM FA at a rate of over 10 mM FA/hour following the aforementioned electrode/charge carrier changes. System design considerations (flow rate, bulk volume) and modeling of the system will also be discussed.

1. Lu X, Leung DY, Wang H, Leung MK, Xuan J. Electrochemical reduction of carbon dioxide to formic acid. ChemElectroChem. 2014 May 13;1(5):836-49.
2. Spurgeon JM, Kumar B. A comparative technoeconomic analysis of pathways for commercial electrochemical CO 2 reduction to liquid products. Energy & Environmental Science. 2018;11(6):1536-51.
3. Pérez-Fortes M, Schöneberger JC, Boulamanti A, Harrison G, Tzimas E. Formic acid synthesis using CO2 as raw material: Techno-economic and environmental evaluation and market potential. international journal of hydrogen energy. 2016 Oct 5;41(37):16444-62.
4. Omosebi, A., Landon, J., Thompson, J., and Liu, K. "Process Intensification for Electrochemical Utilization of CO2." Proceedings of AIChE 2019 Annual Meeting. 11 Nov 2019.

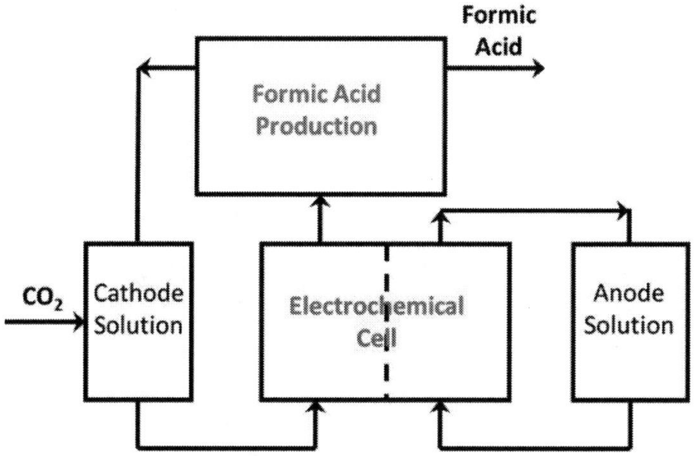

601478 Oxide-Derived Tin Microspheres for Efficient CO_2 Electroreduction to Formic Acid

Abstract:

Rational design and engineering of electrocatalysts for electrochemical CO_2 reduction (CO_2RR) is crucial to effectively, selectively converting CO_2 into carbon-neutral chemicals and fuels. In this work, we have developed low-cost hierarchical tin oxide microsphere electrocatalysts to boost the production of formic acid. Three-dimensional (3D) SnO_2 spheres constructed from small crystalline nanoparticles exhibit high Faradaic efficiencies (70-80%) and high selectivities (77-90%) toward formic acid in a wide potential window of -0.9 V to -1.3 V *vs.* the reversible hydrogen electrode (RHE) at total current densities of 30-100 mA/cm^2. A peak 80.3% Faradaic efficiency is obtained at -1.2 V *vs.* RHE with superior HCOOH partial current density of 50.4 mA/cm^2. Such unique 3D spherical catalysts outperform non-spherical nanoparticles and commercially available SnO_2 benchmarking catalysts due to special hierarchical porous structure, enlarged surface area, larger number of electrochemically active sites, and strongly suppressed H_2 evolution. The spherical oxide-derived Sn catalysts are highly stable up to 36 hours over many start/stop cycles at -1.2 V *vs.* RHE, preserving long-term partial current density of 45 mA/cm^2 and 90% selectivity for formic acid. Our results may provide more guidelines for the introduction of hierarchical porosity into CO_2RR electrocatalysts for large-scale applications.

602882 Designing Catalysts and Reactors Towards Practical CO_2 Reduction

Abstract:

Electrochemical CO_2 reduction, with the energy input from renewable electricity, provides a green and alternative route for the generation of chemicals and fuels. However, its practice is currently challenged at two systematical levels: the lack of selective electrocatalysts to combat the strong completion from water reduction, and the lack of novel reactors for large-scale reaction rates and efficient product separation. In this talk, I will introduce the rational design of both catalytic materials and reactors towards practical CO_2 reduction performances. By dispersing transition metals into isolated single atoms with electronic structures significantly different from their bulk counterparts, we can dramatically suppress the competing hydrogen evolution and deliver an ultra-high CO_2 reduction selectivity of more than 95% under ambient conditions in water. Scaled-up synthesis and efficient reactors demonstrated the potential for practical applications. Furthermore, by designing a novel solid electrolyte reactor, we successfully demonstrated a continuous generation of pure liquid fuel solutions via CO_2 reduction. This technology eliminates the product separation process required in traditional CO_2 reduction electrolyzers, opening up its practical applications in the future.

607263 Carbon Dioxide Utilization through PEM Electrolysis

Abstract:

Electrochemical carbon dioxide utilization can produce carbon-based fuels and chemicals using only three inputs: carbon dioxide, water, and electricity. This approach has the potential to monetize and reduce green house gas emissions by providing a profitable alternative to releasing carbon dioxide into the atmosphere. Large-scale deployment of this technology is key to achieving these benefits.

Opus 12 is developing a polymer-electrolyte membrane (PEM) electrolyzer for carbon dioxide utilization based on existing water electrolyzer designs. PEM electrolyzers have many benefits: 1) They can ramp quickly, allowing use of intermittent low-cost renewable electricity, such as nighttime wind. 2) They operate at high current density, leading to small footprint. 3) They have low capex, thanks to low temperature and pressure operating conditions. 4) They are modular and scalable, allowing for integration with CO_2 emissions sources of diverse volumes. 5) They are operationally simple, with no need for trained operators on site.

This talk will outline Opus 12's progress in scaling PEM carbon dioxide electrolysis, performance metrics and their relationship to process economics, and the market for industrial-scale deployment.

608980 Simultaneous Carbon Utilization and Alkane Conversion Using Solid Oxide Electrolytic Technology

Abstract:

U.S. coal plants have been essential in providing the U.S. with reliable low-cost electricity and will continue to be relied upon as base-load for the foreseeable future. Cost-effectively reducing carbon emissions from the existing U.S. coal-fired generation fleet is necessary for these assets to continue to play an important role in a carbon restricted economy. Carbon utilization is a methodology which could help offset the costs of carbon capture by converting the product CO_2 into valuable products. Carbon monoxide (CO) is an important industrial gas used in manufacturing bulk chemical precursors such as phosgene, commodity materials via carbonylation including aldehydes, ketones, carboxylic acids, anhydrides, esters, amides, imides, carbonates, ureas, and isocynanates. Industrial bulk CO is produced from separating CO from syngas ($CO + H_2$) generated from the steam methane reforming process.

Separation technologies such as cryogenic separation (i.e. cold box), pressure swing adsorption (PSA), membrane separation, and ammonium salt solution absorption are currently used to extract CO. Understandably, these processes are complex requiring large production facilities and huge capital investment. CO_2 electrolysis is a promising solution with several low-temperature solution based electrolyzers being studied. However, solid oxide electrolyzer cells (SOECs) operating at an elevated temperature provide greater cell/electrolysis efficiencies and the ability to use this technology to electrochemically convert CO_2 into CO/O_2 or CO_2/H_2O into syngas/O_2 has been shown over recent years [1–3].

Ohio University (OHIO) with funding from the U.S. Department of Energy's National Energy Technology Laboratory (NETL) Carbon Use and Reuse Program (DE-FE0031709) is developing a process that utilizes solid oxide electrolysis to simultaneously convert CO_2 into valuable products. As a preliminary study, the electrochemical conversion of CO_2 reduction to CO will be evaluated using an SOEC with H_2 as the anode fuel. Nickel and $(La_{0.80}Sr_{0.20})_{0.95}MnO_{3-\delta}$ based electrodes, widely used in solid oxide fuel cells (SOFCs) [4] were chosen as the anode and cathode electrodes, respectively, for the SOEC used in this study. This presentation will discuss these results including experimental performance and process simulations as part of OHIO's carbon utilization project.

[1] F. Bidrawn, G. Kim, G. Corre, J.T.S. Irvine, J.M. Vohs, R.J. Gorte, Efficient Reduction of CO2 in a Solid Oxide Electrolyzer, Electrochem. Solid-State Lett. 11 (2008) B167.

[2] L. Zhang, S. Hu, X. Zhu, W. Yang, Electrochemical reduction of CO2 in solid oxide electrolysis cells, Journal of Energy Chemistry. 26 (2017) 593–601.

[3] Y. Xie, J. Xiao, D. Liu, J. Liu, C. Yang, Electrolysis of Carbon Dioxide in a Solid Oxide Electrolyzer with Silver-Gadolinium-Doped Ceria Cathode, J. Electrochem. Soc. 162 (2015) F397.

[4] R. Mark Ormerod, Solid oxide fuel cells, Chemical Society Reviews. 32 (2003) 17–28.

599367 Examining Routes to *P*-Cymene through Integrating Biological and Chemical Catalysis

Abstract:

p-Cymene is an interesting renewable product for fuel and polymer applications if biological and chemical conversions can be efficiently integrated for its production. An important aspect of efficient integration for the overall processes is determining what is the preferred chemical intermediate for the handoff from biological to chemical conversion. While limonene has been examined previously as the intermediate, recent work has shown that stopping earlier in the same metabolic pathway at 1,8-cineole leads to a significantly higher titer than with limonene. In this study, the direct comparison of converting these potential intermediate reactants to *p*-cymene was performed. Pd on γ-Al_2O_3 catalysts were found to convert efficiently either reactant, as it appeared that the additional dehydration required with 1,8-cineole conversion was facile. To examine the importance of acidity for the catalysts studies were also performed with silica-aluminum or SiO_2 supports. Also, alternative metals, Cu and Ni, as well as stability performance of Pd-based catalysts were examined. By examining the overall biological and chemical production pathway, it was found that the most efficient strategy was to target 1,8-cineole as the biological intermediate that is subsequently converted to *p*-cymene. This study demonstrates the importance of biological and chemical catalysis development informing each other for creating efficient conversion pathways to chemical products.

602506 Assessing the Impact of Allosteric Enzyme Regulations Limiting Ethanol Titer in *Clostridium Thermocellum* Using a Core Kinetic Model

Abstract:

Clostridium thermocellum is a promising candidate for consolidated bioprocessing due to its ability to consume and ferment cellobiose to ethanol. Despite significant efforts, achieved yields and titers fall below industrially relevant targets. This implies that there still exist enzymatic, regulatory and possibly thermodynamic bottlenecks that throttle back metabolic flow. This work aims to reveal some of these factors by elucidating internal metabolic fluxes in wild-type *C. thermocellum* grown on cellobiose via ^{13}C-metabolic flux analysis (^{13}C-MFA). Flux elucidation results confirmed that that both pyruvate, phosphate dikinase and the malate shunt are pyruvate contributors. ^{13}C labeling data interpreted by ^{13}C-MFA alluded to serine generation via the mercaptopyruvate pathway, with candidate genes Clo1313_0176 and Clo1313_0149 identified for encoding this activity in *C. thermocellum*. Using the resultant flux distribution in conjunction with batch fermentation process yield data for various mutant strains, we constructed an updated kinetic model of *C. thermocellum* core metabolism (i.e. k-ctherm101). During kinetic parameterization, we introduced a systematic method for identifying a minimal set of putative regulatory mechanisms required for agreement with the experimental data using the K-FIT kinetic parameterization algorithm. Subsequently, we use k-ctherm101 to explore the effect of removing allosteric regulations on ethanol yield and titer. By exploring all possible simultaneous (up to four) regulation removals we identified combinations that lead to 76% model predicted improvements in ethanol titers. Top strategies include the simultaneous removal of acetyl-CoA inhibition of acetaldehyde dehydrogenase, ethanol inhibition of pyruvate ferredoxin oxidoreductase, and ethanol inhibition of alcohol dehydrogenase combined with nadh inhibition of either alcohol dehydrogenase or acetaldehyde dehydrogenase.

Towards the Potential Coupling of Microbial Biosynthesis and Heterogeneous Catalysis for the Petroleum-Free Production of Alkylated Phenol Compounds

Adam Zuber, *Zhenghong Li, Xiaonan Wang, Ashil Vekaria, Yingxi Lu, Haoran Zhang* and George Tsilomelekis**

Department of Chemical and Biochemical Engineering, Rutgers, The State University of New Jersey, 98 Brett Road, Piscataway, NJ 08854

* Corresponding Authors
Email: g.tsilo@rutgers.edu (George Tsilomelekis), Haoran.Zhang@rutgers.edu (Haoran Zhang)

Amongst variegated alkylation reactions, the tert-butylation of phenol is one of particular importance to industry [1]. It has been estimated that 450,000 tonnes of tert-butyl phenols (TBP) are manufactured annually for production of innumerable chemical commodities [3]. The ortho mono-alkylated product serves a role in pesticides and fragrances; the para counterpart functions as a flavoring agent and petroleum additive, among other compounds. As for the di-alkylated products, 2,4-TBP is utilized in ultraviolet absorbers in polyolefins and PVC stabilizers, while antioxidants and pharmaceutical products make use of 2,6-TBP [2].

In the present study, petroleum-free production of alkylated phenolic compounds from glycerol was achieved by streamlined coupling of microbial biosynthesis and heterogeneous catalysis with the aim to advance biomass conversion to industrially relevant chemical products. First, bacterium *E. coli* was metabolically engineered to convert glycerol to phenol under mild conditions, followed by extraction of phenol from the cell culture using select polymeric resins. Desorption of phenol was achieved by washing with tertiary butyl alcohol (TBA), which serves also as a reactant in the subsequent tert-butylation reaction. The adopted polymeric resins were also evaluated as potential downstream catalysts for the production of tert-butyl phenolic products.

Catalytic activity was evaluated in batch mode at industrially relevant conditions, i.e. low temperature and pressure. Amberlyst ® 15 exhibited low to moderate phenol adsorption capacity but very promising reactivity. Achieving full conversion at 120°C within 4 h, Amberlyst ® 15 obtained over 90% selectivity to mono-alkylated products. Furthermore, full regeneration was accomplished thrice by simply washing with ethanol, highlighting the promise of Amberlyst ® 15 as a heterogeneous catalyst. The holistic approach developed by this study offers a unique opportunity to synthesize end-products directly from glycerol that cannot be achieved in such an efficient manner by using microbial biosynthesis or heterogeneous catalysis approaches alone.

References:

[1] Adam, F., and C. Kueh. J Taiwan Inst Chem Eng, vol. 45, no. 2, 2014, pp. 713–723.
[2] Matloubi MF et al. Trans C Chem Chem Eng 2009;16:81.
[3] Modrogan E, Valkenberg MH, Hoelderich WF. J Catal 2009;261:177.

606618 Optimization of Lignocellulose-Based Production of Biocontrol Agents As Biorefinery Coproducts

Abstract:

Over 80% of *Fusarium sambucinum*, causative of potato dry rot, are now resistant to the fungicide thiabendazole which was once used for disease control. ARS (Peoria, IL) has developed three *Pseudomonas* spp. as biological control agents (BCAs) to serve as alternatives to traditional chemicals for controlling potato dry rot and other maladies, including sprouting, late-blight (incited by *Phytophthora infestans*), pink rot (incited by *Phytophthora erythroseptica*), and leak (incited by *Pythium ultimum*). It is proposed that such *Pseudomonas* spp., which utilize diverse carbon sources, can be developed as valuable coproducts for the lignocellulose biorefining industry. Our recent results suggest that sugars, acetic acid, and furan aldehydes generated from lignocellulose can all be utilized during production of these BCAs. Data will be presented showing BCA viable cell yield and productivity optimization on dilute acid switchgrass hydrolysate as a function of sugar loading, C:N ratio, oxygen transfer rate, phosphate (pH) and growth factor levels. Optimization of the inoculum needed for production of triculture with balanced growth of three *Pseudomonas* strain populations will also be presented, along with a rehydration technique for recovering viable cell activity in dried cell formulations and an evaluation of bio-efficacy in Fusarium dry rot suppression.

608360 Tandem Chemical and Biological Catalysis: Production of 2-Pyrone-4,6-Dicarboxylic Acid Via Lignin Valorization

Abstract:

Lignocellulosic biomass is a bountiful source of renewable carbon for the sustainable production of fuels and chemicals.

Lignocellulosic biomass is composed of primarily three polymers: cellulose, hemicellulose, and lignin. Hemicellulose is a heteropolymer of C_5 and C_6 carbohydrates, whereas cellulose is a homopolymer of β(1→4) linked glucose. These polysaccharides comprise 65-80 wt% of the plant biomass and considered to be easier to upgrade to value-added products (e.g., paper, clothing, and liquid fuels). Lignin, on the other hand, is a heteropolymer of aromatic compounds and is recalcitrant to chemical and/or biological upgrading. As lignin constitutes 20-35 wt% of biomass, and it is the largest source of renewable aromatics on the planet, it is desirable to obtain value-added products from lignin for maximum utilization of the feedstock.

In this work, we have used chemical and biological upgrading in tandem to extract maximum value from lignin. We will show that relatively native lignin can be isolated from lignocellulosic biomass using γ-valerolactone (GVL) and water as the solvent system with sulfuric acid as the catalyst under mild reaction conditions. We demonstrate that this GVL-lignin is successfully depolymerized by hydrogenolysis over a Pd/C catalyst into a mixture of aromatic compounds. This mixture contains molecules with similar chemical structures and presents a challenging separation problem. Indeed, this has been a bottleneck for obtaining value-added products from lignin. We solve this problem by upgrading the mixture of aromatic compounds to a single compound via biological funneling. Specifically, we use engineered *Novosphingobium aromaticivorans* (DSM12444) to transform the mixture of aromatic compounds containing syringyl, guaiacyl, and *p*-hydroxyphenyl substructures to a single product, 2-pyrone-4,6-dicarboxylic acid (PDC). Thus, we show that a complex mixture of lignin hydrogenolysis products can be reduced to a single product leading to easier separation. Additionally, we show that this strategy is agnostic to the biomass type by successfully applying this strategy to lignin extracted from hardwoods (poplar and maple) and grasses (switchgrass and sorghum).

605611 Biological Alcohol Production from CO_2 using Novel Acetogens

Abstract:

The conversion of CO_2 to valuable products such as alcohols can create additional revenues from industrial waste streams, while reducing greenhouse gas emissions. Acetogens such as *Clostridium ragsdalei, C. carboxydivorans and C. ljundahlii have been used to convert syngas components* (CO, CO_2 and H_2) *into alcohols and organic acids via the acetyl-CoA pathway. However, few studies have focused on the use of these microorganisms to convert* CO_2 to various products. In order to use CO_2 as a carbon source, an energy source such as H_2 must be present to allow the microorganism to reduce CO_2 into products. In the current study, *C. ragsdalei strain P11, C. carboxidivorans* strain P7A and three new clostridial strains A, B and C were used to convert CO_2 and H_2 into carboxylic acids and alcohols. Fermentations were performed in 250-mL bottle assays with 50 mL working volume at 37°C and 150 rpm. A gas mix containing H_2:CO_2:N_2 (60:20:20) was fed daily for 360 h. Results showed that all strains grew on CO_2 and produced alcohols and organic acids. However, strains P11 and P7A grew faster and consumed more CO_2 and H_2 compared to the other new strains. There was a lag phase of 48 to 72 h for strains A, B and C. Strains P11 and P7A produced more organic acids than the new strains. However, preliminary results showed that the new strains appear to make more C4 and C6 products compared to strains P11 and P7A. The effects of fermentation parameters such as gas composition and pressure will be presented. Further adaption and improvement of the five strains to reduce lag phase and improve CO_2 and H_2 conversion to alcohols are needed to enhance their potential use in the conversion of CO_2 containing waste streams to biofuels.

610743 Towards Sustainable Circular Economy: Design Framework and Application to Grocery Sacks

Abstract:

Numerous directives and strategies [1,2] aim to establish a Circular Economy (CE) by value retention from a consumer discarded product within the technosphere, thereby implicitly reducing the burden on the ecosphere. Many opportunities to recycle, down-cycle or up-cycle products have been identified each contributing to a different degree of circularity. However, for the CE to be sustainable, that is to obtain a Sustainable Circular Economy (SCE), there is a need to holistically evaluate entire value chains of the product and other down-cycled or up-cycled material. Further, there needs to be a method to systemically find the environmental and economic implications of various value chain pathways and quantify the trade-offs with the establishment of circularity. There is also a requirement to analyze the performance of pathways for multiple stakeholders of circularity along the value chain by means of appropriate CE metrics. We propose a life-cycle design based framework to identify optimal pathways that are best aligned with the goals of SCE from the perspective of multiple stakeholders.

Life-cycle assessment (LCA) is a powerful tool for assessing holistic environmental impacts due to a certain configuration of network flows in a value chain. With a few modifications to this tool such as including the consumption and disposal modules, it can be used to represent circular value chain networks. Further, using mathematical properties of matrices, principles of allocation and displacement [3] and flow conservation, we can use LCA to describe superstructure networks of many alternative value chains. Further, we present the use these life-cycle superstructure networks as basis for a multi-objective optimization problem formulation [4], thereby demonstrating the proposed framework.

In this talk, we will present a novel computational framework to model, assess and design sustainable circular systems by means of a superstructure pathway design approach. The framework is capable of quantifying trade-offs between circularity objectives for multiple stakeholders and environmental objectives. Thus, it allows the governing bodies and other stakeholders to make calculated decisions related to establishing value chains to ensure SCE. It can also be used to find the viability of a novel product substitute in a value chain by including it within the superstructure, and identify opportunities for innovation. We apply the framework to a case study for grocery sacks value chain and consider multiple plastic bag options such as HDPE, LDPE, PP and PLA [5]. Further, we also include the choices related to waste management options at the material recovery facility. Using the framework, we are able to obtain optimal value chain pathways under current technosphere constraints that minimize global warming potential, life cycle cost or maximize circularity metrics. Further, we plot the trade-offs as pareto surfaces and perform sensitivity analyses on various parameters such as reuse of bags and consumer littering behavior. It is observed that a combination of reusable polypropylene bags and compostable polylactic acid bags can lead to a favorable pareto optimal solution for SCE. Numerous other insights on the waste management pathways and effectiveness of reuse of bags are also derived from the framework outcomes.

References

[1] EMF - Ellen MacArthur Foundation. *Report:* What is a Circular Economy?, 2010.

[2] Yuliya Kalmykova, Madumita Sadagopan, and Leonardo Rosado. Circular Economy – From review of theories and practices to development of implementation tools. *Resources, Conservation and Recycling*, 135: 190-201, 2018.

[3] Dieuwertje L. Schrijvers, Philippe Loubet, and Guido Sonnemann. Critical review of guidelines against a systematic framework with regard to consistency on allocation procedures for recycling in LCA. *International Journal of Life Cycle Assessment*, 21(7): 994-1008, 2016.

[4] Adisa Azapagic. Life cycle assessment and its application to process selection, design and optimization. *Chemical Engineering Journal*, 73(1):1-21, 1999.

[5] Didem Civancik-Uslu, Rita Puig, Michael Hauschild, and Pere Fullana-i Palmer. Life cycle assessment of carrier bags and development of a littering indicator. *Science of the Total Environment*, 685: 621-630, 2019.

611322 Identifying Strategies for Systems Scale Transition to Circular Economy Via Physical Supply Use Tables Developed from Bottom-up Models: A Case Study on Illinois Agro Economy

Abstract:

Bio-based production provides versatile opportunity for transition to circular economy because of ample availability of agricultural and animal waste in a region, which still needs attention for proper management. However, understanding agricultural material flows in production system is crucial to achieve circular economy to identify the opportunities for feedstock substitution and auxiliary use. By substituting the waste products as feedstock inputs, the primary sectors of the economy can be more circularized with very low waste flows going to the environment. Hence, it is crucial to identify all possible sources and quantity of agriculture wastes that can be used as a feedstock which can be achieved by a comprehensive material flow accounting for the target economy.

In this work, we propose using computationally developed physical supply use tables (PSUTs) to perform an economy wide material flow analysis to identify and account agricultural wastes that can be used as substitution to existing feedstock, thus providing opportunity for redesign of material flows in economy towards higher circularity. These PSUTs are based on mechanistic engineering models that were developed using fundamental mass and energy balance equations. Using the PSUTs, we identify the potential substitution of feedstock and redesign the system with proposed substitution. Further, with our approach we study the overall impact of this transition in terms of material use efficiency, nitrogen use efficiency and energy efficiency of the system. The case study will demonstrate the potential of this approach to study other questions pertaining to sustainable redesign of material flows towards higher circularity.

603295 *Systems-Level Analysis of Global Plastic Trade: Implications for a Circular Economy*

Abstract:

Since their commercialization in the 1950s, the use of plastics has exponentially increased in popularity. In 1950, 2 million metric tons of plastic were produced globally, compared to around 350 million metric tons in 2018 (Geyer 2017, Plastics Europe 2019). Due to their versatility and customizability, plastics are ubiquitous and used in a variety of consumer and industrial products. However, only about 9% of plastic waste generated has been recycled, with the vast majority of plastic ending up discarded in the environment (Geyer 2017). Improvements in areas like recycling and identifying plastic waste prevention and utilization strategies is essential to shift the vastly linear life cycle of plastics to a circular one. Existing studies have identified top producers and consumers of plastic in the world, but the data is most often aggregated for all plastics and for regions that include multiple countries. Quantifying the origin and destination of specific plastic commodities using systems-based analysis is a key step towards developing strategies for circular economy of plastics.

We present a network model of global plastic trade with a focus on several thermoplastics that account for the majority of global plastic trade. To ensure robust data for this analysis, trade data is compared to other data sources quantifying national plastic production capacity. Import and export data for the same year and commodity are also compared for data verification. The network model is used to identify critical nodes (top exporter and importers) and connections in the network over a 22-year time frame (1994-2016). Conducting this analysis over time provides insight to the evolution of the critical actors and their partnerships in the network, as well as changes in quantity of plastic traded for particular plastic commodities. Correlations between socioeconomic data and alliance/hostility data are computed to gain further understanding to the development of trade partnerships within the network. This network could serve as a starting point for implementing circular economy by identifying target commodities, critical countries and trade relationships to improve circularity in the plastic economy. Preliminary analysis shows that a few countries account for a majority of imports or exports for specific plastic commodities, which influences the resiliency of the network. The temporal evolution of the plastics trade network and its implications for managing end-of-life plastic waste and developing circular economy of plastics will be described. By attaining an in-depth understanding of the network and the relations between its actors, the network model is further utilized to quantify the energy and greenhouse gas emissions embodied in global plastic trade.

610533 A Graph-Analytical Approach for the Development of Circular Economy Networks

Abstract:

Industrial Symbiosis (IS) has proven to be an important tool for improving business sustainability with numerous environmental and economic benefits for all members involved. The breadth of successful examples all over the world and the huge amounts of unexploited waste indicates huge potential. However, many attempts to implement innovative IS practices have failed due to lack of information, knowledge or understanding of the system's complexity. The large amount of unstructured data from different sources renders the analysis a challenging process and the different background knowledge of associated partners makes it difficult to identify and evaluate potential value chains. IS networks are dynamic and require partners who can adapt to the constantly changing environment. In order to capture the complexity of these networks and make fast, accurate and data-driven decisions, companies must implement innovative technologies leveraging the abundance of data available.

This paper proposes graph databases, as the appropriate tools for modeling IS networks and automate decision making. Graph databases have flexible schema, can handle very large sets of semi-structured data, which can describe effectively the varied IS networks and at the same time can perform efficiently complex queries. The proposed model is composed of nodes with 3 labels: Companies, Processes and Materials, which connect to each other with relationships (edges) of different labels. Each node and each edge contains different properties depending on its label. Properties regarding materials can be physical, chemical, biological etc. whereas properties regarding processes can be temperature, pressure, CAPEX, OPEX etc. Properties are powerful tools that allow the user to query, having incomplete knowledge.

With this representation, the user can model large and complex IS networks involving heterogeneous industries and a wide range of by-products exchanged. The graph can be used for designing zero waste Eco-industrial parks, or for identifying potential networks between industrial units at existing industrial parks. The graph can be enriched as more technologies are developed or more practices are applied. Connections can be either implemented or potential, depending on the development stage. Furthermore quantitative data can be used for evaluating possible synergies with mathematical optimization techniques.

The graph can be used for matching waste to process, waste to company, company to company or querying using only desired properties. We have demonstrated examples in agricultural activities, rural and municipal communities and ports, with reliable results. In conclusion, the developed model allows real time analysis about potential IS networks which include several different partners and synergies that would otherwise be extremely difficult to identify.

599092 Integrated Food Facilities: Circular Economy Applied to the Food Industry

Abstract:

Food is one of the three main pillars of the water-energy-food (WEF) nexus. The links among the three are responsible for the sustainability of society.[1] While this nexus is a current trend analyzed and implemented in the chemical and power industries,[2] food production companies are product oriented,[3] that do not typically work as chemical complexes. However, a more integrated analysis can allow energy savings improving the sustainability and the economics of the facility. Food waste is attracting attention lately due to the large volumes generated.[4] By integrating the residues it is possible not only to produce added value chemicals, but also to reduce the consumption of utilities while managing and valorizing the waste generated in their operation. By reusing the waste, the principles of circular economy are implemented within the food industry.

To illustrate the possibilities of circular economy in the food industry two cases of study are presented, orange and coffee. The similarities between both are due to the fact that added value products can be found in the waste and that for waste to be further used to chemicals and power, it is advantageous to remove them. Thus, several pretreatments are evaluated to recover limonene from orange peels (extraction and steam explosion) and tannins, caffeine and oils (extraction with different solvents) from spent coffee grounds. Waste is later digested to obtain a fertilized and biogas that can be further used to provide the thermal and electrical energy for the facility. Gas turbines or combined cycles are considered. Thus the design problem is formulated as a superstructure of alternatives with heat and power integration towards a sustainable integrated food complex.

The circular economy concept reduces the residue and the external energy consumption providing added value production and additional byproducts such as fertilizers. In the case of oranges, The use of steam explosion requires 50% lower production costs but 30% higher investment to treat the waste. The use of hexane consumes 4 times more thermal energy but the processing of the waste requires less digesters. Combined cycles are less profitable due to the additional investment. The lower production cost of the steam explosion allows for smaller facilities to become profitable before the ones using hexane limonene recovery. In the case of coffee waste, the integrated facility allows self integrating the energy required for the operation, consuming 40% of the raw material for that purpose, while producing added value product chemicals such as caffeine and pigments. In addition biogas and digestate are also produced out of the residue from the extraction of the chemicals.

The authors acknowledge the MECD for the research fellowship to AC and PSEM3 GIR. The authors acknowledge the funding received from the European Union's Horizon 2020 research and innovation programme under the Marie Sklodowska-Curie grant agreement No 778168

References

[1] García. D.; You F. The water energy food nexus and process systems engineering: A new focus. *Comp. Chem Eng*. **2016,** 91, 49-67.

[2] Bauer, D.; Philbrick, M.; Vallario, B. U.S. Department of Energy. The Water-Energy Nexus: Challenges and Opportunities. 2014

[3] Fito, P.; LeMaguer, M.; Betoret, N.; Fito, P.J. Advanced food process engineering to model real foods and processes: The "SAFES" methodology. *J Food. Eng*. **2007,** 83 (2), 173-185

[4] Schanes, K.; Doberning, K., Gözet, B. Food waste matters. A systematic review of household food waste practices and their policy implications. *J Clean Prod.* **2018,** 182, 978-991.

609637 Towards Sustainable Resource Integration Networks in Industrial Clusters

Abstract:

Industrial clusters have been pursuing designs and policies that fulfil recent sustainability goals. These policies encourage the implication of improved systems for waste reduction, thus decrease the environmental impact with minimum cost requirement. This simultaneous consideration of the two design aspects is challenging since, mostly, there exist trade-offs between economic and environmental performance. Therefore, there has been a focus on developing methods for multi-objective optimization (MOO) combined with material and energy integration approaches that tackle this issue. Two of the most prominent MOO models are the weighted average method, which converts multi-objective problems into single objective ones, as well as the e-constraint method. However, integration MOO models developed previously rely focus on specific material resources. They only allowed a certain number of resources to be integrated. This work applies holistic resource integration in industrial cluster (material and energy) under multiple objectives. The approach allows the integration of all type of resources, and uses both the augmented e-constraint (an enhanced e-constraint model) and the weighted average method to optimize an industrial cluster. The MOO models allow exploring the trade-offs between economic and environmental performance and represent them in Pareto Optimal sets. A case study of a carbon-neutral industrial cluster is explored and optimized using the MOO model, and an optimal set of solutions is obtained. This method is suitable to generate multiple scenarios to assess policymakers in decision making.

609954 Multi-Objective Optimization of an Electrochemical CO_2 Utilization Process

Abstract:

CO_2 utilization has several advantages including reduction in the level of atmospheric CO_2 through reuse and a cleaner production process for products that are conventionally produced from hydrocarbon-based feedstocks. While many technologies for CO_2 utilization are being investigated, the electrochemical reduction technology has high potential. Due to the direct usage of the electric power in this technology, this process can take advantage of cheap electricity, when available, from renewable energy sources.

For improved commercial feasibility of the electrochemical conversion technology, it is important to increase the energy efficiency, conversion, and selectivity. However there is a strong tradeoff between these objectives. Single pass conversion through an electrochemical cell can be improved if the cell terminal voltage can be increased; however that results in a lower energy efficiency (Feaster et. al). A higher single pass conversion also typically lowers the selectivity. These tradeoffs are difficult to analyze and mitigate in electrochemical cells due to the complex ionic reactions and molecular and ionic transport that take place in the cells. Concentration of the products, anolyte and catholyte chemistry and their concentration, other operating conditions, and the cell design details affect the loss mechanisms in these cells affecting the conversion and yield through complex interaction among the activation overpotential, concentration overpotential, and Ohmic resistance (Delacourt et. al., 2010). While considerable work in this area has been done on the development of the cell material and electrolyte chemistry including several experimental on various types of cell designs, it is difficult to optimize these cells experimentally due to the trade-offs noted above, large number of decision variables, and their complex interactions and lack of measurements that can capture the spatial variation of transport variables. Therefore, a model-based multi-objective optimization study is conducted for improving the economics of the electrochemical CO_2-utilization processes.

The electrochemical CO_2-utilization process that is investigated here produces formate. Studies show that reduction to formate/ formic acid has the highest current efficiencies (around %) (Whipple & Kenis, 2010). This along with the many uses of formate salts (as de-icing agents) and formic acid (as a storage medium for hydrogen gas, replacement for mineral acids in steel picking, and so on) make formate/ formic acid a very good choice as a product. One notable study on an electrochemical cell producing formic acid was conducted by Li and Oloman (Li & Oloman, 2008). They developed a model of trickle-bed reactors with a tin electro-catalyst and a cation membrane separator.

This work extends the model developed by Li and Oloman (Li & Oloman, 2008) by incorporating more detailed and generic chemistry and electrolyte flow models so that the corresponding degrees of freedom can be optimized. The electrolyte chemistry plays a critical role in the performance of these cells. Since CO_2 must be dissolved in the aqueous catholyte for participating in the electrochemical reactions, reaction rates can be enhanced by improved solubility of CO_2 in the catholyte thus reducing the concentration overpotential (Li & Oloman, 2006). However, reduction of water concentration in the catholyte can result in an increase in the Ohmic resistance. The model is developed so that it can also capture the effect of the electrolyte chemistry on the rates of undesired side reactions that produce H_2 and CO. Other than modeling the cathode and anode chambers, ionic transport through the membrane was modeled by considering osmotic drag and migration. The system of partial differential algebraic equations is discretized in space and optimized by a nonlinear programming solver. Both energy efficiency and yield of formate are maximized for desired extent of CO_2 conversion. The multi-objective optimization problem is solved by a lexicographic programming approach (Zykina, 2004). In addition to the cell dimensions, operating conditions such as the catholyte and anolyte flows and their inlet concentrations are optimized for different electrolyte chemistries.

References:

1. Feaster, J.T., Shi, C., Cave, E., Hatsukade, T., Abram, D., Kuhl, K., Hahn, C., Nørskov, J., & Jaramillo, T. (2017). Understanding Selectivity for the Electrochemical Reduction of Carbon Dioxide to Formic Acid and Carbon Monoxide on Metal Electrodes. *ACS Catalysis 2017* 7 (7), 4822-4827

2. Delacourt, C., Ridgway, P. L., & Newman, J. (2010). Mathematical Modeling of CO_2 Reduction to CO in Aqueous Electrolytes. *Journal of The Electrochemical Society, 157*(12)

3. Whipple, D. T., & Kenis, P. J. A. (2010). Prospects of CO2 Utilization via Direct Heterogeneous Electrochemical Reduction. *The Journal of Physical Chemistry Letters, 1*(24), 3451–3458.

4. Oloman, C., & Li, H. (2008). Electrochemical Processing of Carbon Dioxide. *ChemSusChem, 1*(5), 385–391. doi: 10.1002/cssc.200800015

5. Li, H., Oloman, C. Development of a continuous reactor for the electro-reduction of carbon dioxide to formate – Part 1: Process variables. *J Appl Electrochem* **36,** 1105 (2006).

6. Zykina, A.V. A Lexicographic Optimization Algorithm. *Automation and Remote Control* **65,** 363–368 (2004). https://doi.org/10.1023/B:AURC.0000019366.84601.8e

7. Georgopoulou, C., Jain, S., Agarwal, A., Rode, E., Dimopoulos, G., Sridhar, N., & Kakalis, N. (2016). On the modelling of multidisciplinary electrochemical systems with application on the electrochemical conversion of CO2 to formate/formic acid. *Computers & Chemical Engineering, 93*, 160–170. doi: 10.1016/j.compchemeng.2016.06.012

601428 Novel Catalytic Chemical Looping Process Technology for Utilization of CO_2 for Acrylonitrile Production

Abstract:

The present work deals with the novel process of converting CO_2 to CO and acrylonitrile (ACN) using mixed metal oxide oxygen carriers via chemical looping in a PID micro reactor. The first step of the chemical looping process involves the reduction of CO_2 to CO at 500-650°C using the reduced metal oxide followed by ammoxidation of propylene (C_3H_6) and ammonia (NH_3) with the oxidized carrier at 500-600°C using the reduced metal oxide to produce ACN.

X-ray fluorescence (XRF), powder x-ray diffraction (XRD) and Tunneling electron microscopy (TEM) were used to characterize the synthesized catalysts for chemical compositions and crystallinity. Surface area was characterized by N_2-physisorption using the Brunauer-Emmett-Teller theory (BET). For best performing catalysts, oxidation of the reduced catalyst by CO_2 was characterized by CO_2 pulsed chemisorption. CO_2 to CO conversion was found to be 20-30% at 400-500°C and increased to 30-50% at 500-600°C. ACN yields ranging from 6-15% were achieved with ~ 20-40% conversion of propylene and ~ 20-25% of NH_3 conversion. Selectivity for ACN was observed to be 30-40% at 500-600°C. Acrolein was observed as a by-product in all 3 cycles, confirming chemical looping of oxygen from CO_2. ACN is produced primarily in the first cycle. This may be due to carbon deposition on the catalyst surface or loss of Lewis acidic sites, preventing NH_3 binding, selective oxidation, or insertion of nitrogen in the ammoxidation mechanism. Future work will focus on increasing the ACN yield from all the 3 cycles.

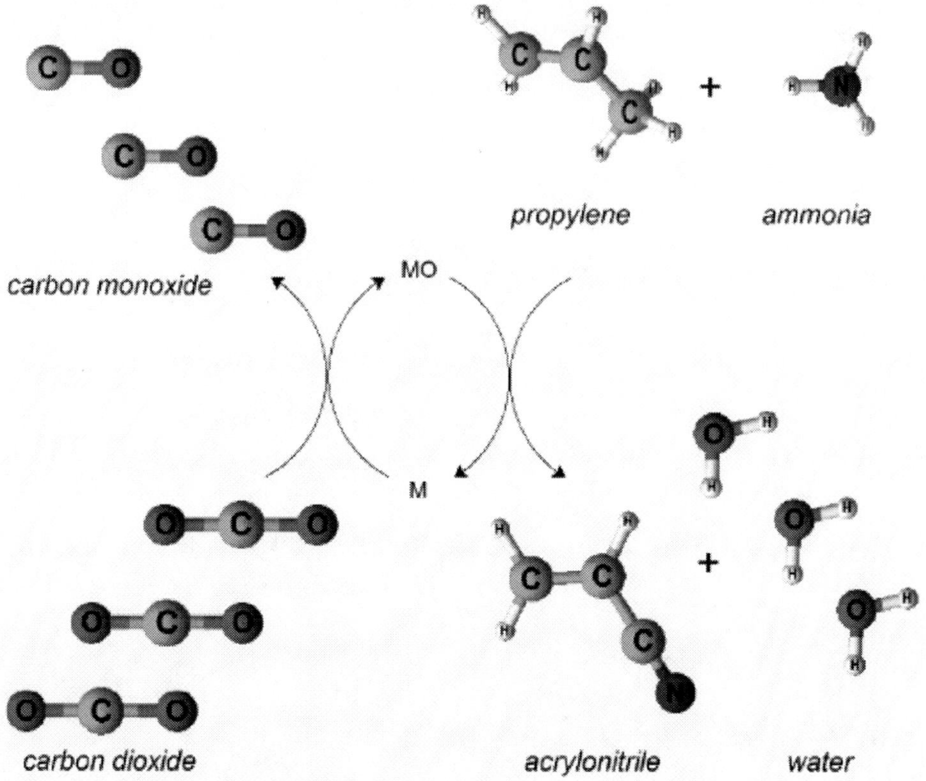

611850 Reforming of Methane with Carbon Dioxide over Cerium Oxide Promoted Nickel Nanoparticles Deposited on 4-Channel Hollow Fibers By Atomic Layer Deposition

Abstract:

Ni nanoparticles were deposited on four-channel structured α-Al_2O_3 hollow fibers by atomic layer deposition (ALD) and CeO_2 was loaded by a liquid phase incipient wetness method to promote Ni catalysts for dry reforming of methane. For Ni/Al_2O_3 prepared by ALD, inactive $NiAl_2O_4$ that originated from the Ni ALD process was incompletely reduced. The introduced CeO_2 was found to weaken the NiO-Al_2O_3 interaction, free NiO from NiO-Al_2O_3 or $NiAl_2O_4$, and improve the reducibility of NiO. The higher reducibility of $NiAl_2O_4$, tuned by CeO_2, further activated the catalyst during DRM, because a larger proportion of $NiAl_2O_4$ was gradually reduced to metallic nickel by the reaction products. The optimal catalytic performance reached a methane reforming rate of 2,410 $Lh^{-1}g_{Ni}^{-1}$ at 850 °C. The CeO_2 promoted catalyst also exhibited an excellent performance after regeneration. In addition, the inhibition effect of CeO_2 on coke formation was observed, due to the enhanced CO_2 dissociative adsorption by CeO_2.

601760 Capture and Conversion of CO_2 into Methane Using Catalytic Sorbent of $NaNO_3/MgO + Ru/Al_2O_3$

Abstract:

A concerted process combining CO_2 capture and utilization can compliment traditional CO_2 capture processes, offering unique advantages, as storage and transportation of CO_2 can be removed, and the produced renewable fuel source can be directly used again in nearby plants. Specifically in case of CO_2 capture & methanation, the energy intensive temperature swing sorption process may also be removed by performing both steps under isothermal conditions. While first generation of catalytic sorbents, such as $Ru/CaO/Al_2O_3$, has previously been shown to be effective in this task, like all first generation materials, it has some drawbacks, including modest CO_2 sorption and methane production capacities.[1] In this work, a catalytic sorbent composed of a physical mixture of $NaNO_3/MgO$ and Ru/Al_2O_3, was examined for combined capture & conversion. A material comprised of 17% $NaNO_3/MgO$ was found to be effective at capturing CO_2 under 10% CO_2/15%H_2O/N_2 flows akin to those in flue gas, showing 7.5 mmol CO_2/g sorption capacity. It was also capable of capturing and desorbing CO_2 over 8 cycles at isothermal conditions at 300 °C by switching the feed gas from CO_2 flow to inert flow.[2]

To evaluate performance of the new catalytic sorbent synthesized, both CO_2 sorption capacity and methane production capacity, equivalent to number of moles of methane produced per mass of catalytic sorbent in one cycle of capture and methanation, were measured. A catalytic sorbent of 2:1 mass ratio of 17%$NaNO_3/MgO$ and 0.5%Ru/Al_2O_3 showed a sorption capacity of 3.5 mmol CO_2/g, and methane production capacity of 1.3 mmol CH_4/g at 300 °C. Both the sorption capacity and methane production capacity are comparable to the highest values reported to date using flue gas relevant CO_2 concentrations (~10% CO_2), making such catalytic sorbents promising materials for further development of combined capture & conversion applications. Ongoing work targets assessment of reaction kinetics and elucidation of reaction pathways using these materials.

Reference:

(1) Duyar, M. S.; Treviño, M. A. A.; Farrauto, R. J. Dual Function Materials for CO_2 Capture and Conversion Using Renewable H_2. *Appl. Catal. B Environ.* **2015**, *168–169*, 370–376.

(2) Park, S.J.; Y. Kim.; Jones, C.W. $NaNO_3$ Promoted Mesoporous MgO for High Capacity CO_2 Capture from Simulated Flue Gas with Isothermal Regeneration. *ChemSusChem* **2020**.

599484 Molten Salt Facilitated Oxidative Dehydrogenation of Ethane with *in-Situ* CO_2 Capture and Utilization

Abstract:

Utilizing CO_2 for chemical production can be an economically attractive option for carbon emission reduction. Among the various CO_2 utilization approaches, oxidative dehydrogenation (ODH) with CO_2 as a soft oxidant offers the potential to produce value-added CO while upgrading light alkanes, such as ethane to ethylene. However, CO_2-assisted ethane ODH is nevertheless an energy intensive process largely due to the limited ethane and CO_2 conversions and hence a high energy cost for product separations. Currently, the highest reported CO_2 conversion in CO_2-ODH of ethane is only 50% at 800 °C. Meanwhile, ethylene yield is generally limited to <20% at lower temperature. Aside from these challenges, CO_2-ODH requires a concentrated CO_2 stream. This adds another energy-consuming step since CO_2 capture from emission sources such as fossil-based power plants are highly energy intensive. To address the challenges from both CO_2 separation and CO_2-ODH, we propose to utilize tailored molten salts as a reaction medium for simultaneous CO_2 capture and utilization. In this scheme, we demonstrate that CO_2 can be captured from a (simulated) power plant flue gas and subsequently utilized via a modified reverse water gas shift reaction with a high CO_2 conversion. Results from temperature programmed reaction (TPR) and in-situ Diffuse Reflectance Infrared Spectroscopy (DRIFTS) suggest that the coupled radical reactions and reverse water gas shift reaction in the molten salt medium was responsible for the high CO_2 to CO conversion and ethylene yield.

596762 National Carbon Capture Center Supporting Development and Scale-up of CO2 Utilization Technologies

Abstract:

Sponsored by the U.S. Department of Energy (DOE) and operated and managed by Southern Company, the National Carbon Capture Center (NCCC) is a cornerstone of U.S. innovation in advancing fossil energy technology solutions. Bridging the gap between laboratory research and large-scale demonstrations, the center works to accelerate the development of technologies to reduce greenhouse gas emissions from natural gas- and coal-based power plants. The facility evaluates next-generation carbon dioxide (CO_2) capture and CO_2 utilization processes from third-party developers, focusing on the early-stage development of the most promising, cost-effective technologies for future commercial deployment. Through pilot testing of more than 60 technologies, the center has directly participated in the reduction of the projected cost of carbon capture from fossil power generation by approximately one-third.

Since its creation in 2009, the NCCC has performed more than 115,000 hours of technology testing, with research now focused bringing on new technologies closer to the marketplace for post-combustion CO_2 capture and CO_2 utilization. Evaluation of developing technologies using actual flue gas under industrial conditions provides critical information on material and process suitability.

The center is broadening its research scope to expand testing of carbon capture technologies for natural gas power generation in addition to its existing coal-fueled testing capability. Integration of the new natural gas flue gas equipment with the existing infrastructure will allow each carbon capture project to receive either natural gas or coal flue gas independently of other projects.

Because reduction of CO_2 emissions is an international issue, the NCCC provides worldwide technology leadership – it co-founded the International Test Center Network, a global coalition focused on research, development and deployment of carbon capture, use and storage (CCUS). Formed in 2012 in collaboration with DOE's Office of Fossil Energy, the ITCN facilitates knowledge-sharing among carbon capture test facilities around the world to accelerate the commercial deployment of carbon capture technologies. The ITCN now has 15 test facilities that are members.

The NCCC has demonstrated its ability to bring on board and manage a multitude of projects to develop the most promising technologies for clean, efficient fossil-fuel energy production with ultra-low CO_2 emissions. Testing at the NCCC also supports research and development of CO_2 utilization technologies that will efficiently, economically and cleanly transform CO_2 into value-added products – partially offsetting CO_2 capture costs from power generation and providing an alternative to conventional manufacturing processes. Operations at the center have resulted in significant scale-ups, process enhancements, and technological breakthroughs. This presentation will detail the center's capabilities, role in supporting development of CO_2 utilization technologies, and future plans.

608981 Catalytic Conversion of CO_2 into Solar Fuels Via Thermochemical Cycle Driven By La-Based Perovskites

Abstract:

Fulfillment of the mounting demand of energy by using solar fuels such as solar H2 or solar syngas, produced via catalytic H2O (WS)/CO2 splitting (CS) reactions, is one of the maintainable and viable approaches for harnessing the renewable and plentifully available solar energy. This process exploits the concentrated solar power heat to drive the high-temperature thermal reduction (TR) of metal oxides (MOs). The solar H2 can be used as a gaseous fuel, whereas; the solar syngas can be converted into multiple types of liquid fuels via the Fischer Tropsch process. Effectively, by applying this solar fuel technology, solar energy can be stored as high-grade chemical energy. This stowed chemical energy is more sustainable since it can be transported and stockpiled for a long time without any degradation. Up to the present time, numerous MOs have been considered for both WS and CS reactions which include volatile i.e., zinc oxide and tin oxide and non-volatile MOs i.e., iron oxide, ferrites, ceria, and doped ceria, and hercynite. Among all the MOs investigated, ceria based oxides appear to be the superlative choice as these oxides possess most of the aforementioned attributes. As an alternative to the ceria based oxides, in recent years, perovskite-based oxides were investigated for thermochemical WS/CS reactions. Among the various perovskite oxides, La-based perovskites were examined considerably for the production of both H2 and CO. Based on the literature review, we understood that the previous studies were focused mainly on La-Sr-Mn perovskites and the redox reactivity of these perovskites were estimated by performing only one or two WS/CS cycles. In this study, the thermal reduction and re-oxidation aptitude of La-Sr-Mn-Co-based (LSMC) perovskite were determined (as an alternative for LSM). An entire spectrum of LSMC perovskites was tested by performing multiple CS cycles. The prime reason to conduct this study was to improve the understanding of the thermochemical community towards the application of LSMC perovskites for WS/CS reactions and to compare the obtained results with the state-of-the-art CeO2 and LSM catalytic material.

610389 Polar Aprotic Solvent and Alkali Metal Ion Effects on the Conversion of C6 Carbohydrates to Alpha-Hydroxy Acids Using Lewis Acid Beta Zeolites

Abstract:

The growing concerns about the limited reserves and adverse environmental impacts of fossil fuels caused increased research efforts for the development of green routes to produce chemicals from biomass. One family of such chemicals obtained from lignocellulosic raw materials includes α-hydroxy acids(AHAs), such as lactic acid(LA). Those platform molecules are promising monomers for the manufacture of biodegradable polymers.

Currently, LA is manufactured using a biochemical route. The process generates high quantities of gypsum waste and has pH and temperature control issues. The use of a catalytic pathway using Lewis acid Beta zeolites, for example Sn-Beta, may overcome those limitations.

Low reaction rates and the generation of undesirable humins limits the conversion of C6 carbohydrates to AHAs using Sn-Beta in water. The conversion decreased from 58% to 29%, after using the same catalyst for three reaction cycles and the AHAs yields also decrease after each cycle, indicating catalyst deactivation. After the regeneration, the initial values were only partially recovered (53%conversion) evidencing an irreversible modification of the zeolite framework.

The use of a polar aprotic cosolvent such as gamma-valerolactone(GVL), increased the catalytic activity as compared to the behavior in water. The use of GVL:water(9:1) also reduced the formation of furans and humins. After three reaction cycles, the conversion only decreased from 88% to 84% and the initial conversion and yields were recovered after calcination in flowing air at 150mL/min. The catalyst is more stable in GVL:water 9:1 compared to the use of only water as solvent.

We also found that the addition of K_2CO_3 to the reactive media improves AHAs production decreasing furans formation in GVL but not in water. This is probably because of the neutralization of Brønsted acids.

604112 Elucidating the Roles of Lattice and Molecular Oxygen during the Oxidative Scission of Methyl Ketones over Supported Vanadium Oxides

Abstract:

At the macro-scale, it is clear that both molecular and lattice oxygen play significant roles in facilitating the scission of ketones over $VO_X/\gamma\text{-}Al_2O_3$. At a mechanistic level, however, their contributions remain unresolved. In order to probe their individual impacts, we have employed Temperature Programmed Surface Reaction (TPSR) spectroscopy. Specifically, we examine trends in product formation during temperature-programmed reactions of chemisorbed 3- methyl-2-butanone over $\gamma\text{-}Al_2O_3$ and $VO_X/\gamma\text{-}Al_2O_3$. Under steady state conditions, this reaction generates two oxidation products, acetone and acetic acid, which are easily resolved by online mass spectrometry.

The role of molecular oxygen was investigated by conducting TPSR experiments over $\gamma\text{-}Al_2O_3$ under both anaerobic (He) and oxidizing conditions (15% O_2 in He). Our results indicate that $\gamma\text{-}Al_2O_3$ is a non- reducible material. Interestingly, despite its non-reducible lattice, oxidative ketone scission does occur over $\gamma\text{-}Al_2O_3$ under aerobic environments. This indicates direct participation of molecular O_2 in the oxidative scission mechanism.

We additionally examined the role of lattice oxygen by performing analogous TPSR experiments over $VO_x/\gamma\text{-}Al_2O_3$ under both anaerobic and aerobic conditions. Results indicate that lattice oxygen from reducible VO_x can also initiate oxidative scission.

A final contrast between oxidative scission over $\gamma\text{-}Al_2O_3$ and $VO_x/\gamma\text{-}Al_2O_3$ is that, although we observe evidence of oxidative scission of 3-methyl-2-butanone on $\gamma\text{-}Al_2O_3$ under aerobic conditions, we only observe production of the acetone fragment. Co-production of acetic acid is only observed over $VO_x/\gamma\text{-}Al_2O_3$. These results suggests that molecular oxygen initiates the scission reaction to form acyl and carboxylate surface species; the former will desorb as acetone without requiring further reduction of the lattice, whereas the latter can only desorb upon reduction of a V-O-X bond. We therefore conclude that both molecular and lattice oxygen participate directly in the oxidative scission of methyl ketones, indicating a mechanism that combines both Langmuir-Rideal and Mars-van-Krevelen pathways.

603787 Lattice Distorted Mnco Oxide Materials As Efficient Catalysts for Transfer Hydrogenation of Levulinic Acid Using Formic Acid As H-Donor

Abstract:

Aqueous-phase hydrodeoxygenation of biomass-derived feedstocks to renewable chemicals and fuels is important for sustainable development of chemical industry [1]. It is well known that, traditional hydrogenation processes using H_2 are usually conducted under harsh temperatures (>150 °C) or pressures (>3MPa), which usually aren't friendly to catalysts and cause huge emission of greenhouse gases. It cannot be ignored that the use of H_2, still derived from fossil fuels, makes biomass conversion costly and unsustainable. In contrast, catalytic transfer hydrogenation (CTH) without external H_2 display both technological and economic advantages, since the reaction is conducted under milder conditions with inexpensive H-donors (alcohols, acids and hydrazine) which result in safe reaction environment, efficient atomic economy and low-cost capital investment [2].

Levulinic acid (LA) is a promising platform molecule produced from the hydrolysis of cellulose. It exhibits great potential for the production of various downstream chemicals, which involves LA hydrogenation to γ-valerolactone (GVL) as the key step. CTH of LA using H-donors represents one of the most sustainable routes compared to conventional hydrogenation [3]. It is known that inexpensive Mn- and Co-based materials have yet to be reported for CTH of LA using formic acid (FA) as H-donor [4].

In this work, we first proposed unique MnCo oxide catalysts for conversion of LA to GVL in FA aqueous medium, which show enhanced stability and activity (80% LA conversion and 80% GVL selectivity, reused at least 5 times). The key finding is that, lattice distortion of MnCo oxides induces electronically coupled $MnCoO_3$ phase at MnO_x-CoO_x interface leading to catalytic performance enhancement [5]. Kinetic analysis confirms an induction period of LA conversion. There exists competitive reactions between LA activation and FA decomposition on catalyst surface. H species accumulated on MnCo catalyst surface are activated first and then react with LA to generate GVL without external H_2.

609176 Selective Hydrogenation of Crotonaldehyde over a Pdsn Catalyst: A First-Principles Study

Abstract:

It remains a challenge to design a catalyst for selective hydrogenation of α, β-unsaturated aldehydes. It is desirable to develop a catalyst that can selectively hydrogenate the carbonyl group and prevent full hydrogenation. Here, we report density function theory (DFT) calculations and detailed kinetic analysis, through which we investigate the hydrogenation reaction of crotonaldehyde on a PdSn surface. We find that the rate of the reaction is determined by dissociation of hydrogen molecules because of its high activation barrier. Meanwhile, the selectively of the hydrogenation products is controlled by the apparent barriers along different reaction pathways - hydrogenation of the C=C and C=O double bonds. We show that hydrogenation of the C=O bond of crotonaldehyde has a lower overall barrier, which leads to kinetically favorable production of the unsaturated alcohol. In addition, further hydrogenation of the unsaturated alcohol is halted because of the higher activation barrier as compared to its desorption from the surface. This work thus provides insights for developing an efficient catalyst for selective hydrogenation.

605049 Catalytic Conversion of Glucose to Tartaric Acid over Synergistic Aupt/TiO_2 Catalysts

Abstract:

Catalytic conversion of biomass to fuels and chemicals is important for establishing environment-friendly chemical processes. Aqueous phase oxidation of sugars and polyols is crucial for sustainable production of value-added carboxylic acids and derivatives. Tartaric acid (TA), a high value-added C_4 dicarboxylic acid, has been widely used for additives of food, dyes and pharmaceuticals. Traditionally, Chemical synthesis method of TA is only via oxidation of maleic anhydride, a two-step reaction generating significant amounts of toxic by-products. Up to date, there has been no detailed study on catalytic synthesis of TA from biomass using heterogeneous catalysts. Therefore, in this work, we proposed a family of AuPt/TiO_2 catalysts for facile TA synthesis (yield: 50%) from glucose in base-free condition at 110 °C and 2 MPa O_2.

In this work, conversion and selectivity of different metal catalysts, including Au, Pt, AuM (M: Pt, Pd, Mn, Cu, Fe) supported by TiO_2, were studied for glucose oxidation. It is found that AuPt/TiO_2 catalyst outperformed other candidates in terms of conversion and selectivity. Compared with monometallic Pt/TiO_2 and Au/TiO_2 catalysts, the bimetallic AuPt/TiO_2 catalyst showed remarkable catalytic performances for conversion of glucose (X ~ 100%) at 110 °C. Meanwhile, selectivity of TA is approximately 50%, suggesting that AuPt/TiO_2 synergistically leads to formation of C_4 products under reaction condition.

Based on the preliminary studies, we also conducted structure sensitivity analysis for oxidation of glucose and gluconic acid to TA, with the aim at understanding possible reaction mechanism on catalyst surface. Experimental studies have demonstrated that C-H and C=O bond activation tend to follow σ-σ and π-π bonding mechanism on catalyst surface. The elrectronic reconfiguration contributes to the synergism of bimetallic AuPt/TiO_2 catalysts for both C-H and C-C cleavage reactions. Meanwhile, kinetic modeling on catalytic conversion and electronic reconfiguration at Pt-Au interface will also be interpreted in details.

606081 Designing Efficient Solvent Recovery Pathways By Integrating Economics and Environmental Metrics

Abstract:

Abstract

Solvents are used widely for industrial processes. Their use and disposal have a significant impact on the environment, directly and indirectly. Most solvents end up in the waste streams of industrial processes (Righi et al., 2018; Tozzi et al., 2018). Their release into the environment has adverse effects on human health, ecosystem quality, climate change, and resources(Goedkoop and Spriensma, 2001). These effects can be quantified as emissions from the air, water, and soil associated with the processing and disposal of hazardous solvents. One of the ways to measure the environmental impacts of hazardous waste materials is through their life cycle assessment (LCA) that can characterize emissions over the entire life of a product. During a life cycle assessment, midpoint impacts characterization such as carcinogens, ecotoxicity, eutrophication, global warming, non-carcinogens, respiratory effects, and acidification are assessed (Jolliet et al., 2003). These midpoint categories are further grouped into three or four main damage assessment or endpoint categories. By assessing the life cycle of a product, various environmental decisions can be implemented to reduce emissions and impacts associated with hazardous solvent wastes. Traditional methods of waste solvent disposal, such as incineration has proved to be less favorable to the environment. It has, therefore, become imperative to develop new ways of hazardous solvent waste disposal, hence solvent recovery.

Although solvent recovery processes improve the cost and sustainability, there has not been an integrated method to simultaneously quantify the environmental impacts and economics of the process. A multi-objective optimization approach, which helps to solve problems with multiple objective functions, is one of the tools that can be implemented (Diwekar, 2013). Previous work on the evaluation of the economic feasibility of the solvent recovery options has already been accomplished (Chea et al., 2020). This work is an extension of the previous study where we incorporate the sustainability matrix into the framework. This is implemented by using a multi-objective optimization approach. The sustainability index used is the Sustainable Process Index (SPI). SPI is an ecological footprint that quantifies the arable area needed to provide goods and services. Every process requires a certain area within the ecosystem to take place. The SPI is made up of seven footprints, quantified as area, namely: (i) area needed for installations and infrastructure, (ii) area for renewable resources, (iii) area for non-renewable resources, (iv) area for fossil carbon usage, (v) areas to embed air, (vi) water, and (vii) soil emissions after the process(Krotscheck and Narodoslawsky, 1996; Narodoslawsky and Krotscheck, 1995). The larger the area needed to process a material into a desired product, the less sustainable the process is in terms of the ecological footprint. This is the basic concept on which the SPI methodology is built upon.

In many studies, we observe that the economic evaluation is at the forefront of process selection, however, when we are targeting recovery from waste streams, as in case of solvent-containing streams, along with meeting the minimum cost requirements we also need to ensure that we are not posing an additional burden on our ecosystem. Thus, we integrated the mathematical models quantifying the SPI of each recovery technology along with its economics and formulated a multi-objective optimization problem for minimizing the costs as well as the ecological footprint. Furthermore, CO_2 and other greenhouse gases (GHGs) were added to the model. This system of equations was modeled using the constraint-based optimization approach in General Algebraic

Modeling System (GAMS) software.

References

Chea, J.D., Lehr, A.L., Stengel, J.P., Savelski, M.J., Slater, C.S., Yenkie, K.M., 2020. Evaluation of Solvent Recovery Options for Economic Feasibility through a Superstructure-Based Optimization Framework. Ind. Eng. Chem. Res. 59, 5931–5944. https://doi.org/10.1021/acs.iecr.9b06725

Diwekar, U., 2013. Introduction to Applied Optimization. Springer Science & Business Media.

Goedkoop, M., Spriensma, R., 2001. 22 June 2001 Third edition 87.

Jolliet, O., Margni, M., Charles, R., Humbert, S., Payet, J., Rebitzer, G., Rosenbaum, R., 2003. IMPACT 2002+: A new life cycle impact assessment methodology. Int. J. Life Cycle Assess. 8, 324–330. https://doi.org/10.1007/BF02978505

Krotscheck, C., Narodoslawsky, M., 1996. The Sustainable Process Index a new dimension in ecological evaluation. Ecol. Eng. 6, 241–258. https://doi.org/10.1016/0925-8574(95)00060-7

Narodoslawsky, M., Krotscheck, C., 1995. The sustainable process index (SPI): evaluating processes according to environmental compatibility. J. Hazard. Mater., Selected papers presented at the Conference on Hazardous Waste Remediation 41, 383–397. https://doi.org/10.1016/0304-3894(94)00114-V

Righi, S., Baioli, F., Dal Pozzo, A., Tugnoli, A., 2018. Integrating Life Cycle Inventory and Process Design Techniques for the Early Estimate of Energy and Material Consumption Data. Energies 11, 970. https://doi.org/10.3390/en11040970

Tozzi, P.V., Wisniewski, C.M., Zalewski, N.J., Savelski, M.J., Slater, C.S., Richetti, F.A., 2018. Life cycle assessment of solvent extraction as a low-energy alternative to distillation for recovery of N-methyl-2-pyrrolidone from process waste. Green Process. Synth. 7, 277–286. https://doi.org/10.1515/gps-2017-0030

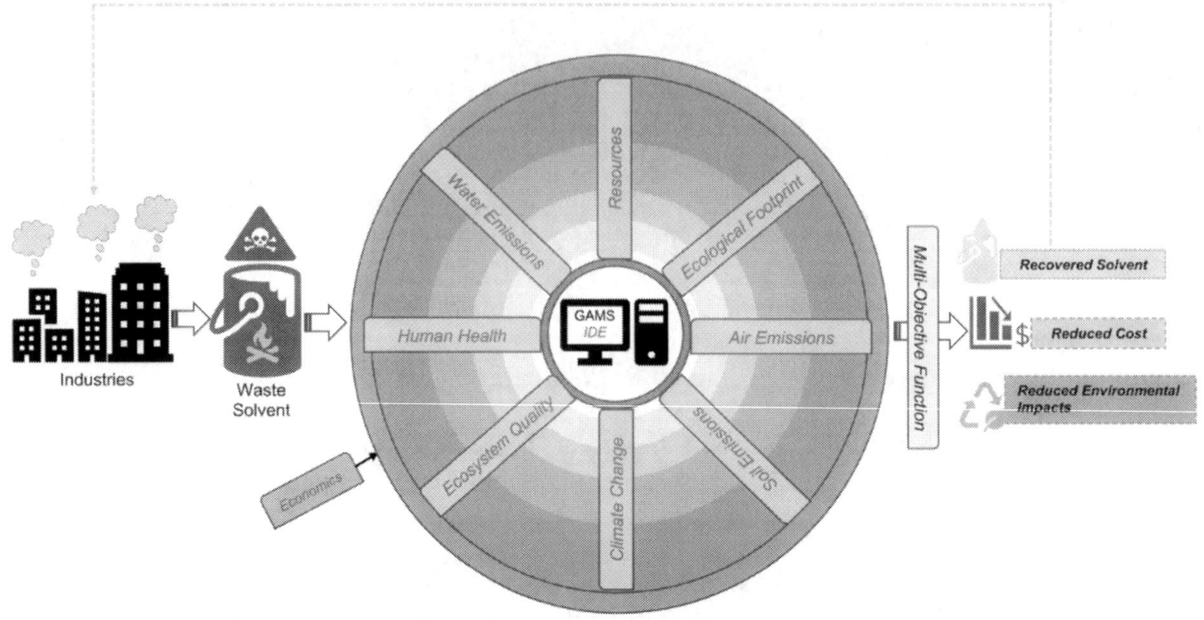

Figure 1. Solvents from waste streams need to be recovered to improve the greenness of the process as well as contribute to the pollution prevention. By incorporating sustainability metric into the economics of the recovery process, a holistic analysis can be performed, which shall aid in informed decision-making.

609380 Modeling Spatial and Temporal Emissions for Animal Farming Using Mechanistic Models

Abstract:

Animal farming is an important sector for economy of several states in USA, however, has high contribution to GHG emissions and waste production as manure. One challenge of tracking emissions from this sector is heterogeneous nature of this sector with varied practices across small scale, large scale and integrated livestock-crop farming. This poses a unique challenge for waste management and also design of waste recycling facilities related to animal farming sector. Hence, there is a need to standardize and automate approach of estimating emissions and waste generation in animal farming sector. We address this challenge using a mechanistic modeling approach for animal growth combined with scaling factors representing scale of operations for farm selected.

We demonstrated here that modeling the production technique of any live animal-based industry is often challenging as the mass transformation process from feed intake to biological body mass growth is very complex. The body mass of animals can vary based on several factors such as quantity and type of nutrient feed, water intake, age, gender, climate and living conditions, etc. To tackle this challenge, we develop a computational model in Python to model the hog farming industry and demonstrated application in Illinois, USA.

The model uses different biomass growth equations for different hog age groups. The nutrient and water intake data were obtained from the United States Department of Agriculture (USDA) databases. The environmental impacts of the hog farming in Illinois were also quantified by integrating environmental impact assessment equations. Formulas derived by The Livestock Environmental Assessment and Performance (LEAP) program of the United Nation's Food and Agriculture Organization were used to calculate the methane emissions of the farm using dynamics of animal growth. The model assumes that the hog farm uses a deep pit manure management system and the pit system collects 100% of the manure, therefore 100% of manure produced is treated by the system. The percentage of volatile solids was assumed to be 75.7% based on literature search. Further, we identify the location of animal farms modeled to show spatial and temporal variation of emissions as geospatial data.

596365 Conflict Resolution of Economic and Environmental Metrics in Dairy Waste Management

Abstract:

The dairy sector in the United States is a multi-billion enterprise that fosters diverse economic and social activities; the total dairy exports of the State of Wisconsin (WI) reached nearly 800 million US dollars in 2018 [1]. However, the dairy industry is facing several environmental challenges; dairy waste management practices generate significant greenhouse gas (GHG) emissions due to methane and nitrogen oxide released during land application of animal manure as fertilizer [2]. Moreover, nutrients contained in manure (nitrogen, phosphorus, and potassium) pollute waterbodies through runoff (e.g., cause harmful algae blooms) [3, 4]. Ammonia pollution in agricultural systems has also become an important concern; evidence shows that ammonia emissions have negative impacts on biodiversity, and it can increase suspended solids in the atmosphere [5,6].

Environmental impacts have drawn the attention of governments and researchers, and various studies have focused on exploring policy incentives to revert these trends [7]. Policy design is affected by inherent conflicts and synergies of diverse environmental impacts and metrics. In this context, decision-making methodologies are needed to uncover these conflicts and synergies. These methodologies leverage the use of life cycle analysis (LCA) to derive appropriate metrics that capture diverse impacts and can leverage the use of supply chain optimization techniques to determine suitable transportation routes for waste and derived products and processing technologies to mitigate environmental impacts [8,9].

In this presentation, we provide a computational framework that integrates LCA analysis and supply chain optimization to analyze environmental impacts arising from livestock waste in the Upper Yahara watershed in WI. Life cycle analyses are conducted for most prevalent management plans in the area, elements including waste collection, anaerobic digestion, solid-liquid separation, storage, and land application, etc. A supply chain optimization model is developed to include an economic metric (system profit) and diverse environmental metrics (GHG, ammonia, fossil energy use, and nutrient pollution). We use sampling methodology to reveal the correlation structure for the economic and environmental metrics. We find that the economic metric is conflicting with only a subset of environmental metrics and thus there is room to mitigate some impacts without sacrificing economics. A utopia-point technique is used to obtain an optimal compromise solution for the system. Finally, a series of case studies are presented to illustrate how policy incentives can be used to manipulate the utopia point and compromise solutions and with this change decision behavior of stakeholders.

References:

[1] United States Department of Agriculture, Economic Research Service, State Fact Sheet

[2] Aguirre-Villegas, Horacio A., and Rebecca A. Larson. "Evaluating greenhouse gas emissions from dairy manure management practices using survey data and lifecycle tools." Journal of cleaner production 143 (2017): 169-179.

[3] López-Díaz, Dulce Celeste, et al. "Systems-Level Analysis of Phosphorus Flows in the Dairy Supply Chain." ACS Sustainable Chemistry & Engineering 7.20 (2019): 17074-17087.

[4] Hu, Yicheng, et al. "Logistics Network Management of Livestock Waste for Spatiotemporal

Control of Nutrient Pollution in Water Bodies." ACS Sustainable Chemistry & Engineering 7.22 (2019): 18359-18374.

[5] Guthrie, Susan, et al. "The impact of ammonia emissions from agriculture on biodiversity." The Royal Society (2018).

[6] Warner, J. X., et al. "Increased atmospheric ammonia over the world's major agricultural areas detected from space." Geophysical research letters 44.6 (2017): 2875-2884.

[7] Sampat, Apoorva M., Gerardo J. Ruiz-Mercado, and Victor M. Zavala. "Economic and environmental analysis for advancing sustainable management of livestock waste: A Wisconsin Case Study." ACS sustainable chemistry & engineering 6.5 (2018): 6018-6031.

[8] Sampat, Apoorva M., et al. "Optimization formulations for multi-product supply chain networks." Computers & Chemical Engineering 104 (2017): 296-310.

[9] You, Fengqi, and Belinda Wang. "Life cycle optimization of biomass-to-liquid supply chains with distributed–centralized processing networks." Industrial & Engineering Chemistry Research 50.17 (2011): 10102-10127.

609732 Optimization of Fuel Composition for Plastic Derived Fuel Oil Produced Via Trash to Tank

Abstract:

A potential method for eliminating waste plastic from the ecosystem is conversion of post-consumer plastic into liquid fuel via pyrolysis. Trash to Tank (3T) is a concept based on the conversion of plastic types, high-density polyethylene, low-density polyethylene, polypropylene and polystyrene, into fuel suitable for use in diesel or kerosene applications. However, the application of 3T via slow pyrolysis in rural regions of developing countries yields varying fuel composition due to the absence of costly distillation techniques required to purify the fuel. Therefore, the outlet fuel composition is mainly affected by input waste plastic type, operating temperatures of the slow pyrolysis reaction, and the addition of catalysts. This contribution studies the impacts of these factors to determine the optimum operating standards for generating either diesel or kerosene fuel. The results of this study will improve 3T application in rural regions by creating specified markets for the plastic derived fuel oil, which in return will minimize waste plastic accumulation in developing countries.

611713 Use of R2 Standard in Sustainable Management of Ewaste

Abstract:

Responsible Recycling or R2 Standard is one of the few Industry Standards created specifically for making eWaste Management sustainable throughout the life cycle of electronic products, especially for post consumer disposal.

The standard initially created by US EPA has now become global standard and is adopted across the world by several organizations including Fortune 500 companies like Dell, IBM, Apple etc. to green their Supply Chain. It is based on Recycle, Reuse and Recover principles and is fully auditable standard.

This presentation provides meaningful and deep insight into:

- Lifecycle (cradle to grave) aspects of Consumer Electronics and its disposal
- Recycle, Reuse and Recovery Principles for managing eWaste
- EHS aspects related to the implementation of R2 standard
- Integration with other ISO standards like related to Quality, Environment, Health and Safety such as ISO 9001, ISO 14001, ISO 45001etc.
- Overview of Focus Materials Management
- Management of Supply chain and Downstream Vendors from cradle to grave approach
- Role of Leadership and Governance in implementing eWaste Sustainability Strategy based on R2 Standard
- Overview of Regulatory Landscape
- Auditing and assessment practices to audit R2 based Sustainable Management System
- Economic and Environmental benefits of implementing R2 Standard
- Case Study discussion
- Summary and Conclusion
- Q & A

597648 Influence of Cr(VI) and Cd(II) on Struvite Crystallization in Simulated Wastewater

Abstract:

Assimilation of residual pollutants such as phosphorus and nitrogen, as nutrients, in swine digestate effluents into a bio-resource is key to promoting sustainable use of resources in farming but surfacing data suggest that heavy metals in such wastewaters cause observable retardation in the struvite growth rate reduces viability of the crystal as a slow-release fertilizer. Struvite crystallization can simultaneously remove and recover phosphorus and ammonia nitrogen both from nutrient-rich wastewater and can be sustainable option to address unwanted nutrient enrichment in receiving bodies of water. In this work, we conducted bulk crystal growth in simulated wastewater. Struvite crystals recovered are characterized using a combination of spectroscopic, and crystallographic techniques. Results from these studies may provide insights on desirable pretreatment on heavy metal rich wastewater which can promote extraction of nutrients through struvite crystallization.

611820 Energy-Efficient Virus Removal with Cotton-Based DIY Water Filters

Abstract:

Fecal contamination of water and associated health burden is a global challenge. Acute diarrhea caused more than a million deaths worldwide in 2015[1]. Although bacteria and protozoa cause diarrhea, human enteric viruses (EVs) are the major cause of water-related diarrhea, with just rotavirus causing 40% of the hospitalizations[2]. Conventional filtration techniques, rapid and slow sand filtration, are not effective against viruses[3,4]. Although membrane filtration techniques such as ultrafiltration, nanofiltration and reverse osmosis (UF, NF, and RO) were shown to be effective against viruses, continuous monitoring is necessary as any defects can hinder the performance drastically[5]. In addition, membrane filtration incurs high operational costs in terms of energy due to their small pore size, limiting their widespread use[6]. Due to the lack of a reliable filtration technology, disinfection and boiling are the available options at household and community scale for protection against viral contamination. Of which, boiling is again energy-intensive[7], and practice of disinfection leads to formation of carcinogenic byproducts[8]. Therefore, there is a need for cost-effective, energy-efficient, and locally accessible virus filtration technologies.

In this work, we show that a simple filter made with unprocessed cotton balls functionalized with *Moringa oleifera* (MO) seed proteins can effectively capture viruses. Cationic proteins from MO seeds have been studied in relevance to indigenous water purification since 1979[9]. Using MS2 bacteriophage as a surrogate virus for EVs, we showed that the proposed filters can achieve ~7 $\log_{10}$ removal of viruses, far exceeding the environmental protection agency requirements of 4 $\log_{10}$ virus removal[10]. In addition, we compared the proposed MO-Cotton filters to commercially available UF and NF membranes to show that MO-Cotton filters can achieve effective virus removal with significantly low-energy requirement. Due to the cost-effective and chemical-free design we believe that MO-Cotton filters can be a potential solution for viral contamination of water worldwide.

References:

(1) Troeger, C.; Forouzanfar, M.; Rao, P. C.; Khalil, I.; Brown, A.; Reiner Jr, R. C.; Fullman, N.; Thompson, R. L.; Abajobir, A.; Ahmed, M. Estimates of global, regional, and national morbidity, mortality, and aetiologies of diarrhoeal diseases: a systematic analysis for the Global Burden of Disease Study 2015. *The Lancet Infectious Diseases* **2017**, *17*, 909-948.

(2) Organization, W. H. Global networks for surveillance of rotavirus gastroenteritis, 2001-2008. *Weekly Epidemiological Record= Relevé épidémiologique hebdomadaire* **2008**, *83*, 421-425.

(3) Yahya, M.; Cluff, C.; Gerba, C. Virus removal by slow sand filtration and nanofiltration. *Water Science and Technology* **1993**, *27*, 445-448.

(4) Shirasaki, N.; Matsushita, T.; Matsui, Y.; Yamashita, R. Evaluation of the suitability of a plant virus, pepper mild mottle virus, as a surrogate of human enteric viruses for assessment of the efficacy of coagulation–rapid sand filtration to remove those viruses. *Water research* **2018**, *129*, 460-469.

(5) Kumar, M.; Adham, S.; DeCarolis, J. Reverse osmosis integrity monitoring. *Desalination* **2007**, *214*, 138-149.

(6) Alzahrani, S.; Mohammad, A. W. Challenges and trends in membrane technology implementation for produced water treatment: A review. *Journal of Water Process Engineering* **2014**, *4*, 107-133.

(7) Gilman, R. H.; Skillicorn, P. Boiling of drinking-water: can a fuel-scarce community afford it? *Bulletin of the World Health Organization* **1985**, *63*, 157.

(8) Richardson, S. D.; Plewa, M. J.; Wagner, E. D.; Schoeny, R.; DeMarini, D. M. Occurrence, genotoxicity, and carcinogenicity of regulated and emerging disinfection by-products in drinking water: a review and roadmap for research. *Mutation Research/Reviews in Mutation Research* **2007**, *636*, 178-242.

(9) Jahn, S. A.; Dirar, H. Studies on natural water coagulants in the Sudan, with special reference to Moringa oleifera seeds. *Water Sa* **1979**, *5*, 90-106.

(10) USEPA: Long Term 2 Enhanced Surface Water Treatment Rule Toolbox Guidance Manual. USEPA, Office of Water Washington, 2010.

611911 Leveraging Reversible Chemistry for Green and Efficient Battery Recycling

Abstract:

Sustainable manufacturing of lithium-ion batteries (LIBs) is critical for future energy storage. Layered oxides (e.g., $LiNi_xCo_yMn_zO_2$, $0<x,y,z<1$, $x+y+z=1$) or NCM is becoming the dominating cathode material in high energy LIBs, which have degradation issues after limited cycling life due to the formation of defects with Li loss and phase changes. Directly resolving these defects to generate new cathodes can not only reduce the high cost but also prevent environmental pollution from disposal of used LIBs. However, currently there is no effective approach to tackle this challenge. We report an effective strategy to resolve the defects of degraded cathode particles to directly regenerate NCM which can achieve very high electrochemical performance. Specifically, we developed a non-destructive approach to successfully renovate degraded $LiNi_{1/3}Co_{1/3}Mn_{1/3}O_2$ (NCM111) and $LiNi_{0.5}Co_{0.2}Mn_{0.3}O_2$ (NCM523) particles. The regenerated NCM particles well maintained their original particle morphology and size distribution. Our method can successfully resolve both the compositional and structural defects of NCM particles. Due to the reconstruction of ideal stoichiometry, low cation mixing and high phase purity, NCM particles regenerated by our approach show high specific capacity, long cycling stability and high rate capability, reaching high performance of pristine cathode materials. This work demonstrates a novel energy-efficient approach to resolve the defects of degraded cathode particles to remanufacture high-performance cathodes, with distinct advantages over traditional hydrometallurgical methods. In addition, our approach can be also extended to resolve the defects of other types of cathode materials after long-time cycling. Therefore, this study builds an important foundation for sustainable manufacturing of energy materials.

604215 Development of a Closed-Loop Process for Separation and Recovery of Li and Co from $LiCoO_2$

Abstract:

Lithium-ion batteries (LiBs) power most of the modern world. The use of LiBs has expanded from portable electronics to electric vehicles, homes, and even to the electrical grid. However, the recycling of LiBs presents new challenges to the battery industry. In this work, we introduce a novel closed-loop process to extract Li and Co metals from lithium cobalt oxide ($LiCoO_2$) using oxalic acid. During the digestion process, Co precipitates as cobalt oxalate dihydrate ($CoC_2O_4 \cdot 2H_2O$), and Li remains in the solution as lithium oxalate ($Li_2C_2O_4$) which provides an efficient separation. The cobalt oxalate is converted back to cobalt oxide by calcination, whereas the Li can be precipitated in the form of carbonate or phosphate from the aqueous oxalate phase. The remaining oxalate in the filtrate after Li precipitation can be recycled using an acid cation-exchange resin process. The used resins can also be recycled, providing a sustainable process for regenerating the oxalic acid. The ability to recover oxalate provides both an economic and environmental advantage compared with other mineral acid (H_2SO_4, HCl, etc.) processes. This presentation will describe the experimental and mechanistic details of the extraction and separation of Li and Co metals and the recovery of oxalic acid. In addition, a techno-economic analysis will be presented comparing our closed-loop oxalic acid process with other conventional hydrometallurgical processes.

605489 Sparged but Not Stirred: Enzyme-Catalyzed Deracemization of Alcohols in a Bubble Column

Abstract:

Enhanced selectivity and lower environmental footprint often are key benefits of enzymatically-catalyzed reactions over chemical synthesis routes. Alcohol dehydrogenases (ADHs) and analogous enzymes are highly selective enzymes for oxidizing polyol substrates (carbohydrates) and for deracemizing or stereoinverting racemic alcohols to enantiomerically pure alcohols, important life science synthons. The required oxidants, nicotinamide cofactors NAD(P)+, are advantageously regenerated to NAD(P)H via NADH oxidase-catalyzed reduction of molecular oxygen to water. We employed a bubble column sparging air into the enzyme-laden solution to transfer the co-substrate oxygen into the reaction medium. We already found that air-sparged but not stirred solutions are beneficial for sustainable reaction rates combined with low enzyme deactivation.[1] We now investigate oxygen transfer, overall reaction rate, and ADH/NADH oxidase deactivation in the deracemization of (R,S)-1-phenylethanol.[2] Four sparged bubble column conditions were tested, two spherical bubble diameters (2.6 and 2.1 mm) and two sparging rates (18 and 5 bubbles/s), and compared against quiescent and stirred solution. Calculated specific mass transfer coefficients kLa ranged from 17 to 87 h^{-1}. Observed enzyme deactivation rates descended from stirred and faster bubble flowrates via slower flow rates to quiescent solution. For each sparged condition, 90% conversion or greater was achieved, with optimal 99% conversion in 6 hours with sparging large air bubbles at slow flowrates. Higher bubble flowrates, i.e. more oxygen transfer per unit time, were found to increase initial enzymatic reaction rates but result in lower average rates and ultimately incomplete maximum conversion. Reaction rates under sparging were 4-8 times faster than in quiescent solution. Lastly, stirring did not significantly aid conversion (41%, 37% quiescent solution after 12h) but caused increased enzyme deactivation (44h half-life, 136h half-life quiescent solution).

References:

[1] M Dias-Gomes[1], B Bommarius[1], SR Anderson, et al., *Adv. Synth. Cat.*, **2019**, *361*, 2574-2581

[2] BR Bommarius[1], SR Anderson[1], JM Woodley, AS Bommarius, submitted

599619 Self-Assembled Metal/Polymer Hybrid Nanoparticles and Their Use As Nanoreactors Towards Cascade Reactions

Abstract:

Confining reactions to hydrophobic microenvironments through the use of self-assembled amphiphilic molecules allows traditionally organic phase reactions to be carried out in a bulk aqueous phase. Replacing the organic solvent with these "nanoreactors" may reduce solvent waste. Flash NanoPrecipitation is a rapid, scalable, and modular method capable of continuously producing such structures. In this work, we use Flash NanoPrecipitation (FNP) as a platform technology to rapidly self-assemble kinetically trapped core-shell polymer nanoreactors that incorporate catalytic gold nanoparticles within tunable hydrophobic microenvironments and investigate their activity towards the reduction of 4-nitrophenol. Localized reagent solubility within the nanoreactors is investigated as the basis for composition-dependent activity differences with corresponding application to the Langmuir-Hinshelwood kinetic model. The partitioning of 4-nitrophenol between nanoreactor core compounds and water was found to vary by as much as two orders of magnitude, and when accounted for in Langmuir-Hinshelwood modeling resulted in individual nanoreactor kinetic activity parameters that fell within two-standard deviations of each other. This suggests that activity differences are the result of reagent solubility and not catalyst activity. Comparison between experimentally observed trends and Hansen Solubility Parameters will be discussed. Application of various nanoreactor systems to a reductive-amination cascade reaction will be presented. Preliminary results show approximately 50% *in-situ* yield of the desired product (4-benzylideneaminophenol). Green chemistry metrics, such as the E-factor, will be analyzed and discussed.

597418 Flow Chemistry-Enabled Investigations of Switchable Hydrophilicity Solvents

Abstract:

The dramatic environmental impacts on society have created the demands for fast development of energy-efficient and environmental-friendly processes towards green and sustainable industrial processes. Currently, most manufacturing processes in chemical and pharmaceutical industries are highly dependent on energy-intensive distillation processes with harmful volatile solvents due to their ease of separation. Recently, switchable solvents have emerged as a promising candidate for liquid-liquid extraction (LLE) processes with a facile reversible tunability of their physiochemical properties triggered by carbon dioxide (CO_2) under a mild condition (*i.e.*, room temperature and atmospheric pressure). In the presence of water and CO_2, switchable solvents react with the formed carbonic acid (H_2CO_3), producing protonated SHS and bicarbonate salts. This reaction can be easily reversed with an inert gas (*e.g.*, nitrogen) or moderate heating (*e.g.*, ~60 °C), thereby providing an ideal characteristic for the next-generation sustainable LLE-based solvent recovery.[1] In particular, Switchable Hydrophilicity Solvents (SHSs) offer a dramatic solubility change in hydrophilic/hydrophobic media upon their protonation/de-protonation, opening a new frontier for energy-efficient chemical processes in various fields from the extraction of bitumen and bio-algal mass and homogeneous catalysis to the recycling of packaging materials and plastics.

Despite the groundbreaking advancements in the discovery and characterization of SHSs over the last decade, the time-, material-, and labor-intensive nature of conventional batch (flask-based) reactors have hindered further exploration of the broad parameter space of SHSs. Conventional batch reactors have significant mass transfer limitations for conducting gas-liquid reactions due to their poorly defined interfacial area (*e.g.*, bubble column provides 50-600 m^2/m^3). Thus, screening of an SHS-based LLE process using a batch reactor might take up to10 h/experiment, further resulting in a limited available chemical database for the design of next-generation SHSs. Furthermore, batch screening methods require 10-1000 mL of chemical consumption for each SHS characterization experiment, which leads to significant waste generation and high cost for parameter space exploration of SHSs. Thus, lack of comprehensive understanding of thermodynamic characteristics of CO_2-triggered hydrophilicity switching of SHSs, mainly due to the above-mentioned drawbacks of conventional batch processes, has negatively impacted the industrial adoption of this exciting class of green solvents.

In this work, we developed and utilized an automated microscale flow chemistry platform utilizing a membrane-based flow reactor for accelerated studies of both continuous (*e.g.*, CO_2 pressure, reaction time, SHS concentration, and flow velocity) and discrete (*e.g.*, the chemical structure of SHS) process parameters associated with CO_2-mediated SHS recovery process. Utilizing a highly CO_2-permeable membrane in a tube-in-tube configuration, offering a significantly higher interfacial area of 4900 m^2/m^3 than conventional batch reactors, we systematically studied CO_2-mediated SHS extraction efficiency and kinetics, *in situ*, as fast as 4 min/experiment (~100 times faster than batch reactor), while utilizing only 4-8 μL of SHS mixture (~2500 times less chemical consumption than batch reactor) for each experimental condition. The single-droplet flow reactor provided a time- and material-efficient approach for comprehensive fundamental and applied studies of the extraction efficiency and kinetics of SHS-based LLE processes, providing accelerated access to the optimized process conditions. Next, the tube-in-tube flow reactor was reconfigured from the single-droplet screening mode to the continuous flow configuration to continuously recover the SHS under the optimized conditions identified by the single-droplet process screening flow reactor.[2] Since the single-droplet and continuous flow reactor configurations share the same intensified mass transfer characteristics, the optimized conditions identified by the screening flow reactor are directly transferrable to the continuous flow reactor. The proposed research will provide a promising solvent utilization technique with valuable insight and robust guidelines for green and sustainable chemical processing.

Reference.

(1) Jessop, P. G.; Mercer, S. M.; Heldebrant, D. J. CO2-Triggered Switchable Solvents, Surfactants, and Other Materials. *Energy Environ. Sci.* **2012**, *5* (6), 7240–7253. https://doi.org/10.1039/C2EE02912J.

(2) Han, S.; Raghuvanshi, K.; Abolhasani, M. Accelerated Material-Efficient Investigation of Switchable Hydrophilicity Solvents for Energy-Efficient Solvent Recovery. *ACS Sustainable Chem. Eng.* **2020**, *8* (8), 3347–3356. https://doi.org/10.1021/acssuschemeng.9b07304.

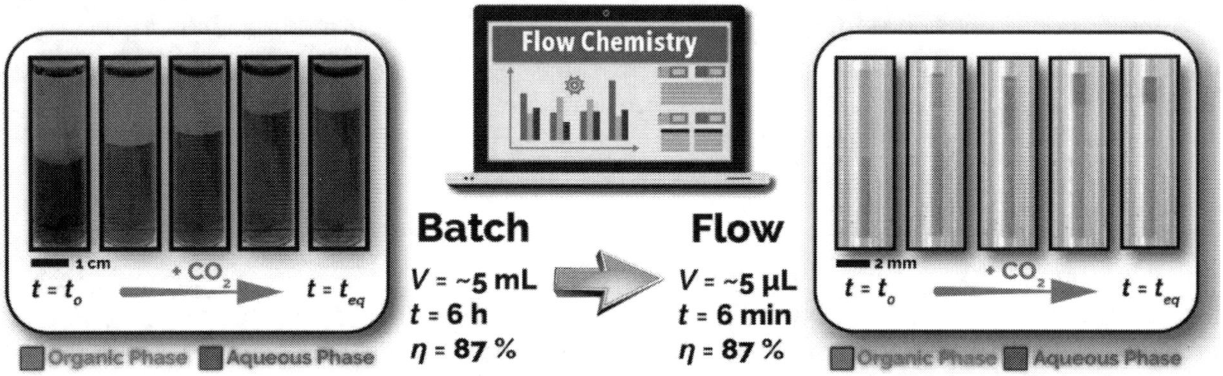

609178 Green Solvents for the Hydroformylation of Long Chain Alkenes

Abstract:

A major trend in the chemical industry is to develop more sustainable products and processes. This is a two-fold process: on one side, the sustainability of existing processes is being improved and on the other, new processes utilizing renewable feedstocks are being developed. For the latter, homogeneous metal transition catalysis (HTMC) is a promising avenue for converting renewable sources into valuable intermediates and end-products. Homogeneous catalysis has many favorable properties including lower reaction temperatures and pressures, which is more suitable for use with renewable feedstocks that require milder reaction conditions than do conventional sources. One disadvantage of using HTMC is the challenging separation and recycle of the dissolved catalyst after the reaction. The catalyst typically consists of a transition metal and an organic ligand, which are frequently highly valuable components, and must be recycled for a sustainable and economic feasible process [1]. As is often the case, large amounts of solvents are used in HTMC for both facilitating the reaction and performing the post-reaction catalyst recovery. In the frame of sustainability, these solvents should also be environmentally benign, preferably having acceptable environmental, health and safety (EHS) criteria.

The reaction and separation can be performed using a novel type of solvent system, a thermomorphic multicomponent system (TMS). In this solvent mixture, the reaction is conducted at a higher reaction temperature in a homogeneous liquid phase, while the separation of catalyst and product takes place using liquid extraction at a lower, separation temperature. This temperature-dependent phase behavior is achieved by using two solvents with a miscibility gap whose critical point lies between the reaction and the separation temperature [2]. The TMS concept addresses the limitation of the widely used Ruhrchemie/Rhône-Poulenc Process for the hydroformylation of short-chain alkenes [1]. Here, the reactant alkenes are dissolved in an aqueous solution and the product aldehydes form a second liquid phase that can be siphoned off. The water-soluble catalyst is almost insoluble in the aldehyde phase, minimizing the catalyst loss. The concept, which is very efficient for short alkenes, cannot be applied for long chain alkenes since the solubility of the reactant in the aqueous catalyst phase is poor, resulting in an inefficient, diffusion-limited reaction. In a TMS, however, the reaction can take place in single liquid phase regime without mass transfer limitations.

The TMS principle has already been successfully applied to the hydroformylation of 1-decene or 1-dodecene using a solvent system comprised of dimethylformamide (DMF) and decane [3, 4]. This system, which is highly efficient from a thermodynamic point of view, suffers from the use of the developmentally toxic solvent DMF. Because of its EHS characteristics, DMF is included on the list of substances of very high concern according to the REACH legislation (Regulation, Evaluation and Authorization of Chemicals) [5]. In order to find a replacement solvent for DMF, a computer-aided TMS selection strategy has recently been developed [6]. Promising TMS components are selected by predicting thermodynamic properties using COSMO-RS and EHS properties using the QSAR software VEGA [7, 8]. Some characteristics considered in this selection procedure are the reactant, product and gas solubilities in the solvent candidates, the boiling points of the solvents, the LLE behavior and 14 different EHS properties (predicted by 33 QSAR models) such as toxicity, persistence, flash point and others.

In this contribution, the screening methodology is briefly introduced, and focus is placed on the experimental validation of the theoretical results. The evaluation of the green solvent candidates is divided into the following steps. Firstly, the boiling point should be checked at ambient conditions. Although, accurate boiling point predictions are available, checking the boiling point first is reasonable because the experiment dependents on one substance and has a low expenditure. Additionally, this step also allows one to evaluate the thermal stability of the solvent, a key element for the process performance. Secondly, the temperature dependency of the liquid-liquid equilibrium (LLE) of the binary solvent mixture is investigated by checking whether the critical point lies within the desired temperature range. This can be done quickly by cloud point experiments and should be the next step in the experimental investigation because of its simplicity and the large reduction of uncertainty in actual LLE behavior, since theoretical predictions thereof are still prone to errors. However, in order to ensure TMS functionality, the liquid phase behavior is also evaluated using a mixture with both the reactant and the product. Therefore, the phase behavior of a mixture of the reactant and the solvents is checked at the reaction temperature, and an analogous procedure is conducted for a mixture comprised of the product and the solvents at the separation temperature. These compositions represent the beginning of the reaction in a batch process, or respectively, for the subsequent separation under the assumption of full conversion. Afterwards, the LLE of the product and the solvents under separation is measured in detail under separation conditions. With these measurements one can then check the feasibility of the product separation. Next, the performance of the reaction within the new solvents is evaluated in terms of solvent stability and adequate reaction rates. Semi-batch experiments are conducted through which samples are taken out of the reactor to evaluate the reaction progress along the reaction time and to check the general functionality. Finally, and perhaps most importantly, the catalyst partitioning between the two liquid phases needs to be investigated, which is the key for an efficient catalyst recycling. This is chosen as the last step because for these measurements an ICP-MS is used and performing measurements in organic solvents using an ICP-MS is cumbersome. However, the use of an ICP-MS is necessary because the leaching of the catalyst into the product phase is extremely low (parts per billion down to parts per trillion), which is not possible with most analytical devices. Here, the best TMS are those with the lowest amounts of catalyst leaching.

Altogether, this experimental approach allows for the comparison of different solvents regarding their applicability as a TMS used in a hydroformylation process and shows the potential for green solvents to replace DMF. The collected data also makes it possible to perform a more rigorous process optimization in subsequent investigations, where decisions about solvent composition can be included. In this manner, a more ecologically beneficial process may be realized, facilitating the development of more sustainable processes for the chemical industry.

Acknowledgment

This work is funded by the Deutsche Forschungsgemeinschaft (DFG, German Research Foundation) - TRR 63 „Integrated Chemical Processes in Liquid Multiphase Systems“ (Gefördert durch die Deutsche Forschungsgemeinschaft (DFG) - TRR 63 "Integrierte chemische Prozesse in flüssigen Mehrphasensystemen" (Teilprojekt B9) - 56091768.). Steffen Linke is also affiliated with the “International Max Planck Research School for Advanced Methods in Process and Systems Engineering - IMPRS ProEng” at the Max Planck Institute for Dynamics of Complex Technical Systems Magdeburg.

References

[1] Arno Behr and Peter Neubert. Applied homogeneous catalysis. Wiley-VCH-Verl., Weinheim, 2012.

[2] Arno Behr, Guido Henze, Dietmar Obst, and Barbara Turkowski. Selection process of new solvents in temperature-dependent multi-component solvent systems and its application in isomerising hydroformylation. Green Chemistry, 7(9):645, 2005.

[3] Kevin McBride, Nicolas Maximilian Kaiser, and Kai Sundmacher. Integrated reaction-extraction process for the hydroformylation of long-chain alkenes with a homogeneous catalyst. Computers & Chemical Engineering, 105:212-223, 2017.

[4] Benjamin Hentschel, Gregor Kiedorf, Martin Gerlach, Christof Hamel, Andreas Seidel-Morgenstern, Hannsjörg Freund, and Kai Sundmacher. Model-Based Identification and Experimental Validation of the Optimal Reaction Route for the Hydroformylation of 1-Dodecene. Industrial & Engineering Chemistry Research, 54(6):1755-1765, 2015.

[5] European Parliament. Registration, Evaluation, Authorisation and Restriction of Chemicals: REACH. https://echa.europa.eu/documents/10162/3f575bea-5a96-4f18- a110-d294d37422c7, 2006.

[6] Steffen Linke, Kevin McBride, and Kai Sundmacher. Systematic green solvent selection for the hydroformylation of long chained alkenes. ACS Sustainable Chemistry & Engineering, UNDER REVIEW, 2020.

[7] Andreas Klamt. Conductor-like Screening Model for Real Solvents: A New Approach to the Quantitative Calculation of Solvation Phenomena. The Journal of Physical Chemistry, 99(7):2224-2235, 1995.

[8] Emilio Benfenati, Alberto Manganaro, and Giuseppina Gini. VEGA-QSAR: AI inside a platform for predictive toxicology. Italy Published on CEUR Workshop Proceedings, (Vol-1107), 2013.

High-temperature solar thermochemistry CO_2 reduction over $NiFe_2O_4/ZrO_2@ZrO_2$ oxygen exchange material

Y. Shuai*[1], B. Guene Lougou*[1,2], H. Zhang[1], B. Jiang[1], A. Mustafa[1], J. Zhao[2], H. Tan[1]

[1] *School of Energy Science and Engineering, Harbin Institute of Technology, Harbin 150001, China*
[2] *School of Chemistry and Chemical Engineering, Harbin Institute of Technology, Harbin 150001, China*

* Corresponding authors E-mail address: (Y. Shuai) shuaiyong@hit.edu.cn (B. Guene Lougou) 15bf02043@hit.edu.cn

Artificial photosynthesis inspired by nature's design concept focusing on harnessing concentrated solar radiant energy flow can be processed for efficient CO_2 reduction into high-grade fuels. Solar thermochemical cycles were firstly investigated by Aldo Steinfeld (1993) [1] to produce renewable fuels and carbon-based energy carrier with CO_2 mitigation. It is an engineering technology that makes full use of the energy of sunlight energy into storable clean transportable fuels and chemicals. Regarding the limitations and challenges associated with the present solar thermochemical technology, high-temperature solar thermochemistry CO_2 reduction is investigated from the perspective of finding advanced oxygen exchange materials for high-efficiency solar fuel production. A composite nanomaterial $NiFe_2O_4/ZrO_2@ZrO_2$ was synthesized by combining impregnation and nanocoating methods. The solar incident radiation flux distribution and thermal-chemical changes inside the reactor have been investigated. The stability of the material's structural and strong synergetic effects of active phase activity could result in higher oxidizer conversion and syngas selectivity. Theoretical description through thermodynamic cycles is carried out for the investigation of the reaction kinetics and redox performance analysis of the solar CO_2 conversion under a high-temperature environment. Regarding new energy industry development, solar thermal chemical reduction of CO_2 has important research significance and broad application prospects. The results provided could contribute to CO_2 reduction technologies and the green environmental protection concept advocated by the world today.

Keywords: High-temperature solar thermochemical; Solar fuels and chemicals; CO_2 reduction; Material structural stability; Redox performance.

INTRODUCTION

While geological CO_2 storage is the most promising way for the sequestration of CO_2 captured, as shown in Fig.1, CO_2 can also undergo a chemical transformation for the synthesis of liquid hydrocarbon fuels and chemicals which are very in line with the green environmental protection concept advocated by the world today. Sunlight can shine on chemical bonds and break it into a gas mixture of H_2 and CO to realize energy-rich fuels. In 2016, the effectiveness of CO_2 re-use is demonstrated by the SUN-to-LIQUID project tested at Madrid's IMDEA[2] which is the first solar thermal chemical kerosene produced from H_2O and CO_2, and able to produce 700 standard litters of high-quality syngas.

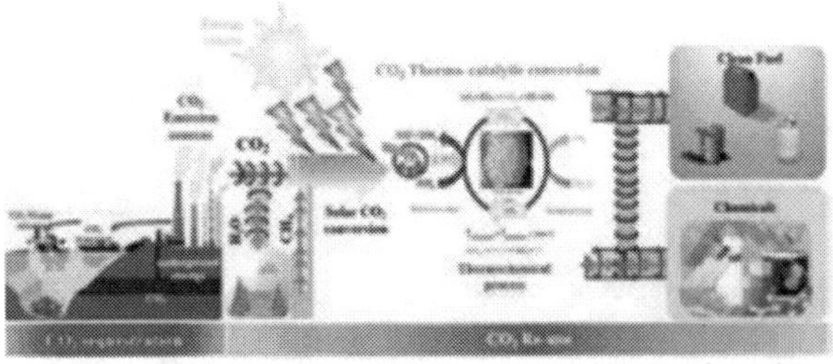

Fig.1. High-temperature Thermochemical Technology for Solar CO_2 Re-use

ANALYSIS

Fig.2-3 describes the laboratory scale solar thermochemical system for energy in sunlight conversion into fuel and chemicals. The collected solar energy is directly focused on the reactor, heating it to high temperatures via complex coupling heat transfer. In this study, the feed gas consists of N_2 gas and CO_2. N_2 gas with a volume flow rate of 120 mL/min is used as a carrier gas and CO_2 with a volume flow rate of 50 mL/min is used as reactive gas flow. The reactor was heated up to 1600 K of maximum reaction temperature. A temperature range of 1000-1600 K is considered for the analysis of two-step solar thermochemical CO_2 reduction. The product gases are directly analyzed by the gas analyzer coupled to the reactor.

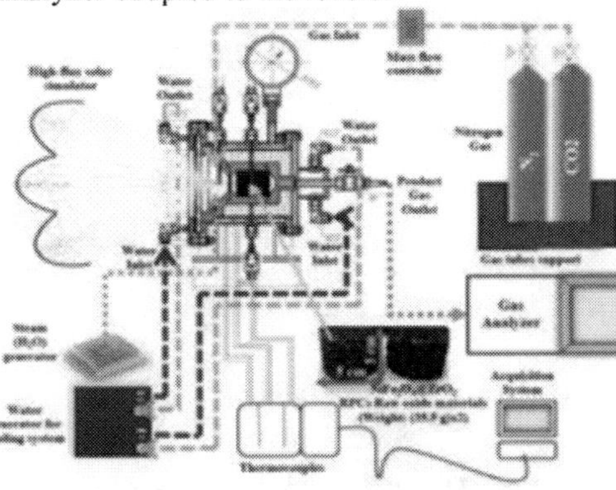

Fig.2. Experimental Setup Describing High-temperature Solar Thermochemical System

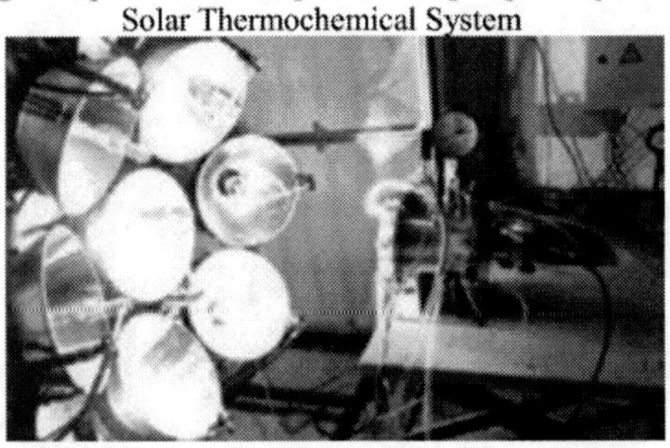

Fig.3. Laboratory-scale solar thermochemical system for energy in sunlight conversion into fuels and chemicals

RESULTS AND DISCUSSION

1. High radiative fluxes transport and thermal-chemical conversion characteristics

The thermal characteristics of the cavity receiver indicate the porous material behaves as a radiant absorber in the cavity receiver as shown in Fig.4a-b. The present reactor efficiency of storing sunlight with safer operating conditions is reported to 85.27% ± 0.85% during thermal charging up to 1787.725 K ± 30.58 K and 76.9% ± 0.11% during thermal discharging step at 1315. 16 K ± 7.53 K[3]. A total of 337.89 mL of CO yield during 46 min of CO_2-splitting over $NiFe_2O_4@ZrO_2$ oxygen exchange material is collected from the gas analyzer under the conditions depicted in Fig.4c. The material thermal stability, thermal resistance, and greater durability are the most important factors regarding high-temperature thermal-chemical conversion enhancement. Besides, the application of the multilayer porous ceramics model has resulted in the improvement of 0.03%-2.43% in solar-to-fuel by using the optimized 1-layer with 4-layers porous ceramic as shown in Fig.4d [4].

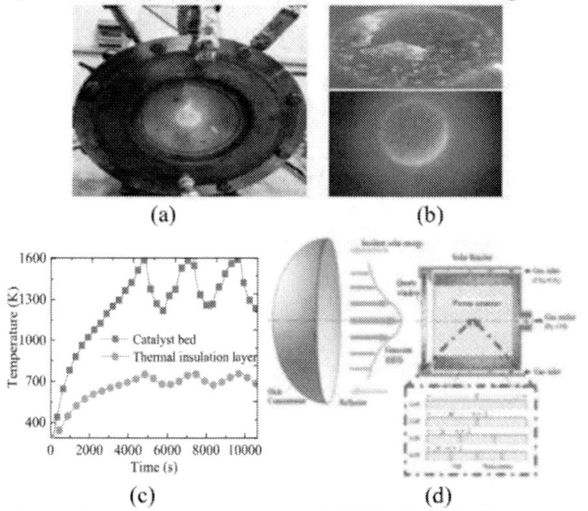

Fig.4. Reactor thermal transport in $NiFe_2O_4@ZrO_2$ oxygen exchange material: (a-b) Heat transfer and thermal characteristics; (c) Cavity receiver thermal characteristic under concentrated sunlight; (d) Solar reaction cavity model of multilayer porous ceramics.

2. Redox thermodynamic analysis of CO_2-splitting

Redox thermodynamics analysis in terms of O_2 exchanges capacity of $NiFe_2O_4@ZrO_2$ and metal encaged Zeolite dopant has been conducted via TGA at 200-1293 °C, 100 sccm of N_2 gas at reduction step, 100 sccm of CO_2 at the oxidation step. SEM and XRD characterization of the materials investigated are depicted in Fig.5. The amount of CO yields up to 205.7 µmol/g with a CO/O_2 ratio of 1.84 can be obtained with the mixing of $40\%NiFe_2O_4/60\%ZrO_2$. Encaged Zeolite materials cannot support the high-temperature environment. However, Co@Zeolite catalyst exhibited remarkable CO_2-splitting activity at very low temperatures under 500 °C which necessity further investigation to confirm their applicability in the thermochemical CO_2 reduction cycles.

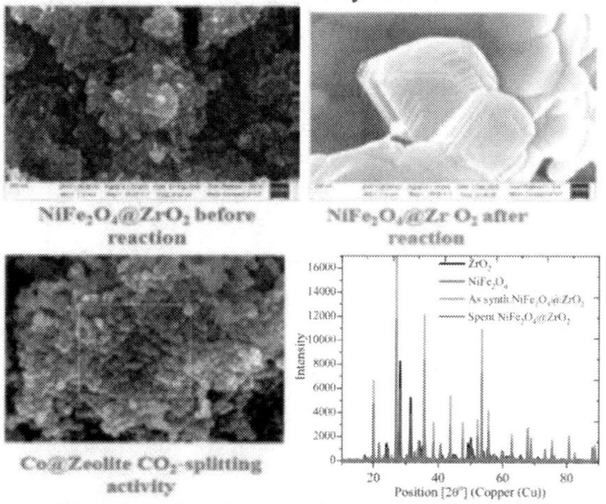

Fig.5. Scanning electron microscopy (SEM) and XRD characterization of the material investigated.

CONCLUSIONS

The reactor typically features solid-gas heterogeneous media. The challenge is enhanced high-temperature heat transfer in such media. Materials-by-design approaches following by fundamental heat and mass transfer models can be explored for understanding and optimizing the complex thermal transport processes and catalytic chemical conversion in a high-temperature environment.

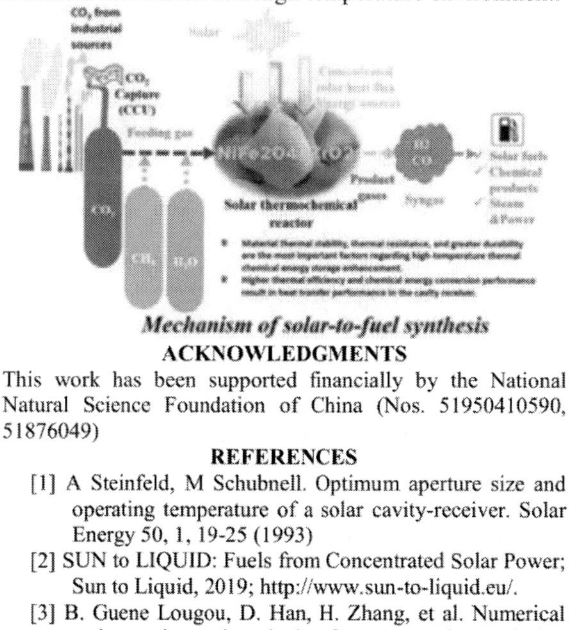

ACKNOWLEDGMENTS

This work has been supported financially by the National Natural Science Foundation of China (Nos. 51950410590, 51876049)

REFERENCES

[1] A Steinfeld, M Schubnell. Optimum aperture size and operating temperature of a solar cavity-receiver. Solar Energy 50, 1, 19-25 (1993)

[2] SUN to LIQUID: Fuels from Concentrated Solar Power; Sun to Liquid, 2019; http://www.sun-to-liquid.eu/.

[3] B. Guene Lougou, D. Han, H. Zhang, et al. Numerical and experimental analysis of reactor optimum design and solar thermal-chemical energy conversion for multidisciplinary applications. *Energy Conversion and Management* **213**, 112870 (2020).

[4] H. Zhang, Y. Shuai, B. Guene Lougou, et al. Effects of multilayer porous ceramics on thermochemical energy conversion and storage efficiency in solar dry reforming of methane reactor. *Applied Energy* **265**, 114799(2020).

609478 Optimal Porosity Distribution of Non-Uniform Macroporous Ceria in Solar Thermochemical Water Splitting

Abstract:

Solar-driven thermochemical water and carbon dioxide splitting is a promising technology which produces synthesis gas in a sustainable way. Previous reactor models showed that ceria-based oxides with macro-porous structure could offer higher surface area and facilitate the endothermal solid-gas reaction by increasing the solar heat absorption. However, non-uniform boundary heat flux and variant fluid concentration along the reactor would lead to large temperature and partial pressure gradient profiles, and . Therefore, a carefully designed non-uniform porosity distribution of the macroporous ceria, matching the heat and flow patterns, could enhance the thermal performance, reduce the concentration difference, increase the reactant use, and essentially improve the overall fuel yield behavior in the solar-driven reactor.

We developed a multi-scale and multi-physics modeling framework for the complete solar-driven two-step redox cycle process, including reduction and oxidation reactions. Based on the volume-averaged conservation equations of mass, momentum, and energy, this model incorporated morphology related effective properties of the porous ceria, coupled the heat and mass transfer interactions, and non-equilibrium local heterogeneous chemistry between solid and fluid phase. The effective transport properties of the porous media were determined by direct pore-level simulations and Monte Carlo ray tracing utilizing the digital 3D micro-structures obtained by X-ray computed tomography.

The porosity distribution was optimized for enhanced reaction rate and fuel yield, considering different operational conditions (temperature, mass flow rate and species concentration). The optimization objectives were to maximize the conversion efficiency and rate of solar energy into chemical energy, considering their constraints given by the conservation equations.

The results indicated that, under certain operation conditions, volumetric radiation absorption could be increased and species concentration could be smoothened and made more uniform within the non-uniform porous reactor. The advancements in the reaction rate and fuel yield for the optimized non-uniform porosity distribution are quantified.

The optimized non-uniform porosity distribution function of the ceria provides a practical opportunity in additive manufacturing, material engineering, and scalable manufacturing.

610468 Thermally-Driven Oxygen Pumping and Sweeping/Separation

Abstract:

One of the key challenges in solar thermochemical (STC) fuels production is to minimize oxygen partial pressures in the thermal reduction step. This partial pressure must be reached by energetically efficient and economically affordable methods—currently unavailable by either vacuum pumping or inert gas sweeping. In an effort to address this unresolved high impact challenge in STC fuels, we have analyzed and experimentally tested a promising new thermally driven approach, featuring high theoretical efficiencies, comparatively low operating temperatures, and exceedingly simple and robust devices—including implementations without moving parts. Our results shows that, even under conservative practical assumptions, the thermally driven pumping and sweeping/separation approach has the potential to substantially exceed state of the art approaches, especially in the low pressure regime, and do so in inherently simple and potentially low cost devices.

604860 Combined Hydrogen and Power Production Systems Using Thermochemical Redox Cycle

Abstract:

Two-step thermochemical redox cycles using metal oxides present a promising approach to produce hydrogen because they utilize intermediate to high-temperature heat. However, their thermodynamic efficiency remains low with the rest of energy supplied to the system emitted to the surrounding or to a heat sink typically at the lowest temperature of thermochemical cycle, that is the temperature for oxidation step (> ~800°C).

We propose a method for increasing system efficiency by adding a bottoming cycle that utilizes this high quality wasted heat. Various conversion processes operating at these temperatures can be used as a bottoming cycle, such as a Rankine cycle or a supercritical CO_2 Brayton cycle for power production, and natural gas steam reforming or biomass gasification for producing fuels. In particular, we analyze three different system concepts for combined hydrogen and power production, having power generation system as the bottoming cycle.

The figure below shows schematics of three system concepts. The first has two loops, hot and cold, to transfer heat from the heat source to thermochemical reactors and the bottoming cycle. In the hot loop, a reactor in its reduction step receives the heat required for increasing the temperature of reactor and the endothermic reduction reaction, from the heat source. In the cold loop, the bottoming cycle receives heat from a reactor in its oxidation step. The remaining thermal energy in the reactor after solid-heat recuperation and the heat generated from the exothermic oxidation reaction are transferred to and utilized in the bottoming cycle. Vacuum pumps and water pumps are used during the reduction and oxidation reactions. In general, excess steam is utilized as input to the reactor during the oxidation step, and it is difficult to recover most of the energy used to generate that steam. The second concept is similar to the first, but it utilizes the water stream used in thermochemical reactors directly as the working fluid of the bottoming cycle to decrease these steam generation related losses. While the first and second concepts have two loops, which need a pump operating at very high temperature (such as graphite or ceramic pump) if using a fluid to transfer heat from the heat source to the reactors, the third concept has only one loop. Because the third concept has only one loop, it does not need such pump and the system is less complicate.

To determine the pros and cons of these three systems, we perform thermodynamic analysis to evaluate the hydrogen production efficiency and power generation efficiency. The first and second concepts achieve higher hydrogen production efficiency than the third concept especially when solid heat recuperation is high. However, water losses in the first concept are large which makes the second system concept more attractive. The second concept reduces water losses but the reactor needs to operate between vacuum and ~ 50 bar, which is the operating pressure of the Rankine cycle. The supercritical CO_2 Brayton cycle, which generally has a better efficiency than the Rankine cycle, cannot be applied to the second concept because the working fluid is steam. The third concept has relatively low hydrogen production efficiency but the upper temperature of the bottoming cycle is relatively high so that it can achieve better power generation efficiency compared to the first and second concepts. Ceria is used for the analysis but the same framework can be applied to the system using other materials. Techno-economic analysis will be provided to compare the levelized costs of hydrogen and electricity produced from these combined hydrogen and power production systems with hydrogen-only production system and power-only production system using the same heat supply.

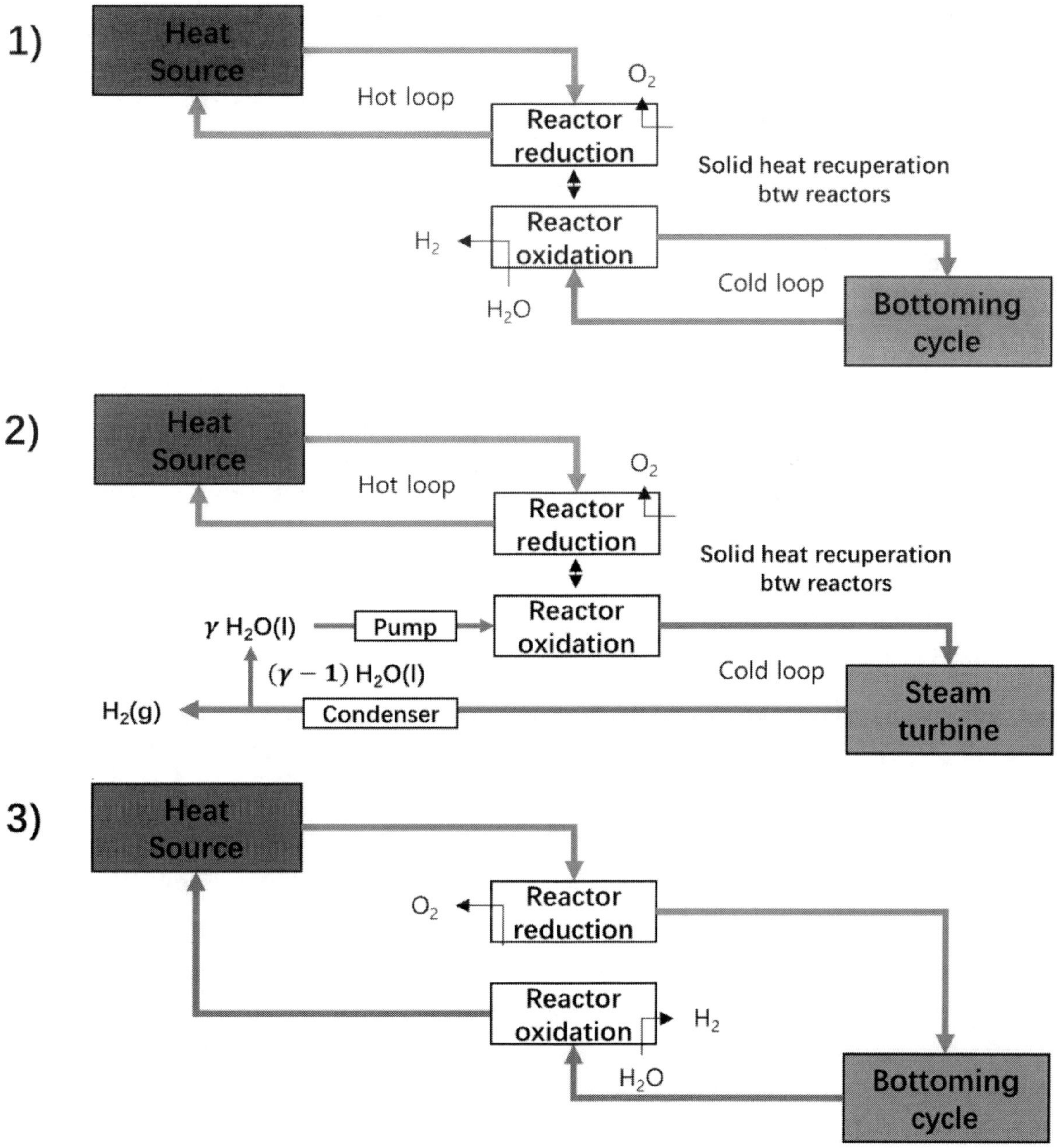

Figure. Three different system schematics of combined hydrogen and power production system

614726 Estimating Chemical Releases and Evaluating Indicators for Manufacturing Processes: Biorefinery Case Studies

Abstract:

During manufacturing and use of chemical substances, the estimation of releases is critical for environmental and human health impact evaluations. Human health impacts, acidification, smog, and many other environmental impacts are directly related to chemical releases. For example, for human health risk assessments, the risk depends upon the amount of chemical released, the exposure people have, and the toxic effects of the chemical. Other environmental impacts are similarly determined through a combination of releases and impact characterizations. Thus, release estimation is a key step in evaluating the effects of manufacturing, processing, using, and disposing of chemicals. This contribution will describe recent efforts to estimate releases, evaluate indicators for manufacturing with case studies for biorefineries that produce cellulosic ethanol via biochemical and thermochemical routes, and provide tools available to aid in these evaluations [1]. In addition, this contribution offers a practical method that enables stakeholders to estimate releases from on-site utilities and common but often neglected unit operations. It also suggests how to guide process designs or improvements that incorporate knowledge from the indicators.

[1] Smith, R.L., Tan, E.C.D., and Ruiz-Mercado, G.J., "Applying Environmental Release Inventories and Indicators to the Evaluation of Chemical Manufacturing Processes in Early Stage Development," ACS Sustainable Chemistry & Engineering, 7, 10937-10950 (2019); DOI: 10.1021/acssuschemeng.9b01961

613423 Keynote: Opportunities and challenges for CST in high temperature industrial processes

Abstract:

The application of concentrated solar thermal energy to high temperature industrial processes is considered to be one of the potential pathways with strong potential to contribute to the decarbonizing of this 'difficult to abate' sector, because the direct use of heat avoids the thermodynamic losses of converting heat to power. However, to realise this potential it is necessary to overcome the challenges of integration (of the solar plant into a complex industrial process), of variability (given the continuous nature of industrial processes) and scale (given that thermal input is typically larger than the output from a single solar field). In addition, the cost of the process must compete against low-cost fossil fuels in the short term and against electrification and hydrogen in the longer term. The paper will review learnings from recent assessments of various potential options with which to overcome these challenges for the alumina and copper production processes, together with some technology developments that offer potential to contribute to the solutions. It will consider the direct use of heat employed in a retrofit context, the local production of fuels and the co-production of fuels and other chemicals, together with hybridization and some greenfield opportunities.

608893 Invited Talk: Grid-Level Thermochemical Energy Storage Using the Magnesium-Manganese-Oxide Redox System

Abstract:

Grid-level energy storage on timescales of 10 to 100 hours is key to a carbon neutral power grid dominated by intermittent renewable generation via wind and solar energy. We are currently developing a thermochemical storage technology that uses a packed bed of porous pelletized Magnesium-Manganese Oxides (Mg-Mn-O). The system is charged with renewable electricity by heating the bed to 1500˚C at absolute pressures around 0.1 bar to thermally reduce the Mg-Mn-O. Energy is recovered by oxidation when passing pressurized air from the compressor of a turbo-generator set through the porous reactive bed. The energy released heats up the air, which is then expanded in the gas turbine of the turbo-generator.

Magnesium-Manganese-Oxides are well suited for the current application because 1) they have excellent cyclability: we have demonstrated over 100 redox cycles with less than 4% loss in activity, 2) they are highly stable: we have demonstrated 500h dwell at 1500˚C and multiple deep cycles between ambient and 1500 ˚C, and 3) they are non-toxic, low-cost, and abundant.

In the current work, we demonstrate Mg-Mn-O thermochemical storage on the bench-scale using a 1 kW/0.1 kWh prototype. The prototype consists of an internally insulated cylindrical steel pressure vessel. A cylindrical, 50 mm diameter packed bed of Mg-Mn-O pellets with an embedded Molybdenum-Disilicide heater is located at the center of the storage unit. The storage unit can be pressurized with air up to 20 bar, it can furthermore operate at reduced pressures down to 0.1 bar. The test system is equipped with flow controllers at the inlets and outlets as well as oxygen sensors at the outlet, enabling closure of the mass and species balance with less than 5% error.

After an initial pressure test, the reactor, filled with 277 g of mostly oxidized Mg-Mn-O pellets, is evacuated to 0.2 bar absolute pressure and it is then heated up. Oxygen evolution due to thermal reduction begins at approximately 1000˚C, it is most vigorous at temperatures between 1250˚C and 1500˚C. Using oxygen sensor and flow sensor derived species balance, the amount of oxygen released during reduction is calculated to be approximately 14g. Subsequently, the reactor is pressurized to 11 bar absolute pressure and the temperature is reduced to 1000˚C. Once 11 bar has been reached, a constant flow of air through the reactor is maintained to drive oxidation of the bed. The outlet oxygen concentration drops below 15%, indicating that oxidation is ongoing. Once the outlet oxygen concentration reaches 19%, oxidation is considered to be complete and the next cycle is started.

We have performed 5 subsequent cycles with an average duration of 3 hours. The average energy storage density based on the packed bed volume obtained from the measured oxygen exchange is 2300 MJ/m^3.

610993 The Thermodynamics of Magnesium Manganate Reduction As Applied to Thermochemical Energy Storage

Abstract:

Thermochemical energy storage shows potential for providing 24 hour a day energy when combined with solar power plants and conventional electricity production means. Operating the system at high temperatures allow for high levels of efficiency to be reached, and the long-duration ability of the technology allows for this type of storage to be used to replace current methods, or to supplement existing infrastructure. $MgMn_2O_4$ is of interest as they are low cost and have high reduction/oxidation temperatures. The energy density of the material determines the storage capacity of any potential systems; therefore we report on the energy density of this material as well as its underlying thermodynamics, using Van't Hoff analysis in conjunction with thermogravimetric data. Density functional theory calculations are used to understand the thermodynamic limitations. We explore the material as a function of Mg:Mn content, and provide novel means of increasing the reactivity. Additionally, material characterization confirms the phase purity of the synthesized material.

611275 Thermal Reduction of Iron-Manganese Oxide Particles in a High-Temperature Packed-Bed Solar Thermochemical Reactor

- Download/View Extended Abstract

Abstract:

The reduction of iron–manganese oxide particles in a high-temperature packed-bed solar thermochemical reactor is investigated using an advanced transient three-dimensional computational fluid dynamics model. The model couples the conductive, convective, and radiative heat transfer, reaction kinetics, and fluid flow in the bed with packed particles and interstitial sweep gases to obtain a detailed description of the transport phenomena in the bed. A reactor prototype that features a reaction tube confining the packed particles and a surrounding diffuse reflective cavity is tested under simulated high-flux solar irradiation to validate the model. The numerically predicted temperature profiles and oxygen generation rates are in good agreement with the experimental data. The validated model is applied to evaluate the thermochemical performance of the reactor. The calculated temperature profiles indicate that uniform temperature distribution in the reactive packed particles is achieved from the onset of the reaction. An energy rate balance analysis shows the instantaneous peak solar-to-thermochemical energy efficiency reaches 9.3%. The optimal operation conditions for the reactor are explored in a parametric study of the sweeping gas velocity and the concentration ratio of incident solar radiation.

612805 Sustainability: Legal and Business Drivers, and Recent Trends

Abstract:

In the United States, achieving sustainability is not specifically mandated by law. So, without a specific mandate, why would domestic governmental and private corporations pursue sustainability goals? In short, to mitigate risk. Environmental and energy law and regulation and compliance, while generally not specifically requiring sustainability, are relevant to sustainability, such that unsustainable and wasteful approaches present both legal and business risk as well as a heightened risk of regulatory noncompliance. Additionally, governmental and private entities are compelled to fulfill fiduciary duties to rate payers and shareholders respectively, while private entities are sensitive to underlying corporate requirements to comply with law and fulfill fiduciary duties, including mitigating material risk that may arise from such unsustainable approaches. Finally, private entities are subject to market sensitivity and potential risk to brand recognition from publicized unsustainable practices. This presentation will review these legal and business drivers for sustainability, as well as recent trends in sustainability arising from recent deregulation, the global approach to plastic waste and our recent experience with COVID-19.

612807 Sustainability: Concepts, Innovation, and Economics

Abstract:

The challenge of the present age is how to sustain the growing human population in a reasonably prosperous and civilized manner within the limits of the one planet, Earth, that we inhabit. For reference, the current human population stands at 7,782,939,000 where I have left the last three digits as zeros because they are changing fast as I write. This is a solvable problem but solving it will require all the human understanding and ingenuity that we possess, and not only technical but also business and innovation. To frame the issue, we will discuss sustainability in a historical context, the measurement of sustainability, and the extended Sustainable Regimes Hypothesis which outlines the conditions necessary for sustainability. We will the discuss the role of innovation and economic policy in finding pathways or solutions out of our dilemma.

612809 The Need for Convergence

Abstract:

It is well-understood that the sustainability challenge cannot be solved by any single conventional discipline. This has given rise to new fields such as sustainable engineering, ecological economics, and industrial ecology, and to frameworks such as life cycle assessment and circular economy. Virtually every corporation is working toward sustainability, and tens of thousands of papers have been published. Despite these efforts, atmospheric greenhouse gas concentrations continue to increase, biodiversity loss continues unabated, and societies continue to become more unequal. We will argue that a root cause behind the unsustainability of human activities is the divergence between disciplines that took place over the last three centuries. This gave rise to disciplines such as economics, ecology, and engineering. The resulting specialization has enabled advances that support human well-being, but it has also resulted in unsustainable development as an unintended side effect. We will provide several examples in support of this argument. With this insight, we will further argue that for meeting the challenge of sustainable development, engineering needs to strive toward convergence with ecology and economics. This requires each discipline to transcend beyond its conventional domain to result in new concepts and principles that will form the foundation of sustainable development. We will identify some of the challenges in such an effort, as well as the unique opportunities for innovation.

612810 Do We Need New Classes of Methods, Tools and Standards?

- Download/View Extended Abstract

Abstract:

Comments from industrial contacts suggest that chemical, petrochemical, pharmaceutical and related industries routinely consider issues concerning sustainability early in their process development. Even though sustainability studies are made at various stages of the process and/or product development, it is not clear if they all use similar sets of standards and/or which currently available methods and tools they commonly employ? Are the available data, models and design-analysis methods sufficient, or, Is there a case for a new set of standards, methods and tools that can help to achieve sustainable development? Should the focus of research be changed to make a greater impact related to sustainability? The answers to the above questions will be presented based on review of published papers, comments from industrial contacts and our own research activities.

600315 Enabling Selective Catalytic Conversion of Biomass-Derived Saturated Furans to C4-C5 Dienes

Abstract:

Conjugated C4-C5 dienes are key industrial monomers, currently recovered from naphtha crackers, with ethylene as the main product. Gradual shift towards cheap light feedstocks has resulted in refineries decoupling the production of ethylene and larger fractions including C4-C5 dienes, accentuating the need to produce this fraction from alternate routes. Previous reports have highlighted a thermochemical route to these dienes from the vapor-phase ring-opening dehydration (dehydra-decyclization) of biomass-derived saturated furans on solid Brønsted acid catalysts including zeolites. However, in addition to a fundamental lack of understanding of the reaction mechanisms and pathways, the diene yields have remained moderately low (50-65 % C basis).

First, combining results from kinetic experiments, DFT calculations, and microkinetic modeling, we highlighted the dehydra-decyclization and the Retro-Prins pathways for THF conversion to butadiene and propene, respectively, on H-ZSM-5. This study established that i) C-O cleavage exhibited the highest degree of rate-control, and ii) the selectivity control descriptor was ratio of C-O/C-C scission rates. We then undertook a detailed catalytic evaluation of zeolites with varying Brønsted acid strength (namely, Al-, and B-substituted zeolites) in different micropore environments (BEA, MFI, and MWW) to provide new insights on the role of heteroatom identity and confining environments in tuning C-O/C-C scission rates for 2-methyltetrahydrofuran (2-MTHF) dehydra-decyclization. While weakly acidic borosilicates selectively catalyzed desirable dehydra-decyclization pathways, micropore environments influenced diene distributions independent of heteroatom identity. Combining these learnings, we reported an unprecedented ~86% (per C basis) 1,3-pentadiene yield on B-MWW.A following techno-economic analysis of proposed furfural-to-butadiene process concluded that the process cannot currently compete with existing petrochemical routes due to high butadiene selling price ($5.43/kg), but we detailed the technological improvements and economic conditions to achieve butadiene selling price below $1.5/kg.

598598 Theoretical Insights into the Selective Conversion of Tetrahydrofuran to 1,3-Butadiene on ZrO_2

Abstract:

1,3-Butadiene is an important industrial chemical used in the manufacture of automobile tires. The current production methods, mainly steam cracking of naphtha and shale gas, are energy-intensive, less selective and produce large amounts of CO_2.[1-2] Renewable biomass-based production of 1,3-butadiene could potentially be a sustainable and energy-efficient alternative due to lower reaction temperatures and green-house gas emissions.[1-2] Out of the various possible catalytic chemical routes, conversion of biomass to butadiene via ethanol, and four-carbon alcohols have been widely explored.[1-2]

A recent method, which has emerged as an alternative, converts tetrahydrofuran (THF) into 1,3-butadiene by dehydra-decyclization over Brønsted acid zeolites.[3-5] Remarkably, we have recently discovered that the dehydra-decyclization of THF to butadiene can be selectively performed (selectivity over 90%) on ZrO_2, a catalyst not typically thought of as Brønsted acidic.

In this paper, we perform Density-Functional theory calculations and microkinetic modelling to gain mechanistic insights into the high performance of ZrO_2. We consider the structural complexity of tetragonal and monoclinic ZrO_2 and investigate the catalytic activity of isolated Lewis acid-base pairs, the possible catalytic activity of Brønsted hydroxyls and the synergy of the two types of sites for two pathways: *(a)* the main reaction to butadiene; and *(b)* the retro-Prins condensation to propene and formaldehyde, the dominant side reaction.

References:

1. D. Cespi, F. Passarini, I. Vassura, and F. Cavani, Green Chem. **18**, 1625 (2016)
2. S. Farzad, M.A. Mandegari, and J.F. Görgens, Bioresour. Technol. **239**, 37 (2017).
3. X. Li, P. Jia, and T. Wang, ACS Catal. **6**, 7621 (2016).
4. O.A. Abdelrahman, D.S. Park, K.P. Vinter, C.S. Spanjers, L. Ren, H.J. Cho, D.G. Vlachos, W. Fan, M. Tsapatsis, and P.J. Dauenhauer, ACS Sustain. Chem. Eng. **5**, 3732 (2017).
5. S. Li, O.A. Abdelrahman, G. Kumar, M. Tsapatsis, D.G. Vlachos, S. Caratzoulas, and P.J. Dauenhauer, ACS Catal. **9**, 10279 (2019).

611768 Active and Selective ZrO_2 Based Catalysts for Diene Monomer Production from Biomass.

Abstract:

The production of diene monomers like butadiene and piperylene from biomass resources has recently attracted interest, serving as a sustainable pathway to produce chemical building blocks vital to polymer production. Cyclic ethers like tetrahydrofuran derived from biomass can be converted to butadiene via a dehydra-decyclization chemistry, where Brønsted acidic materials have been demonstrated to be active catalysts. The main challenge is balancing the activity and selectivity of catalysts, where strongly Brønsted acid catalysts are highly active but facilitate undesired reactions (retro-prins condensation, diene oligemerization). Conversely, weakly Brønsted acidic materials are highly selective, but are far less active. Here we demonstrate that Lewis acidic metal oxide catalysts can potentially overcome this challenge, serving as a highly selective and active dehydra-decyclization catalyst. The screening of a variety of metal oxide catalysts shows that ZrO_2 is an effective dehdyra-decyclization catalyst, achieving near quantitative yield to butadiene from THF at 673 K and moderate weight hourly space velocities (1 g THF g cat^{-1} hr^{-1}). The catalyst is highly selective to dienes across a wide range of conditions (> 90%) and is found to be relatively stable with time on stream. Comparing the different phases of ZrO_2 (tetragonal, monoclinic and cubic), we find the catalytic activity of the catalyst is structure insensitive as evidenced by comparable rates and activation energies. Conversely, the stability with time on stream is highly dependent on the phase of ZrO_2, where tetragonal is found to be the most stable. Despite the deactivation, the initial activity of the catalyst can be regenerated through a mild calcination, suggesting a coking mechanism is responsible for the loss in activity.

607100 The Cascade Reaction of Ethanol to Butadiene over Silica-Supported Silver-Zirconia Catalysts

Abstract:

Silver and zirconia supported on silica is known to be one of the most effective catalysts for the cascade reaction of ethanol to butadiene, also known as the Lebedev reaction. The reaction includes the dehydrogenation of ethanol to acetaldehyde, condensation of acetaldehyde, and appropriate hydrogenation and dehydration reactions to ultimately produce butadiene. This work focused on studying the role of each component (Ag and ZrO_2, respectively) in the Lebedev reaction by modifying the crystallinity of ZrO_2 and the ratio of Ag to ZrO_2 in the reactor. Characterization of the catalysts was accomplished using X-ray diffraction, X-ray absorption spectroscopy, Raman spectroscopy, electron microscopy, and diffuse reflectance infrared Fourier transform spectroscopy (DRIFTS) of adsorbed CO and CO_2. All of the characterization methods were consistent with silica-supported zirconia catalysts presenting the ZrO_2 component in a very highly dispersed phase. Zirconia catalysts, either in bulk (crystalline) or supported (amorphous) forms converted ethanol at 573 K to mostly undesired dehydration products, diethyl ether and ethylene, when Ag was not present. The addition of Ag to bulk ZrO_2 favored the production of acetaldehyde (presumably from the Ag component) without forming an appreciable amount of butadiene (<5%), whereas placing Ag together with silica-supported ZrO_2 exhibited a selectivity of ~35% to butadiene at an ethanol conversion of 20%. The observed change in product distribution dependent on the ratio of Ag to ZrO_2 in the catalyst bed revealed the delicate balance required between the dehydrogenation/hydrogenation function provided by Ag and the acidic/basic function provided by ZrO_2. The spectral features from *in situ* DRIFTS of adsorbed CO and CO_2, used to probe surface Lewis acid and base sites, respectively, may explain the differences in catalytic performance between bulk and supported ZrO_2.

598968 Ethanol Conversion to Butadiene over Isolated Zinc and Yttrium Sites in Dealuminated Beta Zeolite

Abstract:

Isolated Zn and Y sites were introduced into dealuminated BEA zeolite (DeAlBEA) and then investigated for the conversion of ethanol to butadiene. This reaction occurs in three steps – ethanol dehydrogenation to acetaldehyde, reaction of acetaldehyde with ethanol to form crotyl alcohol, and crotyl alcohol dehydration to butadiene. Zn-DeAlDEA and Y-DeAlBEA are particularly effective for one part of the overall reaction. Zn-DeAlBEA exhibits high activity for ethanol dehydrogenation and low activity for to acetaldehyde C-C bonds formation, whereas Y-DeAlBEA exhibits low activity for ethanol dehydrogenation and high activity for crotyl alcohol formation. Crotyl alcohol dehydration is catalyzed by the slightly Bronsted acidic Si-OH groups present in silanol nests of DeAlBEA. Investigations of the mechanism and kinetics of C-C bond formation over Zn-DeAlBEA and Y-DeAlBEA indicate that this process does not proceed via aldol condensation of acetaldehyde but, instead, involves the reaction of acetaldehyde with ethanol. It is also notable that the rate of C-C bond formation over single Y sites in Y-DeAlBEA is 80 time higher than that over isolated silica-supported Y sites. Based on these findings, we propose that the C-C bond coupling over both Zn and Y site present in DeAlBEA require the assistance of the adjacent hydroxyl group. The Si-O-Y(OH)-O-Si (Si-O-Zn-O-Si) together with the adjacent hydroxyl group constitute the active center for crotyl alcohol formation from ethanol and acetaldehyde. Very good EtOH-to-1,3-BD performance was obtained over both the 10Zn-15Y-DeABEA and 25Y-DeABEA catalysts for one-step and two-step processes. When the reaction was performed at 400 °C, the 1,3-BD productivity is 3.6 $g/g_{cat}/h$ with a C_4 olefin selectivity of 62%. Over 25Y-DeABEA catalyst at 400 °C, it was also encouraging to find that the 1,3-BD productivity could reach 6.6 $g/g_{cat}/h$ with a C_4 olefins selectivity of 77.2%.

598971 Kinetic and Mechanism of Acetone Conversion to Isobutene over Hf/Silicalite-1 and Hf/SiO_2

Abstract:

Hafnium (Hf)/Silicalite-1 nanosheets and Hf/SiO_2 catalysts were prepared and investigated for the reaction of acetone to isobutene. The dependence of activity on Hf loadings for the two catalysts differed. For Hf/SiO_2, the rate of isobutene formation increased linearly for Hf weight loadings below 2 wt%, and then increased further but at a diminishing rate. By contrast, no isobutene was formed on Hf/Silicalite-1 for a Hf loading < 2 wt%. However, for higher Hf loadings, the catalyst active increased more rapidly than that observed for Hf/SiO_2. This abrupt activity change is attributed to a change in the structure of the Hf species. The IR spectra of adsorbed pyridine show distinct differences, suggesting differences in the Lewis acid properties of Hf on Hf/SiO_2 and Hf/Silicalite-1. The dependence of the rate of isobutene formation over the two catalysts also differ. Hf/Silicalite-1 exhibits a second order dependence on acetone partial pressure, while the reaction order decreased to 1.6 for Hf/SiO_2. The apparent activation energies also differ for the two catalysts, the activation energy for Hf/Silicalite-1 being significantly lower than that for Hf/SiO_2. Further investigation of the mechanism of isobutene formation revealed that the first step is aldol condensation to form mesityl oxide and the second is the decomposition of this product to form isobutene and acetic acid, which is consumed within the reaction cycle. For Hf/Silicalite-1, isobutene formation is limited by the rate of aldol condensation on Hf sites, whereas for Hf/SiO_2, the decomposition of mesityl oxide limits the rate of isobutene formation.

611632 *Vapor Phase Ethanol Selective Oxidation to Acetaldehyde and Acetic Acid on Gold Catalysts*

Abstract:

The selective oxidation of ethanol to acetaldehyde and acetic acid on gold nanoparticles supported on a metal oxide (Au/TiO_2) was examined via in situ UV-Vis and FTIR spectroscopies and kinetic studies. In situ UV-Vis spectroscopy via a gold plasmon surface resonance sensing technique provided evidence for O_2 adsorption at the $Au-TiO_2$ metal-support perimeter interface. Additionally, in situ Modulation Excitation-Phase Sensitive Detection-UV-Vis and FTIR spectroscopies performed during O_2 modulation at reaction conditions also showed the presence of charge transfer events likely occurring during the spillover of hydrogen arising from ethanol dissociative adsorption at the $Au-TiO_2$ metal-support perimeter. Moreover, kinetic results over a wide range of conditions (473-533 K, 0.1-20 kPa O_2, 0.1-5 kPa ethanol, 0-8 kPa water) showed that ethanol oxidized selectively in vapor phase to mainly acetaldehyde (~50-95%) and acetic acid (~2-30%) with respective apparent activation energies of 47 and 70 kJ/mol. It was also found that acetic acid selectivity was favored at increasing O_2/ethanol partial pressure ratios. Although water presence in the reaction mixture has been implied to promote oxidations on gold catalysts in gas phase and acetic acid formation in ethanol liquid phase oxidation, the present results indicated that in the vapor phase acetaldehyde and acetic acid production rates were mostly insensitive to co-fed water. Average apparent reaction rate orders of O_2, ethanol, and water for acetaldehyde production were ~0.2, ~0.5, and ~0.1, whereas for acetic acid production they were ~0.5, ~0.3, and ~0, respectively. A kinetic model involving ethanol dehydrogenation to acetaldehyde, oxygen adsorption and activation, and acetaldehyde oxidation to acetic acid on a two-site ensemble at the $Au-TiO_2$ metal-support perimeter comprised of Au-O vacancy-Ti-O was found to adequately explain the spectroscopic and kinetic experimental findings. Overall, this study highlights a metal-support synergy which can be explored to further guide the development of new selective oxidation catalysts.

608298 Dissociation, Dissolution, and Diffusion of Nitrogen on V_xFe_y and V_xCr_y Alloy Membranes Studied By First Principles

Abstract:

N_2-selective metal membranes could be used to couple the extraction of N_2 from flue gas and NH_3 synthesis within a single membrane reactor. The larger interstices of BCC metals make them attractive for fast permeation of the large N atom, but N permeabilities reported for pure BCC metals are still impractical. As practical permeabilities could be achieved by alloying these metals, we use density functional theory to study how alloying V, a metal with high N solubility but low N diffusivity, with Fe and Cr impacts N_2 dissociation, dissolution, and diffusion. Although N binding at vacancies was studied, binding energies indicate that these defects may not significantly affect solubility. The studied alloys present V-rich ($o1$) and V-depleted ($o2$) octahedral interstices. Nitrogen binding strength correlates with the number of V nearest-neighbors; hence, binding in $o1$ sites is on average 1 eV stronger than in $o2$ sites. Ab initio thermodynamics suggests that this combination of strong and weak binding sites mitigates the reduction in solubility expected from the alloying at relevant operating conditions. However, the heterogeneity of the interstices generally leads to higher energy barriers for N hopping than those encountered in pure V (1.24 eV). The exception to this observation was the $V_{0.25}Fe_{0.75}$ alloy (1.01 eV). Based on our calculations, at 673 K and 5 bar (N_2 pressure), N solubility and diffusivity in $V_{0.25}Fe_{0.75}$ would be ~3 times smaller and ~53 times larger than pure V, respectively. According to the solution–diffusion model, these findings indicate that an ~18-fold higher permeability would be expected for $V_{0.25}Fe_{0.75}$ relative to V. Permeability is expected to be controlled by bulk diffusion rather than by surface processes, as in all alloys, we find the energy barrier for N_2 dissociation at the alloy surface to be lower than the barrier for bulk diffusion in the same alloy.

609332 Impact of Oxygen on Photocatalytic Nitrogen Fixation

Abstract:

Photocatalytic nitrogen fixation is an approach that may enable the synthesis of ammonia (NH3) from abundant feedstock (air and water) using only renewable energy (sun). Despite the potential sustainability of this approach, few photocatalyst exhibits desirable nitrogen fixation activity in an atmosphere made up of pure nitrogen let alone in air that contain oxygen and other impurities. Separation and purification of nitrogen from air is an energy intensive and expensive process. Therefore, truly sustainable approaches toward achieving photocatalytic nitrogen fixation should catalyze nitrogen reduction in air. The chief challenge with using air is the potential for oxygen reduction (ORR) to compete with nitrogen reduction reaction (NRR). Prior attempts to photocatalyze nitrogen fixation in air on titania, showed that ammonia production rates decreased by approximately 60% when compared with pure nitrogen gas[1]. However, another electrochemical analysis reported that Fe2O3/TiO2 showed negligible change in ammonia production rate in nitrogen and in air[2]. This motivates the need to determine which co-catalyst can be employed to mitigate oxygen reduction reaction activity, to enable photocatalytic nitrogen fixation with air. Here we aim to screen rational metal co-catalyst for activity for nitrogen and oxygen reduction reaction using a rotating ring disk. We then demonstrate the photocatalytic activity through a series of photocatalytic experiments.

[1] Hirakawa, Hiroaki, et al. "Photocatalytic conversion of nitrogen to ammonia with water on surface oxygen vacancies of titanium dioxide." *Journal of the American Chemical Society* 139.31 (2017): 10929-10936.

[2] Manjunatha, Revanasiddappa, et al. "Electrochemical ammonia generation directly from nitrogen and air using an iron-oxide/titania-based catalyst at ambient conditions." *ACS applied materials & interfaces* 11.8 (2019): 7981-7989.

598500 Manipulating Structures and Chemical Bonding in Nitrides Using First-Principles Methods for Ammonia Production

Abstract:

In this work, an alternative approach to the Haber-Bosch process based on a cyclic process is investigated for smaller-scale, intermittent NH_3production at atmospheric pressure. Manganese (Mn) and Mn nitrides are uniquely advantageous for the proposed cyclic step catalysis scheme. The elemental N is incorporated into Mn substrate, as lattice N, which will be reduced subsequently by H_2 to yield NH_3. Currently, the lattice N utilization (< 8%) and NH_3 yield remain too low.

On the nitrogen fixation side of the N conversion cycle, the nitridation of Mn exhibits strong temperature dependence. XRD analyses revealed a transition from the N-lean Mn_4N phase to a N-rich Mn_2N (and MnN) phase as temperature increases. Above 800°C, the N-rich nitride phase becomes unstable and decomposes into Mn_4N and N_2. This transition was also confirmed by the calculated free energies from Density Functional Theory (DFT) calculations. On the NH_3 formation side of the cycle, an Eley Rideal-Mars-van Krevelan pathway is assumed to study the hydrogenation and extract of lattice N. The formation of NH_3 tends to be strongly endothermic on most of the nitride models considered. The diffusion of sublayer lattice N onto the top layer is the kinetic controlling step. Based on DFT calculations, we showed that early hydrogenation NH_x intermediates still follow linear scaling relationships a behavior similar to transition metal surfaces. From kinetic modeling, we showed that single-atom doping with Fe, Co, Ni will promote NH_3 formation, while Cr, and Mo dopant hinder this process. These molecular-level insights based on well-defined sites and lattice structures are critical for the identification of promising materials, and ultimately, accelerate material discovery and synthesis.

603654 Enhancing Potential of Catalysts for Plasma Nitrogen Oxidation

Abstract:

N_2 fixation is a central process in the production of fertilizers and chemicals. The Birkeland-Eyde (B-E) process, an early attempt to fix nitrogen synthetically, used an electric arc thermal plasma to reach the high temperatures necessary to push the N_2/O_2 equilibrium in air towards NO. B-E proved too energy intensive and low yield to compete with the Haber-Bosch (H-B) process. Models and experiments both indicate that the energy costs to oxidize N_2 to NO can be reduced in a lower-temperature, non-thermal plasma (NTP). Models suggest that energy consumption can be reduced by more than 2.5-fold relative to the H-B process. Realizing this theoretical limit is challenging for NTP alone, partly because of the low NO production rate. Empirical evidence suggests that NO production can be increased through coupling of the NTP with a catalyst, but the nature and role of the catalytic enhancement remains unknown. Here we use microkinetic models to compare N_2 oxidation rates and conversions in a non-thermal plasma, catalyzed thermally, and with the two combined. We included important active species produced in a NTP and parametrized the catalytic model with DFT-computed data for a stepped Pt surface. The turnover frequency at zero conversion over Pt increases substantially over the thermal-only chemistry with the incorporation of dissociated N and O and vibrational excited N_2 and O_2 at levels expected to be present in the plasma. Similarly, at finite conversion, as simulated in a continuous stirred-tank reactor model, both dissociated and vibrational excited N_2 and O_2 increase the NO production on Pt. We discuss implications for application of plasma catalysis to N_2 oxidative fixation.

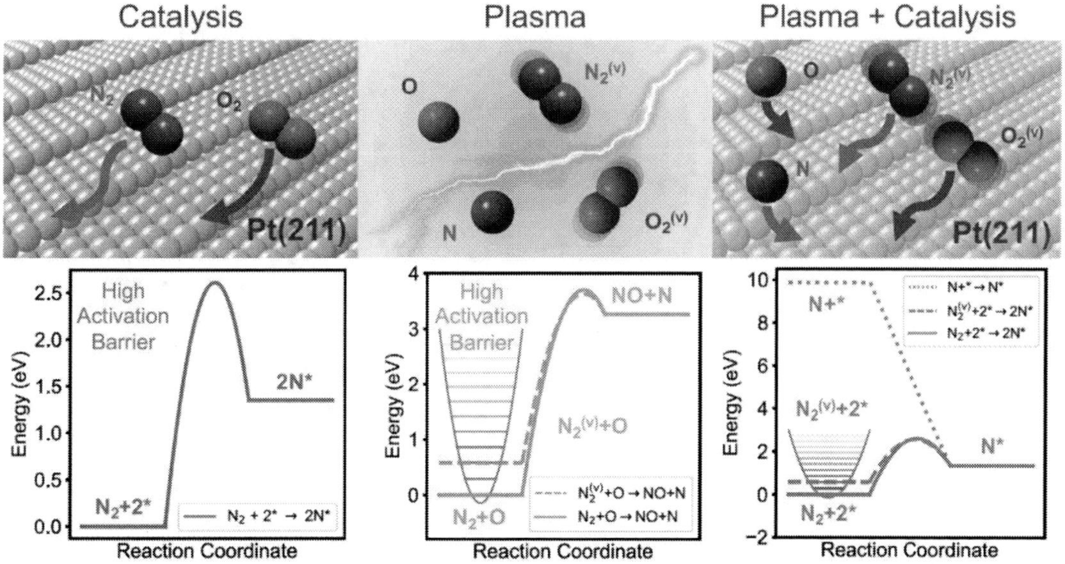

Scheme 1: Schematic representation of active species and N_2 activation processes in thermal catalysis, non-thermal plasma and non-thermal plasma coupled with catalysis.

607801 Theoretical Insight into the Surface Dynamics of Lithium-Mediated Nitrogen Activation

Abstract:

The electrochemical synthesis of ammonia, an important chemical which is conventionally synthesized industrially by the highly centralized Haber-Bosch process, holds promise for more sustainable and decentralized production. A lithium metal mediated reduction of nitrogen, involving applying a strongly reducing potential to a metal electrode in a nonaqueous electrolyte containing a lithium salt, has reproducibly demonstrated electrochemical production of ammonia[1,2,3]. The reduction of the lithium ions from the electrolyte to lithium metal or other solid lithium phase at the catalytic surface is hypothesized to play a dominant role in the activation of nitrogen, but the composition and structure of the catalytically active phase is not fully understood presently. Initial theoretical studies have shed light on the possible catalytic role of lithium metal as well as lithium nitride and hydride phases. Due to the ease of diffusion of species within lithium and its high reactivity, understanding the composition and properties of the catalytically active phase and the reaction dynamics within this phase are active areas of research.

In this work, density functional theory based molecular dynamics simulations, and enhanced sampling methods such as metadynamics, are used to simulate the reaction dynamics at finite temperature. We gain physical insight into the role played by structural fluctuations in nitrogen activation. We calculate the rates of these steps and investigate their dependence on the dynamic structure and composition of the active phase. This work provides new insight into the dynamics of nitrogen activation in lithium-medicated nitrogen activation, and also demonstrates the value of dynamic simulations and enhanced sampling methods for studying chemical reactions in complex environments.

[1] A. Tsuneto et al. Journal of Electroanalytical Chemistry. 1994, 367, 1-2, 183-188.

[2] N. Lazouski et al. Joule. 2019, 3, 4, 1127-1139.

[3] J. A. Schwalbe et al. ChemElectroChem. 2020, 7, 1542.

610777 Cyclic Stabilization Method for Lithium Mediated Electrochemical Nitrogen Reduction

Abstract:

Lithium mediated nitrogen reduction is a proven method to electrochemically synthesize ammonia (NH_3). However the process has so far been unstable, and the continuous deposition of lithium limits the practical applicability of the process. In addition, very little is understood about the mechanism. To address these issues, this report sets out to achieve two goals. Firstly, we present a general molecular-level model to describe the mechanism of lithium mediated N_2 reduction to NH_3. Despite its simplicity, this model can explain a number of mechanistic features that have been observed experimentally, such as the dependence of faradaic efficiency (FE) upon N_2 partial pressure and EtOH concentration. Based on this understanding, we develop a new strategy for increasing the FE by cycling the applied potential between a regime where Li deposition is favored and one where Li dissolution is favored. This strategy leads to a significant improvement to both the system's stability and catalytic efficiency. We show experimentally that the Li cycling process is stable for up to (at least) 125 hours of operation and yields a substantial increase in FE from 21% during constant Li deposition to 37% during potential switching. Importantly, we additionally measure an energy efficiency of 7.1% that is well beyond the previously highest reported value of 1.5% in the literature.

606864 Theoretical Insights into the Potential-Dependent Competition between Electrochemical Nitrogen Reduction Reaction and Hydrogen Evolution Reaction

Abstract:

The electrochemical N_2 reduction reaction (NRR) produces ammonia cleanly and sustainably at ambient conditions. The challenge in realizing NRR is the suppression of the hydrogen evolution reaction (HER), thus understanding the competition between NRR and HER is critical. One of the biggest obstacles for developing catalysts for NRR is a premature decrease of NRR activity with electrode potential before reaching the limiting potential which can facilitate NRR exothermically. Undesirably, the NRR activity has been consistently shown to reach a maximum quickly and then decrease with increasing reduction potential, but the exact mechanistic origin is unclear. In this talk, we discuss the potential-dependent competition between NRR and HER using the constant electrode potential model and the microkinetic modeling. We find a potential-dependent crossover between the H and N_2 binding energies, resulting in the crossover between the H and N_2 coverages. The simulated NRR activity vs. electrode potential trend is in great agreement with the experimentally observed trend. We find that the crossover originates from the larger charge transfer in *H formation than that of *N_2 and *NNH formation. Consideration of various catalysts shows that the larger charge transfer of *H than *N_2 and *NNH formation, and the potential-dependent activity crossover is a general trend, thus presenting a fundamental challenge to realizing NRR.

602734 Renewable NH_3 As Alternative Energy Fuel for Ocean Exploration

Abstract:

Renewable NH_3 as Alternative Energy Fuel for Ocean Exploration

Jian Liu, Yuyan Shao, Vanda Glezakou, Robert Cavagnaro, and Li-Jung Kuo

Pacific Northwest National Laboratory, Richland, WA 99352, USA.

Wind and wave energy constitute an abundant renewable energy source readily available to the United States. However, their practical use is limited by their fluctuating power output, expensive transmission line infrastructure and complicate integration systems to the existing grid. We propose a modular energy storage solution to convert wave or tidal energy into a carbon-neutral NH_3 fuel using an electrochemical method for ocean exploration and applications at sea. Currently, state-of-art electrocatalytic conversion using precious metals favors the competing hydrogen evolution reaction leaving insufficient vacant sites (Faraday efficiency<1%) for nitrogen reduction to occur. A high surface carbon catalyst with pyridinic N derived from metal-organic frameworks is developed under the guidance from the theoretical simulation. The enhanced N_2 adsorption and dissociation and the potential ability to suppress the competitive hydrogen evolution reaction will help to make the electrochemical synthesis of NH_3 from seawater and air more viable. An economic assessment of making and using NH_3 prepared through this route as the alternative energy fuels will be discussed. The performance of the electrocatalyst will be characterized using the rotating ring disk electrode method and compared with that of the conventional electrocatalysts.

611810 Mixed-Phase Perovskite Materials for H_2 Generation and Hybrid Charge Storage

Abstract:

Perovskite (ABO_3) materials have applications both in fuel production and energy storage. Although mixed metal oxide-based systems have been thoroughly investigated, the information available on the use of perovskite for H_2 generation is extremely meagre. In this study, Mn and Ti- based perovskite materials (as mixed phase) were synthesized using surfactant templating assisted sol-gel technique, which were analyzed using powder x-ray diffraction, BET, SEM/EDX, and utilized for thermochemical water-splitting process for H_2 generation. This material was loaded inside Inconel reactor and multiple thermochemical cycles were performed. H_2 generation was continuously monitored using online H_2 sensor and GC. As calcined perovskite materials in the form of mixed phase were utilized to fabricate supercapacitors, which were characterized for their gravimetric capacitance. In addition, supercapacitors were also fabricated using carbon electrode made from the HTL derived biochar. The cyclic voltammetry curves obtained at 5 V and 2 A were used to determine the specific capacitance and compared with commercial supercapacitors.

599875 Hydrogen and Ammonia for Renewable Energy Storage: The Economics of Location, Scale, and Demand Type

Abstract:

Generation of electrical power from intermittent sources of renewable energy such as the wind or the sun is a promising approach to improving energy sustainability. Energy storage is necessary to balance this intermittent renewable generation with electrical power demands. Batteries are the most commonly used energy storage technology, but their capital cost and relatively low energy density makes them unsuitable for longer-term, high-capacity energy storage. Hydrogen, produced via water electrolysis and later converted back to electrical energy using a fuel cell, is another potential energy vector. The main limitation of hydrogen-based energy storage is the storage infrastructure itself; high-pressure gaseous storage necessitates expensive types and amounts of material, and hydrogen liquefaction requires significant additional energy. Ammonia production using electrolysis-derived hydrogen and nitrogen obtained via electrically driven air separation has recently received worldwide attention [1-3] as a potential route to decrease these storage costs, as ammonia is liquid at ambient temperature and moderate pressures. Ammonia can be used to generate electrical power through traditionally fossil-fueled technologies (e.g. internal combustion gensets, fuel cells) with minor modifications. Potential limitations of ammonia-based energy storage are (i) the capital investment needed for additional process units (e.g. nitrogen production, ammonia production), (ii) the additional energy needed to transform the hydrogen to ammonia, and (iii) the potentially lower power generation efficiency of ammonia as compared to hydrogen which along with (ii) can lower overall power-to-power efficiency.

In this work, we elucidate this trade-off between storage cost, chemical production cost, and energy efficiency to determine the best chemical-based energy storage systems for a variety of locations, power demand scales, and power demand types. We perform this investigation through a mixed integer linear programming (MILP) optimal combined capacity planning and scheduling model, which minimizes the levelized cost of energy supply (LCOE) for islanded (not grid connected) renewable energy systems. The model optimally selects and sizes the units in the system, while simultaneously scheduling the operation of these units (i.e. on/off, chemical production rates, storage inventories, power generation rates) during each distinct period of a scheduling horizon which captures both diurnal and seasonal variation in intermittent renewable generation and power demand. The modeling framework generates variable length scheduling periods by clustering consecutive hours of full year time series data for wind capacity factor, solar capacity factor and power demand. This temporal aggregation makes the model computationally tractable, allowing for high throughput computational studies. In our previous work [4], we used our capacity planning and scheduling model to determine the minimum LCOE of using hydrogen and/or ammonia as energy storage for 1 MW residential renewable energy supply systems in 15 American cities which represent climate-demand regions throughout the United States. We found that ammonia is generally more economical than hydrogen as a single method of energy storage and that in every location, cost savings can be achieved by using both synergistically. In the present study, we expand the scope to optimally size and schedule hydrogen and ammonia renewable energy storage systems in each location for power demand scales up to 100 MW, with various combinations of commercial and residential demand profiles at each scale. The results provide a more comprehensive understanding of the economics of chemical-based energy storage and reveal the most promising applications for ammonia energy storage.

References

[1] Klerke, A., Christensen, C. H., Nørskov, J. K., & Vegge, T. (2008). Ammonia for hydrogen storage: challenges and opportunities. *J. Mat. Chem., 18(20),* 2304-2310.

[2] Zamfirescu, C., & Dincer, I. (2008). Using ammonia as a sustainable fuel. *J. Power Sources, 185(1)*, 459-465.

[3] Lan, R., Irvine, J. T., & Tao, S. (2012). Ammonia and related chemicals as potential indirect hydrogen storage materials. *Int. J. Hydrogen Energy, 37(2),* 1482-1494.

[4] Palys, M. J., & Daoutidis, P. (2020). Using hydrogen and ammonia for renewable energy storage: A geographically comprehensive techno-economic study. *Comput. Chem. Eng, 127*, 106785.

610513 Hydrothermal Liquefaction of Bagasse for H_2, Bio-Oil and High Value Products

Abstract:

Hydrothermal Liquefaction of Bagasse for H_2, Bio-oil and High Value Products

Vinod S Amar, Anuradha Shende, Bharath K Maddipudi and Rajesh V Shende* Department of Chemical and Biological Engineering, South Dakota School of Mines and Technology, Rapid City, South Dakota 57701, United States of America

Corresponding Author: *Rajesh Shende, Department of Chemical and Biological Engineering, South Dakota School of Mines and Technology, Rapid City, South Dakota 57701, United States of America (E-mail: *Rajesh.Shende@sdsmt.edu*)

Abstract

Beer bagasse (spent brewer grain), a by-product generated in breweries, is rich in fiber and proteins. Being a low-cost biomass substrate, in this research, we investigated hydrothermal liquefaction (HTL) of bagasse for the production of H_2, bio-oil and other value-added products. HTL of beer bagasse was performed at 250-300°C under non-catalytic and catalytic conditions with bagasse to solvent (B:S) ratios of 1:5, 1:10, and 1:30. While HTL reaction was in progress, the product gases were analyzed using GC. After HTL reaction, biocrude was separated from the biochar and analyzed using HPLC, GC-MS and TOC analyzer. The HTL derived light bio-oil (LBO) and heavy bio-oil (HBO) were blended with biodiesel prepared from the spent cooking oil. HHVs of biochar, LBO, HBO, and blended fuels were calculated based on their elemental analysis. The results obtained with H_2, bio-oils, blend fuels, and value-added products from the HTL of beer bagasse will be presented.

600663 CuO-Cu_2O Redox Thermochemical Storage for High-Temperature Solar Process Heat

Abstract:

The integration of concentrating solar thermal technologies with thermal energy storage enables dispatchable power generation and around-the-clock operation of high-temperature industrial processes such as cement manufacturing, mineral and metallurgical extraction, waste recycling, and thermochemical fuels production. In particular, thermochemical energy storage (TCS) utilizes endothermic/exothermic reversible chemical reactions to store and release solar process heat at temperatures exceeding 600°C and with relatively high gravimetric heat capacity. We report on the experimental demonstration of a lab-scale TCS reactor prototype for storing heat at 1000°C. It consists of a packed-bed reactor filled with CuO-based granules undergoing reduction-oxidation (redox) cycles with a gravimetric heat capacity of about 470 kJ/kg, as measured by thermogravimetry. Two series of 30 consecutive redox cycles were performed by swinging either the oxygen partial pressure of the gas inflow (e.g. with air or nitrogen) or the air inflow temperature, yielding stable conversion in both cases. During the heat release, a spontaneous stabilization of the outflow temperature of the system at around 1025°C (equilibrium temperature of the CuO/Cu_2O system in air) was observed, which is beneficial for the downstream process. The influence of the gas flow rate and gas inflow temperature on the TCS system performance was parametrically investigated.

596060 Downstream Process with Countercurrent Chromatography to Recover Bioproducts from Sustainable Feedstocks

Abstract:

Separation of bioproducts produced from renewable feed streams is challenging in downstream process due to complex feed composition and low feed concentration. Current technologies such as ion exchange, adsorption, and liquid-liquid extraction are often faced with large solvent consumption, limited choice of resins, additional pretreatment and regeneration steps, and limited pH operation range. CounterCurrent Chromatography (CCC) is an emerging approach that uses at least two immiscible liquids for both mobile and stationary phase and is a potentially attractive separation method to recover bioproducts from complex process streams. Although there are multiple advantages of CCC for its liquid stationary phase in the context of natural products, there is little attention on employing CCC into downstream process for commodity bioproducts and biochemicals. This talk will discuss the feasibility of incorporating CCC into downstream process to produce high-value products from lignocellulosic feedstocks. Separation of bioproducts from lignocellulosic feedstocks such as alkaline pretreatment liquor and reductive catalytic fraction oils will be presented as demonstrations. The energy and solvent demands of the system correlated with the physicochemical properties of solvent systems, CCC operation mode, and feed composition will be discussed.

610487 Separation of Lignocellulosic Sugars from Pyrolytic Bio-Oil Using Simulated Moving Bed

Abstract:

Anhydrosugars are the major carbohydrate products of autothermal pyrolysis of lignocellulosic biomass that can be upgraded into biofuels and bio-based chemicals. Our research group has developed a fast pyrolysis process, namely autothermal pyrolysis, which can produce high yields of bio-oil by thermal depolymerization of pretreated biomass. A significant fraction of the heavy ends of bio-oil contain sugars, mainly levoglucosan, but also xylose as the major carbohydrate products. The anhydrosugars in bio-oil are acid-hydrolyzed to produce glucose, which can serve as a fermentation substrate for manufacturing common renewable fuels such as ethanol.

The bio-oil contains some phenolic monomers generated from the thermal depolymerization of lignin during the pyrolysis process. These phenolic monomers end up in the hydrolyzed, neutralized, and filtered water-soluble sugar fraction of the bio-oil due to a moderate solubility of phenol and other substituted phenolic monomers in water. The presence of phenolic species in the aqueous extraction of pyrolytic bio-oil poses significant toxicity challenges in employing fermenting microorganisms downstream and therefore must be removed from the aqueous solution of bio-oil sugars prior to fermentation. Our previous report shows effectiveness of hydrophobic polymeric resins in preferential adsorption of phenolic species compared to levoglucosan and glucose from an aqueous fraction of bio-oil.

In this work, we have developed a simulated moving bed technology (SMB) to employ continuous scale separation of sugars from phenolic species for an aqueous bio-oil fraction using the principle of adsorption chromatography. Sugars were separated from phenolics in the SMB processing a hydrolyzed, neutralized and filtered water-soluble fraction of bio-oil containing 225 g/L glucose, 10 g/L xylose and 21.1 g/L phenolics at the rate of 4 mL/min. Although it is a binary separation, the SMB is modeled to separate two classes of chemicals (sugars and phenolics) rather than two individual chemical species. The SMB uses a 3-1-1-1 configuration of six columns each having 53 cc bed volume and can operate at 0.5-8.0 bed volumes per hour (BV/h) flow rates. The columns are packed with hydrophobic resin Dowex L493 to adsorb the phenolic compounds preferentially on the resin surfaces as they are more non-polar compared to the sugars in aqueous solution. Ethanol is used to desorb the adsorbed phenolic species on the column surfaces while DI water is used to purge any adsorbed sugars on columns into a purified sugar solution. Water is also used to regenerate the columns after each cycle of feeding, desorption and purging before the next cycle of feed flow begins. The entire unit periodically switches the entry point locations of feed, purge, desorption, and regeneration steams to simulate a continuous flow of solid beds of adsorbent resin against the liquid solution of water-soluble sugars. The purified sugar product contained 154 g/L glucose and 0.8 g/L phenolics after 6 h of operation at 4 BV/h using 20 min switch interval time and 1:1 ethanol/water consumption. We have further explored the effects of switch interval time and flow rates of the streams on the separation efficiency of the SMB unit to optimize its design.

603910 Black Liquor Gasification Chemical Recovery and Combined Cycle Power at Pulp and Paper Mills Biorefinery

Abstract:

In the areas of chemicals and allied products and energy process engineering, it is of particular interest to design processes with low energy requirements (energy intensive) (preferably renewable materials, e.g., biomass) and the use of minimum raw materials or use of a byproduct as feedstock (materials intensive) and minimum number of equipment (i.e., lean manufacturing). This is the basis from which the area of process intensification (PI) arises. The main objective of PI is to propose production and manufacturing processes in plants as small as possible and with low total energy, materials, water, and other resources.

Pulp and paper are primarily made out of wood fibers originating from natural forests or pulpwood plantations. Recycled fiber and other fiber sources such as agricultural residue are also utilized and recycled fiber is becoming more commonly used in pulp and paper making. There are two significant pulping technologies available that differ greatly in terms of process, i.e., mechanical and chemical pulping. Approximately 30% of the total pulp production in European Union is from mechanical pulping while the rest is produced by means of chemical pulping. North America has major pulp and paper industry, about 21% of the total pulp produced is from mechanical pulping and rest is produced chemically.

A pulp mill that produces bleached kraft pulp generates 1.7-1.8 ton of black liquor (measured as dry content) per ton of pulp. Black liquor thus represents a potential energy source of 250-500 MW per mill. As modern kraft pulp mills have a surplus of energy, they could become key suppliers of renewable fuels in the future energy systems. Today, black liquor is the most important source of energy from biomass. It is thus of great interest to convert the primary energy in the black liquor to an energy carrier of high value. A key advantage of black liquor compared to biofuels and fossil fuels is that it is already at the mill; the handling infrastructure already exists and there are no collection and transport costs.

An extended case study is presented in which black liquor, a byproduct of wood-pulping process, is utilized to make chemicals and biofuels, in particular methanol, dimethyl ether and syngas. Different feedstock, use of byproduct and alternatives and principles of PI are commented (Das, 2020).

Reference: Das, T.K. (2020). Industrial Environmental Management: Engineering, Science and Policy, Wiley, Hoboken, NJ.

600048 From Biomass to Power Using Integrated Compact Pyrolysis, Combustor, and Stirling Engine

Abstract:

The feedstock logistic challenges, high feedstock procuring and shipping costs, high capital and operating costs, and a stringent requirement of making gasoline and diesel blends are the major barriers to commercialize biomass conversion pathways such as pyrolysis to make renewable fuels. These barriers can be overcome by developing portable technologies with a low weight, small footprint, and a high energy efficiency that can upgrade biomass to power. In this work, the viability of converting wood chips and plastic to power via integrated compact catalytic fast pyrolysis, combustor, and stirling engine technologies was determined through simulation and modeling analysis. The catalytic fast pyrolysis component consists of auger reactor followed by a catalytic upgrading reactor and the air-cooled condenser that upgrades wood chips and plastic to a high energy dense fuel. The ASPEN simulations followed by sizing analysis were performed to determine the size, weight, and net energy efficiency of producing 42.5-Watt equivalent of power generation from wood chips and plastic using the integrated portable technology. The impact of catalytic fast pyrolysis oil yields, oil quality, condenser type and its efficiency, type of fuel combustor, and combustor efficiency on the size, weight, and net energy efficiency of upgrading biomass to power was determined.

602795 Techno-Economic Analysis and Life-Cycle Assessment of Emerging Technologies for Bioprocessing Separations

Abstract:

Limited availability, rising costs, and environmental concerns about fossil fuels have generated considerable interest in finding alternative, renewable sources including biomass, which can be converted into a number of biofuels and bioproducts. In comparison to petroleum-based products, high processing costs, mainly associated with bioprocessing separations, limit widespread implementation of biofuels and bioproducts. Bioprocess-related separations are also complicated, regardless of the conversion pathway, due to the dilute nature of products and the chemically complex mixtures that result from biomass deconstruction. In many cases, bioprocessing separation approaches lack a technology baseline, or definition of the state of technology (SOT). This work focuses on evaluating the technology readiness of novel separations technologies for the conversion of biomass into biofuels and bioproducts and will address three key topics: 1) SOT description, 2) economic viability analysis of both SOT and innovative separations, and 3) environmental impact assessment of both SOT and novel processes. The current SOT for lignin valorization, dilute carbon recovery, and impurity removal is used to identify potential opportunities for improvement and provide a baseline for comparison with the emerging technologies. Detailed techno-economic analysis (TEA) and life cycle assessment (LCA) are applied to understand the key drivers and challenges related to the economic feasibility and environmental impacts, respectively.

603386 Bio-Based Pyrrolidone from Succinic Acid: Process Design and Life Cycle Assessment

Abstract:

Bio-based chemicals are a possible pathway to reduce the carbon emissions of the chemical industry. Thus, many bio-based process routes are currently under development. One of these process routes is the bio-based production of succinic acid, which is already established on an industrial scale. However, succinic acid is limited in its applications and thus in its market. The market for succinic acid has been recently expanded by the development of a new process route from succinic acid to the high-value chemical N-vinyl-2-pyrrolidone (NVP) by Haus et al. 2019. The market for NVP is expected to grow by 8.5 % p.a., reaching a total volume of 2.75 billion dollars by 2024. In combination with its current market price of NVP of around 5000 $/tonne, this makes NVP an attractive target as a bio-based chemical. However, for new bio-based chemicals, it is essential to evaluate the potential to reduced environmental impacts compared to its fossil-based counterpart, since a reduced impact is not guaranteed.

In this work, we thus conduct a comparative Life Cycle Assessment of n-vinyl pyrrolidone to examine the environmental advantages and drawbacks of bio-based NVP production. The functional unit of the LCA study is 1 kg of NVP and we apply cradle-to-grave system boundaries. The production system for the bio-based NVP includes the cultivation of four potential bio-feedstocks (corn, corn stover, sugar beet, and sugar cane), the fermentation of sugar to succinic acid, and the newly developed processes for the conversion of succinic acid to NVP. The production system for the fossil-based NVP includes the production of maleic anhydride based on butane, the conversion of maleic anhydride to gamma-butyrolactone, and the conversion of gamma-butyrolactone to NVP. Since no data on the conversion of gamma-butyrolactone to NVP is available in commercial databases, we modelled this process step in commercial flowsheeting software based on available patents.

To generate the inventory for bio-based NVP production, we design a process for the production of bio-based NVP and generated a flowsheet model based on lab results. The bio-based NVP production is based on a two-stage reaction: In the first reactor, succinic acid is amidated with ethanolamine and then hydrogenated with a ruthenium catalyst in a water phase to form N-(2-hydroxyethyl)-2-pyrrolidone (HEP). Succinic acid is converted to HEP with 70 % yield. Various by-products are formed, including an uncharacterized oligomer. After the first reactor, HEP is purified. For this purpose, HEP is first separated from the oligomer and water in an extraction column using dichloromethane as solvent. Afterward, HEP is separated from dichloromethane and by-products in a distillation column. HEP is then vaporized and dehydrated to NVP in a gas-phase reaction at 350 °C in the second reactor. This reaction takes place with above 99% conversion and 95% selectivity. By-products and educts are separated via distillation.

The LCA shows that the bio-based production of NVP reduces the environmental impact significantly compared to fossil-based production. Depending on the used feedstock, global warming impacts are reduced by 45-60 %. Production of succinic acid has the largest contribution to the overall environmental impact, followed by feedstock production. Increasing the selectivity of HEP in the first reaction step could further improve the bio-based process.

Thus, the bio-based production of NVP from succinic acid increases not only the economic potential but also reduces carbon emissions compared to its fossil-based counterpart and thus shows a large potential as a new bio-based production route.

612486 GIS Integrated Optimization Model for Location of Biogas Combustion Power Plants Under Carbon Offsets in Wisconsin

Abstract:

There is an increasing interest in low carbon renewable energy generation in the US. Several technologies have been developed to convert biomass into energy, and anaerobic digestion is one such technology to convert food waste and animal manure into power by biochemical conversion and combustion. Many studies have looked at the optimization of the biomass supply chains in combination with environmental impacts. However, there are very few studies in the literature on determining optimum location of biopower plants fed by food waste and manure. This study evaluates the optimum locations, sizes, and the number of plants for biopower production in Wisconsin using both mixed integer linear programming (MILP) programming and GIS network analysis (ArcGIS V10) tools. The main objective of the study is to maximize the profits for biopower facilities accounting for both the profits from supply chain and carbon credits. Furthermore, the model predicts the cost-effective supply chain networks between feedstock locations and biopower plants. From a predetermined set of 24 candidate locations, the model predicts facility capacities among a discrete set from 0.5 MW to 20 MW, and matches sources of dairy manure and urban food waste to feed the facilities. Other decision variables include tipping fee for food waste, manure cost, and carbon credits of \$15/ton CO_2 savings, where greenhouse gas savings for avoiding food landfilled and conventional manure management were taken from Ankathi et al. (2018). Inclusion of a carbon credit increased the profitability of biopower production and boosted biopower production capacity from 50 to over 80 MW in WI. The systems analyses presented will also include the net carbon footprint savings of the WI biogas power production compared to business as usual.

Ankathi, S.K., Potter, J., Shonnard, D.R., (2018) Carbon Footprint and Energy Analysis of Bio-CH_4 from a Mixture of Food Waste and Dairy Manure in Denver, CO, *Environmental Progress / Sustainable Energy*, 37(3), 1101-1111, DOI 10.1002/ep.12762

604128 Sustainability and Cost Tradeoffs of Epoxidized Sucrose Soyate: A Bio-Based Thermoset Resin

Abstract:

In this study, we assess the sustainability and cost of a bio-based thermoset resin derived from soybean oil, Epoxidized Sucrose Soyate (ESS). This bio–based resin has gained considerable attention due to its high functionality, significant mechanical properties and promise to provide a potentially sustainable alternative to petroleum-based thermoset resins in a variety of applications. We carried out a comparative life cycle assessment (LCA) of ESS and a functionally similar petroleum-based commercial resin. We model the large scale production of ESS using laboratory scale experimental data (with 98% conversion) using Aspen plus software, and use the results to assess the sustainability and cost of producing ESS resin in a commercial scale. The LCA results demonstrated that Epoxidized Sucrose Soyate (ESS) is more sustainable and environmentally beneficial compared to the conventional petroleum-based resin and thus can be a viable alternative. ESS displayed almost a 93% reduction in global warming. In addition to conducting LCA, we assess the sustainability of ESS according to green design principles including atom economy, use of renewable resources and process efficiency. To assess the economic feasibility of commercial production of ESS commercially, we perform techno-economic analysis (TEA). Coupling TEA with process simulation allows us to detect hotspots in the production process and optimize it to make the resin more competitive in the market. Moreover, we conducted sensitivity analysis to address uncertainty in the selling price of the ESS resin. The combination of environmental and cost performance data of the ESS provides critical decision-making information regarding the sustainability and cost-effectiveness of this resin.

607330 Carbon and Energy Implications of Fast Pyrolysis Biorefineries with Blended Feedstocks and Decentralized Supply Chain Design in the Southeastern United States

Abstract:

Variations in feedstock cost, quality, and availability are common risks for traditional biomass supply chains that rely on centralized design and a single type of feedstocks.[1] Two risk mitigation strategies have been suggested by the literature: using blended feedstocks[2] and decentralized preprocessing facilities.[3] Many Life Cycle Assessment (LCA) have been developed to either investigate the benefits of producing biofuels from varied feedstocks or decentralized biomass supply chains.[4,5] However, few of previous studies have investigated the environmental implications of using both blended biomass and decentralized preprocessing sites, nor did they quantified the impacts of variations of biomass quality, blending ratios, and preprocessing technologies.

This study addressed this knowledge gap by developing a cradle-to-gate LCA model for fast pyrolysis biorefineries using blended feedstocks with decentralized preprocessing depots.[6] Pine residues and switchgrass are two types of biomass included in this study. The primary tasks for depot preprocessing include size reduction, moisture reduction, and feedstock blending and densification. This study explored two different preprocessing technologies, conventional pelleting processing (CPP) and high moisture pelleting process (HMPP).[4] The life-cycle inventory (LCI) data of depot preprocessing and biorefining were generated by process simulations. A scenario analysis was conducted to evaluate the impacts of depot sizes (i.e., 250, 500 oven dry tonne (ODT)/day), biorefinery capacities (i.e., 1,000, 1,500, 2,000 ODT/day), blending ratios (i.e., pine residue/switchgrass ratio: 100%/0%, 75%/25%, 50%/50%, 25%/75%), and allocation methods for surplus electricity (energy allocation and system expansion). This study mainly focused on Global Warming Potential (GWP) and primary energy consumption.

The results showed that the life-cycle energy consumption and GWP of decentralized systems with depots are 0.7–1.1 MJ/MJ biofuel produced and 43.2–76.6 g CO_2 eq./MJ, respectively. The largest contributors to those results are biorefinery processing and depot preprocessing. Compared to the traditional centralized system without depots, the decentralized system reduces the energy consumption in biorefinery but increases the overall life-cycle environmental impacts due to the drying stage in depot preprocessing. However, by blending more switchgrass in the feedstock, both energy and GWP of depot preprocessing can be largely reduced. For example, increasing switchgrass blend to 75% would lower the energy and GWP of the decentralized system to a level similar to the centralized system using 100% pine residues. For the two preprocessing technologies, HMPP has lower energy consumption and GWP. In addition, allocation methods for the surplus electricity have large impacts on the results but do not change the comparative conclusions.

References

(1) Festel, G.; Würmseher, M.; Rammer, C. Scaling and Learning Effects of Biofuels Conversion Technologies. *Energy Technology* **2014**, *2* (7), 612–617.

(2) Edmunds, C. W.; Reyes Molina, E.; André, N.; Hamilton, C.; Park, S.; Fasina, O.; Adhikari, S.;

Kelley, S. S.; Tumuluru, J. S.; Rials, T. G.; et al. Blended Feedstocks for Thermochemical Conversion : Biomass Characterization and Bio-Oil Production From Switchgrass-Pine Residues Blends. *Frontiers in Energy Research* **2018**, *6* (August), 1–16.

(3) Lamers, P.; Roni, M. S.; Tumuluru, J. S.; Jacobson, J. J.; Cafferty, K. G.; Hansen, J. K.; Kenney, K.; Teymouri, F.; Bals, B. Techno-Economic Analysis of Decentralized Biomass Processing Depots. *Bioresource Technology* **2015**, *194*, 205–213.

(4) Iglesias, L.; Laca, A.; Herrero, M.; Díaz, M. A Life Cycle Assessment Comparison between Centralized and Decentralized Biodiesel Production from Raw Sunflower Oil and Waste Cooking Oils. *Journal of Cleaner Production* **2012**, *37*, 162–171.

(5) Hiloidhari, M.; Baruah, D. C.; Singh, A.; Kataki, S.; Medhi, K.; Kumari, S.; Ramachandra, T. V.; Jenkins, B. M.; Thakur, I. S. Emerging Role of Geographical Information System (GIS), Life Cycle Assessment (LCA) and Spatial LCA (GIS-LCA) in Sustainable Bioenergy Planning. *Bioresource Technology* **2017**, *242*, 218–226.

(6) Lan, K.; Ou, L.; Park, S.; Kelley, S. S.; Yao, Y. Life Cycle Analysis of Decentralized Preprocessing Systems for Fast Pyrolysis Biorefineries with Blended Feedstocks in the Southeastern United States. *Energy Technology* **2019**, 1–13.

607050 Integrating Machine Learning and Process Simulation to Estimate Energy and Greenhouse Gas Emissions of Activated Carbon Production

Abstract:

Activated carbon (AC) is a carbonaceous material that can be produced from coal and biomass. It has been used as fertilizer additives in soil amendment, adsorbents in water pollutant removal, catalysts in flue gas cleanup [1]. In 2015, the worldwide demand for AC was 12.8 million ton [2] and keeps growing. Depending on the feedstocks and technologies, the environmental footprints and resource consumption of AC production have large variations. For example, with steam activation, the cradle-to-gate global warming potential (GWP) of AC produced from hard coal is 18.28 kg CO_2-eq./kg AC, while the GWP of AC from wood chip is only 8.6 kg CO_2-eq./kg AC [3]. Various technologies (e.g., physical activation and chemical activation) applied to specific feedstock will also lead to different environmental impacts. It is critical to understand the environmental impacts of different combination of feedstocks, technologies and operational parameters in AC production. Such understandings could significantly enhance the feedstock selection, process optimization, and supply chain design for sustainable production of AC.

Several researches have evaluated the environmental impacts of AC production through Life Cycle Assessment (LCA) [3–7]. However, most studies have focused on AC produced from specific feedstock (e.g., coconut shell) and operational parameters, and thus their results cannot be extended to other feedstocks with different operational conditions. In this work, we integrated machine learning and process simulation to estimate the primary energy and Greenhouse Gas (GHG) emissions of AC produced from different feedstocks with varying operational conditions. Artificial neural network (ANN), a machine learning technique, was trained by a large dataset collected from literatures and used to predict the yield of steam AC production [8]. The trained ANN model was integrated with pyrolysis kinetic model [9] and Aspen Plus process simulation to estimate the overall energy and mass balance that were used to estimate overall primary energy consumption and GHG emissions [10].

The results indicated large variations in primary energy consumption and GHG emissions across 73 different woody biomass species (43.4-277 MJ/kg AC and 3.96-22.0 kg CO_2-eq./kg AC). A sensitivity analysis was conducted and the results showed the large impacts of biomass composition (e.g., hydrogen and oxygen content) and operational parameters (e.g., activation temperature). Such understandings will be helpful for feedstock selection and process optimization. For example, the results showed that higher hydrogen content (and H/C molar ratio) of biomass increases the primary energy consumption, and higher oxygen content (and O/C molar ratio) increases the energy recovery ratio. This conclusion will be useful for selecting suitable biomass for energy-efficient AC production. The process-based environmental data generated from this work can be used a transparent data source for future LCA studies related to AC or AC applications (e.g., wastewater treatment). Although this study focuses on AC made from woody biomass, the integrated modeling framework can be applied to other types of biomass or bio-based products.

References

[1] Dias JM, Alvim-Ferraz MCM, Almeida MF, Rivera-Utrilla J, Sánchez-Polo M. Waste materials

for activated carbon preparation and its use in aqueous-phase treatment: A review. J Environ Manage 2007;85:833–46. https://doi.org/10.1016/j.jenvman.2007.07.031.

[2] González-García P. Activated carbon from lignocellulosics precursors: A review of the synthesis methods, characterization techniques and applications. Renew Sustain Energy Rev 2018;82:1393–414. https://doi.org/10.1016/j.rser.2017.04.117.

[3] Gu H, Bergman R, Anderson N, Alanya-Rosenbaum S. Life Cycle Assessment of Activated Carbon From Woody Biomass. Wood Fiber Sci 2018;50:1–15.

[4] Bayer P, Heuer E, Karl U, Finkel M. Economical and ecological comparison of granular activated carbon (GAC) adsorber refill strategies. Water Res 2005;39:1719–28. https://doi.org/10.1016/j.watres.2005.02.005.

[5] Arena N, Lee J, Clift R. Life Cycle Assessment of activated carbon production from coconut shells. J Clean Prod 2016;125:68–77. https://doi.org/10.1016/j.jclepro.2016.03.073.

[6] Hjaila K, Baccar R, Sarrà M, Gasol CM, Blánquez P. Environmental impact associated with activated carbon preparation from olive-waste cake via life cycle assessment. J Environ Manage 2013;130:242–7. https://doi.org/10.1016/j.jenvman.2013.08.061.

[7] Kim MH, Jeong IT, Park SB, Kim JW. Analysis of environmental impact of activated carbon production from wood waste. Environ Eng Res 2018;24:117–26. https://doi.org/10.4491/eer.2018.104.

[8] Liao M, Kelley SS, Yao Y. Artificial neural network based modeling for the prediction of yield and surface area of activated carbon from biomass. Biofuels, Bioprod Biorefining 2019;13:1015–27. https://doi.org/10.1002/bbb.1991.

[9] Anca-Couce A, Scharler R. Modelling heat of reaction in biomass pyrolysis with detailed reaction schemes. Fuel 2017;206:572–9. https://doi.org/10.1016/j.fuel.2017.06.011.

[10] Liao M, Kelley S, Yao Y. Generating Energy and Greenhouse Gas Inventory Data of Activated Carbon Production Using Machine Learning and Kinetic Based Process Simulation. ACS Sustain Chem Eng 2020;8:1252–61. https://doi.org/10.1021/acssuschemeng.9b06522.

609416 Spatial Life Cycle Analysis of Soybean-Based Biodiesel Production in Indiana Using Process Modeling

Abstract:

Spatial Life Cycle Analysis (LCA) studies are vital for decision making as conventional LCAs utilize data that is representative of an average system with regards to life cycle stages, thus providing only an aggregate picture. For example, all regions may not use the same level of energy efficient production technology to produce a product and/or may use different raw materials to produce the same product. Such variations can lead to significant changes in the net impacts of different life cycle stages of a product. Therefore, it is important to account for the differences in impacts due to spatial variations in LCA studies. However, it is challenging to find spatial/ regional level life cycle inventory (LCI) data in the excising LCA databases. The more spatial desegregation desired, the more challenging it is to find data. To overcome this challenge, in this study, process modeling techniques were used to fill in the regional life cycle inventory data for performing a spatial LCA of biodiesel production from soybean in both north and central Indiana. Process models were developed for all the life cycle stages in the production process of soybean oil and soybean-based biodiesel for both central and northern regions of Indiana. The spatial LCI data for the soybean farming was available from the United States Department of Agriculture's (USDA) regional datasets.

Once all the LCI data was acquired, the data was fed to SimaPro to integrate with upstream processes to perform a full scale LCA of producing a liter of biodiesel from both north and central Indiana biodiesel producers. The impact categories studied were global warming potential (kg CO_2 eq) and freshwater eutrophication (kg P eq) using the 2008 US TRACI impact assessment method. The spatial LCA study made it possible to compare two different spatial scenarios of biodiesel production in two different regions of Indiana. With the analysis performed, it is possible to build a county level environmental impact distribution maps for soybean farming in Indiana. The results in the study show that the spatial distribution of environmental impacts of biodiesel production is sensitive to at least four factors: (1) fertilizer and water application rates for feedstock, (2) land usage, (3) distance to the processing plant, and (4) the available options for soybean farmers to sell their harvested soybeans. While the technology for biobased processing can be same, it can have different spatial impacts based on these factors.

604207 Advanced Manufactured Carbonate Materials for CO_2 Utilization in Algal Biomass Production

Abstract:

The ability to easily and cheaply transport carbon dioxide (CO_2) from point-sources such as power plants to distributed utilization sites of widely varying scales will enable broader application of captured CO_2. In the case of algal biomass cultivation, the use of carbonate materials for CO_2 capture, transport, and delivery to algae has the potential to: 1) eliminate the requirement for co-location of algal production facilities with power plants or costly, low-volume pipelines, 2) develop a means of inorganic carbon transport, storage, and CO_2 mass transfer tuned to algal productivity levels, and 3) reduce carbon capture costs by eliminating the need for energy-intensive CO_2 stripping and compression processes.

Lawrence Livermore National Laboratory has developed advanced manufactured materials that consist of aqueous sodium carbonate, which captures CO_2 as sodium bicarbonate, encapsulated within a CO_2-permeable polymer to increase the surface area and improve carbon capture rates by an order-of-magnitude compared with carbonate solution, shown in Figure 1. Advanced manufacturing enables the use of these otherwise kinetically-limited but inexpensive and environmentally-benign solvents for carbon capture, storage, and delivery to algal ponds. In collaboration with Sandia National Laboratory we have demonstrated the biocompatibility and ability of these carbonate-based CO_2 capture materials to deliver CO_2 and control the media pH in algal cultures up to 100L. In addition to experimental data we will show results from a techno-economic analysis and life cycle assessment of this technology.

Figure 1: Image of silicone-sodium carbonate composite printed into a mesh structure using advanced manufacturing, and scanning electron microscopy images of the cross-section of a filament of the composite showing the distribution of carbonate particles.

This work was performed under the auspices of the U.S. Department of Energy by Lawrence Livermore National Laboratory under Contract DE-AC52-07NA27344. (LLNL-ABS-808071)

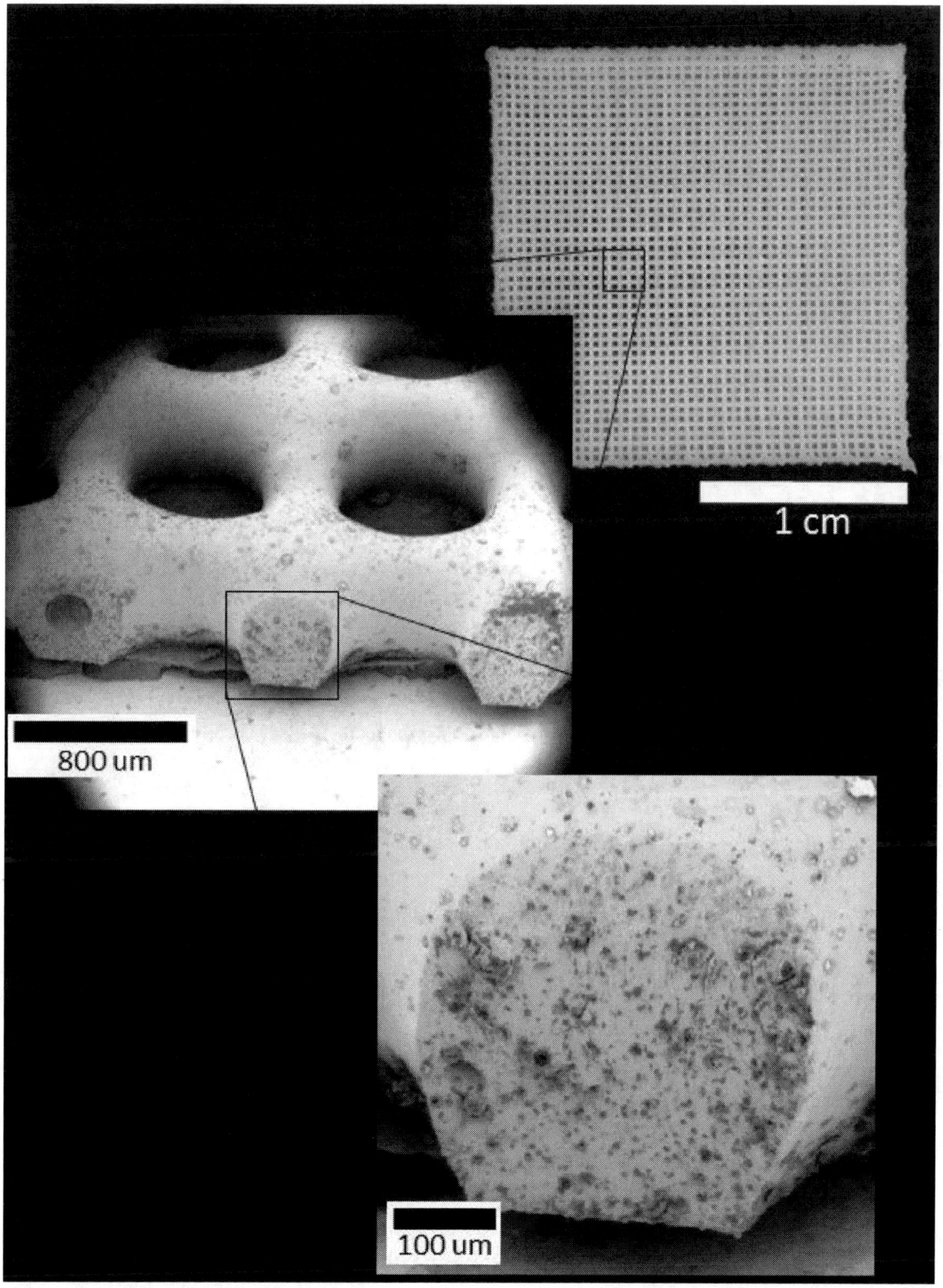
1 cm
800 um
100 um

601533 NH_4oh Looping with Membrane CO_2 Absorber and Distributed Stripper for Enhanced Algae Growth

Abstract:

The University of Kentucky Center for Applied Energy (UK CAER) has devised a unique, integrated CO_2 capture and utilization technology. CO_2 from coal-fired power generation flue gas is first captured at half the operating cost of a typical aqueous CO_2 capture system (CCS), distributed in an aqueous stream and then fixed by algae in bioreactors where the algae production is increased by 50% over that with a typical intermittent nutrient feeding system. Lower CCS operating cost is achieved by eliminating the flue gas pretreatment step for cooling and SO_2 removal, eliminating steam extraction from the power generation steam cycle for solvent regeneration, and eliminating CO_2 compression. Higher algae production is achieved by continuous, just-in-time nutrient feed to the bioreactors directly from a distributed solvent regenerator, which maintains the bioreactor pH for optimum growth.

The process starts with a uniquely configured membrane absorber, where the flue gas is indirectly contacted with an ammonium hydroxide (NH_4OH) solvent. Dissolved NH_3 is attractive for both CO_2 capture and as an algae nutrient. For CO_2 capture it is inexpensive, has a low regeneration energy, is thermally- and oxidatively-stable and has a viscosity near that of water, which makes is easy to transport. Numerous studies have shown that the scrubbing capacity of NH_3 is approximately 0.9-1.2 kg of CO_2/kg of NH_3, with a CO_2 removal efficiency of ~99% and half the solvent regeneration energy than that of 30 wt% MEA[1, 2, 3]. NH_3 is attractive as an algae nutrient due to its low cost.

The rich NH_4OH solvent is pumped to a set of distributed regenerators which are co-located with the algae bioreactors. Solvent pumping, transport and distribution reduces the balance of plant (BOP) cost compared to a typical aqueous CCS related to the flue gas duct and boost fan required to transport the flue gas. The energy required for the distributed solvent regeneration is supplied by solar-thermal panels eliminating the need for steam extraction from the power generation steam cycle. After solvent regeneration, the product stream contains both the CO_2 captured from the flue gas and volatized NH_3 from the solvent. This product stream is fed directly to the bioreactors, eliminating the need for compression of the CO_2 stream. The relative amounts of CO_2 and NH_3 in the product stream are adjusted and controlled by a controlling the regeneration conditions (pressure and temperature). The continuous feed of the right ratio of nutrients overcomes the typical inhibition of algae growth resulting from frequent pH swings in the bioreactor due to unbalanced (intermittent) feeding systems for CO_2 and N. Also, because the regenerators will operate at pressure and be located in close proximity to the bioreactors, there is no worry about pressure drop when sparging the gas into the algae. Sparging produces small bubbles which is beneficial for mass transfer efficiency.

One known challenge when using an NH_4OH solvent is high NH_3 emission. Hydrophobic membranes are used for CO_2 capture using an aqueous NH_3 solution[4, 5] without the direct contact between flue gas and aqueous solution. Additionally, UK CAER CO_2 capture and utilization process manages NH_3 slip in three extra measures. First, NH_3 slip is minimized by working with minimal species partial pressure, which is proportional to the concentration in the liquid. Hence, lowering the capture solvent concentration will lower the NH_3 partial pressure. Second, UK CAER's previous work has demonstrated that the addition of Zn^{2+} into NH_3 solutions to chelate the NH_3 can reduce NH_3 volatility. Third, the configuration of the membrane CO_2 absorber utilizes condensed water from the flue gas to continually wash the gas-side of the membrane to reduce fouling and recapture NH_3 slip.

Additional details about the UK CAER unique, integrated CO_2 capture and utilization technology will be presented along with technology development plans.

1. Diao, N., Q. Li, and Z. Fang. 2004. Heat transfer in ground heat exchangers with groundwater advection. *International Journal of Thermal Sciences*. 43: 1203-1211, <https://doi.org/10.1016/j.ijthermalsci.2004.04.009>
2. He, Q., M. Chen, L. Meng, K. Liu, and W. Pan. 2004. Study on Carbon Dioxide Removal from Flue Gas by Absorption of Aqueous Ammonia. Western Kentucky University. <https://www.semanticscholar.org/paper/Study-on-Carbon-Dioxide-Removal-from-Flue-Gas-by-of-He-Chen/4994d1230f70482e0b2fa8abbecd9a1daedfd5e0>
3. Yeh, A.C., and H. Bai. 1999. Comparison of ammonia and monoethanolamine solvents to reduce CO_2 greenhouse gas emissions. *The Science of the Total Environment*. 228: 121-133, <https://doi.org/10.1016/S0048-9697(99)00025-X>
4. Villeneuve, K., D. Roizard, J.C. Remigy, M. Iacono, and S. Rode. 2018. CO_2 capture by aqueous ammonia with hollow fiber membrane contactors: Gas phase reactions and performance stability. *Separation and Purification Technology*, 199: 189-197, <https://doi.org/10.1016/j.seppur.2018.01.052>
5. Toro Molina, C., and C. Bouallou. 2016. Carbon dioxide absorption by ammonia intensified with membrane contactors. *Clean Techn Environ Policy* 18, 2133–2146 (2016) <https://doi.org/10.1007/s10098-016-1140-0>

609009 Enhancing Monoethanolamine Regeneration at Low Temperatures Via in-Situ Precipitation of Solid Carbonates

Abstract:

Climate change is one of the biggest risks facing mankind and scientific evidence correlates the rise in global temperature to the concentration of CO_2 in the atmosphere. Carbon dioxide gas represents a majority of 82% of the total greenhouse gas emissions in the United States with nearly 33% of total emission coming from power generation plants that burn coal and natural gas. Currently, Carbon Capture and Sequestration (CCS) with amine solvents stands as the most effective and industrially applicable chemical separation technology demonstrated at a commercial scale. Aqueous solutions of monoethanolamine (MEA) are highly regarded as ideal chemical absorbents for CO_2 capture due to their high absorption capacity (reaching 0.5 mol CO_2/mol MEA), fast reaction kinetics, and high degree of regenerability. However, amine scrubbing processes suffer from the parasitic energy demand where half of the operational energy is devoted to thermally regenerate the rich amine solvent and the other half is to compress the captured CO_2.

Precipitating CO_2 as a solid phase represents an economical and effective way of regenerating CO_2-rich amine solutions. Particularly, reacting carbonate ions in solution with calcium ions (from various sources) takes advantage of the availability of large amounts of dissolved CO_2 in amine solutions and the spontaneous mineralization reaction to $CaCO_3$ – the most thermodynamically stable form of carbon. In this work, we show that CO_2 loading of 22 vol% MEA solutions can be significantly reduced from 0.28-0.42 to 0.034-0.05 mol CO_2/mol MEA by mixing the loaded solutions with stoichiometric amounts of CaO, $Ca(OH)_2$ and $CaCl_2$ at low regeneration temperatures. In our experiments, $Ca(OH)_2$ and CaO demonstrated an outstanding mineralization property by removing 86.9 mol% and 83.6 mol% of the absorbed CO_2, respectively, with almost no gas desorption. On the other hand, $CaCl_2$ was found to mineralize 72.1 mol% of the absorbed CO_2 while 13.4 mol% was desorbed as a gas. The presence of chlorine ions was found to reduce the alkalinity of the MEA solution which leads to increased gaseous CO_2 desorption during regeneration. In addition, our results demonstrate that partially mineralizing CO_2 and generating solid $CaCO_3$ in solution improves the kinetics of the subsequent absorption/desorption cycles. Specifically, the maximum absorption and desorption rates for the solid containing solutions showed 8.8 and 20% increase in the maximum rates of absorption and desorption, respectively, over the solid free solutions.

599191 Life Cycle Analysis Tools for Evaluating the Environmental Footprint of Carbon Utilization Projects

Abstract:

Capturing carbon dioxide (CO_2) emissions from power and industrial sources and using that CO_2 to make useful products is an emerging area of research that would benefit from a consistent and unbiased framework, such as life cycle assessment/analysis (LCA), to better understand the environmental impacts and net life cycle GHG reductions compared to the current state of the alternative in the marketplace. From a methodological perspective, CO_2 utilization systems are complex due to the intrinsic links established between the power sector and the utilization sector (e.g., biofuels, cement, chemicals, etc.). In their 2019 report, "Gaseous Carbon Waste Streams: Utilization Status and Research Needs," the National Academies identified life cycle assessment benchmarking as a key research priority to "facilitate consistent and transparent assessments of the net greenhouse gas emissions of carbon utilization technologies." Technology developers and LCA analysts could benefit from guidance that establishes best practices for CO_2 utilization LCA to allow for unbiased assessment of potential benefits and opportunities for further research.

LCA is uniquely suited to employ in the analysis of CO_2 utilization technologies because of the methodical approach that forces a cradle-to-grave understanding of the system. It is not uncommon to see CO_2 utilization analyses that focus mainly on the utilization technology and apply a simplified approach to the upstream CO_2 source. We contend that a robust treatment of the upstream CO_2 source is imperative in any CO_2 utilization LCA, because of the important link between the source of the CO_2 and the use of the CO_2 in the overall environmental impact. This holistic view is necessary to understand the net benefits of the entire system, not just a subset. Conducting LCA guidance work early is important in the development of these emerging technologies, because it allows time to implement change while technologies are still nascent. Additionally, U.S. Department of Energy (DOE) and the federal government is increasingly requiring LCA as part of funding for primary research and tax incentives like "45Q" for CO_2 capture, utilization, and storage projects (H.R. 1892, 2018).

In the interest of supporting the creation of useful and consistent LCAs of CO_2 utilization projects, the DOE National Energy Technology Laboratory has developed a CO_2 utilization LCA toolkit. The primary audience for the toolkit is the group of principal investigators that have been awarded a grant through a DOE Carbon Use and Reuse funding opportunity announcement. However, we believe that there is wider application beyond those grant recipients. The toolkit comprises a guidance document, reporting templates, openLCA contribution analysis tools, LCI data, and subject matter expertise. The guidance document should be considered as a companion to the requirements established in ISO 14040/14044, providing specific guidance on methodological issues for scoping and completing the LCA. The guidance, data, and tools are publicly available and free. This presentation will provide an overview of each part of this first-of-a-kind toolkit, it's utility to the LCA community, and value in assessing emerging technologies.

References

H.R. 1892, 115 Cong. (2018) (enacted). Retrieved from https://www.congress.gov/bill/115th-congress/house-bill/1892

National Academies of Sciences, Engineering, Medicine. (2019). Gaseous Carbon Waste Streams Utilization: Status and Research Needs. Washington, DC: The National Academies Press.

595898 First Operation of 5 kW_{CH4} Local Carbon Recycling System: Product Gas Quality of CO_2 Methanation

Abstract:

An industrial sector makes up more than one-third of global CO_2 emission and therefore efforts for industrial CO_2 mitigation are to be further intensified and accelerated. Companies and organizations with awareness of its importance and urgency have launched initiatives, declaring their strategies as well as specific and challenging targets [1-3]. Application of on-site renewable energy, e.g., solar and wind, in industrial factories is one of the most promising approaches to reduce CO_2 emission significantly. However, its fluctuating nature in power necessitates energy storage within the premises when surplus power arises. Also, due to spatial constrains of factories, its storage media needs to have high volumetric energy density.

This study explored a local carbon recycling system (LCRS, Figure 1). The process begins by generating electricity from on-site solar cells and/or wind turbines. Thereby H_2 is produced, electrolyzing water. Concurrently, CO_2 is captured from flue gas of furnaces. Then CH_4 is synthesized by CO_2 methanation reaction ($CO_2 + 4H_2 \leftrightarrow CH_4 + 2H_2O$). The product gas is subsequently stored in a pressurized tank and, as is necessary, returned to furnaces as a fuel. The system realizes energy storage with CH_4 which has a high volumetric energy density, also enabling to utilize presently existing natural-gas furnaces with minimum retrofitting.

LCRS needs to reduce operating expense by enhancing its energy efficiency, which requires the load reduction of auxiliaries such as vacuum pumps, compressors and heaters. The previous studies [4,5] developed a two-staged methanation reactor which realizes high CO_2 conversion (≧99%) at low pressure of 200 kPa (G) and thermally self-sustained conditions. The reactor can recover the redundant heat in the process via thermal oil at the same time. A CO_2 separator was also developed in another study [6]. The separator repeatedly switching physical adsorption and desorption of CO_2, swinging temperature and pressure. It features exploitation of methanation heat for temperature swing and H_2 sweeping for reducing vacuum pump load.

This study built up a 5 kW_{CH4} LCRS integrating the previously developed methanation reactor and CO_2 separator, and conducted its first operation. While a burner, which emulates a furnace in a factory, burnt the product gas of methanation, the CO_2 in the flue gas was continuously captured and converted to CH_4. The LCRS was started up from a cold state, sequentially implementing the followings: (1) heating of the reactor and the separator, (2) flue gas provision to the separator and feed gas provision to the reactor, and (3) provision of the methanation product gas to the burner. The result shows that the system can complete all the startup procedure in 53 minutes and realize the methanation product gas quality of more than 93% CH_4 concentration at steady state.

[1] Toyota Motor Corporation. “Challenge 3 Plant Zero CO_2 Emissions Challenge”, retrieved Feb. 2, 2020, from https://global.toyota/en/sustainability/esg/challenge2050/challenge3/.

[2] Denso Corporation. “EcoVision”, retrieved Feb. 2, 2020, from

https://www.denso.com/global/en/csr/environment-report/ecovision/

[3] Toyota Industries Corporation. "Reduction in the volume of CO_2 Emissions" retrieved Feb. 2, 2020, from https://www.toyota-industries.com/csr/environment/process/co2/.

[4] Sayama, Shogo, et al. "Energy Efficiency Improvement of CO_2 Methanation Process by Using a Thermally Self-Sustained Two-Stage Reactor: Preliminary Evaluation of a Reactor Concept", Kagaku Kogaku Ronbunshu, 2020, 46(3).

[5] Sayama, Shogo, et al. "Local Carbon Recycling System: Performance Evaluation of a Thermally Self-sustained Two-staged Methanation Reactor", SCEJ 85th Annual Meeting, 2020.

[6] Kunitomi, Seiichi, et al. "Local Carbon Recycling System: Energy Consumption Reduction of CO_2 Adsorber", SCEJ 85th Annual Meeting, 2020.

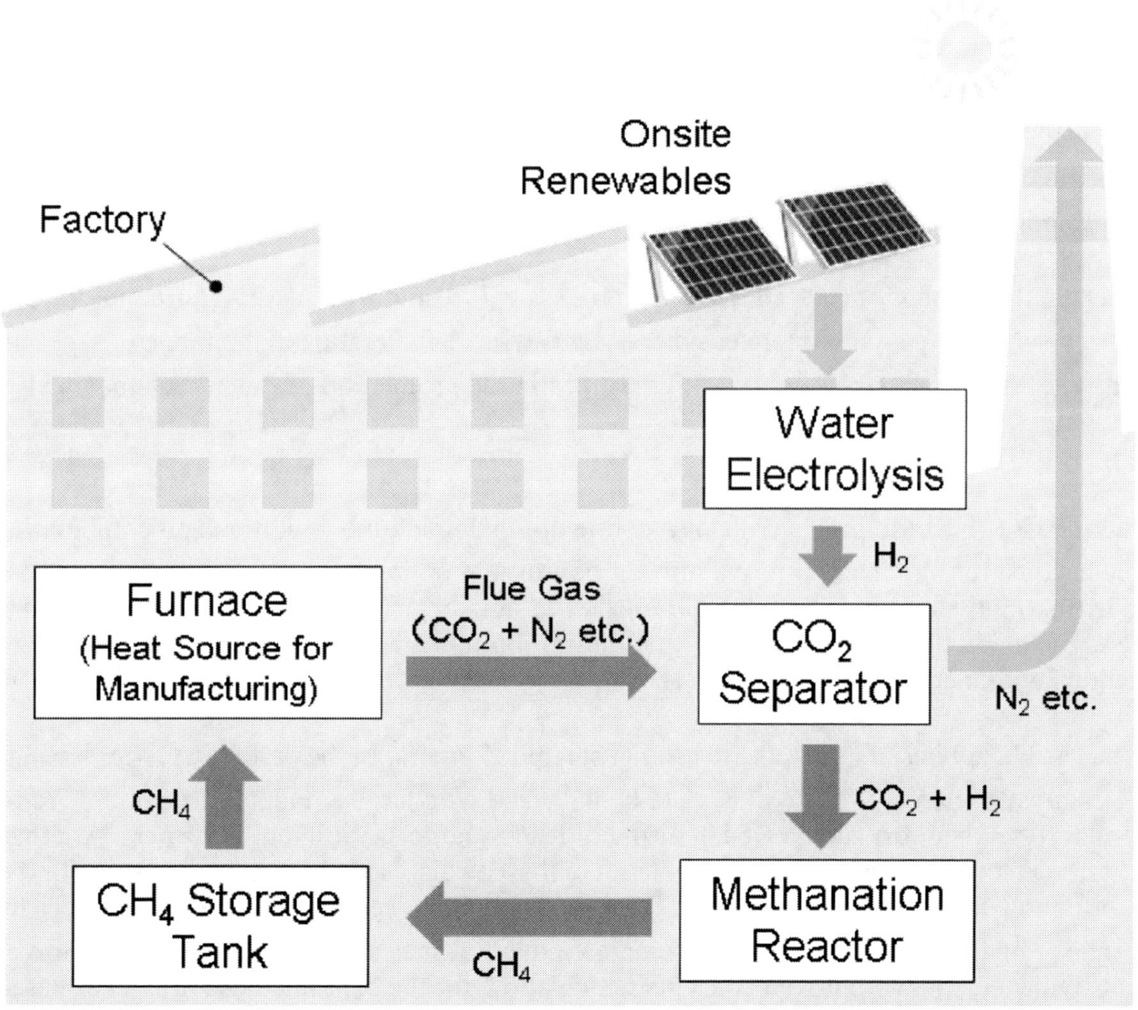

599069 Industrial Scale CO_2-Reforming for Production of CO-Rich Synthesis Gas at Low Steam to Carbon Ratio

Abstract:

1. Introduction

Important bulk chemicals such as hydrogen, ammonia, and methanol are typically produced in a multiple step process. The first step is normally conversion of natural gas, or a similar feedstock, to produce hydrogen or synthesis gas (a mixture consisting of mainly hydrogen and carbon monoxide). This is followed by the actual synthesis and purification. The synthesis gas production step is often carried out by reforming of the feedstock with mixtures of steam and carbon dioxide (henceforth referred to as CO_2-reforming).

CO_2-reforming is an environmentally interesting process as it offers a way of utilizing CO_2, which is a polluting greenhouse gas, and in many industries often is considered as a waste product. CO_2-reforming is a process which can be designed with overall negative CO_2-emissions, or in other words can be designed to utilize more CO_2 than what is produced. It is therefore expected to play an important role in combination with CO_2 capture technologies.

An increase in the amount of CO_2 added to the process will result in an increased fraction of CO in the produced synthesis gas. In principle, synthesis gas with H_2/CO ratios in the range 0,5 – 3 can be produced. These CO-rich gasses are typically utilized in the production of functional chemicals.

2. CO_2-reforming in industry

When talking CO_2-reforming, it is critical to define optimal operating conditions, in combination with a suitable catalyst to avoid carbon formation. Carbon formation in a typical steam reforming plant is dictated by thermodynamics. In the typical reformer design, no affinity for carbon formation is allowed anywhere in the catalyst bed. In Fig. 1 the carbon limits are illustrated for various feed compositions with methane, steam, and CO_2.

In most cases, Nickel-based catalysts are used in steam reformers. In the case of CO_2-reforming over a Nickel-catalyst, the process gas will have to be balanced with large amounts of steam to circumvent the potential for carbon formation. This in turn will lead to larger, inefficient reformers. Alternative technologies exist in the market, such as noble metal based catalysts which have a lower affinity for carbon formation and a higher activity for CO_2-reforming. However, high raw material prices make this alternative unattractive for extensive use on an industrial scale. Therefore, there is incentive to continuously develop new improved concepts which allow for the continued use of Nickel-based catalysts for CO_2-reforming.

3. New high temperature reactor for production of CO-rich synthesis gas

A new reformer technology has been developed, where preheated CO_2 is added directly downstream of a main reformer and then equilibrated in an adiabatic reactor. Supported by a detailed understanding of the underlying thermodynamic and kinetic mechanisms, the technology utilizes the high temperature of the synthesis gas from the main reformer to circumvent carbon formation. This is illustrated in Fig. 2 for an example where the main reformer is operated at steam to carbon ratio of 1.0.

In theory, this technology can be utilized to tailor a synthesis gas to practically any H_2/CO ratio, using a cost-efficient nickel based catalyst, while maintaining a low steam to carbon ratio in the main reformer – all this, without the risk of carbon formation. In principle the adiabatic reactor may be added downstream of any reformer type and can be used for both grassroot and revamp cases.

Bench scale experiments have been carried out to demonstrate carbon free operation with the new reactor technology at significantly lower steam-to-carbon ratios than is possible with stand-alone CO_2-reforming.

A series of economic and process calculations have been carried out to illustrate the benefits of the new technology. Selected results include:

- The size of a steam reformer can be reduced by up to ca. 20%;
- The synthesis gas composition is shifted towards CO with high selectivity relative to the converted CO_2;
- The solution can be tailored to produce synthesis gas covering a large range of H_2/CO ratios.

4. **Conclusion**

In conclusion, the new high temperature CO production reactor is a promising solution for producing synthesis gas with a high content of CO at a low steam to carbon ratio. The technology can be used to retro-fit an existing unit towards more CO production or included in new projects. It is an excellent match in cases where excess CO_2 is available. In the presentation, the new technology will be described and the benefits will be illustrated with specific examples.

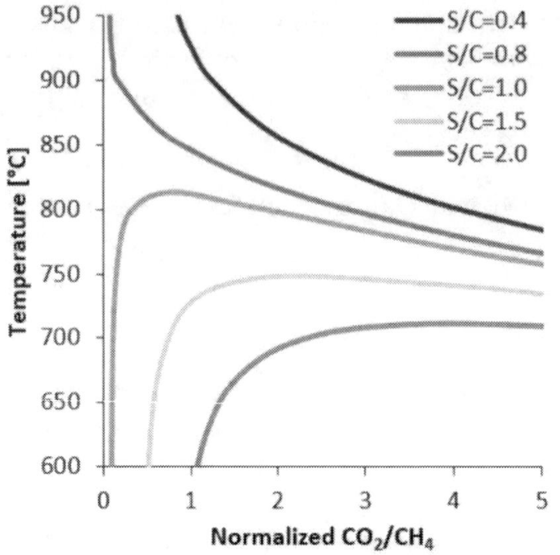

Fig. 1: Thermodynamic evaluation of carbon limits as a function of normalized H_2O/CH_4 (S/C), CO_2/CH_4, and temperature. A point placed below a curve in the graph will have a potential for carbon formation

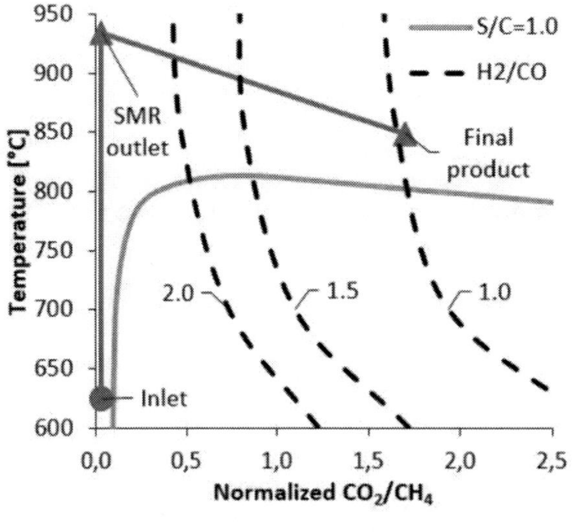

Fig. 2: A working line illustrates how the carbon limit can be circumvented by doing high-temperature CO_2 addition. The main reformer (in this case a Steam Methane Reformer) is operated at H_2O/CH_4 (S/C) of 1.0.

600317 Effects of CO_2 Methanation on Economic Feasibility and Global Climate Change in Coal-Fired Power Plants

Abstract:

The continued use of fossil fuels due to industrial development, economic and population growth has increased the concentration of CO_2 in the atmosphere that causes climate change. The carbon capture and storage (CCS) technology has been developed to mitigate the increase of CO_2 in the atmosphere. The CO_2 capture and utilization (CCU) technology has been proposed to resolve the limitation of the storage capacity of the captured CO_2. This study presented a techno-economic feasibility of 500 MW_e coal-fired circulating fluidized-bed (CFB) power plants including CCU. An amine absorber unit (AAU) and a CO_2 processing unit (CPU) were used to capture CO_2, in air- and oxy- combustion CFBs, respectively. Only 2% and 5% of the captured CO_2 were methanized with H_2 because of the excessive consumption of electricity in water alkaline electrolysis (AEL) that produces H_2. The effects of CO_2 methanation with H_2 on the global climate change were investigated for coal-fired power plants around the world.

598979 Calcination-Free Production of Calcium Hydroxide at Sub-Boiling Temperature

Abstract:

Calcium hydroxide ($Ca(OH)_2$), also known as portlandite, slaked lime, or hydrated-lime, is a widely used product in a variety of industries: as a soil stabilizer in road and building construction, as a bleaching agent in the Kraft paper process, and as a flocculent in the water treatment industry. According to the United States Geological Survey (USGS), global lime ($Ca(OH)_2$ and CaO) production reached 420 million metric tons in 2018. $Ca(OH)_2$ has drawn attention in the context of climate change mitigation because it can be used as a CO_2 sorbent via the formation of a stable calcium carbonate ($CaCO_3$) solid following the reaction: $Ca(OH)_2 + CO_2 \rightarrow CaCO_3 + H_2O$ a reaction which can offer cementation, e.g., in concrete. But, the production of $Ca(OH)_2$ features a carbon intensity on the order of 0.77 tons of CO_2 per ton of $Ca(OH)_2$ produced.

This work establishes a new route to synthesize $Ca(OH)_2$ by low-temperature precipitation from an alkaline solution saturated in the solid. This calcium-rich alkaline solution can be obtained by dissolving Ca-bearing wastes such as slags in water. The dissolution of slag releases calcium ions in a wide range of concentrations below the saturation level of $Ca(OH)_2$ in water. Therefore, the slag leachate is concentrated using reverse osmosis (RO) by such that a less concentrated stream permeates through the membrane while dissolved salts and other impurities become concentrated in the retentate stream. Such concentration occurs on account of a combination of size exclusion, charge exclusion, and physico-chemical interactions between the ions in solution and the functionalized semipermeable membrane. Once the solution is concentrated, precipitation of $Ca(OH)_2$ from the saturated leachate can be induced by a temperature-swing process since the solubility of $Ca(OH)_2$ in water decreases with temperature. This allows for the production of $Ca(OH)_2$ crystals via a low-temperature, aqueous route.

The proposed process presents several advantages compared to traditional $Ca(OH)_2$ production. First, the raw materials constitute industrial by-products and waste, that are available at large scale, globally. Second, the process eliminates the CO_2 generation associated with the calcination reaction. Finally, a detailed account of the mass and energy balances discloses that the operation of this process is particularly advantaged by the use of low-grade waste heat (e.g., $T < 100°C$), e.g., from industrial operations (e.g., power plants, steel plants, etc.) to operate the temperature swing process. This ensures that the CO_2 emissions associated with the proposed $Ca(OH)_2$ production process are very low; i.e., on the order of 0.27 tons of CO_2 per ton of $Ca(OH)_2$ produced. Taken together, this work highlights the opportunities, advantages, and knowledge gaps that need to be overcome to create low-temperature calcination-free pathways for the industrial-scale production of portlandite and other divalent hydroxides.

601618 “Sustainabilty – the Need for Scalable Solutions & a Framework to Make Decisions”

Abstract:

Sustainability had taken a new dimension in recent times with employees and investors challenging industries to be more sustainable. An overview of CO_2 generation sources globally and the USA will be presented and a case will be made for the need to scaleable solutions. The scalability and mutual interaction of these solutions have resulted in trade-offs. In this talk, the need for a frame-work for decision making and the role scientists and engineers can play is discussed.

606476 The Impact of Policy Instruments on Deployment of Low-Carbon Heating in the UK: A Whole System Optimisation Study

Abstract:

As part of the UK's commitment to emitting net zero greenhouse gases by the year 2050, the country must decarbonise the space heating sector, which currently contributes 17% of these emissions. This will involve retrofitting 24 million residential buildings, currently using natural gas, to a low carbon form of heating. The main options to achieve this are either hydrogen based, such as hydrogen boilers, or electricity based, such as electric heat pumps and electric resistive heating. Each option involves considerable costs along the supply chain, including upgrading infrastructure and installing new equipment in homes.

One of the key challenges facing policymakers is that, at present, natural gas heating is often the cheapest form available to households. In addition, considerable investment will be required in hydrogen production and transportation before it becomes available to consumers. Consequently, the transition away from natural gas heating is likely to require financial incentives. Each of the strategies is likely to present different costs to different stakeholders and so policies will be required targeting different points of the supply chain to incentivise the uptake of low-carbon heating.

The aim of this study is to assess how policy instruments, designed to promote the uptake of renewable heating, might impact both the heating technology mix and the supporting infrastructure when considering the least cost solution to decarbonisation. The Value Web Model, a mixed integer linear programming model for integrated energy value chains, was developed to optimise the deployment of heating technologies alongside energy generation, transportation, and storage to meet energy demands in the UK. This model considers natural resource availability (including wind, solar, land, and natural gas) alongside energy demand with high spatial and temporal resolution to determine not only which technologies should be used, but where they should be located. The results will show how different incentives applied at different points along the supply chain and in different amounts impact both the penetration of the technology they are aimed at and the overall technology mix. This work will highlight the impacts of incentives on the wider energy system and add to the discussion, relevant to policymakers and researchers, on how to decarbonise energy systems at the lowest cost.

611169 The Role of Chemical Engineering in Sustainable Energy Policy

Abstract:

Technological advancements are often made long before their application is mandated by policy. Conversely, the way a technology is regulated can make or break its adoption. Sustainability of energy involves balancing our energy needs with the well-being of the environment and the community. This talk will examine examples of policy successes and how chemical engineering can contribute to sustainability of energy and energy policy.

600820 Review of US Environmental Protection Agency General Exposure Scenario Modeling for Evaluating Chemical Safety

Abstract:

Generic Scenarios were initially developed by the US Environmental Protection Agency (USEPA) as a part of the chemical risk assessments process required under the Toxic Substances Control Act (TSCA) used by the USEPA to evaluate the hazards associated with the modeled chemical. From these humble beginnings, Generic Scenario documents have evolved into Organisation for Economic Co-operation and Development (OECD) Emissions Scenario Documents (ESDs) produced to describe a chemical's sources, production processes, pathways and use patterns. This presentation provides a review of exposure modeling performed in the development of existing Generic Scenario documents. This review was conducted to understand the evolution of the Generic Scenario process, with a goal of simplifying the modeling approaches used to estimate exposure levels and human health effects related to manufacture and use of chemicals currently required under the Lautenberg Chemical Safety for the 21st Century Act.

597556 Effectively Communicating Sustainability and Engineering within Federal Policy

Abstract:

As engineers and scientists we wonder why it seems that science is void from the decisions made be policy makers and governmental entities. There are scientists and engineers in government, but we have to be willing to put in the work to be effective, and those implementing policy have to be willing to listen. The ideas presented in this talk will give insight into how varying levels can effectively communicate scientific and sustainable ideas to those in policy. Part of this talk will be describing the process for becoming a Science and Technology Policy Fellow of the American Association for the Advancement of Science, as well as the roles of a fellow in the Executive branch of the federal government and how sustainability can be communicated to an array of non-scientific audiences.

603063 Making Amino Acids from Biomass Via Chemical Catalysis

Abstract:

Current amino acid production is largely based on microbial conversion processes. L-Glutamate, L-lysine, L-threonine, and L-isoleucine production using *Corynebacterium glutamicum* are all well-established in industry, but these bio-based processes are not without problems. Given the challenges encountered in bio-based amino acid production, there are substantial opportunities for chemists to develop robust chemo-catalytic processes from renewable sources such as biomass. Under the concept of biorefinery, a number of catalytic processes to transform various types of biomass or biomass components, such as cellulose, hemicellulose, lignin, lipids and chitin, into value-added chemicals have been developed in the last two decades, which has been a testament to the power of chemical catalysis in biomass valorization but chemical conversion of biomass into amino acids is rare.[1]

Considering these, we developed a two-step protocol to make several amino acids from sugars.[2] The first step prepares α-hydroxy acids by catalytic transformations of polysaccharides, lignin, and their derivatives using procedures reported in the literature. Following that, the hydroxyl groups are replaced by amine groups to access a series of α-amino acids. To this end, a supported Ru catalyst with high efficiency and recyclability was developed for the desired conversion, allowing six amino acids to be generated from their corresponding α-hydroxy acids. Based on the catalytic system developed in the study, a new two-step chemical process to convert glucose into alanine in 43% yield was demonstrated. Following that, we developed a photo-catalytic pathway to enable the same transformation but reduce the reaction temperature from 200 degree to near room temperature. More recently, we developed one-pot conversion of waste glycerol into alanine[3] and three-step transformation of hemicellulose into proline.

[1] Yan, N.; Wang, Y. Chem, 2019, 5, 739.

[2] Deng, W. et. al. PNAS, 2018, 115, 5093.

[3] Wang, Y. et. al. Angew. Chem. Int. Ed., 2020, 59, 2289.

607235 Development of Solid Base Activated Carbon Catalysts for Ketone Condensation and Aviation Fuel Synthesis

Abstract:

Acid hydrolysis and fast pyrolysis of biomass can generate a range of aldehydes (e.g., furfural) and carboxylic acids (e.g., acetic acid). The acids can be converted to linear ketones (e.g., acetone) via catalytic ketonization and furfural to cyclic ketones (e.g., cyclopentanone or CP) via hydrogenation and rearrangement using supported catalysts. Recent work indicates that the bio-derived ketones can be converted to aviation fuel by aldol self-condensation followed by hydrodeoxygenation. The cyclic hydrocarbons derived from self (CP only) or cross (CP+linear ketones) condensation would have high densities and volumetric heating values due to strong ring strain. Studies have shown that anionic clays, particularly hydrotalcites, have a high density of base sites generating higher condensation conversion and yield compared to other catalysts, yet little work has studied hydrotalcite supported catalysts and cross condensation reactions. In this work a Mg/Al-hydrotalcite on activated carbon (AC) catalyst is synthesized and a series of self and cross condensation reactions with acetone and cyclopentanone are studied. It is theorized that the AC support will significantly increase surface area and base active sites and thus increase reaction rates and yields (relative to typical base catalysts). The solid base catalyst could potentially be recoverable, reusable, and used in continuous processing. Results will be presented for a range of reaction conditions and compared to a traditional base catalyst.

604067 A Pathway Towards Carbon Neutral Fuels: Synthesis of I-Butanol with Renewable Methanol

Abstract:

The Institute of Electrochemical Process Engineering (IEK-14) of the Jülich Research Center is taking part in the project C³-Mobility, exploring different pathways for the synthesis of carbon neutral fuels from green methanol. Unlike methanol and ethanol, pure C4 alcohols comply with fuel regulations.

A lot of research nowadays deals with the upgrading of ethanol into 1-butanol, because it is more favourable as fuel than ethanol[1]. A disadvantage of the used *Guerbert* Reaction is the low selectivity and that methanol will react only as electrophile[2].

We therefore decided to build upon the C-methylation protocol of Siddiki et al for the synthesis of iso-butanol[4]. In a typical experiment, 10% ethanol is reacted in a methanolic NaOH solution (N_2-Atmosphere, 423K, autogenous pressure, 4h). Samples are taken every 60 minutes and measured by GC-MS . Triple determination was applied to verify the quality of the measurements. We tested different commercially available catalysts from *Alfa Aesar* (5% Rh, 5% Pd, 5% Ru 5% Pt, 1% Ir support: carbon), that have been examined by XRD and BET. The preliminary results show that iso-butanol is the only detectable liquid final product in differing amounts for different catalysts.

Wu et al. correlated the ethanol upgrading experiments of Riittonen et al with the d-Band center and found that the conversion decreased with a downshift of the d-Band center[3]. Our findings differ from those in a way, that the highest rate of formation for iso-butanol could be achieved with 1% Ir/C instead of the expected 5% Pt/C or 5% Ru/C. More detailed characterization of the catalysts is underway and includes the characterization of surface functionalities like base and acid centers, as well as the structural analysis of metal particles. Further insight into possible reasons for this difference in results will be the subject of the presentation.

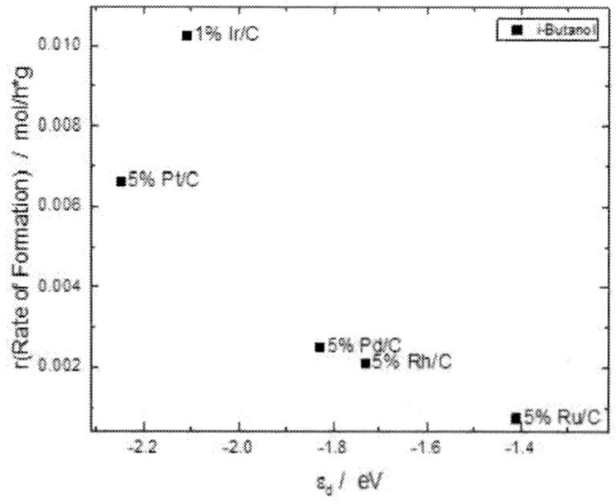

Table 1: Rate of formation for i-butanol at 150°C in correlation to the d-band center in eV per gram catalyst

Catalyst	r(Rate of Formation) mmol/h*g	ε_d eV
5% Rh/C	2.1±0.3	-1.73
5% Pd/C	2.5±0.7	-1.83
5% Ru/C	0.8±0.08	-1.41
5% Pt/C	6.6±0.7	-2.25
1% Ir/C	10.2±2.1	-2.11

Table 1: Rate of formation for i-butanol at 150°C in correlation to the d-band center in eV per gram catalyst

609589 Hybridization Potential of Carbon Capture and Utilization: A Smoother Transition Towards a Decarbonized Ethylene Production.

Abstract:

Meeting the Paris Agreement will require global climate change mitigation efforts, among which carbon capture and utilization (CCU) emerges as an appealing strategy as it can bring value to industrial activities while utilizing captured CO_2. With this spirit, at present, direct electro- and thermo- catalytic routes are being investigated extensively to produce a myriad of valuable CO_2-based chemicals. The main difference between these routes is that the former activates the CO_2 molecule directly with electricity, whereas the latter is based on the hydrogenation of CO_2 in tandem with electrocatalytic water splitting.

So far, however, these two leading CCU technologies, electro- and thermo- catalytic routes, have been mostly assessed as stand-alone systems. In addition, even though significant high costs characterize both routes compared to the fossil-based alternative, the hybridization potential of fossil and CCU technologies, from a process perspective, remains mostly unexplored. Hence, a comparative analysis of these routes and their hybridization is considered critical for establishing future research directions and developing effective environmental regulations.

In our work, we focus on ethylene, an important building block of the chemical industry, and cover this research gap considering various power (i.e., wind, solar, and nuclear) and CO_2 sources (i.e., capture from natural gas or coal power plants, or directly from the air). Our results show that CCU can help to significantly reduce the carbon footprint of ethylene and the impacts on human health, ecosystem quality, and resource depletion can be significantly reduced. The thermo-route is currently economically and environmentally superior when compared to the electro-route, yet the latter shows the highest potential for improvements, being at an earlier stage of development with respect to the thermo-route. The substantial energy needs, for both routes, make the production of CO_2-based ethylene, at present, economically unappealing, i.e., a 1.7 to 3.9-fold decrease in costs would be required to become competitive with the fossil-based alternative. In order to overcome this barrier, we apply optimization tools and the concept of hybridization to smooth the transition towards more sustainable chemicals. In particular, we developed a Linear Programming (LP) model considering the whole range of power and CO_2 sources to quantify the optimal hybrid for different decarbonization levels. Accordingly, we find that it is possible to produce cradle-to-gate carbon-neutral ethylene with only a 1.3-fold increase in cost by optimally combining fossil and CCU routes.

610814 Climate-Resilient Process Design Using the Flexibility Analysis Approach

Abstract:

The impacts of climate change can be fatal in various ways. We have seen the increased fluctuation of precipitation and temperature, which causes frequent natural disasters, such as flooding, drought, severe storms, and hurricanes. These are the direct impacts of climate change primarily related to the water cycle, which is one of the supporting services from ecosystems. The change in supporting ecosystem services will affect other ecosystem services (i.e., provisioning, regulating, and cultural services) as well as economic and societal activities that rely on ecosystem services. Numerous studies indicated that the knock-on (indirect) impacts of climate change on economic activities are significant as well. For example, Martinich and Crimmins (2019) estimated that the overall economic damages in the U.S. range in the hundreds of billions of dollars annually by 2100 under a high emission scenario. However, mitigation strategies under a low emission scenario will result in substantial economic benefits. Therefore, it is critical that we need to investigate both vulnerability and adaptation of technological systems to climate change.

Currently, most work on the climate change impacts has been conducted on agricultural and energy sectors. However, studies on manufacturing systems are still lacking. Designing flexible/resilient manufacturing systems to climate change is necessary to make them adapt to uncertain climate disturbances and changes in ecosystems and economic markets. Additional costs will be involved with adaptation strategies by showing trade-offs with the benefits of avoiding climate change impacts. In this work, we address climate-resilient process design (CRPD) problems using a flexibility analysis (FA) approach.

The main motivation for FA is to guarantee the operability (the capability of feasible operation) of manufacturing processes under changes and disturbances (uncertain parameters) at the design stage of processes [2]. Uncertain parameters can be physical properties such as inlet temperature, flow rate, and concentrations; and economic properties such as price and demand of products. The flexibility of a given design can be calculated by maximizing the range of unknown parameters that ensure a feasible operation. Also, by including a cost minimization function and solving a multi-objective optimization problem, optimal designs with desired flexibility and cost can be identified.

FA is suited for exploring CRPD since future climate projections are uncertain. Numerous global climate models (GCMs) have been developed to project future climate data. Based on the climate data, future ecosystem provisioning services, such as water supply, can be simulated using hydrological modeling or ecosystem modeling tools. For example, water is an important resource for any manufacturing system since it is used for cooling and steam-heating processes. Therefore, those projected and simulated data can be considered as uncertain parameters for the FA study. The preliminary study on a heat exchanger network (HEN) shows that its flexibility needs to be increased to adapt it to future climate disturbances (warmer water temperature). Also, the study exhibits the increased investment cost as a trade-off with higher flexibility.

Although FA has not yet been applied to investigate the flexibility of manufacturing systems to the climate change impacts, there are apparent research opportunities to explore adaptation strategies using FA. FA methodology has been established for decades to address flexible process design problems. Therefore, applying FA for CRPD could be readily acceptable and applicable to engineering disciplines. Moreover, since FA is based on the mathematical engineering model, it is extendable for integrating with other mathematical models such as supply chain models and economic models. Thus, the flexibility of processes can be examined while accounting for supply chain networks and the planetary boundary. Considering that climate change has indirect impacts on many sectors, incorporating a broad system boundary in the model could be particularly valuable. However, the increased computational challenges will be a trade-off.

References

[1] Martinich, J., & Crimmins, A. (2019). Climate damages and adaptation potential across diverse sectors of the United States. Nature climate change, 9(5), 397-404.

[2] Grossmann, I. E., Calfa, B. A., & Garcia-Herreros, P. (2014). Evolution of concepts and models for quantifying resiliency and flexibility of chemical processes. Computers & chemical engineering, 70, 22-34.

611243 Towards Integration of Design and Operation of Techno-Ecological Synergistic Systems

Abstract:

Over the last few decades, the engineering community has dedicated an immense amount of effort to minimize its impact on environment. Innovations like integrated mass and heat exchange network, waste reduction algorithms and eco-industrial parks are such examples that have helped reduce the impact of the chemical industry on environment [1]. Life cycle assessment and optimization methods were developed to avoid shifting of impacts and ensure individual unit scale reductions are reflected across the supply chain and product life cycle. Although these innovations have enhanced quality of human life and ecosystems, environ- mental issues like lake eutrophication, global climate change and air pollution in major cities across the global and in US persist. The currently developed impact reduction methods failed to determine the appropriate level of reduction in order to avoid ecological overshoot.

To address this fundamental challenge, there is a need to shift engineering design paradigm to explicitly consider capacity of ecosystems in technological designs. The framework of Techno-Ecological Synergy (TES) is one such approach that enables the explicit inclusion of ecosystem capacity in design and operation of manufacturing process by proposing sustainability metrics across local, regional and global scales [2]. As opposed to previously developed methodologies, TES aims to reduce emissions through a synergistic improvement in process efficiency and ecological capacity by investing in ecosystems as unit operations. Increased reliance on ecosystems for efficient functioning of technological systems incentivizes their expansion and protection while providing economically and environmentally win-win solutions.

Inclusion of ecosystems like forestry, wetlands, etc. as unit operations in design and optimization of technological systems are accompanied with their novel set of design and operational challenges. While the design challenges are often spatial placement and service shed boundary delineation problems, the temporal nature of operational challenges can be partitioned across three scales: (i) long-term planning, (ii) mid-term scheduling and (iii) short-term control problems. Typically, the planning scale would be concerned with an increase in ecosystem capacity over span of years such growth of forest or maturation of wetland. The month-to-month variation in the ecosystem capacity due to seasonal variation is accounted by the mid-term scheduling problem to maximize the operational performance of the plant. Lastly, the minutes-to-hour scale local control decisions ensure satisfaction of critical sustainability and safety constraints. Along with above challenges, the TES design and operation problem also need to immunize against the inherent uncertainty in predicting weather on which the ecosystem's productive is dependent.

Bakshi and co-workers have applied the framework of TES in design of sustainable biofuel supply chain [3], biodiesel manufacturing [4] and watershed management to demonstrate a possibility of win-win solutions using ecosystems as unit operations. The applications have largely ignored operational challenges involved with inclusion of ecosystems. Previously, Shah and Bakshi [5, 6] solved an integrated design and planning problem over a span of 20 years to demonstrate that a TES design for air quality regulation case study can lead to environmentally, societally and economically win-win solutions. They considered a forest ecosystem along with technocentric selective catalytic reducer (SCR) to design and operate a chloralkali production facility under monthly aggregated NO2 sustainability requirement. The sustainability constraint was based on extensive mass-based emission requirement, while the practical human well-being constraints are based on intensive property like concentration in form of Air Quality Index (AQI).

In this work, we expand the integrated design and planning framework to explicitly ac- count for AQI constraints on hourly scale as well as incorporate the shorter-term scheduling scale in the long-term design and planning problem. We define and solve the integrated de- sign and operation problem at various scales namely hourly, weekly and monthly to demonstrate the effect of aggregation on the optimal design and operation. We further bridge the scheduling and planning operational scale using an approximate dynamic problem formulation, where the smallest scale hourly decisions are approximated using a surrogate deep neural network model. Availability of hourly scale modified intensive resultant concentration data allows calculation of accurate hourly scale damages and benefits to human life due to NO2 emissions. This work is a step toward a fully integrated design and operation framework for engineering design in the presence of dynamic ecosystem services.

References

[1] B. R. Bakshi. Toward Sustainable Chemical Engineering: The Role of Process Systems Engineering. Annual Reviews in Chemical and Biomolecular Engineering, Accepted, 2019.

[2] B. R. Bakshi, G. Ziv, and M. D. Lepech. Techno-ecological synergy: A framework for sustainable engineering. Environmental science & technology, 49(3):1752– 1760, Feb 3, 2015.

[3] T. Ghosh, X. Liu, and B. R Bakshi. Including ecosystem services in sustainable process design across multiple spatial scales. In Computer Aided Chemical Engineering, volume 44, pages 1837–1842. Elsevier, 2018.

[4] M. Charles, G. Ziv, G. Bohrer, and B. R Bakshi. Connecting air quality regulating ecosystem services with beneficiaries through quantitative serviceshed analysis. Ecosystem Services, 41:101057, 2020.

[5] U. Shah and B. R. Bakshi. Accounting for nature's intermittency and growth while mitigating no 2 emissions by technoecological synergistic design-application to a chloralkali process. Journal of Advanced Manufacturing and Processing, 0(0):e10013, Mar 28, 2019.

[6] U. Shah and B. R Bakshi. Quantification of physical and monetary benefits of forest ecosystem: A case study for net positive impact manufacturing. In 2019 AIChE Annual Meeting. AIChE, 2019.

599104 Multiscale Analysis for Power-to-Gas-to-Power Facilities. Technical and Social Issues of Substituting Base Load

Abstract:

Solar and wind can easily provide the world's energy demand. However, the variability of both sources makes it difficult to use them to provide base load, where nuclear and coal have been the resources of choice (Simla and Stanek, 2020). In the transition towards a more sustainable energy production sector, the development of technologies for the storage of wind and solar energy to be used at demand are being the focus of the research (Johlas er al, 2020). Six major technologies are considered for energy storage, chemical storage, electrochemical technologies, electrical systems, mechanical systems, and thermal storage and thermodynamic cycles (Matos et al., 2018). The selection of the technology depends on the purpose and capacity of the storage as well as the energy source. Power to gas is typically suited for large power storage such as the case of substituting the base load that so far is provided by the coal and nuclear power plants .The expected future hydrogen economy and the possibilities of using it for power (Gahleitner 2013) or synthesis (Davis and Martin, 2014) have developed electrolyser technology that is capable of responding to the variability of energy sources, reducing the cost of renewable hydrogen (Proost, 2019). However, a number of challenges must be addressed in terms of power harvesting, storage and power production.

A multiscale analysis is developed to evaluate the feasibility of storing renewable based hydrogen in the form of metal hydrides, and its further use in fuel cells to produce power. The renewable power to power ratio is computed at process level, where water splitting, oxygen and hydrogen production and purification, metal hydride production and decomposition and hydrogen use in a fuel cell are considered. Next, a macroscopic supply chain study is performed to identify the solar and wind energy harvesting technology and the region to produce the power required to later generate electricity that substitutes current base load in a region. The two problems are is modelled and optimized using a mathematical approach. The case study of Spain is considered where coal facilities are closing down in June 2020. Economic and social aspects are considered for the allocation of not only PV panels and wind turbines, but also the new power facilities aiming at mitigating the impact of the energy transition in the regions affected and helping develop the economy of the entire country.

The production of power from MgH_2 is advantageous due to the decomposition of water in the production of the MgH_2 so that the yield to power of the facility is above 1 at the cost of further need to process MgO to recover Mg via Carboreduction. Solar panels are to be installed to the South-East while the distribution of wind turbines is divided into that same region and the north of the country. However, aiming at job generation or GDP growth, alternative solutions are found. Multiobjective optimization is the way to find trade-offs. The power production cost for the system results in a competitive value of 0.084€/kWh, but the investment required adds up to 13000 €/kW, a high value compared to current non sustainable technologies.

References

Davis, W., Martín, M ., 2014a. Optimal year-round operation for methane production from CO_2 and Water using wind energy. Energy 69, 497-505

Gahleitner, G (2013) Hydrogen form renewable electricity: An international review of power to gas pilot plants for startionary applications. Int. J. Hydrogen Energ 38, 2039-2061

Johlas, H., Witherby, W., Doyle, J.R. (2020) Storage requirements for high grid penetration of wind and solar power for the MISO region of North America: A case study. Renew. Energ. 146, 1315-1324

Matos, CR., Carneiro, J.F., Silva, PP. (2019) Overview of Large-Scale Underground Energy Storage Technologies for Integration of Renewable Energies and Criteria for Reservoir Identification. J Energy Sotrage 21 241-258

Proost, J. (2019) State-of-the art CAPEX data for water electrolysers and their impact on renewable hydrogen price settings. Int. J hydrogen Energ. 44, 4406-4413

Simla, T., Stanek, W. (2020) Reducing the impact of wind farms on the electric power system by the use of energy storage. Renew. Energ. 145, 772-782

611225 Incorporating Process Safety in Heat Exchanger Network Synthesis

Abstract:

Although there have been documented instances of heat exchanger failures resulting in property damage, plant shutdowns and loss of life [1], safety has not been always the primary objective in heat exchanger network synthesis (HENS) [2-3]. In many cases, pressure-related process safety metrics, including the potential for overpressure, are neglected. If inadequately protected, they can result in devastating consequences. This work incorporates the safety of exchanger tube ruptures due to overpressure scenarios in HENS. Tube ruptures are non-steady state overpressure events whereby the low pressure side is pressurized by the high pressure side [4]. This pressurization can occur in as short as milliseconds. The severity of pressurization increases with large differences in shell and tube side pressure [5]. Using dynamic simulations that capture the over-pressurization during a tube rupture, the severity of this event in a network can be mitigated by using a combination of stream-matching and/or overpressure protection equipment (e.g. relief valve) [6]. However, the dynamic nature of tube rupture increases the complexity of designing safer heat exchange networks. To this end, we develop an optimization-based formulation that accounts for the tradeoff between the safety and cost of HENS. Unique safety profiles are generated for all plausible integration alternatives. A new piecewise approximation is used to obtain the minimal breakpoints needed to accurately capture the relationship between temperature and safety, based on the work of Rebennack and Kallrath [7]. This approximation is able to accommodate a wide range of pressures, temperatures, and fluid phase with significant accuracy. We illustrate the incorporation of process safety using several case studies. Specifically, our results demonstrate that optimization-based methods can significantly increase the safety of a heat exchanger network with minimal increase in cost.

[1] Grim, L. (2017). CSB Investigation: Williams Geismar Olefins Plant Reboiler Rupture and Fire. In *ASSE Professional Development Conference and Exposition*. American Society of Safety Engineers.

[2] Yee, T. F., & Grossmann, I. E. (1990). Simultaneous optimization models for heat integration—II. Heat exchanger network synthesis. *Computers & Chemical Engineering*, *14*(10), 1165-1184.

[3] Ponce-Ortega, J. M., Jiménez-Gutiérrez, A., & Grossmann, I. E. (2008). Optimal synthesis of heat exchanger networks involving isothermal process streams. *Computers & Chemical Engineering*, *32*(8), 1918-1942.

[4] API Standard 521. (2014). Pressure-Relieving and Depressuring Systems.

[5] Hellemans, M. (2009). *The safety relief valve handbook: design and use of process safety valves to ASME and International codes and standards*. Elsevier.

[6] Harhara, A., & Hasan, M. M. F. (2020). Dynamic modeling of heat exchanger tube rupture. *BMC Chemical Engineering*, *2*(1), 1-20.

[7] Rebennack, S., & Kallrath, J. (2015). Continuous piecewise linear delta-approximations for univariate functions: computing minimal breakpoint systems. *Journal of Optimization Theory and Applications*, *167*(2), 617-643.

607709 Effluent Reduction in Pulp and Paper Manufacturing: White Water Recycle in Bleaching Section

- Download/View Extended Abstract

Abstract:

The pulp and paper industry is one of the largest consumers of water in the US industrial sector. Water is an integral part of paper manufacturing, starting from wood preparation to fiber conditioning and classification. Over the past thirty years, mills across the United States have made great strides to reduce their freshwater usage. In recent years, there has been an increase in the industry for the adoption of waste minimization strategies covering various aspects and avenues in mill operations. Of the various process areas in a typical integrated bleached Kraft mill, the paper machine and pulp production areas account for around 75% of the total mill freshwater consumption, making it a low hanging fruit for water conservation.

In this work, we used **W**indows **G**eneral **E**nergy and **M**aterial Balance **S**ystems (WinGEMS) to develop a process simulation model of an integrated softwood line of a Kraft pulp mill. The model includes the bleaching section, the stock preparation section, and the paper machine section. The simulation model estimates the material flows, including Non-Process Elements (NPEs) and water, and the energy flows. We investigated the impact of replacing freshwater use in the final bleaching stage shower washer with white water from the paper machine on the overall freshwater consumption and on the NPEs flowrates.

Steady-state equations for the Elemental Chlorine Free (ECF) bleaching sequence $D_0EopD_1EpD_2$ are included in the simulation. Industry and literature data were used to verify the ECF bleaching simulation model, and the difference was in the range of 1.8 to 5.5% for final stage brightness and the post first extraction pulp kappa number, respectively. The simulation model prepared can predict changes in the kappa number, brightness and the chemical oxygen demand (COD) values with changes in the operating parameters. Simulation of the stock preparation section includes the filler ($CaCO_3$) addition, cleaners and screening compound blocks. The paper machine section is simulated with the approach system, headbox and sheet forming section as compound blocks. The next step includes the use of Visual-MINTEQ (V-MINTEQ) software to determine the distribution of Ca^{2+} ions in the paper machine white water. This comprehensive model for the bleaching, stock preparation and paper machine section along with Ca^{2+} ion distribution allows the assessment of the extent and impact of white water recycling from the paper machine section into the bleaching section without compromising the final paper brightness.

610078 On a Systematic Approach to Simultaneously Integrate Water and Energy Utilities in Multi-Product Biorefineries

Abstract:

Recent negative prices of crude oil put more stress on green products to minimize their production cost and environmental footprint. Water, solvent, and energy, known as utility requirements (mass and energy), are the obvious contributors to both economic and environmental sustainability. Reuse/recycle can reduce the cost but, at the same time, increases the complexity of the network. Regeneration can save mass utilities and reduce the effluents, provided that the regeneration cost is lower than the utility savings. The optimal use of each utility is a complicated design problem, and numerous systematic approaches exist in the literature. However, existing methods consider only the simultaneous use of water and energy (or one main component and energy), handling any other substance as a contaminant. Moreover, multiple technological options result in different process designs. Different design configurations create different thermal profiles for the energy network and different effluents for the mass network. The higher the chemical/biological load of the discharged stream, the higher its treatment cost. In addition, intermediate storage is another degree of freedom, which has an impact on the thermal profile of the network and, thus, on energy consumption. As a result, degrees of freedom are not only the stream allocations, but also the mass and energy flows in the network, and the design of the auxiliary units. Decisions need to consider the cost for utilities, for treatment, and for investment in auxiliary units and, at the same time, to keep the complexity of the network low. The challenge is to coordinate all the individual problems into a single formulation.

This work introduces the concept of interacting superstructures for the integrated optimization of utility networks and process design. The method takes into account multiple components (contaminants) and does not pivot on a central component (e.g., water, H2). Alternative design options exist for regeneration and storage. The objective function evaluates the role of each component and balances the trade-offs between reuse, regeneration, and discharge. The problem is nonlinear by nature, but assumptions and piecewise linearization techniques are adopted to keep it linear. The optimization model is applied to design the utility network of a real-life biorefinery process. Results show that the utility network optimization (UNO) model managed to reduce up to 70% of the total annual cost, compared to the initial design, by reducing mainly the utility cost for both mass and energy. UNO identified that, in the particular case study, 62% of the thermal flows are a degree of freedom and managed to allocate mass and thermal flows properly. The UNO model finds practical application in grassroots and retrofit systems, exclusively for the design of the utility network but not for the optimization of the process or the supporting unit design. It can be used to analyze how the uncertainty of different costs affect the process and find flexible designs, able to absorb price fluctuations, simply by changing the mass flows. Future work will incorporate options for the energy system, including heat pumps and generation of different steam qualities. The integration of detailed treatment systems and the analysis of total site networks is also left as a future challenge.

Acknowledgments

Financial support from CIMV and the Marie Curie project RENESENG [FP7-607415] is gratefully acknowledged.

609151 Developing Performance-Advantaged Chemicals Utilizing Bioprivileged Molecules: Organic Corrosion Inhibitors

Abstract:

Biomass-derived molecules hold the promise to create not only drop-in replacement chemicals but also novel performance-advantaged chemicals. One path for the development of these performance-advantaged chemicals is through utilization of bioprivileged molecules, which are intermediates that can be used to generate a library of compounds for testing. In this work, a bioprivileged molecule, triacetic acid lactone (TAL), was used for the creation of an organic corrosion inhibitor molecular library. A number of compounds generated from the TAL library demonstrated promising inhibition efficiency for mild steel in both sulfuric acid and hydrochloric acid solutions as measured from electrochemical impedance spectroscopy. The interaction of the corrosion inhibitors on the mild steel surface was also examined by SEM, XPS, and adsorption studies. By coupling the library synthesis, corrosion inhibition performance and quantum chemical calculations, the structure/function performance of the molecules was elucidated. The work demonstrates how bioprivileged molecules can be used as the basis for identifying novel molecules that can lead to enhanced performance in end use applications.

610864 Advanced Robust Catalysts for Producing Stable Biomass Pyrolysis Oils for Upgrading to Hydrocarbon Fuels

Abstract:

Biomass feedstocks contain inorganic compounds generally classified as ash. The ash consists of compounds of potassium, calcium, magnesium, silicon, phosphorous and other elements. These elements have been reported to influence both the pyrolysis reactions as well as destabilizing the pyrolysis oils during storage. The inorganic elements have also been reported to deposit on catalyst surfaces during *in situ* catalytic pyrolysis leading to the eventual deactivation of acidic catalysts such as zeolites. The deposition of inorganic elements and their effect on formulated red mud (FRM) catalyst during *in situ* catalytic pyrolysis of pinyon juniper wood was investigated. The inorganic elements were measured for the fresh, coked, and regenerated catalysts. The BET specific surface area of the FRM catalyst decreased from 76 m^2/g for the fresh catalyst to 53 m^2/g for the stable regenerated catalyst. After three regenerations, the BET specific surface area stabilized at 53 m^2/g and remained constant after all other regenerations. Potassium, calcium, magnesium, and phosphorous were deposited on the catalyst. Potassium deposition was linear with the number of regenerations while magnesium and calcium depositions were initially rapid but leveled off after three regenerations of the catalyst. Phosphorous deposition was almost linear, but the data were more scattered compared to potassium. The potassium deposition was attributed to physical phenomenon whereas calcium and magnesium depositions were more akin to chemical reactions related to the loss of BET surface area of the catalyst. The deposition of these elements on the surface of the catalyst did not deactivate it. After each catalyst regeneration, the oil yield was not significantly affected and the oil oxygen content and viscosity decreased slightly. This clearly showed that formulated red mud is a robust catalyst suitable for *in situ* catalytic fast pyrolysis of biomass.

610050 One-Step Conversion of Bioderived Hydroxy Acids to a Renewable Alkanediol

Abstract:

Among biomass-derived compounds that can serve as building blocks for the chemical industry, hydroxy acids are particularly attractive as it can be used for the direct production of renewable chemicals such as dienes, diacids, diols, etc.[1] Visolis aims to develop a renewable process to generate a high value industrial monomer, Mevol, from a hybrid process combining fermentation of renewable feedstock to a biological, C^6-hydroxy-acid intermediate and its catalytic upgrading to Mevol. Mevol is our first target commercial product in the specialty chemical space due to substantial industry interest in renewable polyols with applications in coatings and in the production of soft polyurethanes. Commercialization of bio-based Mevol will allow for expansion of our renewable C^6hydroxy-acid platform to other green chemicals.

In collaboration with PNNL, Visolis has demonstrated the one-step conversion of C^6-hydroxyacid to Mevol over a water-tolerant, acid-resistant hydrogenolysis catalyst operating at lower pressure and low temperatures with yields exceeding 90% for over 300 hours time on stream. In contrast, conventional and commercially available catalysts are poisoned by these impurities. Understanding the catalyst's level of tolerance to impurities will relax purification constraints on the reactant feed leading to a simpler, more economical process for scale-up and commercialization and the lower H_2pressure substantially reduce CapEx requirement for plant construction. Our platform thus takes the best aspects of both approaches - the ease of obtaining renewable synthons through fermentation and subsequent catalytic upgrading of synthons - thereby achieving significant process efficiencies for renewable chemicals production allowing us to be cost-competitive with petrochemical based manufacturing.

References:

[1] C. Angelici, B. M. Weckhuysen, P. C. A. Bruijnincx, ChemSusChem 6 (2013) 1595-1614.

600404 Nitrogen Behavior in Glycine Decomposition Under Hydrothermal Condition

Abstract:

Nitrogen recovery from biomass feedstock is useful in terms of biorefinery to obtain fertilizer component together with energy in hydrothermal conversion of biomass. However, actual biomass contains various forms of nitrogen, and its behavior is not easy to understand. Thus, in this study, glycine, the simplest amino acid, was employed as model compound, and behavior of nitrogen was studied. We used a laboratory scale continuous reactor, and glycine aqueous solution was delivered together with preheated hot compressed water. Mixing of these two flows at the entrance of the reactor enabled rapid heating of the feedstock to the reaction temperature. At the exit of the reactor, the reactor effluent was mixed with cooling water to achieve rapid cooling. In this way, precise reaction time was determined. We further minimized the dead space of the reactor. The effluent was collected and analyzed for nitrogen products, which were mainly ammonia and methylamine. This is the first result of measuring the reaction characteristics with sufficient accuracy in supercritical water. One of the interesting discussion is whether the reaction rate and selectivity are different between subcritical and supercritical condition. If they are the same, they may be considered as radical reaction, while if not, they may be considered as ionic reaction. Comparison of our result with previous study by Sato, conducted under subcritical conditions, the reaction was found to be radical reaction. The finding of this study provides important information for nitrogen recovery from biomass feedstock using hydrothermal conditions.

592664 High-Efficiency Hydrogen Production By Decoupled Water Splitting in an Electrochemical - Chemical Cycle

Abstract:

As the human population grows, a transition towards clean and renewable energy sources becomes critical in curbing the negative effects of fossil fuel combustion. To facilitate this transition, renewable sources must be combined with energy storage and transportation infrastructures. Hydrogen production by water splitting has long been considered a promising solution to mitigating the intermittency issues of renewable energy, while also providing a high energy fuel for transportation. However, water splitting technologies still face challenges to improve their performance, efficiency and cost, especially when combined with renewable power sources. In conventional electrolyzer technologies, hydrogen is produced simultaneously with oxygen in the same cell, and a membrane or diaphragm is essential for product separation. Nevertheless, a dangerous mix of hydrogen and oxygen may still form under partial load conditions, which are inherent to renewable power source. Furthermore, the four-electron oxygen evolution reaction (OER) that takes place at the anode requires large overpotentials of over 400 mV, resulting in significant efficiency losses. Here, we present a strategy for high-efficiency decoupled water splitting, wherein the hydrogen and oxygen are produced in two separate steps in an electrochemical – thermally-activated chemical (E-TAC) cycle. The conventional OER catalyst anode is replaced by a cobalt-doped nickel hydroxide [$Ni_xCo_{1-x}(OH)_2$] anode; in the first step, hydrogen is produced at the cathode while the $Ni_xCo_{1-x}(OH)_2$ anode is oxidized to $Ni_xCo_{1-x}OOH$ in an efficient one-electron reaction. This step is followed by a spontaneous chemical step that reduces the anode to its initial composition while oxidizing water and releasing oxygen under mild heating. This E-TAC process enables high-efficiency water splitting in a membrane-free cell without the risk of product crossover.

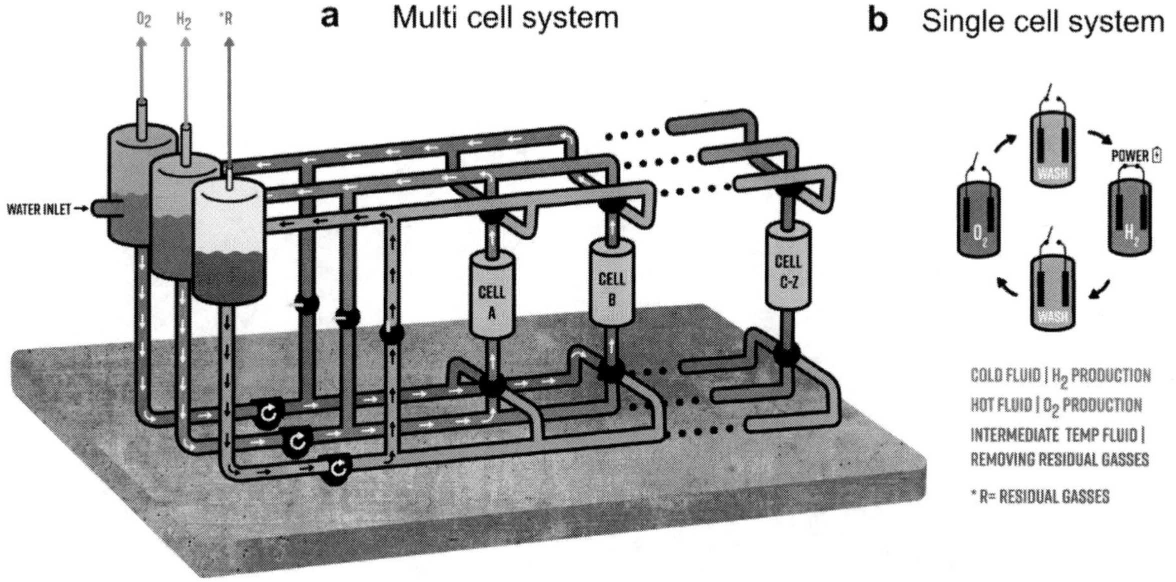

605497 Efficiency Limits of Underwater Solar Cells: An Excellent Application for Wide-Band-Gap Semiconductors?

Abstract:

Operation of underwater vehicles and autonomous systems is currently limited by the lack of long-lasting power sources. These systems could potentially be powered using underwater solar cells,[1–3] but the material requirements to achieve their full potential are not well-understood. Using detailed-balance calculations, we show that underwater solar cells can exhibit efficiencies from 55% in shallow waters to more than 65% in deep waters, while maintaining a power density > 5 mW cm^{-2}.[4] We show that the optimum band gap of the solar cell shifts by ~0.6 eV (1.8 to 2.4 eV) between shallow and deep waters and plateaus at 2.1 eV at intermediate depths, independent of geographical location. This wide range in optimum band-gap energies opens the potential for the vast library of wide-band gap semiconductors to be used for high-efficiency underwater solar cells.

1. Jenkins, P.P., Messenger, S., Trautz, K.M., Maximenko, S.I., Goldstein, D., Scheiman, D., Hoheisel, R., and Walters, R.J. (2014). High-Bandgap Solar Cells for Underwater Photovoltaic Applications. IEEE J. Photovoltaics *4*, 202–206.
2. Walters, R.J., Yoon, W., Placencia, D., Scheiman, D., Lumb, M.P., Strang, A., Stavrinou, P.N., and Jenkins, P.P. (2015). Multijunction Organic Photovoltaic Cells for Underwater Solar Power. 2015 IEEE 42nd Photovolt. Spec. Conf. PVSC 2015, 1–3.
3. Kong, J., Nordlund, D., Jin, J.S., Kim, S.Y., Jin, S.-M., Huang, D., Zheng, Y., Karpovich, C., Sertic, G., Wang, H., et al. (2019). Underwater Organic Solar Cells via Selective Removal of Electron Acceptors near the Top Electrode. ACS Energy Lett. *4*, 1034–1041.
4. Röhr, J.A., Lipton, J., Kong, J., Stephen, A., and Taylor, A.D. (2020). Efficiency Limits of Underwater Solar Cells. Joule *4*, 1–10.

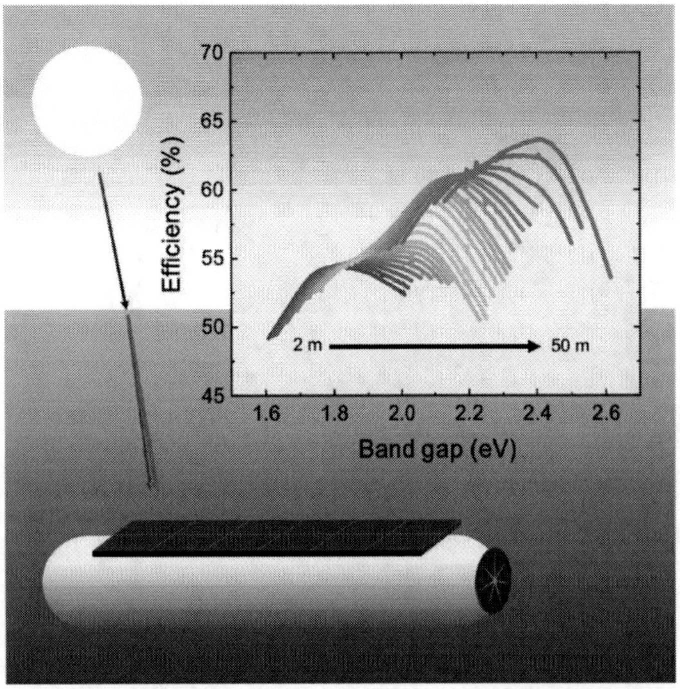

609924 Microvascular Carbon-Carbon Composite for Concentrated Solar Power Gas Receivers

Abstract:

Concentrated solar power systems offer a sustainable energy source by integrating it with thermal energy storage systems to produce electricity on demand. The technology uses mirrors to concentrate sunlight into a central receiver, which is used to heat a working fluid. The state-of-the-art approach to constructing gas receivers is the use of nickel-alloy modules with microchannels, showing promise by achieving nearly 90% global receiver efficiency based on simulated solar testing with super-critical carbon dioxide as working fluid (sCO_2) [1]. While the initial achievements of this technology are impressive, these receiver designs do, however, suffer from technical challenges that must be overcome before adoption. An alternative to these metallic systems is the use of high-temperature carbon-carbon (c-c) composites with novel microvascular (microVasc) composites technology [3]. In this work, these composites are made by first preparing prepregs with sheets of PAN-based carbon fibers and phenolic resin with the sacrificial polymer micro-channel (Polyacrylonitrile - PLA) between them, which is later removed by heat treatment under vacuum. The prepreg are then carbonized under an inert atmosphere (N_2) at 1000 °C. Due to gas evolution, this process lowers density and increases porosity. Therefore, to obtain the desirable mechanical and heat transfer properties, the composite is densified using phenolic resin under the same conditions. The c-c composites will then be evaluated for their mechanical and heat transfer properties. In addition, multiple densification cycles will be necessary in order to obtain the desired properties. Lastly, hydraulic and leakage testing of the channels will be necessary to prove their efficacy with flowing of sCO_2.

References

[1] Zada, K. R., Hyder, M. B., Drost, M. K. & Fronk, B. M. J. Sol. Energy Eng. 138, 61007 (2016).

[2] Chung, Deborah DL. Carbon fiber composites. Elsevier, (2016).

[3] Esser-Kahn, A. P. *et al. Adv. Mater.* 23, 3654–3658 (2011).

598561 On the Economics of a Thermal Management System for Industrial-Scale Natural Gas Adsorption Storage

Abstract:

One of the major strategies to lower CO_2 emissions in power generation is to use renewables such as wind and solar photovoltaics. A major challenge to achieving reliable and sustainable supply of power is that renewables are intermittent and require the use of energy storage technologies. Even with modest improvements to current battery technology, electrical energy storage is too expensive. An alternative solution to ensure sustainability of power supply with relatively low CO_2 emissions is the use of stored natural gas. Natural gas storage can ensure a continuous power supply when renewable energy production is insufficient to meet demand. Natural gas can be stored as compressed gas (CNG) at 3000 psig, in liquid phase (LNG) at -162 ^{o}C, or in adsorbed phase (ANG) at moderate pressure (60 bar) using commercially-available activated carbon adsorbent. CNG and LNG storage technologies are energy intensive processes due to the extreme storage conditions (high pressure and cryogenic temperatures, respectively). ANG storage, however, has the potential to store gas at moderate conditions with a relatively high density. The amount of gas stored using adsorbents is greater than twice that of compressed gas phase at the same operating conditions. The feasibility of ANG storage at an industrial scale critically depends on the cost of the adsorbent material and involved equipment. The adsorption process is exothermic - heat is released when gas is stored in a porous medium, and heat needs to be supplied during the release of the adsorbed gas. Due to temperature changes in the adsorption bed during the adsorption and desorption stages, the usable adsorption capacity is reduced by more than 20%. The adsorption capacity can be maximized by operating under isothermal conditions. Controlling the bed temperature requires the thermal management of the adsorption bed with a heating/cooling system in a shell and tube configuration heat exchanger vessel. As a result, the capital expenditure of the technology is significantly increased compared to running the adsorption/desorption process at non-isothermal conditions and thus a compromise in cost/storage capacity is required. In this work, the economics of an industrial-scale adsorbed natural gas storage (~140 MSCF) for different vessel configurations is explored. The benefits of increasing the usable adsorption capacity via thermal management of the adsorption bed are also addressed. The usable methane storage capacity for the different systems is calculated using dynamic theoretical modeling of the adsorption/desorption cycle. Transient, three-dimensional CFD modeling is used to verify the adsorption bed configuration. The risk adjusted net present value of these systems is estimated to economically compare the gas storage system using a thermal management system, operating at non-isothermal conditions, and storage via compression at the same operating conditions are presented.

606452 Trading Off the Costs and Benefits of Electric and Hydrogen Technologies in Decarbonising the Heat Sector: Whole-System Value Chain Optimisation

Abstract:

Space heating and cooling account for 17% of the UK's greenhouse gas emissions whilst gas based central heating can be found in nearly 24 million homes in the UK, which is 86% of all residential buildings. In order to reach zero carbon emissions by 2050, as required by the Climate Change Act, these homes will need to switch to some form of low carbon heating. This task is very challenging because of the wide variety of building stock and infrastructure present throughout the UK.

Although a low carbon heating strategy is yet to be set by the UK government, electrification of heat, such as heat pumps and electric storage heating, is one of the main options. Contrasting these two technologies, heat pumps provide higher efficiency with lower electricity use and lower grid upgrade requirements, whilst electric storage heaters have lower equipment and installation costs and allow users access to off-peak electricity. The CO_2 emissions associated with both technologies depend largely on the emissions of the electricity grid, although heat pumps use less electricity to provide the same amount of heat and so may be better placed to help the UK meet its carbon budget. Electrification of heat also needs to be compared with other low-carbon heating strategies, such as renewable hydrogen produced via power-to-gas technologies and injected into the gas grids.

In this study, a value chain optimisation of the UK energy system was conducted using the Value Web Model (VWM). The VWM is a mixed-integer linear programming model equipped to model the production, conversion, and transportation of energy, alongside the development of technologies and infrastructure to meet energy demands with high spatial and temporal resolutions. The aim of this study is to determine the role of different heating technologies in a carbon neutral energy system. The results from the whole-system value chain optimisation are discussed in terms of the costs and benefits of each of the heating strategy options with respect to the consumers and producers. The effects of different policy instruments such as feed-in-tariffs and renewable heat incentives will be discussed. This presentation will discuss future UK scenarios involving these technologies and the opportunities and uncertainties associated with each. The decarbonisation of heating is an important next step in the transition to net zero and this work will provide evidence regarding an effective long-term decarbonisation strategy for the UK.

611643 Subnational Supply Chain Analysis of Cobalt in the US to Guide Sustainable Manufacturing of Batteries for Renewable Energy: A Multi-Regional Input-Output (MRIO) Approach.

Abstract:

The increasing demand for mineral-based resources that face supply risks calls for managing the supply chains for these resources at the regional level. Cobalt is a widely used cathode material in Li-ion batteries, which form the major portion of batteries used for renewable energy storage — a necessary technology for overcoming the challenge of intermittency, thus making renewable energy more reliable and energy generation more sustainable. This necessitates understanding cobalt's supply risks and identifying the stocks of cobalt available for future use via recycling or mining. We address these needs in this work using multi-regional input-output (MRIO) analysis. MRIO analysis can be used to obtain the trade network of critical materials and offers a more expedient way to identify sub-regional material flows and stocks. However, tracing the flows of minor materials, such as cobalt, requires further sectoral disaggregation beyond the industrial classifications defining the input-output sectors. Commodities containing cobalt will be further disaggregated from a MRIO table of the United States and the subnational trade structure of cobalt observed from the MRIO will be presented. The presentation will highlight the methodology adopted for disaggregating a subsector of an economic input-output table beyond the level of aggregation defined by the most detailed NAICS codes. Results from mining the MRIO network for cobalt will highlight the geographical distribution of cobalt use, availability, and stocks as domestic waste that will become available as a secondary feedstock for battery production.

610821 Can Combined Heat and Power (CHP) be a Friend to Renewable Energy?

Abstract:

As the share of variable renewable energy generations rises on U.S. grids, the need for greater grid flexibility rises. Using existing combined heat and power (CHP) capacity could permit adding grid flexibility at lower cost. This study evaluates the potential use of CHP based on capacity adequacy of CHP systems. The capacity adequacy is estimated using annual- and hourly-averaged capacity factors. The potential is examined for various sectors including utility, independent power producer, commercial, and industrial sectors across seven independent system operators (ISOs) and regional transmission organizations (RTOs). The result shows that existing CHP systems can provide grids with a considerable amount (0.7-8.5 GW) of capacity, which accounts for 1-8% of the peak electric demand and 2-14% of the median demand in grid regions. This high level of available capacity is mainly attributed to consistently low capacity factors (<0.6) across regions and sectors. The result also revealed a consistently low utilization of CHP systems in manufacturing facilities, and an hourly-varying utilization of CHP systems in office buildings. Such a great potential of built-in flexibility in grids can minimize non-trivial incremental cost that would incur with additional power capacity.

606505 Distributed Modular Green Ammonia Synthesis with Absorption Separation: Understanding Rate Limitations, Insights Toward Improvement

Abstract:

We have been investigating a design that may better enable distributed, small-scale, modularized ammonia synthesis - particularly green ammonia made with hydrogen from sustainable wind energy. Ammonia can be separated from unreacted hydrogen and nitrogen by absorption in magnesium chloride; we have found ways to ensure the uptake/regeneration cycle is reproducible over many cycles. While we have generally found that the absorption (uptake) can be quite rapid, but the desorption (regeneration) can be rate limiting for production.

In this presentation we investigate the desorption (regeneration) step, showing that the process appears to involve two steps: (i) reactive decomposition of the ammine salt to release ammonia, and (ii) diffusion of the released ammonia through a shell of salt depleted of ammonia. The experiments and analysis, performed with lab-scale colums (grams of absorbent), can be used to design cycling performance for pilot-scale columns (kilograms of absorbent) used in a prototype module for ammonia production on the order of kilograms per day. The insights also suggest improved absorbent materials that might avert diffusion rate limitations during regeneration.

602142 Alkali & Green Catalyst Activated Geopolymers Using Coal and Wood Fly Ash

Abstract:

Geopolymers are greener construction material alternatives to the commonly used Ordinary Portland cement(OPC) in terms of the carbon footprint apart from possessing suitable physiochemical & mechanical properties. Geopolymers however do need an alkaline activator, such as NaOH or KOH, which increases overall cost and manpower as it requires skilled personnel for process handling. Here, wood fly ash, a by product of wood combustion, which is alkaline in nature is used as an internal source of alkaline activator for geopolymerisation along with coal fly ash. The high calcium content of the ash mixture produces geopolymer–cementitious hybrid (GCH), which is a mixture of geopolymers and alkali activated cement (calcium hydrates). GCH is characterized for its morphology, size, porosity, surface groups, mineral content and also subjected to leaching tests as to check if heavy metal leaching from wood fly ash has been controlled. Various factors such as water-to-ash ratio (w/a), method of curing, type of alkaline activator, sieving have been considered here through factorial experimental analysis with the leaching patterns of various elements such as Pb, Zn, Cr, As. Hg, Se being studied. The dominant factor for leaching was found to be the type of alkaline activator (i.e. Wood Fly Ash versus Na_2SiO_3) with the best formulation being 0.3DI_p_s (0.3 represents the w/a ratio, DI represents use of wood fly ash as the pure source of alkaline activator, "p" denotes that samples were precured and "s" denotes sieving) when compared to the Denmark leaching standards. This study points to the production of alkaline activated materials for construction without the use of an external activator where the final product exhibits controlled leaching as required especially when used in applications involving exposure to moisture or water bodies. To avoid the GCH formation, a follow up study is planned to produce purer geopolmers where the ratio of coal fly ash to wood fly ash will be varied (70-30,60-40 wt%). A green catalyst will be used to pretreat the coal fly ash as to increase it surface area and reduce the calcium content to prevent formation of calcium hydrates. The treated coal fly ash along with wood fly ash will be subject to green catalyst based geopolymerisation, with the formed structure tested for its geopolymeric nature through chemical analysis. Concrete would be cast out of the formed geopolymers using them in certain levels (eg: 5%) as cement substitutes along with OPC. Mechanical strength of the cast concrete would be analysed through standard testing methods with leaching properties of the final structure also studied through standard leaching techniques as to check for its suitability in different applications with pure OPC made concrete was used as control.

605469 Systematic Study of Charge Transfer between Catalytic Au Nanoparticles and Support and Its Implication for CO Oxidation

Abstract:

Catalytic activity of gold nanoparticles (NPs) supported on various oxides has been observed for various reactions. Strong metal-support interaction (SMSI) between metal NPs and supports is one of key factors explaining improved catalytic performances of these particles. Various studies of the mechanisms of SMSI contributing to enhanced activity suggested that electron transfer from support to NPs plays an important role.

In this work, we deposited Au NPs on SiO_2/Si substrates (n-type, Fig 2. (a)) using Helium Nanodroplet Deposition (HNDD) method (Fig 1.), which can produce very clean Au NPs with narrow size distribution (~2 nm). By designing the novel Au/ SiO_2/ Si system, we tuned a charge transfer between Au NPs and Si to understand its effect on CO oxidation. We found that the onset temperature for CO oxidation has decreased with increased charge transfer.

Fig 1. shows the simplified structure of HNDD system. A beam of helium nanodroplets expanded from low temperature nozzle and then captured vaporized Au atoms, which then formed NPs within the droplets. These nanoparticles were then deposited onto Si wafers modified with variable thicknesses of SiO_2 layers.

The structure of Au/ SiO_2/ Si system and the proposed mechanism are depicted in Fig.2 (a) - (b). In the absence of SiO2 layer, a heterojunction is formed based on the work function difference between Au and *n*-silicon. More specifically, positive charges tend to accumulate at the surface of Si with the band bending to higher energy while negative charges tend to accumulate at interface of Au NPs. By varying the thickness of SiO_2 layer, the amount of negative charges can be finely controlled. To verify this mechanism, the samples were characterized by XPS and TPD. XPS data for Si 2p and Au $4f_{7/2}$ (Fig.2 (c) and (d)) binding energies show that peak positions moved toward lower binding energy with thinner SiO_2 layer, indicating more charges accumulate on both sides of SiO_2 layer as it becomes thinner. Fig.2 (e) demonstrates the onset temperature of CO oxidation as a function of the SiO_2 thickness. The onset temperature of CO oxidation decreased as the thickness of SiO_2 layer became thinner, coinciding with shifts in BE.

By creating a model of Au/ SiO_2/ Si, we systematically studied charge transfer between Au NPs and support, and its effect on CO oxidation. This study highlights the importance of this mechanism on catalytic reactions, which can guide a development of the next generation of catalysts for environmental protection.

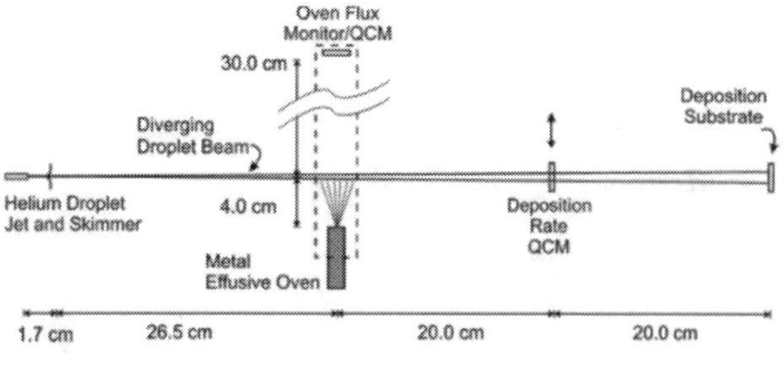

Figure 1. Helium Nanodroplet Deposition System.

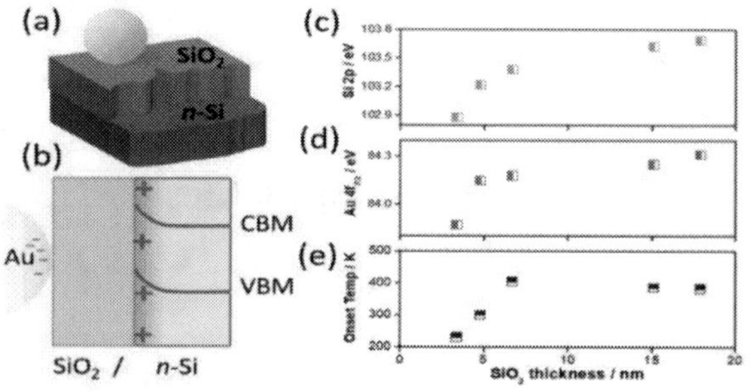

Figure 2. (a): Structure of Au/ SiO_2/ Si system. (b): Bend binding in Au/ SiO_2/ Si system. (c) - (d): XPS peak position of Si 2P and Au $4f_{7/2}$ vs. SiO_2 thickness. (e): Onset temperature of CO oxidation vs. SiO_2 thickness.

605361 Toxicity Study of Powdered and Cast Incinerated Sewage Sludge Ash - Chemical Treatment, Leaching Patterns and Marine Toxicity

Abstract:

Disposal of ash obtained from thermal treatment of solid waste has been a cause of concern due to its toxic content and potential leaching of these toxins into the environment, such as if it is used in land reclamation applications. In this study, bottom ash from the incineration of sewage sludge was subject to acid treatment to facilitate removal of heavy metals. A phosphate treatment was also explored to fix heavy metal from leaching. The treated samples were examined for heavy metal leaching using a standard protocol (EN 12457). A relatively mixed leaching profile was observed, with different heavy metals leaching out more or less compared to raw untreated ash. Malonic acid, an organic acid, was found to remove more heavy metals compared to nitric acid, a strong inorganic acid. When evaluated for marine eco-toxicity, acid treated ash was found to be more toxic as compared to untreated raw ash, with malonic acid-treated samples being the most toxic; whilst phosphate treated ash displayed a similar toxicity profile to that of untreated raw ash. Raw ash was compared also against concrete sand, a commonly used land reclamation material. There was relatively low toxicity toward microalgae, barnacles and brine shrimp. However, raw ash triggered toxicity responses in the sea urchin egg fertilization assay indicating presence of toxic substances not present in sand. The raw ash was tested against human cell lines also, where it showed size and dose-dependent toxicity, with greater toxicity at higher doses and smaller sizes. This study demonstrates the importance of evaluating the toxicity of post-treatment of ash if it will be used for environment applications such as coastal reclamation. Appropriate treatments are required to minimize the potential leaching of harmful contaminants. Further studies will be pursued with to investigate the effect on mechanical properties if fresh or aged sewage sludge ash is used as a substitute for cement as compared to Ordinary Portland Cement concrete tiles. Further studies on the leaching patterns in seawater and effect on microbial communities will also be pursued.

605869 Dynamic Simulation and Model Predictive Control of Post-Combustion CO_2 Capture Process with Advanced Flash Stripper

Abstract:

Amine-based post-combustion CO_2 capture is an option that is readily applicable to power plants to mitigate their negative environmental impact. However, high operating cost due to significant regeneration energy stands as the main obstacle for its widespread adoption. Advanced flash stripper (AFS) is an intensified process designed to reduce the energy consumption of the conventional stripper. The capture process with AFS can reduce the regeneration energy by about 20%, but the complex heat integration scheme in AFS may reduce its operational flexibility and controllability. According to a previous study on regulatory control of the AFS process, it is revealed that achieving a major control objective (e.g.CO_2 capture rate) is highly challenging with decentralized PID controllers.

This study aims to suggest an effective control structure for the CO_2 capture process with AFS under a number of operational scenarios. For this, a dynamic model for the capture process with AFS using 30wt% Monoethanolamine (MEA) solvent is built in the commercial software platform of *gPROMS*, to examine dynamic characteristics of the process. Step response analysis is performed using the developed model and the result shows that the process exhibits slow dynamics with strong interactions among its variables, due to the heat integration and recycle. Then, the use of a multivariate control technique like model predictive control (MPC) is considered to handle the complex process dynamics. Closed-loop dynamic behavior with different control structures of MPC and decentralized PID controllers are simulated and compared. Our results indicate that the MPC can give superior control performance, robustness and convenience of operation over the decentralized control, under the various studied scenarios of changes in the inlet flue gas flow rate and CO_2 capture rate set-point.

611828 Evaluation of an Industrial Bio-Catalyst for Reducing Hydrogen Sulfide of the Gas Stream from Domestic Wastewater Plant Treatment

- Download/View Extended Abstract

Abstract:

Several substances are sources of unpleasant odor emissions coming from wastewater treatment plants (WWTP) which hydrogen sulfide (H_2S) is the most representative molecule. Chemical scrubbing is the conventional and predominant process used for treating gas streams with these kinds of emissions that affect the air quality and displeasure of inhabitants of the nearby neighborhoods. The aim of this research work was to evaluate the removal of hydrogen sulfide of a gas stream from WWTP of a city in the north of Colombia by using a commercial Bio-catalyst (BOC). Structural, design, and operation changes of the scrubber equipment were made in order to implement the BOC as a scrubbing solution. A central composite design was used to study the effect of the BOC volume rate and the damper valve (air entrance control) opening percentage on the removal grade of the H_2S. The results of the experimental analysis showed that both variables had a significant effect on the removal of H_2S. Optimal conditions of operation were found to be at BOC volume rate of 17.74 ml/min and 20% of valve opening percentage with a removal percentage of H_2S of 87.43%. This work was able to show the application of the BOC as a scrubbing solution for removing H_2S of a gas stream from WWTP. In addition, the BOC has some taking advantage of the conventional process due to is biodegradable and environmentally friendly (low toxicity and non-corrosive).

602163 Invited Talk: Numerical and Experimental Investigation of Aerosol Particles Reactions Driven By Concentrating Solar Power

Abstract:

Solar energy conversion to chemicals and fuels receives progressively more attention. Many of the possible conversion routes incorporate particles or could be transformed into particle-phase reactions. When the particles are diluted into a fluid stream, usually a carrier inert gas or a reacting gas mixture, belong to the large family of aerosol processes. An aerosol process is the central theme of this presentation; specifically, one where solid or liquid particles are directly irradiated with concentrated solar power.

Aerosol phase reactors with a directly irradiated reaction zone can employ direct solar power with potentially higher solar to chemical energy efficiencies than the indirect solar reactors. Many studies investigate particle reactors such as fluidization and moving, but only a few of them focus on aerosol reactors. On the other hand, aerosol phase processes are very similar to solid fuel combustion and gasification, and free-board reactions. Therefore, in this study, we investigated the particle phase reactions of the oxygen evolution step of the sulfur-ammonia water-splitting cycle where ammonium and potassium-based sulfate particles decompose and convert to pyrosulfates with the gradual release of ammonia and sulfur oxide gases.

Particles of different compositions were synthesized via evaporation, crystallization, and drying of precursor aqueous solutions of potassium sulfate, potassium pyrosulfate, and ammonium sulfate. The onset temperatures, reaction energies, and exact mechanisms were extracted using bench-scale experiments with a combination of differential scanning calorimetry (DSC), thermogravimetric analysis (TGA), Fourier transformation infrared spectroscopy (FTIR), and mass spectroscopy (MS). The experimental observations were employed to develop a mathematical model and the results to fit the model parameters. After the successful verification and validation of the model, it was applied to predict the evolution of the particles exposed to concentrating light. Finally, more experiments were conducted into our high flux solar simulator to compare the numerical with experimental results. Overall, the numerical model can predict the particle phase reactions accurately. However, the numerical predictions are very sensitive to the irradiance levels. In other words, the predictions of the particle model can vary significantly depending on the assumed irradiance of the concentrating light, which implies that further work is essential in order to estimate the overall solar-to-chemical energy conversion efficiency.

607157 Reaction Screening for Thermochemical Energy Storage

Abstract:

A practical way to manage intermittency of solar energy in concentrating solar power (CSP) plants is to integrate them with thermal energy storage (TES). CSP plants integrated with TES is estimated to contribute to 11% of the global electricity generation by 2050.[1] The commercially deployed CSP plants are typically integrated with two-tank, molten-salt sensible heat storage unit, which delivers heat to the power cycle at low temperature (565 °C), resulting in low solar-to-electricity efficiency. Furthermore, due to low energy density of molten-salt, large amount of material is needed, which can become too expensive for large-scale systems. On the other hand, thermochemical energy storage (TCES), which is the reversible conversion of solar-thermal energy to chemical energy, allows high operating temperatures, high storage density and low heat loss over long periods.

Today, TCES is at an early stage of development. Previously, several fluid phase and solid-gas reactions have been studied. Recent analyses of fluid-phase reactions show that they suffer from low energy efficiency and high cost due to the need to store reacting gases.[2] However, solid-gas reactions are more promising because they require less gas storage and the products can be easily separated. Some work has been done to identify promising reaction candidates[3,4] based on equilibrium temperature and energy density. However, it still remains unclear which reaction chemistry leads to the lowest storage and electricity production costs. Accordingly, the goal of this work is to (i) evaluate solid-gas reactions belonging to the oxidation, hydration, and carbonation family via optimization models and (ii) rank material properties based on their effect on energetic and economic performance of CSP-TCES plants.

To this end, we developed a general process synthesis model for CSP with solid-phase TCES systems. The model is formulated as a two-stage stochastic programming model to account for variability in daily and seasonal solar irradiance, which are represented as modes and scenarios, respectively. The model requires plant characteristics (e.g., weather, operation time, and capacity) and reaction properties (e.g., reaction enthalpy and entropy, and specific heat capacity of each component) as inputs. Daggett in southern California is selected as the plant location with 100 MW capacity and its weather data is obtained from National Solar Radiation Data Base.[5] Component heat capacity, reaction enthalpy and entropy are estimated using Barin database.[6] The solution of the optimization model yields design (e.g., solar field area, receiver size, reactor and storage tank sizes) and operating conditions (e.g., stream flow rates, charging and discharging duration) that leads to the minimum levelized cost of electricity production. Other performance metrics such as energy efficiency and storage costs are also computed.

We take three major steps to create a ranked list of promising reaction candidates. First, we create a comprehensive list of reaction candidates belonging to oxidation, hydration, and carbonation family based on stoichiometry. Second, to maintain computational tractability, we perform a preliminary screening, and choose the reactions with positive enthalpy and equilibrium temperatures between 400 °C and 1500 °C. Third, the model is solved for the reaction candidates obtained after preliminary screening and they are ranked based on the cost of electricity production. Furthermore, we also formulate a material optimization problem by including reacting material properties as the decision variables in the process synthesis model. This analysis provides insights in the material and reaction properties that are most suitable for CSP-TCES plants and serve as guidelines for the development of novel TCES systems.

References

1 IEA publications, *Energy Technology Perspectives 2017 - Catalysing Energy Technology Transformations*, 2017.

2 X. Peng, T. W. Root and C. T. Maravelias, *Green Chem.*, 2017, 2427–2438.

3 P. Pardo, A. Deydier, Z. Anxionnaz-Minvielle, S. Rougé, M. Cabassud and P. Cognet, *Renew. Sustain. Energy Rev.*, 2014, **32**, 591–610.

4 L. André, S. Abanades and G. Flamant, *Renew. Sustain. Energy Rev.*, 2016, **64**, 703–715.

5 NREL, National Solar Radiation Data Base, https://rredc.nrel.gov/solar/old_data/nsrdb/1961-1990/, (accessed 8 April 2020).

6 I. Barin, *Thermochemical Data of Pure Substances*, VCH Verlagsgesellschaft, FRG., 1989.

608666 Thermal Energy Storage Using the Boudouard Reaction

Abstract:

This study presents a firm, indefinite storage of solar heat via chemical conversion of simple carbon-based chemicals using both the forward and reverse Boudouard reaction [1-2]:

$C + CO_2 \leftrightarrow 2CO$ ($\Delta H = 171$ kJ/mol)

Depending on the operational needs, the energy storage reaction may be conducted to optimize performance at temperatures ranging between ca. 900K and 1100K (~600°C to ~800°C) by modifying the initial loading ratio of C and CO_2 into the system. The process involves a fluidized bed reactor in which process solids reside where the heat is coupled into or out of the power cycle working fluid through the use of bidirectional heat pipes, and with the gas reactants/products blown through packed beds of the respective zeolite material (copper doped zeolite 5A for CO [3-4], and simple ZSM-5 for CO_2 [5-6]). In this mode, when excess heat is present, the process absorbs some of the heat and stores it as chemical energy in the storage tanks, and when insufficient heat is supplied by the solar field, the process reinjects heat directly into the power cycle.

Boudouard reactors at 5kWh and 10kWh scales have been designed and constructed. Some tests on the reactor have been conducted to examine its performance, including longevity, specific energy, charge and discharge temperatures and exergetic efficiency. Meanwhile, the simulation model has been developed based on the structure of the fluidized bed Boudouard reactor and optimized by matching the experimental data for the prediction of similar applications.

Reference:

[1] Boudouard, O. Influence De La Vapeur D'eau Sur La Réduction De L'anhydride Carbonique Par Le Charbon *C. R. Hebd. Acad. Sci.* **1905**, *141*, 252–253.

[2] Hunt, J.; Ferrari, A.; Lita, A.; Crosswhite, M.; Ashley, B.; Stiegman, A.E. Microwave-Specific Enhancement of the Carbon–Carbon Dioxide (Boudouard) Reaction *J. Phys. Chem. C* **2013**, *117*, 26871–26880.

[3] Dulaurent, O.; Courtois, X.; Perrichon, V.; Bianchi, D. "Heats of Adsorption of CO on a Cu/Al2O3 Catalyst Using FTIR Spectroscopy at High Temperatures and under Adsorption Equilibrium Conditions" *J. Phys. Chem. B*, **2000**, *104*, 6001–6011.

[4] Saha, D.; Deng, S. Adsorption Equilibria and Kinetics of Carbon Monoxide on Zeolite 5A, 13X, MOF-5, and MOF-177 *J. Chem. Eng. Data* **2009**, *54*, 2245–2250.

[5] Wang, Q.; Luo, J.; Zhong, Z.; Borgna, A.CO2 capture by solid adsorbents and their applications: current status and new trends *Energy Environ. Sci.* **2011**, *4*, 42-55.

[6] Choi, S; Drese, JH; Jones, CW Adsorbent Materials for Carbon Dioxide Capture from Large Anthropogenic Point Sources *ChemSusChem* **2009**, *2*, 796-854.

608659 Opportunities to Enhance Sustainability of Copper Production with the Hys Cycle

Abstract:

We report on the potential to integrate the hybrid sulphur cycle (HyS) with copper production through the coproduction of H_2, O_2 and H_2S Figure 1 presents schematic diagrams of the HyS cycle and the copper production process. O_4. According to national greenhouse and energy reporting data (NGER), mining was the second-largest contributor to Australia's CO_2 emissions in 2018-2019 at 28.5% [1]. Given that many mines in Australia are located in remote areas, where fuel costs are relatively high, and solar resources are typically available. There is strong motivation to seek opportunities to lower both cost and carbon footprint, in current mineral processing plant by retrofitting with renewable energy.

The HyS cycle, initially developed by the Westinghouse Electric Corp for H_2 and O_2 production through the splitting of water from nuclear heat source [2], has been found to offer strong potential for low-cost H_2 production. The cycle was then further developed by the German Aerospace Centre (DLR) using solar thermal energy [3]. The HyS cycle was estimated to achieve a solar to hydrogen efficiency of 30%, which is almost twice that of the commercially available PV-electrolysis route [3-4].

This work has been undertaken through the Australian Solar Thermal Research Institute (ASTRI) using Aspen Plus. Figure 1 presents schematic diagrams of the HyS cycle and the copper production process. As shown in Figure 1(a), the HyS cycle is a chemical looping process with sulphuric acid as one of the circulated components. In the first reactor, sulphuric acid is thermally decomposed at temperatures of 600-1200 °C, and a mixture of O_2, SO_2 and H_2O in the gas phase is produced. Gaseous SO_2, after separation from oxygen, is sent to the second reactor, where, it is electrolysed with water to produce H_2 and H_2SO_4. The hydrogen product leaves the cycle while the sulphuric acid is returned to the first reactor to complete the loop.

As shown in Figure 1(b), the copper production process, comprises two main pathways, which depend on the composition of the mineral feedstock, as follows:

1. Pyrometallurgy, in which the minerals are heated reacting sulphide ores with oxygen-enriched air to produce copper and SO_2 gas in a series of pyro-furnaces (smelters, converters and fire refiners). The oxygen is typically provided at present using air separation units (ASU).
2. Hydrometallurgy, in which the oxide ores are chemically treated to produce pure copper. Since the hydrometallurgy process requires sulphuric acid, usually the sulphur dioxide from the pyro-pathway is sent to the sulphuric acid plant. The sulphuric acid product is then utilised to dissolve the copper-containing ores for subsequent precipitation in electrowinning cells.

The sulphur dioxide product from the copper pyrometallurgy process has potential to feed into the HyS cycle to generate sulphuric acid to be utilised in hydro-pathway while the H_2 and O_2 products from the HyS cycle have potential to be used in the copper pyro-pathway to displace ASU and/or fossil fuels. The potential of this linkages will be reported in the paper through the following scenarios:

1. the HyS cycle is scaled to produce enough hydrogen for the pyro-furnaces;
2. the HyS cycle is sized to meet the oxygen demands in the pyro-pathways; and
3. the HyS cycle is sized based on the production rate of the SO_2 from the furnaces. In this scenario, the sulphuric acid would be supplied from the HyS cycle rather than the acid plant with any excess acid being used in the hydrometallurgy pathway or sold as a product.

Figure 1 (Caption): Schematic representation of a) HyS cycle b) Copper production flowsheet. HyS cycle produces hydrogen (fuel), oxygen and sulphuric acid that can be used in the copper production process.

References

1. National Greenhouse and Energy Reporting Data (NGER) website, accessed: 28.04.2020. http://www.cleanenergyregulator.gov.au/NGER/National%20greenhouse%20and%20energy%20reporting%20data/Data-highlights/2018-19-published-data-highlights
2. Brecher L.E. and Christopher K.W., "Electrolytic decomposition of water", Westinghouse Electric Corp., Patent

3,888,750, June 10, 1975.
3. Niehoff, A. G., Botero, N. B., Acharya, A., Thomey, D., Roeb, M., Sattler, C., & Pitz-Paal, R. (2015). Process modelling and heat management of the solar hybrid sulfur cycle. International Journal of Hydrogen Energy, 40(13), 4461-4473.
4. Thomey, D., de Oliveira, L., Säck, J. P., Roeb, M., & Sattler, C. (2012). Development and test of a solar reactor for decomposition of sulphuric acid in thermochemical hydrogen production. International journal of hydrogen energy, 37(21), 16615-16622.

1a) HyS Cycle

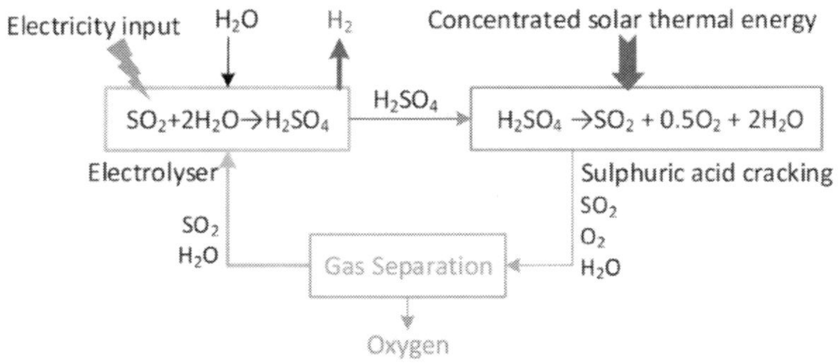

1b) Copper production flowsheet

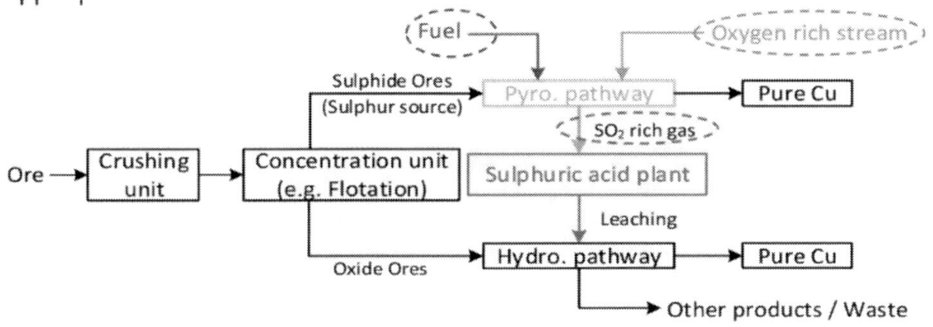

618118 Sponsored Technology Workshop - Instrumentation for Energy Device Research; from Material Studies to Cells to Full Stacks up to 1000V - Gamry Instruments

Abstract:

This workshop will focus on the instrumentation Gamry has to support battery and energy device research from material studies to cells to full stacks. We will discuss applications of an electrochemical quartz crystal microbalance for materials studies, potentiostats for materials and whole cell studies and then we will showcase our newest instrument capable of performing EIS on a 1000 V stack of cells.

613373 Legal Drivers for Waste Plastic Law and Regulation

Abstract:

Environmental law and policy are designed to mitigate the harmful effects of pollution. Yet, plastic generally tends to fall outside the scope of stringent environmental regulation, depriving entities producing or distributing plastic from specific critical incentives that could best support their development of meaningful sustainable approaches. Familiar approaches to designing environmental law and policy is not specifically designed to address persistent synthetic materials such as the current suite of synthetic commercial plastics distributed in the market today. Environmental programs generally focus on addressing pollution by mitigating exposure to chemicals that pose a threat of acute and chronic toxicity including carcinogenicity. The synthetic macro and microplastics in our environment, while likely leaching some chemical pollutants, have observably caused harm through their persistence in the environment resulting from their inert nature. Through their characteristics of fracturing into smaller and smaller particles while not dissolving, reacting or degrading, synthetic plastics may result in significant acute and chronic harm from physical obstruction of biological processes, whether as macro, micro or nanoparticles. This harm is generally not mitigated through conventional environmental regulation focusing on mitigating chemical toxicity. Ongoing research that better defines this harm, and provides reproducible parameters, metrics and nomenclature for such harm, will support the development of new approaches to mitigating such harm. These new approaches can support better application of existing law and policy, while also informing development of more tailored authority to address this type of environmental pollutant. These new approaches also support litigation and other methods of distributing the burden of risk and harm better defined from this research once the harm can be quantified. This presentation will summarize the current status of global policy approaches, including the United States, and how both ongoing efforts to qualify and quantify harm, and these new legal and policy measures, serve to provide significant incentive for all plastic dependent industry sectors to seek more aggressive approaches to keeping plastic out of the environment.

613246 Public Health Perspectives Associated with Waste Plastics

Abstract:

Each stage of the plastic lifecycle poses major risks to human health. Worldwide, most people are exposed to plastics at all stages of their lifecycle. Plastic pollution is pervasive worldwide infiltrating our environment, wildlife and marine animals, our food supply, and our human health. Health problems associated with waste plastics include: various forms of cancer; diabetes; several organ malfunctions; impacts on skin, eyes, and other sensory organs; birth defects, among other impacts. Additionally, there are harmful impacts associated with plastic additives, including bisphenol-A, heavy metals (e.g., cadmium and lead); flame retardants, phthalates, perfluorinated substances (PFAS), endocrine disrupting compounds, and other chemicals Plastic production has increased from 2×10^6 metric tons in 1950 to 3.80×10^8 metric tons in 2015.

Direct environmental exposure to plastics is a result of extraction and transport, refining and manufacture, consumer products and packaging, waste management, fragmentation and micro-plastics formation, cascading exposure as plastics degrade, and ongoing environmental exposures. Humans are exposed to microplastics and associated toxic chemicals through inhalation, ingestion, and direct skin contact throughout the life cycle of plastics. Public health is exacerbated by the fact that more than 2/3 of all plastic produced remains in the environment, as debris in the oceans and waterways, solid waste and litter on land, micro- or nano-particles in the air or agricultural soils, microfibers in water supplies, or microparticles in the human body, etc. As a result, plastics have permeated our air, our soil, our water, and our bodies.

This plenary presentation addresses public health perspectives associated with waste plastics, and offers approaches for minimizing waste plastic formation.

613477 Keynote: Solar Chemistry in Mexico and the National Laboratory of Solar Concentration and Solar Chemistry

Abstract:

Mexico is a country that has abundant solar energy resources, in particular, in the northwestern region of the country, it has an average annual DNI values of over 7.5 kWh/m2/day, which are among the best on the planet. Therefore, it is expected that solar concentration technologies will be deployed in the country that take advantage of this energy resource; in addition to CSP, those related to thermochemical processes such as the production of solar fuels or the development of new materials will be expected.

The article presents the most important investigations that have been carried out in Mexico in the last decade, particularly due to the impulse generated by the establishment of the National Laboratory of Solar Concentration Systems and Solar Chemistry in the country.

605907 Invited Talk: Reaching Beyond 1'500°C with the Synhelion Absorbing Gas Solar Receiver: Results of Experimental Campaign

Abstract:

We report about the development and testing of a novel high-temperature solar receiver for generating process heat beyond 1'500°C. The receiver operation principle is similar to the "greenhouse effect" and exploits the ability of molecular gases such as water vapor and carbon dioxide to absorb a significant fraction of longer wavelength thermal radiation while being mostly transparent to terrestrial solar radiation. The receiver is constituted by an absorbing cavity with a windowed aperture. The cavity is filled with a streaming participating gas flowing from the aperture towards the back of the cavity. The incoming solar radiation enters the aperture, passes with minimal absorption through the gas, and is successively absorbed by the black surfaces of the cavity. The surfaces in turn thermalize and re-radiate with a blackbody spectrum. The gas absorbs to a large extent the thermal re-emission and is heated up, acting as heat transfer fluid (HTF). At the same time, the gas shields the aperture from the re-emission [1]. The left side of Figure 1 depicts schematically the system. The receiver works with radiation only as heat transfer mechanism and does not require, a priori, any convective contribution.

To assess the potential of the approach we modelled the system with the most accurate method available, spectral line-by-line photon Monte Carlo raytracing (MC-LBL) with the absorption coefficients derived from the HITEMP 2010 spectroscopic database [1]. With a solar irradiation flux of 1'200 kW/m2 at the aperture and a gas inlet temperature of 1'000 K, the receiver efficiency (defined as the ratio of thermal power absorbed by the HTF to the solar power incident on the receiver aperture) remained above 80% for gas outlet temperatures up to 1'800 K for receivers operated with steam [1].

To demonstrate in practice the system capabilities and to validate the models developed we built a large 250 kW input power receiver prototype operating with steam. The cylindrical cavity of the receiver prototype had an internal diameter of 0.8 m and a clear aperture diameter of 0.56 m. We tested the setup between summer 2019 and spring 2020 at DLR's Synlight solar simulator facility. The system was operated "on-lamp" for almost 150 hours. Steam outlet temperatures in excess of 1'550°C were reached. The system after test is shown on the right side in Figure 1.

Fig. 1: (Left) Schematic illustration of the Synhelion absorbing gas receiver. (Right) Photograph of the receiver front taken after testing.

References

[1] G. Ambrosetti and P. Good, A novel approach to high temperature solar receivers with an absorbing gas as heat transfer fluid and reduced radiative losses. Solar Energy 183, 521–531 (2019).

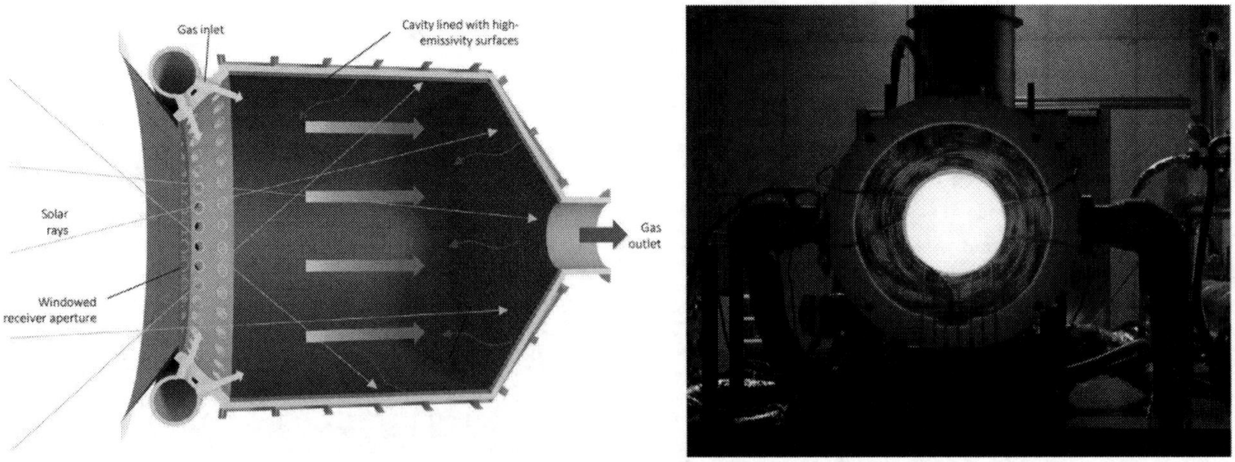

609254 CFD Modeling and Performance Evaluation of the Synhelion Absorbing Gas Solar Receiver for Process Heat at 1500°C and Beyond

Abstract:

1. Introduction

In the field of renewable energies, solar is the only energy source that can be exploited, with proper concentration techniques, to fulfill the needs of applications requiring thermal energy at very high temperature levels, e.g. higher than 1'000°C, such as thermochemical syngas production or industrial process heat (i.e., cement pyroprocessing). However, the maximum operating temperature of commercial solar receivers is currently lower than 873 K due to major limiting factors such as: heat transfer fluids (HTF), materials, and heat losses [1]. The Swiss company Synhelion SA is developing a new generation of solar receivers, for large-scale installations, suitable to operate at temperature levels higher than 1'500 °C [2]. The experimental demonstration will be detailed in [3]. As opposed to conventional commercial receivers, which mainly rely on indirect heating of the HTF, the proposed solution is based on the direct absorption of thermal radiation by a gaseous HTF. The latter must have specific spectral properties: it has to be almost transparent to the high-radiation intensity wavelengths of the solar spectrum and, at the same time, mostly opaque (i.e. strong absorption bands) in the wavelength range of thermal radiation. Water vapor or carbon dioxide are among the suitable HTF candidates. In the present paper, the absorbing gas receiver concept will be introduced along with a detailed description of the numerical model developed to evaluate its performance. Various simulations campaigns were carried out with the aim of evaluating the effect of important parameters, such as: (i) HTF mass flow rate, (ii) HTF entrance angle and (iii) gravity, on the receiver thermo-fluid dynamics behavior. The major findings obtained will be presented and discussed.

2. The Synhelion absorbing gas solar receiver

The Synhelion absorbing gas receiver, schematically depicted in figure 1a, is a cavity-type receiver. Concentrated solar radiation enters the receiver from the windowed aperture, travels along the whole cavity length to be then absorbed by highly-absorptive surfaces located at the back of the cavity. This, in turn, causes the temperatures of these surfaces to sensibly increase thus re-emitting longer wavelength thermal radiation into the cavity. At the same time, the HTF flows also from the aperture towards the back of the cavity absorbing the thermal radiation emitted by the absorptive surfaces. As a result, the temperature of the HTF increases, as it flows through the cavity, reaching the maximum value as it approaches the innermost section in the vicinity of the receiver outlet section. The fact of being mostly opaque in the wavelength range of thermal radiation, also enables the HTF to effectively shield the absorptive surfaces from aperture minimizing hence the radiative heat losses towards the external environment.

3. Numerical model development and validation

A computational fluid dynamics (CFD) model was developed, exploiting Fluent code from ANSYS, with the twofold aim of clearly understanding the physical phenomena, as well as their interactions, occurring into the system and creating a useful tool for assisting the design phase of the receiver. From the numerical modeling standpoint, the major challenge is that radiative heat transfer in participating media has to be accounted for. Based upon a tradeoff between results accuracy and computational efforts, this heat transfer mechanism was described through the discrete ordinates (DO) method coupled with the weighted-sum-of-gray-gases (WSGG) model accounting for the variations of the participating media radiation properties.

At the time this study was conducted, no experimental data were available; the CFD model was hence validated against benchmark results of a numerical model, developed by Synhelion, based on Monte Carlo line-by-line that represents the most accurate methods to model radiative heat transfer in participating ia [4]. The validation process allowed to observe that, despite a slight overestimation of the receiver performance (of about 5%), the CFD model with simplified treatment of radiative heat transfer is able to replicate the behavior of this type of receiver.

4. CFD simulations campaigns – Results and discussion

Once validated, the CFD model was exploited to run a series of simulations campaign with the aim of evaluating how different operating conditions affect the Synhelion absorbing gas receiver. The most representative operating parameters evaluated were: (i) HTF mass flow rate, (ii) orientation of the HTF inlet velocity vector and (iii) gravity (upward- and downward-facing receiver). Moreover, two scenarios of receiver operating pressure, ambient pressure and 10 bars respectively, were also considered and compared. The beneficial effect of operating at higher pressure is the increased density of the HTF, which leads to a downsizing of the receiver proportional to the operating pressure increase. The reference receiver geometry assumed for the simulations, is 2D axisymmetric (figure 1b) with cavity dimensions of 16 m diameter and 16 m length, and 0.72 m diameter and 0.72 m length, in the case of unpressurized and pressurized receiver respectively. For both the cases, 600 kW/m^2 concentrated solar flux on the absorptive surfaces was assumed with water vapor, the HTF under investigation, inlet temperature of 1000 K. The cavity is modelled as ideally insulated and therefore, the only source of heat loss is assumed to be through the aperture by means of thermal radiation only. The internal surfaces of the cavity are considered to have emissivity equal to 1 (black surfaces) and a fully diffusive emission of thermal radiation. Besides the thermo-fluid dynamics behavior, the overall receiver performance were also quantified by computing its thermal efficiency, defined as the ratio of the power removed by the HTF divided by the total input power.

Effect of HTF mass flow rate

This first CFD simulations campaign was devoted to evaluate the receiver performance operating with different HTF mass flow rate values. The reference HTF mass flow rate ranges considered were: 38 kg/s – 472 kg/s and 0.07 kg/s – 0.92 kg/s in the case of unpressurized and pressurized receiver respectively. The graph in figure 1c summarizes the resulting receiver thermal efficiency as a function of the HTF outflow temperature. The higher the HTF outflow temperature, the lower the HTF mass flow rate thorough the receiver. A consideration that can be drawn, is that reducing the HTF mass flow rate leads to a decrease of the receiver thermal efficiency due to higher heat losses. It is also interesting to observe the different evolution of the receiver performance with the efficiency that rapidly decreases moving away from the highest and the lowest HTF mass flow rate values considered. In between, the thermal efficiency curves show an inflection point in proximity of the HTF outflow temperature of about 1'500 K.

Effect of HTF entrance angle

Five different orientation of the HTF inlet velocity vector, as shown in figure 1b, were simulated in order to evaluate the effect of the HTF entrance angle on the receiver performance. The resulting receiver efficiency is reported in the graph of figure 1d. Based upon the obtained results, it can be concluded that inclination angles lower than about 30° from the aperture have a negligible effect on the receiver performance. An inclination angle of about 45° has a beneficial effect if the receiver operates at high HTF mass flow rates; conversely, it shows a detrimental effect in the case of lower HTF mass flow rate values. At higher inclination angles, the receiver efficiency seems to be generally higher; however, the receiver behavior is affected, at the same time, by the combined effect of HTF inclination angle and inlet velocity since the same inlet section was assumed for all the simulations.

Effect of gravity

In the case of large HTF density gradient, a gravity-induced buoyancy flow could take place affecting the flow field, and temperature distribution, into the cavity. Two evaluate the importance of gravity on the receiver performance, an additional CFD simulations campaign was performed assuming an upward- and downward-facing receiver. With these receiver orientations is possible to maintain the same 2D computational domain. The unpressurized receiver was assumed as reference. According to the results obtained, shown in figure 1e, the effect of gravity on the receiver performance is not negligible especially in the case of low HTF mass flow rates. As the HTF mass flow rate increases, the receiver performance are gradually less dependent on gravity up to a certain condition in which the effect of gravity can be safely neglected.

References

[1] C. Ho, Advances in central receivers for concentrating solar applications“, Solar Energy 152, 38-56, (2017)

[2] G. Ambrosetti and P. Good, A novel approach to high temperature solar receivers with an absorbing gas as heat transfer fluid and reduced radiative losses. Solar Energy 183, 521-531 (2019).

[3] G. Ambrosetti, P. Good, P. Furler, L. Geissbühler, D. Rutz, S. Ackermann, Reaching Beyond 1’500°C With The Synhelion Absorbing Gas Solar Receiver: Results of Experimental Campaign, Abstract submitted for 2020 AIChE Annual Meeting.

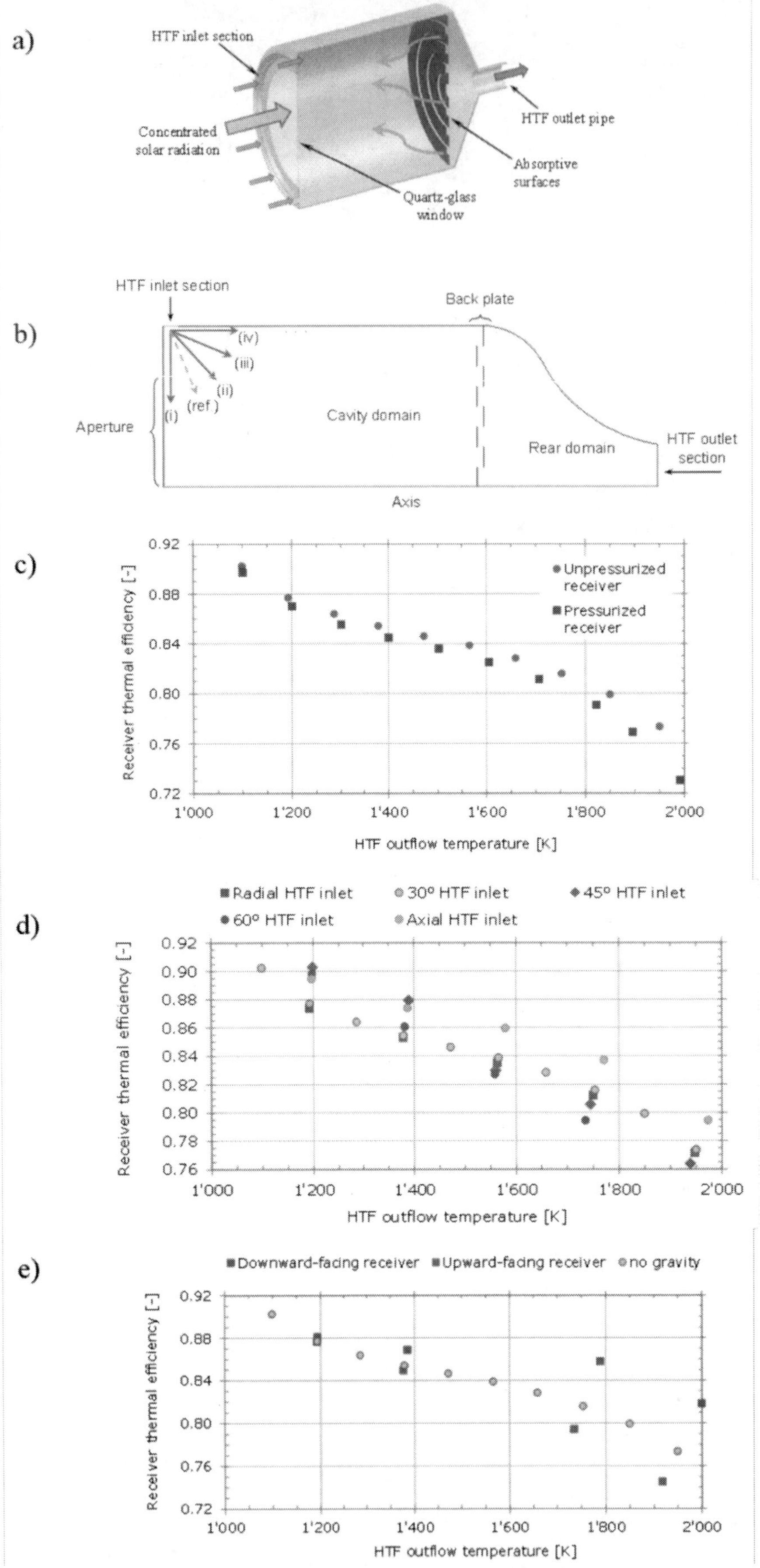
a)
HTF inlet section
Concentrated solar radiation
HTF outlet pipe
Absorptive surfaces
Quartz-glass window
b)
HTF inlet section
Back plate
(iv)
(iii)
(ii)
(ref)
(i)
Aperture
Cavity domain
Rear domain
HTF outlet section
Axis
c)
Unpressurized receiver
Pressurized receiver
Receiver thermal efficiency [-]
0.92
0.88
0.84
0.80
0.76
0.72
1'000
1'200
1'400
1'600
1'800
2'000
HTF outflow temperature [K]
d)
Radial HTF inlet
30° HTF inlet
45° HTF inlet
60° HTF inlet
Axial HTF inlet
Receiver thermal efficiency [-]
0.92
0.90
0.88
0.86
0.84
0.82
0.80
0.78
0.76
1'000
1'200
1'400
1'600
1'800
2'000
HTF outflow temperature [K]
e)
Downward-facing receiver
Upward-facing receiver
no gravity
Receiver thermal efficiency [-]
0.92
0.88
0.84
0.80
0.76
0.72
1'000
1'200
1'400
1'600
1'800
2'000
HTF outflow temperature [K]

611661 Optical Analysis of a Novel Solar Thermochemical System with a Rotating Tower Reflector and a Receiver–Reactor Array

Abstract:

We propose a novel concept of a rotating tower reflector (TR) in a beam-down optical system to alternate concentrated solar irradiation of an array of solar receiver–reactors realizing multi-step solar thermochemical redox cycles. Heat recovery can be implemented in the receiver–reactor array, which is reported to result in dramatically improved thermal-to-chemical efficiency of receiver–reactors. Here, optical and radiative characteristics of the proposed optical system are explored analytically and by Monte-Carlo ray-tracing simulations.

We study the effects of system geometrical and optical parameters on the optical and radiative performance. TR axis is required to be tilted for accommodating the receiver–reactor array, resulting in reduced optical efficiency. We demonstrate that the annual optical efficiency of a baseline system with the receiver–reactor located south of the tower decreases from 46% to 37% for the axis tilt angle of TR increasing from 2° to 20°. Locating the receiver–reactor array south of and as close as possible to the tower offers the most promising optical configuration. The optical analysis conducted in this study provides a general formulation to enable predictions of thermal-to-chemical efficiency of the receiver–reactor array and overall solar-to-chemical efficiency of the solar thermochemical plant.

606896 Numerical Study on a Novel Solar-Driven Interface Evaporation Device Combined with Thermoelectric Material

Abstract:

In the continuous development of solar energy, solar-driven evaporation has attracted much attention due to its huge industrial potential. In recent years, many kinds of solar-driven interface evaporation devices have been proposed by confining the conversion of solar-thermal energy to the air/liquid interface. However, it can't work during night time and thus the evaporation efficiency is still limited. In this work, we discussed the possibility of combining thermoelectric technology with interface evaporation and proposed a novel solar-driven interface evaporation device combined with thermoelectric material. The design of the key components including thermoelectric structures and evaporation structures are introduced. A one-dimensional numerical model is developed to evaluate the evaporation performance of the novel design. We also compare the evaporation performance of the novel design with the traditional solar-driven interface evaporation device.

611754 Solar Pyrolysis Production of Self-Activated Bio-Derived Carbons for Supercapacitors

Abstract:

Supercapacitors (SC), also known as ultracapacitor or electrochemical capacitors, are energy storage devices whose characteristics place them in the gap between high power density of traditional capacitors, and high energy density of rechargeable batteries. In particular devices that store energy mainly by ion adsorption at the electrode/electrolyte interface are called electric double layer capacitors (EDLC). Carbon materials with high surface area, adequate micro-meso pore size distribution, good electrical conductivity, and chemical stability, are the most common electrode materials used in EDLCs. In order to increase the sustainability of the production of these devices, bio-derived carbons (BDCs), sometimes called biochars can be used, in particular, those obtained from agricultural wastes that do not compete with food supply. The production of bio-derived carbons useful for supercapacitors by means of solar pyrolysis is reported in this work. Two abundant lignocellulosic wastes, agave angustifolia leaves, and pruned tomato plant, were transformed by solar pyrolysis to BDCs. The process was carried out in the IER-UNAM High flux solar furnace (HoSIER), by using a glass spherical reactor. We studied different solar pyrolysis temperatures ranging from 450°C to 1560°C to produce the BDCs and correlate their physicochemical properties with their ability to store energy. XRD was used to determine the effect of solar pyrolysis temperature on the obtained BDCs. Additional structural properties of BDCs were analyzed by SEM, Raman spectroscopy and physisorption. Elemental analysis and EDAX were used to determine the chemical composition of wastes, and their effect on BDCs. It was found that BDCs from tomato plant waste had higher surface areas with well-developed microporosity without any extra activation process, confirming a self-activation process during pyrolysis. Electrochemical properties of BDCs were analyzed by cyclic voltammetry in 3 electrode cells, and the ones showing better performance were tested in 2-electrode supercapacitor cells. These cells exhibit specific energies and power values similar to a commercial carbon designed for supercapacitors.

611315 Non-Isothermal Conditions in Countercurrent Thermochemical Reactors for Fuel Production

Abstract:

The efficiency of two-step thermochemical water and carbon dioxide splitting cycles based on non-stoichiometric, redox active, metal oxides is critically dependent on the selection of the operating conditions and the design of thermochemical reactors. Thermodynamically, the reduction step is favored at high temperature and low oxygen partial pressure, while the re-oxidation step is favored at low temperatures and excess oxidant (water or carbon dioxide). Countercurrent reactors can maximize the chemical potential utilization of the material offering high efficiencies. However, as several authors have highlighted, an oversimplified thermodynamic analysis of these reactors could lead to a violation of the second law of thermodynamics. Some previous works have analyzed these reactors based on isothermal reduction or oxidation conditions, where chemical equilibrium ($\Delta G = 0$) is reached only at one point along the flow path. The present work proposes a method to determine the chemical equilibrium at each infinitesimal point along the flow path, leading to non-isothermal conditions based on both mass conservation and Gibbs chemical equilibrium criteria. Our results provide higher flexibility on the design and use of countercurrent reactors in thermochemical cycles being especially relevant to solar driven reactors where the temperature distribution is not homogeneous in the reduction or oxidation reactors. Additionally, material heating may be more rapid than the reduction reaction contributing to the inhomogeneity at least for reduction.

610092 Achieving High Temperature and Robustness in Thermochemical Reactor Prototypes

Abstract:

Solar thermochemical reactor prototypes rely on simulators to achieve high temperatures and rapid temperature swings. Simulators as heaters are both costly and bulky, and often play a significant role in limiting reactor designs. To enable faster reactor prototyping and materials testing, we have developed a novel heater that can reach both the temperatures necessary in thermochemistry, and has outstanding temperature cycling capabilities. The heater is also compact and inexpensive. Here, we report on the testing campaign on a prototype heater, demonstrating sustained temperature in excess of 1550 °C, and thousands of heating/cooling cycles at 400 °C/min. The robust performance confirms the feasibility of the heater concept and its potential in thermochemical reactor prototypes, as well as much more broadly in a range of high-temperature applications.

611860 Numerical Modelling of Ceria Undergoing Reduction in a Particle–Gas Counter-Flow: Effects of Chemical Kinetics Under Isothermal Conditions

Abstract:

A numerical model is employed to simulate a single tube reactor featuring a downward particle flow counter to an upward inert gas flow for ceria reduction in the dilute flow regime. The coupled phenomena of mass and momentum transfer as well as chemical kinetics are simulated assuming isothermal operation for the reactor. The model predicts the reduction extent under varying reaction kinetics as well as design and operational choices. The reduction extent is found to increase with the reaction rate constant until achieving the thermodynamic upper limit at a certain critical value. This critical rate constant signifies a transition from a chemical kinetics limited conversion to a gas advection limited conversion. The effect of the reactor length and the particle size on reaction extent is studied for a range of realistic cases. An empirical correlation is further developed to quantify the effects of particle and gas flow rates on reduction extent at both slow and fast kinetics. The present work offers insights to help guide reactor design and operation towards achieving the maximum reduction extent. Future effort will expand the two-phase numerical flow model to account for non-isothermal reactor conditions.

600783 Porous Silica-Encaged Ultrafine Metal Nanocatalysts for Solar Thermochemical Methane Reforming

Abstract:

Methane reforming is an important industrial process for hydrogen and synthesis gas production. For example, 95% of hydrogen produced globally in 2018 (70 million tonnes) was obtained via methane reforming. However, methane reforming processes are highly endothermic. To obtain high methane conversion to meet industrial production requirements, the process proceeds at high temperatures, typically in excess of 700 °C and in presence of catalysts. Solar-driven thermochemical processes provide an alternative and sustainable route for methane reforming at high temperature. Supported Ni-based materials are the most commonly used catalysts for this process. Catalytic activity of catalysts depends on the particle size of the metal species. The performance improves with the decrease of metal particle sizes due to increased accessibility to active sites. However, rapid deactivation due to the severe sintering of metal species during the reactions under high temperatures was observed in previous studies. Porous silica, especially those with uniform porous structures such as mesoporous silica and microporous zeolite, have been regarded as ideal supports to encapsulate ultrasmall metal species inside the matrix due to the excellent confinement effect of ordered porous systems. Porous silica-encaged metallic catalysts possess superior hydrothermal/thermal stability under high temperatures. In this work, we have developed a novel and universal synthesis strategy for preparing mesoporous silica/zeolite-encaged ultrafine bimetallic Ni-CeO_x and Ni-Rh nanocatalysts, and investigated their catalytic activity for the solar-driven methane reforming application. Additionally, the particle size, the location and fine structure of obtained nanocatalysts, and the long-term/recycling stability of the catalysts during the solar-driven methane reforming processes are investigated.

609964 Correction of the DRIFT in Solar Tracking Systems through Closed LOOP Control

Abstract:

Two-axis solar tracking systems in solar concentration applications require greater precision in the automatic tracking of the apparent position of the Sun, this in order to keep the focal area in a fixed position with respect to the focal axis of the concentrator system, since that a slight variation translates into a deviation of the concentrated sun image from the area of interest. This work focuses on the discussion of the methods implemented to solve the problems of drift and correction of elevation angles and azimuths in solar tracking systems.

The proposed method is based on an open loop control system in which the angles of the solar vector (azimuth and elevation) are calculated using a solar position algorithm. For the closed-loop control system, two types of feedback were implemented, the first refers to an algorithm for image processing obtained by a digital camera that directly observes the sun, thus allowing a position correction to be obtained in both axes of the monitoring system.

The second feedback sensor is based on an adjustable peephole which consists of a cylindrical body with four cylindrical tunnels longitudinally angled and distributed orthogonal to each other at the same radial distance, which house an LDR photoresistor that provides a voltage. proportional to the sunlight incident on them. This voltage is translated into a position correction angle on the two tracking axes.

To carry out the evaluation of the drift correction, three two-axis solar tracking systems were used. A 36 m2 Heliostat (E3) and a parabolic mirror solar tracker (2M) located in the Hermosillo Solar Platform, Sonora, Mexico. The third monitoring system used is a Fresnel-type solar concentrator for the characterization of photovoltaic cells (FRESNEL) located at the IER-UNAM in Temixco, Morelos, Mexico. The results obtained from the comparison of the open-loop and closed-loop control by analyzing the drift of the concentration spot of the three monitoring systems are discussed.

REFERENCES

[1] M. E. H. Chowdhury, A. Khandakar, B. Hossain, and R. Abouhasera, "A low-cost closed-loop solar tracking system based on the sun position algorithm," J. Sensors, vol. 2019, 2019, doi: 10.1155/2019/3681031.

[2] V. S. S. Kumar, "Automatic Dual Axis Sun Tracking System using LDR Sensor," Int. J. Curr. Eng. Technogy, vol. 4, no. 5, pp. 3214–3217, 2011, doi: 10.14741/ijcet/4/5/2014/22.

[3] M. H. F. B. M. Fauadi, F. A. Jafar, L. O. Yee, M. N. Bin Maslan, and S. H. Yahya, "The 8th International Conference on Robotic, Vision, Signal Processing & Power Applications," Lect. Notes Electr. Eng., vol. 291, pp. 327–334, 2014, doi: 10.1007/978-981-4585-42-2.

[4] R. Pérez-Enciso, E. Brito-Bazan, C. A. Pérez-Rábago, C. A. Arancibia-Bulnes, D. Riveros-Rosas, and C. A. Estrada, "Correction of the concentrated sunlight spot's drift of the IER-UNAM's solar furnace," Appl. Therm. Eng., vol. 75, pp. 1187–1191, 2015, doi: 10.1016/j.applthermaleng.2014.05.045.

[5] K. Azizi and A. Ghaffari, "Design and manufacturing of a high-precision sun tracking system based on image processing," Int. J. Photoenergy, vol. 2013, 2013, doi: 10.1155/2013/754549.

609852 Process Simulation of Lignocellulosic Material Pyrolysis to Producing High Added-Value Chemicals

Abstract:

Sustainable processes and proper waste management have become increasingly priority for industries and governments. The current global scenario points out to the critical demand for renewable energy alternatives to drive the reduction of the market dependence with regards to the fossil fuels and their effects on climate change as a consequence of the greenhouse effect. In recent years, the biomass derivatives (such as corn, straw, bagasse, and many others) have become natural sources which the chemical composition is especially convenient for the production of fuels and chemicals. Biomass is a renewable resource derived from forestry, agriculture, agricultural crops, aquatic plants and food processing.

The biomass chemical and structural composition is highly dependent on the plant genetic factors, environmental influences and their chemical interactions. Derived-biomass materials have shown a high energy content. To converting the biomass into chemicals and fuels, fast pyrolysis technology has been found to be efficient because high heating rates favor the formation of liquid product.

The liquid phase derived from the biomass pyrolysis, conventionally termed bio-oil, is a complex mixture of organic compounds. The bio-oil is usually composed of aliphatic and aromatic hydrocarbons, organic acids, alcohols, esters, ethers, aldehydes, ketones and sugars.

Many of the chemical compounds found in the bio-oil are very useful for the chemical industry, including 5-hydroxymethylfurfural (HMF), 1,3-cyclopentadione, 2-methoxy-4-vinylphenol, levoglucosan, 4-hydroxy-3-methoxycinnamaldehyde and 2-methoxy-4-propylphenol. As an example, the 1,3-cycle-pentadione is found in widespread synthetic applications that include the synthesis of prostaglandins, antibiotics, herbicides, among other biologically active components. Its synthesis requires controlled and selective alkylation process. 2-methoxy-4-vinylphenol is an aromatic compound used as a flavoring agent, being one of the compounds responsible for the wheat natural aroma. The 4-hydroxy-3-methoxycinmaldehyde has antioxidant potential suitable for use in antioxidant and anti-radical research studies of ferulate activities. The 2-methoxy-4-propyl-phenol is applied to the enhancement of smoke, clove, spicy cinnamon and pepper, vanilla and fruit nuances.

Due to the complexity, experimental tests on biorefinery units tend to be expensive and time consuming. Process simulation is an effective way to study the pyrolysis process without the expenses and delays generated by laboratory-scaled tests. Studies on pyrolysis process modelling have scarcely provided full description of the biomass conversion kinetics. Most current studies have utilized kinetic parameters calculated based on experimentally obtained results of material conversion. Despite recent studies on integrated processes for chemical and energy use of biomass, a limited number of works has presented the pyrolysis process simulation, focusing on the high-added value products to replace non-renewable products. Based on this scenario, the present work presents a study on the thermal conversion of lignocellulosic residues through the fast pyrolysis process, from a biorefinery point-of-view, aiming to obtain HMF, levoglucosan and *p*-coniferaldehyde.

The simulation design was structured through elementary compositional simulation, using PRO/II® software. The process is based on the concept of biorefinery, coupling a fast pyrolysis unit to a separation process. The process was divided into three main stages: pre-treatment, pyrolysis and separation. The pyrolysis process modeling was developed based on the classic kinetic multi-stage model and the Diebold concept for the production of active cellulose. The result shows the viability of the production and recovery of HMF from an alternative source to petroleum derivatives.

608467 Upgrading Chemically Depolymerized Lignin into Value-Added Products: Maximizing Feedstock Utilization

Abstract:

Lignocellulosic biomass is a bountiful source of renewable carbon for the sustainable production of fuels and chemicals.

Lignocellulosic biomass is composed of primarily three polymers: cellulose, hemicellulose, and lignin. Hemicellulose is a heteropolymer of C_5 and C_6 carbohydrates, whereas cellulose is a homopolymer of $\beta(1\rightarrow4)$ linked glucose. These polysaccharides comprise 65-80 wt% of the plant biomass and considered to be easier to upgrade to value-added products (e.g., paper, clothing, and liquid fuels). Lignin, on the other hand, is a heteropolymer of aromatic compounds and is recalcitrant to chemical and/or biological upgrading. As lignin constitutes 20-35 wt% of biomass, and it is the largest source of renewable aromatics on the planet, it is desirable to obtain value-added products from lignin for maximum utilization of the feedstock.

In this work, we have used chemical and biological upgrading in tandem to extract maximum value from lignin. We will show that relatively native lignin can be isolated from lignocellulosic biomass using γ-valerolactone (GVL) and water as the solvent system with sulfuric acid as the catalyst under mild reaction conditions. We demonstrate that this GVL-lignin is successfully depolymerized by hydrogenolysis over a Pd/C catalyst into a mixture of aromatic compounds. This mixture contains molecules with similar chemical structures and presents a challenging separation problem. Indeed, this has been a bottleneck for obtaining value-added products from lignin. We solve this problem by upgrading the mixture of aromatic compounds to a single compound via biological funneling. Specifically, we use engineered *Novosphingobium aromaticivorans* (DSM12444) to transform the mixture of aromatic compounds containing syringyl, guaiacyl, and *p*-hydroxyphenyl substructures to a single product, 2-pyrone-4,6-dicarboxylic acid (PDC). Thus, we show that a complex mixture of lignin hydrogenolysis products can be reduced to a single product leading to easier separation. Additionally, we show that this strategy is agnostic to the biomass type by successfully applying this strategy to lignin extracted from hardwoods (poplar and maple) and grasses (switchgrass and sorghum).

610980 Oxidative Stability and Biodegradability of Bio-Based Lubricants Using the Petrooxy and Bio-Kinetic Methods

Abstract:

The key properties for bio-based lubricants are their biodegradability, cold flow behavior and oxidative stability. In order to be considered as a candidate to obtain bio-based lubricants, a material from renewable sources must provide properties which overcome the disadvantages usually observed for vegetable oils in this application, poor oxidative stability and high pour point. In this study, the oxidative stability and biodegradability of bio-based lubricants were compared to petroleum-based lubricants and vegetable oils through the PetroOXY and Bio-kinetic methods, respectively. Chemical modifications of castor oil fatty acids were carried out using 2-ethylhexanol. The acetylation reactions were performed with acetic anhydride (1:2) using KOH (5 % wt.). Furthermore, commercial antioxidants and phenolic compounds (saturated and unsaturated cardanol), obtained from cashew nut shell liquid, were added to the synthesized samples. The results show that the chemically modified bio-based products exhibited improved oxidative stability (up to 6 times) and depressed pour point (-42 °C) when compared to fresh castor oil. Overall, the addition of antioxidants increased 6 to 20 times the oxidative stability of the bio-based products. The biodegradability tests showed that synthesized samples have half-life around 30 days.

611044 Effects of Preprocessing Parameters on Material Attributes and Flow Behavior of Loblolly Pine

Abstract:

The inhomogeneous, elastic and cohesive nature of biomass poses several feeding and handling challenges during granular flow which create issues when transporting material through biorefinery unit operations (hoppers, silos, screw conveyors, etc.). It is therefore essential to study the mechanical and physical properties of these materials and accurately characterize bulk flow to ensure uniform handling and prevent issues of segregation, agglomeration, plugging, ratholing, stagnant material, arching, etc. This work is aimed at understanding and solving the flow challenges that biorefineries are faced with, through extensive experimental investigation of the mechanical properties and flow behavior (resulting from shear-failure of the bulk sample) of industrially relevant mature loblolly pine wood. Shear tests are conducted on a rotary shear tester (Schulze-style tester) at the different consolidation stresses (approximately 1 – 10 kPa) that the material might experience in an industrial scale hopper to obtain mechanical properties of the sample. This project investigates the linkages between material attributes (particle size and shape, mechanical properties arising largely from the biological cellular structure and biopolymer composition, and moisture content manifesting as enhanced particle cohesion due to capillary forces, preprocessing method, etc.), and resulting flow performance. Step wise explanatory regression of the shear test data provided a preliminary prediction estimate for apparent cohesion (output flow response) with respect to the preprocessing and material attributes (prediction parameters) discussed above. Smaller particles are seen to have higher cohesion and lower flowability. This is hypothesized to be an observational result of the interparticle forces exceeding gravitational forces in finer particles. Further, the measured properties depend on the consolidation forces. These results indicate that material might stagnate and cause mechanical locking at certain particle sizes and consolidation stresses, preventing flow. These observations imply careful consideration of operational conditions and material properties on equipment design and operation is essential to avoid material flow issues.

598193 Optimal Economic and Environmental Design to Produce Bio-Jet Fuel

Abstract:

Environmental regulations show more and more focus on reducing the effects of using fossil fuels. The aviation sector's commitment to society has led to the development of the process to replace these fuels with fuels from renewable sources. The design of these processes must be aimed at being efficient, profitable, and environmentally friendly. Bio-jet fuel is a fuel that can be obtained by different technological routes. The process of converting alcohol to bio-jet fuel known as ATJ (alcohol to jet) shows a particular interest because obtaining the raw material is in a promising development showing good results for its industrial scale implementation. In order to evaluate this technology, its study is proposed through the modeling and simulation of the process. The ATJ process consists of four main stages, dehydration, oligomerization, hydrogenation, and separation. These processes are modeled in the Aspen plus software, in order to obtain the necessary mass and energy balances to evaluate the process performance and calculate the proposed objective functions. In order to design the process, a DETL optimization algorithm is used, which is programmed in Visual Basic and linked to the simulation software. To guarantee an optimal design of the process, an economic function measured through the total annual cost (TAC) and an environmental function, calculated through the eco-indicator 99 (EI99) methodologies based on the life cycle analysis, is used. For the optimization of the process, the ethanol feed varied, the degrees of freedom of the process and the variation of the distribution of components in the reactor was considered to guarantee the properties that the bio-jet fuel produced should normally carry.

After the optimization process, the results show a trade-off between the proposed objectives, considering a utopian point in which it is considered that there is a balance between the objectives, the following information is generated: the optimum capacity of the plant is 445.12 kg/h of bio-jet fuel, with an optimal ethanol feed of 1239.67 kg/h, for these conditions the optimal TAC is 1.150×10^6 \$/year with an environmental impact of 2.793×10^8 points/year, with a sale price of 1.876 \$/kg of bio-jet fuel, the energy consumption of the process is 10.69 MJ/kg of jet fuel. This cost of sale would be reduced if the sale of the other generated components such as diesel is considered. Multi-objective optimization considering economic and environmental aspects of the transformation of alcohol and its subsequent upgrading to jet fuel proved to be an efficient tool for process evaluation.

595964 Autonomous Monitoring of Wellbore Integrity Applying Time Reverse Nonlinear Elastic Wave Spectroscopy (TR NEWS) and Fiber Optic Sensing and Communication

Abstract:

We present an integrated approach to monitoring long-term wellbore integrity based on measurement of passive and active seismo-acoustic signals and stress changes in the near-wellbore region. Our goal is the development of an autonomous system that can be deployed in wells for long- term (e.g., decades), unattended monitoring both wellbore integrity and associated stress changes. Our signal target is threefold: (i) small seismo-acoustic signals associated with moving (leaking) fluids; (ii) the stress field along the wellbore where changes in stress may presage damage and leakage; (iii) nonlinear properties in the near-wellbore region that arise when damage develops and progresses. We envision an autonomous system consisting of both wellhead and downhole components based on the combined signals of piezoelectric and/or optical sensors integrated with innovative strategies for both source generation and data analysis. Our approach is based on our extensive work in four areas: (i) the physics of time reversal (TR) to focus acoustic energy, allowing interrogation of material properties locally; (ii) nonlinear elastic wave spectroscopy to probe damage in materials; (iii) the measurement and interpretation of seismo-acoustic signals, including recent work focused on extracting very small signals; and (iv) the characterization of in situ strains and stresses. The final system would entail autonomous, passive monitoring for changes in the local characteristics of both stress/strain field and seismo-acoustic noise, with periodic use of active monitoring (TR NEWS) to probe local changes in elastic properties tied to damage.

600714 Optimal Selection and Location of Nitrogen Recovery Systems for Intensive Pig Farming

Abstract:

Farming is an industry in current raise to meet the increasing demand of meat products due to the increase in the life standard of large population sectors. Spain is one of the main producers of swine products, accounting more than 25 million of pigs, representing 20% of animals of the pig farming industry in Europe. Within Spain, Castilla y León shows a farming scenario in which, while cattle farming is mainly extensive, with a low density of animals and a low stocking rate of organic waste per area unit, intensive swine is carried out in facilities with a high density of animals where manure must be properly managed due to the high rate and spatial concentration of the organic waste generated. However, Castilla y León accounts for 4 million pigs located in intensive breeding facilities. Intensive farming practices are associated to environmental and public health issues including soil degradation, atmospheric pollution associated with the emission of ammonia, and the runoff of nitrogen and phosphorus to waterbodies, resulting in nutrient pollution and eutrophication [1]. These effects are reflected in the fact that 31.3% of towns in Castilla y León [2] have nitrate concentrations in water over the threshold stipulated by the European Union (50 mg/L), most of them related with nitrogen releases from the incorrect management of manure fertilizer application in crops.

To improve the current manure management practices, which actually are limited to the application of manure in the lands around the facilities, or its treatment in ponds, the installation and allocation of nitrogen recovery process are evaluated in this work. The main drivers for operating costs of manure processing systems are the economies of scale and the potential coupling of nitrogen recovery processes with energy production systems, such as anaerobic digestion with biogas upgrading systems in form of electricity or bio-methane, providing additional income sources to the system. Nevertheless, the treatment of several farms has to be centralized in the same facility to reach processing capacities with size enough to be cost-effective, although transportation costs of organic waste as well as the associated carbon footprint, can negatively impact the operation of the treatment systems. The model proposed is divided into two levels. In a first level, the available processes for nitrogen recovery from swine waste are evaluated though mathematical modelling. The optimal processes and operating conditions are determined for different sizes of swine breeding facilities [3, 4]. In the second level, surrogate models for the capital and operating costs as function of the processing capacity are developed from the results obtained in the previous stage. These surrogate models are embedded in a supply chain problem determining the optimal allocations of the facilities and flow of waste from the livestock facilities to the processing centers [5]. The final goal of the work is to assess the economic feasibility of the implementation of a cooperative network for swine waste processing

The combination of real farms data regarding size and location with the optimization model developed for the selection and allocation of nitrogen recovery systems allows us to work with a real problem, identifying and addressing some of the challenges to mitigate nutrient pollution associated intensive swine farming, reaching the most cost-effective solution feasible for the region studied. In addition, the formulation developed in the present work can be easily adapted to other regions with available data for farming facilities.

References

[1] Backus, G. B. C., Van Wagenberg, C. P. A., Verdoes, N., 1998. Environmental impact of pig meat production. Meat science, 49, S65-S72.

[2] Consejo Económico y Social de Castilla y León, 2019. Informe Previo sobre el Proyecto de Decreto por el que se designan las Zonas Vulnerables a la contaminación de las aguas por nitratos procedentes de fuentes de origen agrícola y ganadero, y se aprueba el Código de Buenas Prácticas Agrarias.

[3] Huang, H., Xiao, D., Liu, J., Hou, L., Ding, L., 2015. Recovery and removal of nutrients from swine wastewater by using a novel integrated reactor for struvite decomposition and recycling. Scientific reports, 5, 10183.

[4] Liu, Y., Kwag, J. H., Kim, J. H., Ra, C., 2011. Recovery of nitrogen and phosphorus by struvite crystallization from swine wastewater. Desalination, 277(1-3), 364-369.

[5] Taifouris, M. R., & Martin, M. (2018). Multiscale scheme for the optimal use of residues for the production of biogas across Castile and Leon. Journal of Cleaner Production, 185, 239-251.

603810 Used Cooking Oil Supply Chain Analysis, Optimization and Integration in the City of Bogotá, Colombia.

Abstract:

Global population growth involves increasing challenges regarding the sustainable development of our society, and in particular regarding the correct management of residues mainly in overpopulated areas. Without proper management regulations and practices, residues can cause a cascade of environmental, social and economic problems. Among the different urban residues, used cooking oil (UCO) is a highly problematic food waste which is typically disposed through syphons and drains. These disposal practices generate a variety of problems including infrastructure damage, flooding, vectors proliferation, ecosystems pollution, and even illegal collection and redistribution. A main reason for the mismanagement is the large economical cost involved in the mitigation of such practices.

The effects of UCOs mismanagement are sharply affecting densely populated urban centers mainly in developing countries. Particularly in Bogotá, the capitol city of Colombia, with nearly 10 million people in the metropolitan area, UCOs represent a major issue. According to recent estimations the city generates at least 45 kt UCO/yr. and the local government expends nearly 6 to 8 million dollar per year on corrective actions and maintenance related to UCOs mismanagement. Thus, urgent actions are required to promote the correct UCOs handling and possible exploitation. In this regard, this work deals with the study, characterization and optimization of a UCOs collection chain in the city of Bogotá for their further use as oleochemical feedstock.

First, data collection characterizing the UCOs logistic chain in the city of Bogotá was carried out. This task involved the location and nature of generators, available volumes, UCOs characteristics, characterization of collecting practices and routes, and the inventory of resources consumption during the collection process. Then, a model of the collection scheme was constructed and implemented in Python, using a Vehicle Routing Problem approach using weighted Hamiltonian paths. Weighting factors were defined according to economic and environmental indicators calculated as costs and equivalent-CO_2 emissions, respectively. Using a Genetic Algorithm with combinations as individuals, a reduced set of preferred collection routes were identified as optimal candidates. Finally, the scheme was integrated with the location of a hypothetical biorefinery facility with a potential portfolio of UCOs derived byproducts. Then, location of the facility and product portfolio distribution were simultaneously optimized using an Evolution Strategy algorithm. As result, an optimized collection scheme integrated with a facility whose optimal product portfolio and production capacity was identified.

609309 Engro's Innovative Approach for Sustained Plant Operation with 25+ Years Life of Reformer Radiant Tubes

Abstract:

Being the energy intensive unit, primary reforming furnace is the host for massive heat generation needed to cater the heat requirement of endothermic reforming reaction for the production of syn gas. With being among the high temperature zone, the reliability of furnace equipment is imperative to sustained plant operation as the upset not only is a risk to business sustainability but the failure of furnace equipment poses catastrophic risk to the safety of assets and personnel. The reliable operation of furnace equipment becomes more challenging as the plant ages.

This paper describes the challenges faced in sustaining plant operation where the furnace tubes surpassed double the recommended life. The double terrace induced draft furnace designed by foster wheeler is a primary reforming unit at a vintage ammonia plant commissioned 1967. The furnace radiant tubes were last replaced in 1992 and have seen a life of two hundred thousand operating hours against the recommended life of one hundred thousand hours.

The secret of double the recommended useful life lies in the stringent inspection regimes and best operating practices. The efforts to ensure the tubes in operable conditions range from apt temperature monitoring regimes to tubes inspection in terms of creep measurement and diametric growth. The paper describes in detail the reliability measurements along with the challenges faced to sustain extended life of reformer tubes spanning over 28 years.

609410 Designing Sustainable Supply Chains in Light of Planetary Boundaries: A Case Study of Bioethanol Production from Sugarcane in Argentina

Abstract:

With the increase of international trade and the quest for sustainable development, supply chains (SCs) worldwide are facing significant transformative changes to integrate sustainability concerns into their design and operation. Traditionally, the methodological approach to design sustainable supply chains has been through Life Cycle Optimization (LCO) approaches, which consists of incorporating Life Cycle Assessment (LCA) principles into mathematical programming models. In essence, the optimization models allow representing the SC problem through a set of equations, while LCA allows quantifying all the material and emissions inputs and outputs throughout the whole lifetime of the SC which are converted into relevant environmental impacts. This LCO approach undoubtedly provides valuable insights into the decision-making process to identify SC configurations showing improved sustainability performance. However, the question that arises is whether the environmental impact of these optimal SC is low enough to be considered truly sustainable, and if not, by how much the environmental pressure should be further reduced. This kind of insight is out of the reach of the traditional LCO approach, which calls for new approaches coupling LCA with absolute measures and ultimately incorporating absolute sustainability assessment into the design of future SCs.

In this context, the planetary boundaries (PBs) concept provides a theoretical basis to evaluate systems referring to the absolute carrying capacity of the Earth, i.e., to assess whether the Earth is able to cope with environmental burdens from human activities. In essence, the PBs framework identifies a set of critical environmental processes and their respective maximum perturbation limits which altogether define the so-called safe operating space (SOS), which can be used as the absolute sustainability reference. In this work, we incorporate the PBs concept into the design of sustainable SCs by relying on a recently proposed set of life cycle assessment (LCA) characterization models which allow expressing all impacts and burdens of the SCs in terms of the control variables of the PBs. The proposed modelling framework allows finding the optimal solution for the whole network minimizing the transgression of the PBs, which need to be previously translated to the regionally-operating scale of the SC using distributive fairness principles. The overall approach, therefore, allows quantifying the performance of SCs in terms of maximum thresholds, thereby providing information on their absolute sustainability level.

To demonstrate how the proposed framework would work in practice, we apply it to a case study which considers the optimal design of the bioethanol production from sugarcane in Argentina to satisfy an increased blending mandate in the transportation sector. To do so, we modified our previously developed mixed-integer linear programming model by incorporating the PBs framework. Our results show that, due to the stringent downscaled PBs imposed, there is no SC configuration capable of satisfying all planetary boundaries concurrently. However, the transgression to critical PBs such as climate change and acidification would be significantly reduced. We hope that our work opens up new avenues for research and business applications to operationalize the design of future SCs in light of PBs towards decision-making aligned with absolute sustainability.

599198 Stoachastic Techno-Economic Analysis for an Integrated Strategy to Coproduce Jet Fuel Range Alkenes and Pentanediols

Abstract:

Cellulosic biofuels is an attractive renewable fuels, because the raw material, lignocellulosic biomass, one of the most plentiful resource and there is no concern about carbon dioxide emission. However the problem in the use of biofuel is a high price due to high feedstock cost and technological immaturity [1, 2]. In addition, the recent drop in oil prices also opens up a price gap between biofuels and fossil fuels [3]. To overcome this bottleneck, a new approach to process economics through simultaneous production of valuable chemicals has been proposed [1, 4]. Recently, an integrated strategy coproducing jet fuel range alkenes (JFA) and pentanediols (1,2-pentanediol and 1,5-pentanediol; PeDs) from cellulose and hemicellulose fraction of lignocellulosic biomass is proposed, the fuel price assumed by deterministic techno-economic analysis is comparable to current biofuel price. Although the result of deterministic analysis shows that the integrated strategy is economic feasible, an uncertainty in these major economic parameters could lead uncertainty in the minimum selling price (MSP). An uncertainty analysis introducing Monte Carlo simulation can be used to assume more reliable production cost and MSP.

We performed the stochastic analysis for the integrated strategy coproducing JFA and PeDs from lignocellulosic biomass based on the deterministic techno-economic results by Monte Carlo simulation using @Risk software. Major cost parameters which have large impact on the MSP of JFA were identified by deterministic techno-economic analysis, and then uncertain input variables were sampled from distributions of the parameters assumed based on the previous price data reported in literature. Next, Monte Carlo simulation with 10,000 iteration was performed to show probability distribution of the MSP of JFA.

Compared to the deterministic analysis, the stochastic analysis can provide probabilistic results on economic feasibility of the integrated strategy when the market conditions are varied. And it is also used as a reliable information for risk-averse investment. Although this study only considered the price uncertainties, we will consider the technical uncertainties such as conversion and selectivity of the catalytic reactions provide more reliable information to the decision makers.

Literature cited:

[1] Huang K, Won W, Barnett KJ, Brentzel ZJ, Alonso DM, Huber GW, et al. Improving economics of lignocellulosic biofuels: An integrated strategy for coproducing 1, 5-pentanediol and ethanol. Applied Energy. 2018;213:585-94.

[2] AlternativeFuelDataCenter. Alternative fuels report. 2017.

[3] Macrotrends. Crude Oil Prices. 2016.

[4] Bond JQ, Upadhye AA, Olcay H, Tompsett GA, Jae J, Xing R, et al. Production of renewable jet fuel range alkanes and commodity chemicals from integrated catalytic processing of biomass. Energy & Environmental Science. 2014;7:1500-23.

604234 Recycling of Cathode of Lithium-Ion Batteries: A Unique Application of Oxalate Chemistry

Abstract:

Lithium-ion batteries (LiBs) are used commercially in numerous applications ranging from portable electronics to electric vehicles. The worldwide battery technology market is expected to reach $150 billion by 2025. However, at the end of their life span, LiBs must be sustainably recycled such that they do not end up in landfills. In this work, a closed-loop, sustainable, and green process for separation and recovery of Li and Co from $LiCoO_2$ using oxalic acid is being developed. Oxalic acid is unique in terms of its reducing, chelating, and precipitating properties. During digestion of $LiCoO_2$ in oxalic acid, the Co directly precipitates as cobalt oxalate dihydrate ($CoC_2O_4 \cdot 2H_2O$), and the Li remains in solution ($Li_2C_2O_4$) providing an efficient separation. The $CoC_2O_4 \cdot 2H_2O$ can be converted to cobalt oxide (Co_3O_4) by calcination in air. The Li can be precipitated from the aqueous oxalate phase in the form of a carbonate or phosphate. The remaining oxalate in the filtrate after Li precipitation can be recycled using an acid cation-exchange resin process. The Co_3O_4 and Li_2CO_3 can be used to produce new $LiCoO_2$ using a solid-state synthesis process. This poster will provide an overview of the entire process, which includes metal extraction, precipitation, and thermal conversion with specific attention towards the regeneration of oxalic acid.

607565 Improve the Thermal Processability of Poly(3-hydroxybutyrate-co-3-hydroxyvalerate) (PHBV) Bioplastic with Natural Rubber

Abstract:

The plastic industry is looking for bio-based and biodegradable alternatives to improve its sustainability. Poly(3-hydroxybutyrate-co-3-hydroxyvalerate) (PHBV) is a promising bio-based and biodegradable thermoplastic made from bacterial fermentation of carbon-rich resources. However, industrial applications of PHBV has been limited by its brittleness and especially by its poor processability. PHBV has highly linear molecular structures, its polymer chains are easily untangled under strain, resulting in low melt strength and low resistance to stretching. This causes issues in processes that require high melt drawing or stretching flow, such as film blowing and blow molding. PHBV's high degree of crystallinity and low thermal stability also makes its thermal processing changeling as it degrades at temperatures close to its melting points.

Natural rubber was incorporated into PHBV through reactive extrusion to improve the thermal processability of PHBV. Peroxide and coagent were used to improve rubber particle dispense and compatibility between PHBV and rubber. The PHBV/rubber blends had higher complex viscosity and more significant shear thinning behavior than pristine PHBV, indicating enhanced chain entanglements. The PHBV/NR blends also showed enhanced strain-hardening in their extensional viscosity. The presence of natural rubber limited PHBV degradation during heating due to rubber dissipating heat more efficiently than PHBV. The difference between Tp (peak degradation temperature) and Tm (melting temperature) of PHBV increased from 127 °C to 142 °C, suggesting expanded range of PHBV processing temperature and broadened PHBV processing window.

This new rubber enhanced PHBV bioplastic material with improved processability is promising to be thermally processed into bio-products such as food packages to improve the sustainability of the industry.

611680 Risk-Centric Optimal Synthetic-Pathway Generation System Using Monte Carlo Tree Search and Open Data of Reactions and Material Safety

Abstract:

In R&D, Acquainting chemical engineering data such as chemicals, reactions, safety and danger is considered to be an essential part of the process and experiment design. Today, Although a number of data related to chemical engineering are operated privately or publicly(accessible on the Internet), data is not only distributed in a wide range or differ in format and also has various types of reactants used in each reaction path, making it difficult to select the suitable path. Thereby this leads in a small scale to a number of trial and error, and in a big scale to economics loss because of wasting materials and time, moreover, Serious safety problems. Therefore, It is an essential part in the planning stage before the research is conducted that the appropriate synthetic pathway design which can prevent safety accidents.

Currently, there is a commercial S/W reaction path design such as Chematica, but it is difficult to use universally in cost aspects(high cost) and specific conditions(cost optimization). Therefore, this study proposes a system that generates a path optimized for safety with Graph algorithm considering reaction energy and material stability(or risk). Concretely, this study proposes a smart path search/generation system based on the automatic collection of open source and open data using database construction and Monte Carlo Tree Search(MCTS).

In the data mining step, we conducted automatic Web Crawling and Web Scraping which can be continuously updated to build system DB, and collected public data such as Google patents, PubChem and ChemSpider. So far, we have collected a large amount of synthetic data, including about 4.8 million compound data and about 2.2 millions synthetic data.

Then, After the data which collected for the synthetic pathway design was constructed as a data networks, apply them to the search algorithm. In general path search algorithms such as Depth First Search (DFS) and Breath First Search (BFS), the time complexity is O(|V|+|E|) when implemented as an adjacency list, and the time complexity is O(|V|^ 2) when implemented as an adjacency matrix. This because there is an explosive increase in the operation speed in proportion to the increase in data. In order to improve this, this study tried to shorten the operation time by applying MCTS algorithm. The synthetic pathway is proposed by searching for which path is generated in a specified depth through data such as reaction, risk, energy, etc. At this time, the weight of MCTS is determined based on degree of risk of reaction, reactant and product. The risk of the reaction that can be derived from the enthalpy change are calculated using the Benson Group Increment Theory(BGIT). Hazard of the reactant and product is classified based on the related public data like NFPA 704 code, Protective Action Criterias(PACs), Acute exposure Guideline Levels(AEGLs) and Emergency Response Planning Guidelines(ERPGs). Consequently, the generated synthetic pathways are represented as Knowledge graph and string data tree types. The result of this study are as follows: 1) Knowledge graph results are classified by color. 2) String result shows the synthesis information, chemicals and risk each node's paths and paths.

The system driving test was conducted using some DB data. And compared with other S/W that shows suitable results in a specific condition or designs a direct synthetic pathway, the proposed system has advantage in automatically generating open-source-synthetic-pathways applicable to a wide range based on large open data. Moreover, It is expected highly meaningful about safety through the proposed system that helps to prevent safety accidents and efficiently manage additional information when we refer to the path generated.

CHARACTERIZATION OF FACTORS AFFECTING THE GAS WELL DELIVERABILITY OF THE JIAOSHIBA SHALE GAS FIELD BY MULTIVARIATE STATISTICAL ANALYSIS

Yizhong Zhang*[1,2]*, Maolin Zhang*[1]*, Bin Ju*[3]*, Jincai Shen*[4]*, Lan Ge*[4]*, Fanhua Zeng*[2]**

**Corresponding author. Email address: Fanhua.Zeng@uregina.ca.*

[1] Hubei Cooperative Innovation Center of Unconventional Oil and Gas, Yangtze University, Hubei, Wuhan, China, 430100

[2] University of Regina, Regina, SK, Canada, S4S 0A2

[3] School of Oil and Natural Gas Engineering, Southwest Petroleum University, Chengdu, Sichuan, China, 610500

[4] SINOPEC Chongqing Fuling Shale Gas Exploration and Development Co., Ltd., Chongqing, China, 408014

Abstract

The Longmaxi shale gas in the Jiaoshiba Shale Gas Field is the first and the largest successfully developed shale gas in China. The analysis of key factors of deliverability of gas wells plays a significant role in efficiently developing shale gas reservoirs. However, there are few methods analyzing the key factors of deliverability of shale gas wells by considering both geological characters and engineering parameters. In this study, the reservoir characteristics, drilling and completion data, fracturing data (including the well logging interpretation of horizontal section, the track and layer section that the well bore passed through, and fracturing parameters) and production data of 86 gas wells in the Wufeng-Longmaxi shale, Jiaoshiba Shale Gas Field in Eastern Sichuan Basin, China are statistically analyzed and ordered by multivariate statistical analysis. The gas well deliverability is described by absolute open flow potential and 9 factors (physical parameters of reservoir, including density of rock, porosity, total organic carbon, brittle mineral content of rock and drilling and fracturing engineering quality, including horizontal length of gas well, horizontal length ratio through different stages in the formation, number of fracturing stages, liquid volume, and sand amount) that may affect the deliverability of shale gas wells are considered. The results reveal that the absolute open flow potential has a close relationship with porosity, total organic carbon, brittle mineral content and total liquid volume. The multivariable nonlinear regression models of absolute open flow potential with different key factors considered are built, and the prediction results are reliable by goodness-of-fit test acceptable with real gas well deliverability testing data. The research methods and results in this study can provide a reference basis for reasonable production rate allocation, design, and adjustment of development plan for the shale gas reservoirs.

Keywords: Gas well deliverability, Longmaxi shale gas, Jiaoshiba Shale Gas Field, Multivariate statistical analysis, Multivariable nonlinear regression

1. Introduction

Following the increased demand of petroleum energy and the developed science and technology, the industry nowadays has focused on the unconventional resources. Shale gas reservoirs are one of the most popular kind of unconventional resources with extremely low permeability, tight porous structure, large specific area and high heterogeneity, which have attracted attentions of researchers worldwide [1-3]. Shale gas has self-generated, self-contained, organic-rich, nanoscale pore structure, and other unusual characteristics and it could be considered as a source as well as a reservoir [4-5]. In addition to free gas and small amount of dissolved gas, adsorbed gas is largely existed in the tight formation, which makes it difficult to produce owing to the complexity of the reservoir [6-8]. The Sichuan-Chongqing region in southwest China is one of the fastest shale gas production growing regions in the world. The annual shale gas deliverability of China has reached 2×10^{10} m^3 with the annual shale gas production of 1.5×10^{10} m^3 [9]. As one of the main contributors of natural gas production growth in China, the gas production of Jiaoshiba shale gas reservoir has exceeded the daily production of 1.67×10^{7} m^3/d since 2013 [10]. The large production of commercial shale gas flow of the Longmaxi Formation (e.g. more than 2×10^{5} m^3 /d of JY1 well) has made the Jiaoshiba area one of the most important shale gas field in China.

Compared with typical shale gas reservoirs in North America, the shale gas in Sichuan Basin has lower TOC value, lower porosity, lower gas saturation, smaller thickness and higher buried depth, which makes its exploration more complicated [9]. In both the early test stage and current production stage, there are large differences among deliverability of the shale gas wells in the Jiaoshiba area. Therefore, the deliverability evaluation is the basis of reasonable shale gas production and prediction, and the research on deliverability evaluation and its related reservoir and engineering parameters of shale gas wells is particularly important. Previous studies have focused on the impacts of reservoir quality including sedimentary environment [11], structure [12-13], mineralogy and lithology [14-17] and engineering quality including drilling and completion techniques [12, 18] and multi-stage hydraulic fracturing performance [19-20] on the shale gas production. Their results revealed that both the reservoir quality and engineering quality will affect the produced gas rate of the shale gas wells. Most of researches did analysis based on the relationship between the interested variable and its single impacting factor, comprehensive multivariate statistical analysis with consideration of correlation and collinearity among factors needs to be further studied. Moreover, instead of the produced gas rate or gas bearing content, the absolute open flow (AOF) potential may be a better parameter to describe the deliverability of gas well. The AOF potential is a hypothetical gas flow rate that may be produced if the flowing pressure at the bottom of the well (BHFP) were equivalent to the atmospheric pressure, i.e. the maximum flow rate that the gas well may produce [21]. In this study, the well deliverability of Jiaoshiba shale gas is described by the AOF potential based on BHFP and the impacts of reservoir and engineering characteristics on the AOF potential will be comprehensively analyzed. The data of over 86 wells of Fuling Jiaoshiba shale gas field is analyzed and the impacts of the reservoir quality (i.e. density of rock, porosity, total organic carbon, brittle mineral content of rock) and the engineering quality (i.e. horizontal length of gas well, horizontal length ratio through different formation stages, number of fracturing stages, liquid volume, and sand amount) on the AOF are quantified. The most important factors affecting deliverability of the gas wells in the Jiaoshiba area will be obtained by hierarchical clustering analysis method. The regression correlation between the well deliverability and the important parameters will be built. The objective of this study is to find a new method to evaluate the key factors that affect the well deliverability of shale gas reservoir by multivariate statistical analysis with consideration of correlation and collinearity among factors and build multivariable nonlinear regression models based on these key factors, so that the well deliverability of shale gas reservoir could be predicted.

2. Shale gas reservoir characters

The Longmaxi shale gas in the Jiaoshiba area located in the Eastern Sichuan Basin in the southwest region of China is the first and the largest successfully developed shale gas reservoir in China [22]. The shale gas bearing formation is the Upper Ordovician Wufeng Formation to Lower Silurian Longmaxi Formation with a thickness of about 200–300 m, area of 272.44 km^2 and buried depth of less than 3500m. Comprehensively considering the lithology, electrical and physical properties, geochemical and gas-bearing characteristics, the gas bearing shale section in the Upper Ordovician Wufeng Formation to Lower Silurian Longmaxi Formation could be vertically divided into three parts with stable horizontal stratigraphic distributions for each member (i.e. Long1, Long2 and Long3 from bottom to top). Long1 member is rich in fossils (e.g., radiolarian, calthrop, and graptolite) and is composed of black mudstone and argillaceous siltstone. The Long2 member consists of argillaceous siltstone and fine-grained sandstone. The Long3 member is composed of dark-gray mudstone and gray shale with fewer fossils [5, 23].

Previous studies suggested that Long1 member consisted of one 3-ord sequence, and could be divided into three system tracts (i.e. the transgressive system tract (TST), early highstand system tract (EHST), and late highstand system tract (LHST)) [10, 15, 16]. As shown in **Fig.1** [5], this study is targeted in the Long1 member, which is abundant of organic matter with pores and could be divided into 3 stages (i.e. stage1, stage2 and stage3 from bottom to top) based on physical properties with gas bearing increasing from top to bottom. The Long1 member is of good brittleness, and the brittle mineral content gradually increases from top to bottom. The natural gas in the layer belongs to over-mature dry gas, in which methane is the main component without containing hydrogen sulfide, carbon monoxide and hydrogen.

3. Samples

The AOF potential is a hypothetical gas flow rate that may be produced if the flowing pressure at the bottom of the well (BHFP) were equivalent to the atmospheric pressure [21]. The AOF potential may be a better parameter to describe the deliverability of the gas wells as it gives the maximum flow rate that the gas well can produce. As shown in Fig.2, the AOF potential data of and 9 factors that may affect the deliverability of gas wells (i.e. horizontal length, number of fracture stages, total liquid volume, total amount of sand, rock density, porosity, TOC, brittle mineral content of rocks, horizontal length ratio through stage 1 and stage 3 of Long1 member) are collected (data is offered by SINOPEC Chongqing Fuling Shale Gas Exploration and Development Co., Ltd., Chongqing, China). Though **Fig.2** shows certain relationship between AOF potential and single impacting factor, the correlation and collinearity among different factors are not considered. According to the data, the average reservoir pressure of the Long1 member is 37.69MPa with pressure coefficient 1.55. The reservoir temperature is 85.99°C. The pay zone thickness is between 20m-200m and the most favorable thickness is about 38m. The kerogen type of the organic matter is mostly type II_1 to II_2. The total organic carbon (TOC) contents of the shale range from 1%-8% with average value of 5% and it is increasing from top to bottom. The vitrinite reflectance (R_o %) of organic matter ranges from 2.5%-3.7% with average value of 2.9%. The porosity of the shale is between 1.88% to 7.06%, with average value of 4.15% and the permeability of the shale is less than 0.001mD. The horizontal wells in the sample have lateral length between 1500m to 1800m, well spacing between 500m to 600m, and the distance between adjacent clusters between 20m to 25m. All the horizontal wells are multi-stage hydraulic fractured with number of fracturing stages between 15 to 29. The fluid volume per stage is between 1500m^3 to 1800m^3, and proppant volume per stage is between 50m^3-65m^3. The fractured half-lengths range from 300m to 350m, and propped half-lengths range from 250m to 270m. The fracture heights range from 40m to 50m, and main fracture width range from 5mm to 10mm.

4. Methods

4.1. Normality test

When working with multivariable regression problems, some assumptions should be followed. The regression process of independent variables and the target variable are assumed to follow Gauss–Markov theorem (i.e. the ordinary least squares (OLS) estimator has the lowest sampling variance within the class of linear unbiased estimators, if the independent variables have no perfect collinearity, and the errors in the regression model are uncorrelated with equal variances and expectation value of zero) [24]. Thus, the first step of the multivariate statistical analysis is to test whether the sample variables follow normal distribution. The Q–Q (quantile-quantile) plot is applied, which is a probability plot that shows a graphical comparison between the samples and normal distribution by plotting their quantiles against each other [25]. If the points in the Q–Q plot approximately lie on the 45-degree line (i.e. $y=x$), the distribution of the sample being compared are similar enough to normal distribution. If normality test of a variable behaves not well, certain transformation (i.e. logistic transformation, logarithm transformation, index transformation, power transformation, etc.) of the variable and nonlinear regression model should be considered.

4.2. Standardization

Before the key factor analysis, the calculation of correlations between each parameter is essential. Since the units, scales and ranges of the factors are different, the first step is to standardize all the parameters based on the normality test results. The method applied in the study is the Fisher Z-transformation method. The standard score (Z score) for each data point is its difference from sample mean divided by the sample standard deviation.

$$Z = \frac{X - \overline{X}}{s} \tag{1}$$

where

X is the data point.
$\overline{X}$ is the sample mean.
s is the sample standard deviation.

4.3. Correlation analysis

After standardization, the correlations between each factor are calculated. This is to eliminate the repeated consideration of correlation and collinearity in the later multivariable regression. In other words, if two variables are strongly related, only the principle variable between them is considered in later analysis. The sample data has a number of n wells, and each well has a number of k variables including a dependent variable (X_k) which is Z scores for AOF potential and $k-1$ independent variables (X_1, X_2, ..., X_{k-1}) which are Z scores for the factors affecting AOF potential. The sample data could be shown in matrix form as follows

$$X = \begin{bmatrix} x_{1,1} & x_{1,2} & \cdots & x_{1,k} \\ x_{2,1} & x_{2,2} & \cdots & x_{2,k} \\ \cdots & \cdots & & \cdots \\ x_{n,1} & x_{n,2} & \cdots & x_{n,k} \end{bmatrix} \tag{2}$$

Then the correlation matrix for all variables would be

$$R = \begin{bmatrix} r_{11} & r_{12} & \cdots & r_{1k} \\ r_{21} & r_{22} & \cdots & r_{2k} \\ \cdots & \cdots & \cdots & \cdots \\ r_{k1} & r_{k2} & \cdots & r_{kk} \end{bmatrix} \tag{3}$$

$r_{ij}(i,j=1,2,...,k)$ is the correlation coefficient for $x_{ij}(i=1,2,...,k)$ and $x_{ij}(i=1,2,...,k)$, and the calculation equation is

$$r_{ij} = \frac{\sum_{m=1}^{k}(x_{im}-\overline{x_i})(x_{jm}-\overline{x_j})}{\sqrt{\sum_{m=1}^{k}(x_{im}-\overline{x_i})^2\sum_{l=1}^{k}(x_{jl}-\overline{x_j})^2}} \tag{4}$$

The value of correlation coefficient should be between -1 and 1. If the correlation coefficient is close to 1, the involving two variables have a strong positive relationship, and if it is close to -1, the involving two variables have a strong negative relationship. If the correlation coefficient is close to 0, this suggests that the involving two variables may be independent to each other.

4.4. Hierarchical clustering analysis

As long as correlations among factors are obtained, clustering method could be applied to find the key factors while eliminate the factors with small effect on AOF potential or strong correlation with other factors. Hierarchical clustering analysis (HCA) is a method to find the hierarchy relationship among variables, so that the variables could be divided into clusters [26]. HCA is performed on the normalized sample data set of the 86 well data including AOF potentials and 9 different factors, using centroid linkage clustering (Eq.5) as linkage criteria. The method will try to link the two most related factors in the data set to form a new cluster at each step (**Fig.3** agglomerative hierarchical clustering). The new merged cluster is then considered as a variable in the relation map and agglomerative approach is again used. Finally, a tree-hierarchy structure with linkage lines that join all the factors would be built.

$$D_{uv}^2=(\overline{X}_u-\overline{X}_v)'(\overline{X}_u-\overline{X}_v) \tag{5}$$

, where $\overline{X}_u$ is the centroid for cluster C_u. $\overline{X}_v$ is the centroid for cluster C_v.

4.5. Multivariable nonlinear regression

Once the main factors that affect the AOF potential are found, other factors are either unessential to AOF potential or have strong correlation to the main factors, which could be eliminated. The next step is to linearize the main factors, including logistic transformation, logarithm transformation, index transformation, power transformation, etc. Then, a set of multivariable linear regressions is carried out to find the optimal regression model. If there are k variables, the linear regression for the variables is

$$y_i=\alpha+\beta_1X_{i1}+\ldots+\beta_kX_{ik}+\varepsilon_i=\alpha+\sum_{j=1}^{k}\beta_jX_{ij}+\varepsilon_i, i=1,2,\ldots,n \tag{6}$$

Least squares could be used to find the intercept α and coefficients β_j (j=1,2,...,k) , which is to find the optimums for α and β_j (j=1,2,...,k) when the sum of squared residuals, *SSe*, is minimized. The squared residual r_i^2 is defined as the difference between real value and predicted value of the dependent variable.

$$r_i=y_i-\hat{y}_i=y_i-f(x_i,\alpha,\beta) \tag{7}$$

$$SSe=\sum_{i=1}^{n}r_i^2 \tag{8}$$

$\hat{y}_i$ is the estimation value of dependent variable y_i. $a,b_j(j=1,2,\ldots,k)$ are the estimation values of intercept and coefficients $\alpha,\beta_j(j=1,2,\ldots,k)$. Let the partial derivatives with respect to $a,b_j(j=1,2,\ldots,k)$ equals 0.

$$\begin{cases}\dfrac{\partial SS_e}{\partial a}=0 \\ \dfrac{\partial SS_e}{\partial b_j}=0 \end{cases} j=1,2,\ldots,k \tag{9}$$

It could be also proved that $a,b_j(j=1,2,\ldots,k)$ is the unbiased estimation of $\alpha,\beta_j(j=1,2,\ldots,k)$.

5. Results and Discussions

5.1. Normality test

Fig.4 shows the Q–Q (quantile-quantile) plots for the 10 variables of the 86 wells. The results reveal that the points of most variables in the Q–Q plots approximately lie on the 45-degree line, which means they follow normal distribution well. The only variable may not behave with good normality is horizontal length ratio through stage 1 and stage 3 of Long1 member. This variable may need transformation (i.e. logistic transformation, log transformation, logistic transformation, etc.) in the later regression study. Thus, the multivariable nonlinear regression would be needed in this study.

5.2. Standardization and correlation of the variables

The matrix of correlation coefficients of all variables is shown in **Table 1**. **Table 1** suggests that there is no strong linear relation between the AOF potential and single one factor. Relatively, the AOF potential is positively related to the TOC, the brittle mineral content, and the total liquid volume. It is negatively related to rock density. Some factors show strong correlation with each other, and the number of fracture stages and total liquid volume have the highest correlation coefficient of 0.946. The drilling and fracturing engineering factors, including the number of fracturing stages, the length of horizontal well, the total liquid volume and the total sand amount, are highly linear related. Among the geological factors, the rock density is highly negative to TOC, brittle mineral content, and the TOC and the brittle mineral content of rock are highly linear related. The collinearity among some factors is very strong and may affect AOF potential modelling, which should be eliminated in the multivariate statistical analysis of AOF potential.

5.3. Hierarchical clustering analysis

According to **Table 1**, the number of fracturing stages and total liquid volume have the highest correlation coefficient, which is 0.946. These two factors could be firstly clustered together, and the central point of this cluster is calculated by centroid method. Then the correlation analysis is done again after applying the whole cluster as a variable. This progress is continued until all the variables converged to one cluster. The result of hierarchical clustering analysis with linkage lines is shown **Fig.5**, and **Table 2** gives the importance order of the effect of each factor on the AOF potential.

It is suggested in the hierarchical clustering analysis results that the order of the factors that affect AOF potential is: a. porosity, b. TOC, c. brittle mineral content of rock, d. horizontal length ratio through stage 1 and stage 3 of Long1 member, e. number of fracturing stages, f. total liquid volume, g. horizontal length, h. total sand amount and i. rock density. The AOF potential firstly has closest relationship to porosity. It then has closer relationship to the cluster of TOC, brittle mineral content of rock, horizontal length ratio through stage 1 and stage 3 of Long1 member. These four factors belong to the first cluster. The other four factors including number of fracturing stages, total liquid volume, horizontal length and total sand amount belong to the second cluster. Rock density belongs to the third cluster. Thus, the main factors that affect AOF potential are porosity, TOC, brittle mineral content of rock and number of fracturing stages.

5.4. Multivariable nonlinear regression

After the main clusters in the hierarchical clustering analysis are obtained, the key factors that affect the AOF potential are found. The independent variables in the multivariable nonlinear regression models could be chosen according to the cluster results. With consideration of the correlation and collinearity among factors, one factor in each cluster with smallest linkage distance is chosen in the regression model, and four models after testing (**Table 3**) are regressed in this study.

With trials of nonlinear transformation of the involving variables, the linear regression of the transformation is ideal, and all the coefficients have passed the student t test for significance. The four models are shown as follows.

Model 1:

$$Y' = \ln Y, X_1' = \ln X_1, X_2' = (\ln X_2)^8, X_3' = X_3 \tag{10}$$

$$Y' = -13.067 + 1.509X_1' + 0.012X_2' + 0.867X_3' \tag{11}$$

Model 2:

$$Y' = \ln Y, X_4' = \ln X_4, X_2' = (\ln X_2)^8, X_3' = \ln X_3 \tag{12}$$

$$Y' = -1.756 + 1.465X_4' + 0.011X_2' + 1.043X_2' \tag{13}$$

Model 3:

$$Y' = \ln Y, X_4' = \ln X_4, X_5' = X_5^{\frac{1}{100}}, X_2' = (\ln X_2)^8, X_3' = \ln X_3 \tag{14}$$

$$Y' = -1.603 + 1.434X_4' + 0.226X_5' + 0.011X_2' + 0.848X_3' \tag{15}$$

Model 4:

$$Y' = \ln Y, X_1' = \ln X_1^{\frac{1}{10}}, X_3' = X_3^{\frac{1}{100}}, X_2' = (\ln X_2)^8, X_6' = X_6 \tag{16}$$

$$Y' = -12.556 + 5.211X_1' + 0.149X_5' + 0.012X_2' + 0.035X_6 \tag{17}$$

It is shown from **Fig.6** that the total AOF potential of 86 wells is $5420.48\times10^4 m^3/d$ and the total AOF potential of 86 wells predicted by the 4 models range from $5110.24\times10^4 m^3/d$ to $5607.74\times10^4 m^3/d$. The average AOF potential of single well is $63.03\times10^4 m^3/d$ and the average AOF potential of single well predicted by the 4 models range from $59.42\times10^4 m^3/d$ to $65.21\times10^4 m^3/d$. The coefficient of determination (R^2) in the figure is the proportion of the variance in the dependent variable Y that is predictable from the independent variables $X_j (j = 1, 2, \ldots, k)$. The R^2 values show that all the 4 models could be relatively preferable models to predict the AOF potentials based on different factors. In total, the prediction value of model 1 and model 4 are closer to the real AOF potential value with higher coefficients of determination (R^2) and less prediction errors.

6. Conclusions

In this study, instead of the produced gas rate, the absolute open flow (AOF) potential is analyzed as a better parameter to describe the deliverability of gas well. The reservoir parameters, drilling and completion data, fracturing data (including the well logging interpretation of horizontal section, the track and layer section that the well bore passed through, fracturing scale and monitoring parameters) and sample production data of 86 gas wells in the Wufeng-Longmaxi shale, Jiaoshiba shale gas field in Eastern Sichuan Basin, China are considered in the multivariate statistical analysis. The hierarchical clustering analysis could give the orders of the key factors that affect gas well AOF potential as well as eliminate the collinearity among the factors, which gives the basis of multivariable nonlinear regression. The results suggest that the order of the factors affects AOF potential is: a. porosity, b. TOC, c. brittle mineral content of rock, d. horizontal length ratio through stage 1 and stage 3 of Long1 member, e. number of fracturing stages, f. total liquid volume, g. horizontal length, h. total sand amount and i. rock density. The estimation results of the AOF potential by the multivariable nonlinear regression models based on the key factors from hierarchical clustering analysis have small differences to that of the actual values of the sample with goodness-of-fit test acceptable. The predictions of AOF potential of the models are reliable, and the feasibility of this multivariate statistical analysis is verified in the study of shale gas well deliverability. The research results can provide a reference basis for design of drilling and completion, the reasonable production rate allocation, design and adjustment of development plan of the shale gas reservoirs.

Acknowledgement

The authors would like to acknowledge the support from National Natural Science Foundation of China (NSFC Grant Number: 520004032) and the Thirteen-Five National Science and Technology Major Projects of China: 1. Study on Shale Gas Fracturing and Gas Production Regularity in Wufeng Formation and Longmaxi Formation in Fuling Area (Sub-Issue Number: 2016ZX05060-026) and 2. Research on the efficient development technology and policy of inter-salt shale oil reservoir (Sub-Issue Number: 2016ZX05049005-010).

References

1. Zolfaghari, A., Dehghanpour, H., & Xu, M. (2017b). Water sorption behaviour of gas shales: II. Pore size distribution. *International Journal of Coal Geology, 179*, 187-195.
2. Zhang, Y., Yao, S., Zhang, M., Zhou, X., Mei, H., & Zeng, F. (2018). Prediction of adsorption isotherms of multicomponent gas mixtures in tight porous media by the oil–gas-adsorption three-phase vacancy solution model. Energy & Fuels, 32(12), 12166-12173.
3. Zhang, Y., Zhang, M., Mei, H., & Zeng, F. (2019). Study on salt precipitation induced by formation brine flow and its effect on a high-salinity tight gas reservoir. *Journal of Petroleum Science and Engineering, 183*, 106384.
4. Dong, T., Harris, N. B., Ayranci, K., Twemlow, C. E., & Nassichuk, B. R. (2017). The impact of composition on pore throat size and permeability in high maturity shales: Middle and Upper Devonian Horn River Group, northeastern British Columbia, Canada. *Marine and Petroleum Geology*, 81, 220-236.
5. Wang, C., Zhang, B., Hu, Q., Shu, Z., Sun, M., & Bao, H. (2019). Laminae characteristics and influence on shale gas reservoir quality of lower Silurian Longmaxi Formation in the Jiaoshiba area of the Sichuan Basin, China. *Marine and Petroleum Geology, 109*, 839-851.
6. Cao, T. T., Song, Z. G., Wang, S. B., & Jia, X. (2015). A comparative study of the specific surface area and pore structure of different shales and their kerogens. *Science China Earth Sciences, 58*(4), 510-522.
7. Chalmers, G. R. L., & Bustin, R. M. (2007). The organic matter distribution and methane capacity of the lower cretaceous strata of northeastern British Columbia, Canada. *International Journal of Coal Geology, 70*(1-3), 223-239.
8. Chalmers, G. R. L., & Bustin, R. M. (2008). Lower cretaceous gas shales in northeastern British Columbia, Part II: Evaluation of regional potential gas resources. *Bulletin of*

Canadian Petroleum Geology, 56(1), 22-61.

9. Ma, X., Li, X., Liang, F. , Wan, Y. , & Guo, W. . (2020). Dominating factors on well productivity and development strategies optimization in Weiyuan shale gas play, Sichuan Basin, SW China. *Petroleum Exploration and Development, 47*(3), 594-602.
10. Shu, Y., Lu, Y., Chen, L., Wang, C., & Zhang, B. (2020). Factors influencing shale gas accumulation in the lower Silurian Longmaxi formation between the north and South Jiaoshiba area, Southeast Sichuan Basin, China. *Marine and Petroleum Geology, 111*, 905-917.
11. Wang, Y. , Dong, D. , Li, X. , Huang, J. , Wang, S. , & Wu, W. . (2015). Stratigraphic sequence and sedimentary characteristics of Lower Silurian Longmaxi Formation in the Sichuan Basin and its peripheral areas. Natural Gas Industry B, 2(2–3), 222-232.
12. Hu, M., Huang, W., & Li, J. (2018). Effects of structural characteristics on the productivity of shale gas wells: A case study on the Jiaoshiba Block in the Fuling shale gasfield, Sichuan Basin. *Natural Gas Industry B, 5*(2), 139-147.
13. Zhang, H., Wang, J., & Zhang, H. (2016). Investigation of the main factors during shale-gas production using grey relational analysis. *The Open Petroleum Engineering Journal, 9*(1).
14. Wei, C., Wang, H., Sun, S., Xiao, Y., Zhu, Y., & Qin, G. (2012, January). Potential investigation of shale gas reservoirs, Southern China. In *SPE Canadian Unconventional Resources Conference*. Society of Petroleum Engineers.
15. Wang, C., Zhang, B., Lu, Y., Shu, Z., Lu, Y., Bao, H., ... & Chen, L. (2018). Lithofacies distribution characteristics and its controlling factors of shale in Wufeng Formation-Member 1 of Longmaxi Formation in the Jiaoshiba area. *Petroleum Research, 3*(4), 306-319.
16. Wang, C., Zhang, B. Q., Lu, Y. C., Shu, Z. G., Lu, Y. Q., Bao, H. Y., ... & Chen, L. (2018). Lithofacies distribution characteristics and main development controlling factors of shale in Wufeng Formation-Member 1 of Longmaxi Formation in Jiaoshiba area. *Acta Pet. Sin., 39*(6), 631-644.
17. Kim, G., Lee, H., Chen, Z., Athichanagorn, S., & Shin, H. (2019). Effect of reservoir characteristics on the productivity and production forecasting of the Montney shale gas in Canada. *Journal of Petroleum Science and Engineering, 182*, 106276.
18. Curtis, M. E., Cardott, B. J., Sondergeld, C. H., & Rai, C. S. (2012). Development of organic porosity in the Woodford Shale with increasing thermal maturity. *International Journal of Coal Geology, 103*, 26-31.
19. Cipolla CL, Lolon EP, Erdle JC, Rubin B. Reservoir modeling in shale-gas reservoirs. *SPE reservoir evaluation & engineering*. 2010 Aug 1;13(04):638-53.
20. Zhu W, Ma D, Qi Q, Song Z, Yue M. Multi-zone coupling productivity of horizontal well fracturing with complex fracture networks in shale gas reservoirs. Natural Gas Industry B. 2018 Feb 1;5(1):80-9.
21. Mohammed, K. R., Stigliano, H., Vicente, M. G., & Angulo, R. J. (2010, January). Comparison of Different Deliverability Testing Techniques in Low Permeability Gas Reservoir and Its Impact on the Absolute Open Flow (AOF) Estimation. In *SPE EUROPEC/EAGE Annual Conference and Exhibition*. Society of Petroleum Engineers.
22. Guo, X., Qin, Z., Yang, R., Dong, T., He, S., Hao, F., ... & Liu, K. (2019). Comparison of

pore systems of clay-rich and silica-rich gas shales in the lower Silurian Longmaxi formation from the Jiaoshiba area in the eastern Sichuan Basin, China. *Marine and Petroleum Geology*, *101*, 265-280.

23. Yang, R., He, S., Hu, Q., Hu, D., Zhang, S., & Yi, J. (2016). Pore characterization and methane sorption capacity of over-mature organic-rich Wufeng and Longmaxi shales in the southeast Sichuan Basin, China. Marine and Petroleum Geology, 77, 247-261.
24. Henderson, C. R. . (1975). Best linear unbiased estimation and prediction under a selection model. *Biometrics, 31*(2), 423-447.
25. Gnanadesikan, M. B. W. . (1968). Probability plotting methods for the analysis for the analysis of data. *Biometrika, 55*(1), 1-17.
26. Rokach, Lior, and Oded Maimon. (2005). "Clustering methods." Data mining and knowledge discovery handbook. Springer US, 321-352.

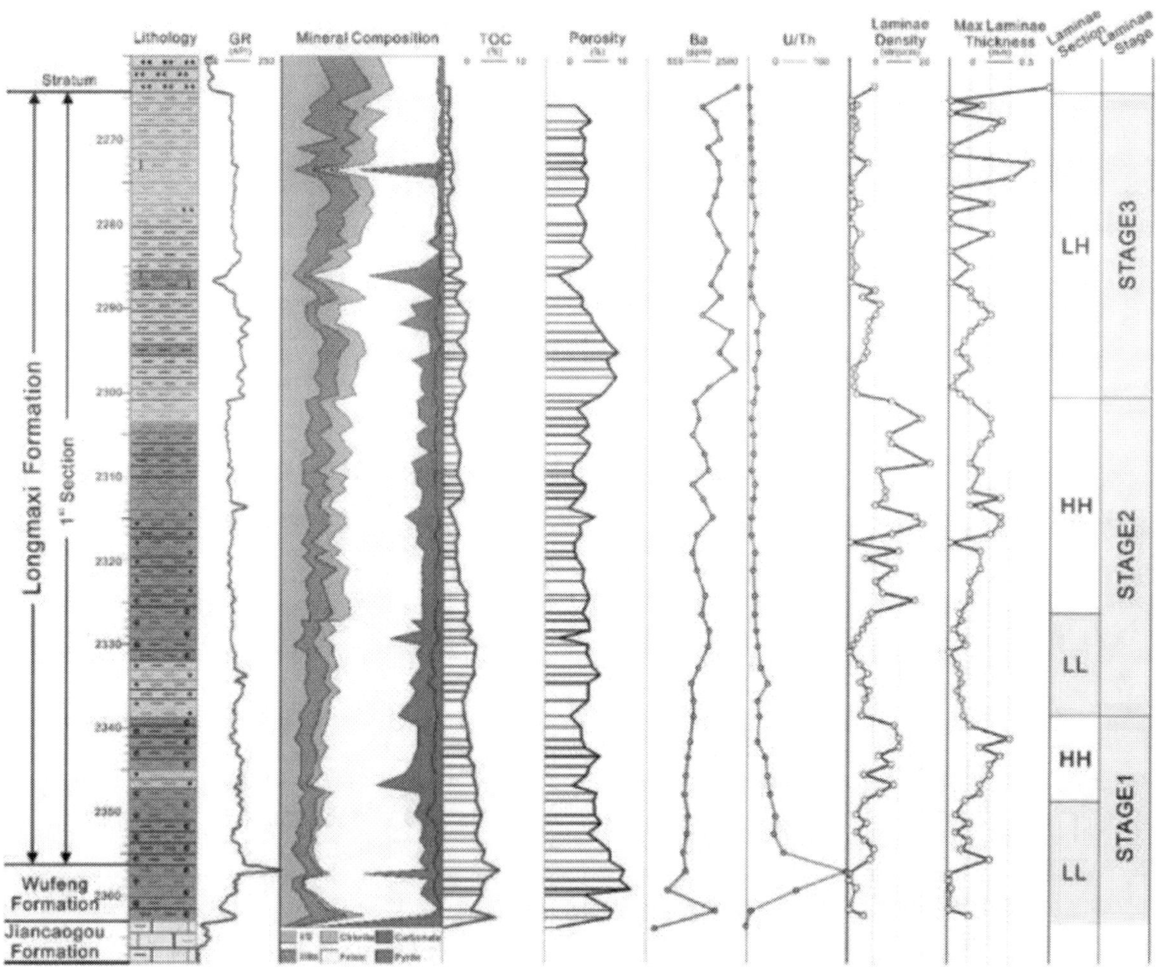

Fig.1 Lithofacies, gamma ray log, mineral content, TOC content, porosity, geochemical proxies, laminae density, and laminae stages of the Long1 member in well A (Wang et al., 2019).

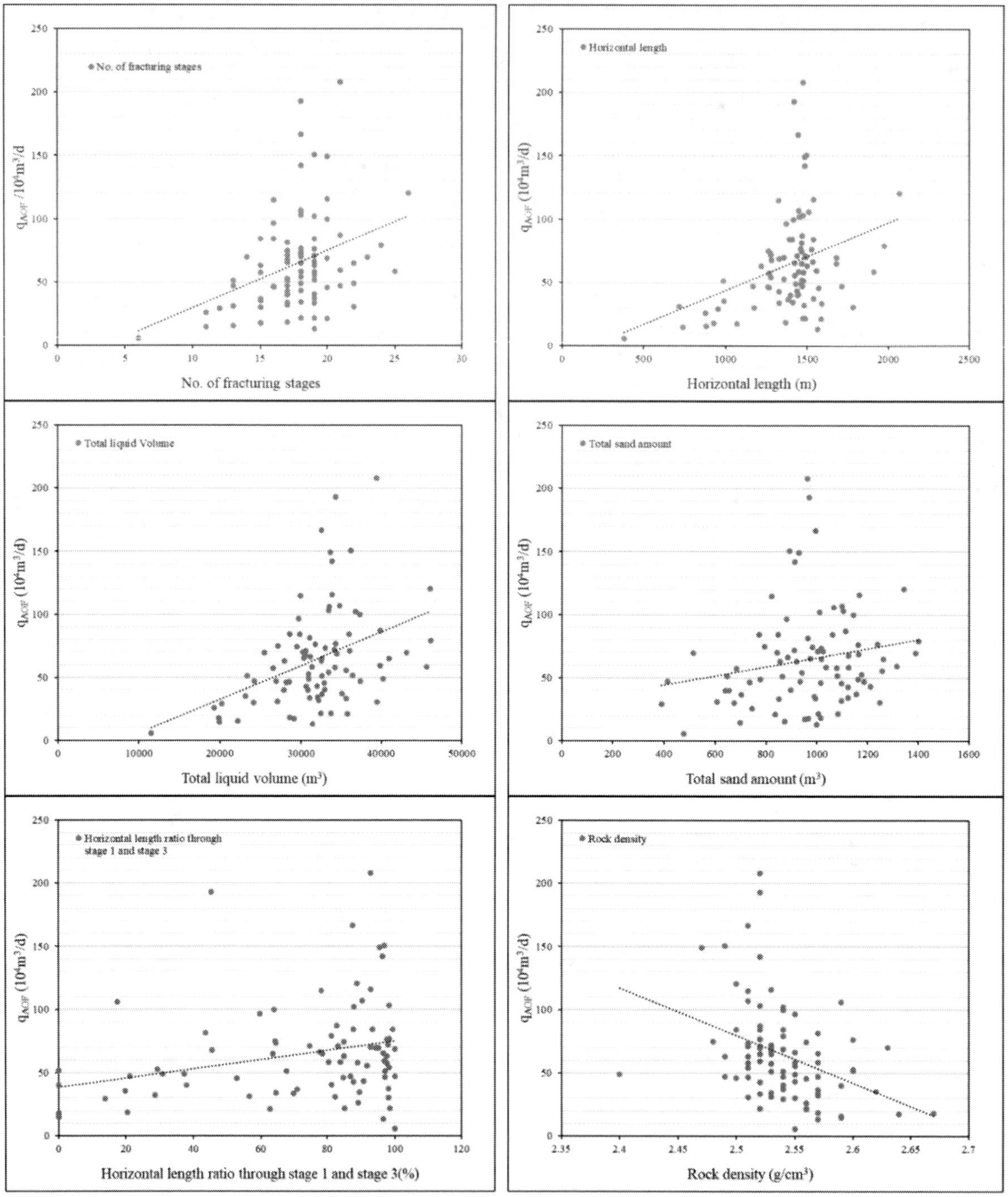
No. of fracturing stages
q_{AOF} /10⁴m³/d
No. of fracturing stages
Horizontal length
q_{AOF} (10⁴m³/d)
Horizontal length (m)
Total liquid Volume
q_{AOF} (10⁴m³/d)
Total liquid volume (m³)
Total sand amount
q_{AOF} (10⁴m³/d)
Total sand amount (m³)
Horizontal length ratio through stage 1 and stage 3
q_{AOF} (10⁴m³/d)
Horizontal length ratio through stage 1 and stage 3(%)
Rock density
q_{AOF} (10⁴m³/d)
Rock density (g/cm³)

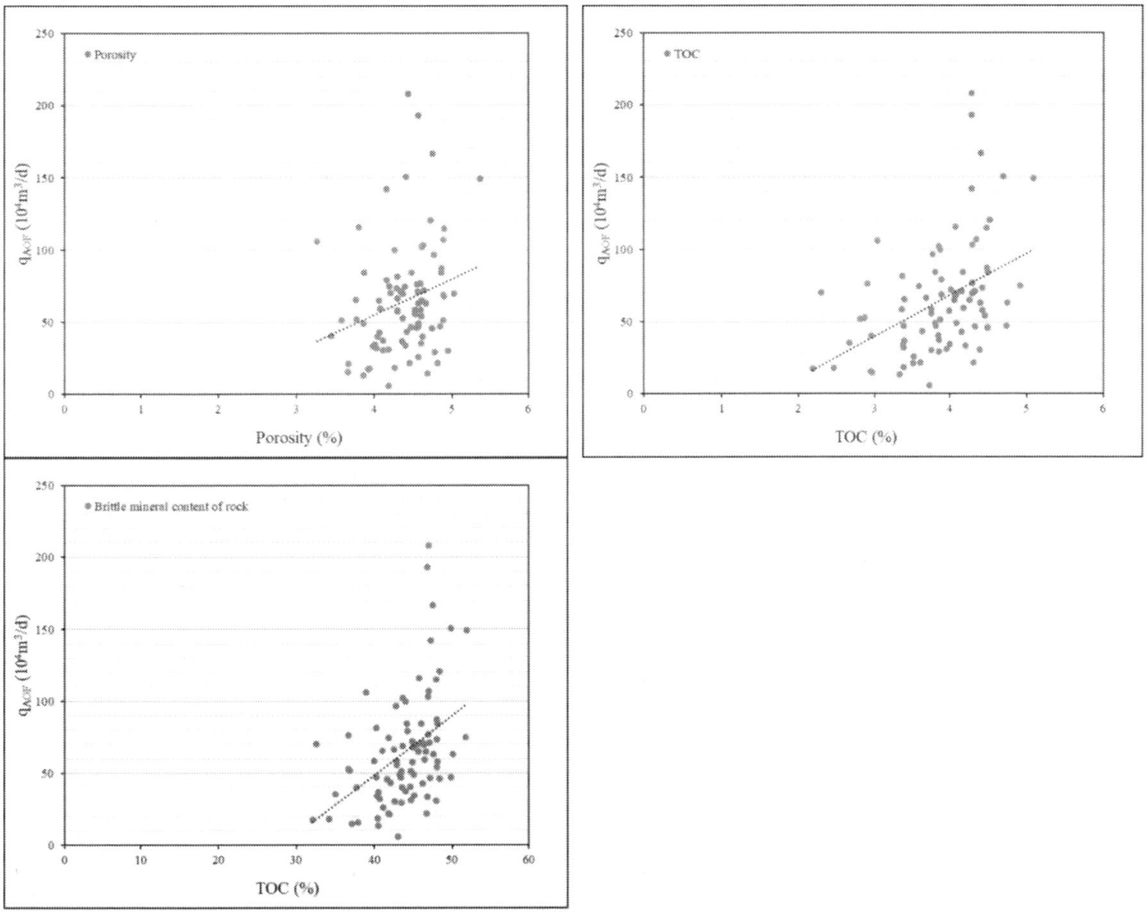

Fig.2 The scatter plots with single linear regression line of sample data of 86 gas wells in Jiaoshiba area with the AOF potential against 9 factors that may affect the deliverability of gas wells (i.e. number of fracture stages, horizontal length of well, total liquid volume, total sand amount, rock density, porosity, TOC, brittle mineral content of rocks, horizontal length ratio through stage 1 and stage 3 of Long1 member)

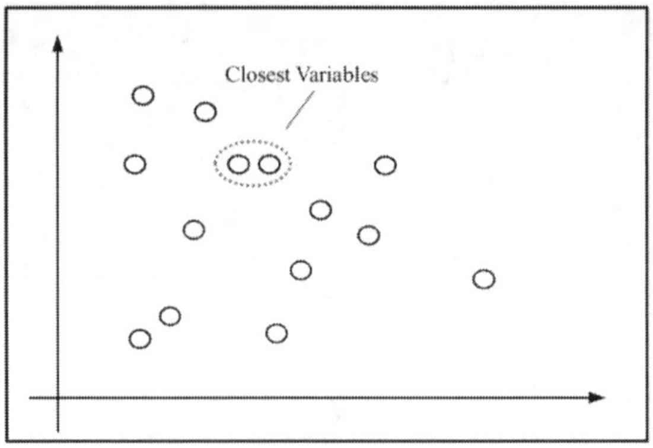

(a)

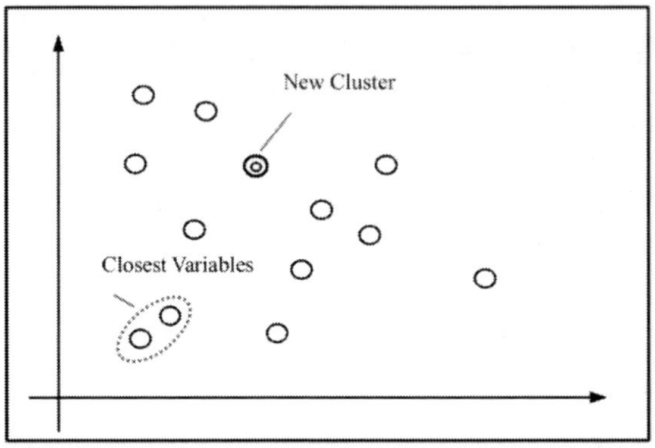

(b)

Fig.3 Hierarchical Clustering Analysis: (a) Clustering the variables by closest relation; (b) The center of new cluster is calculated by centroid method

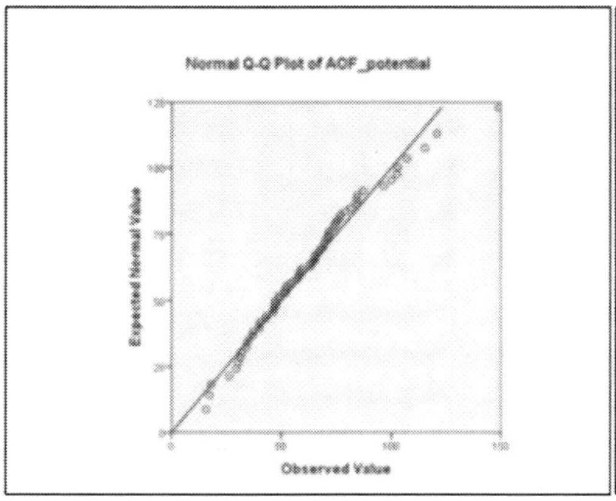
Normal Q-Q Plot of AOF_potential
Expected Normal Value
Observed Value

a.

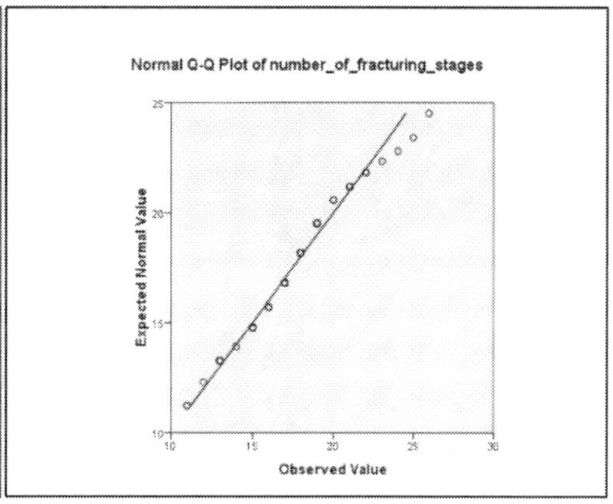
Normal Q-Q Plot of number_of_fracturing_stages
Expected Normal Value
Observed Value

b.

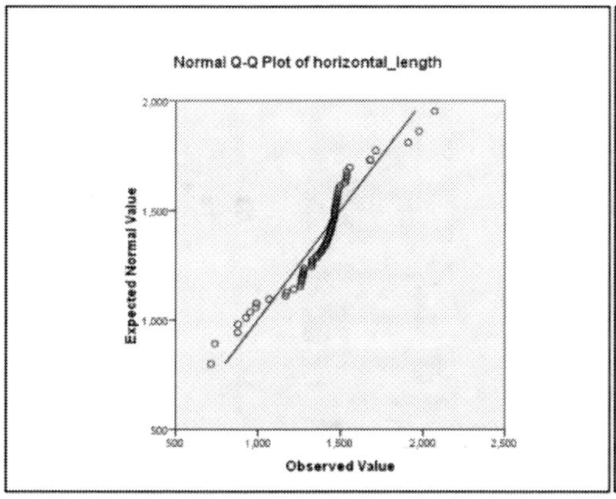
Normal Q-Q Plot of horizontal_length
Expected Normal Value
Observed Value

c.

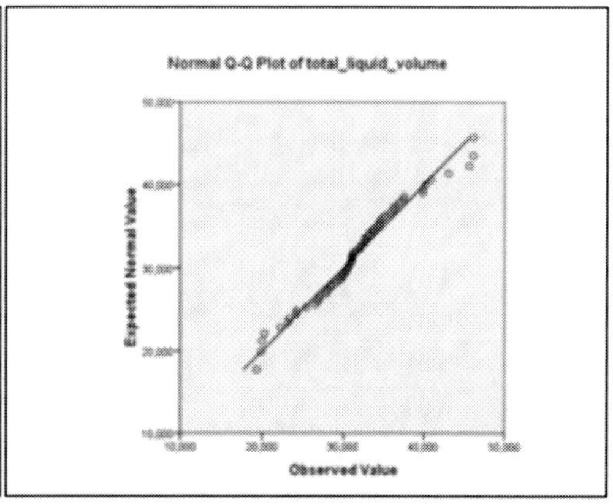
Normal Q-Q Plot of total_liquid_volume
Expected Normal Value
Observed Value

d.

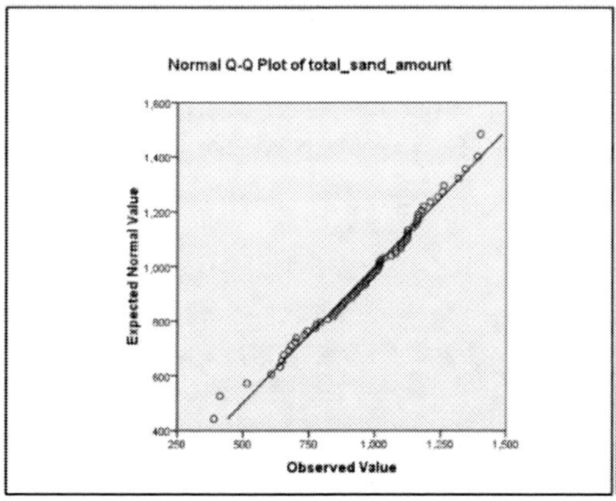
Normal Q-Q Plot of total_sand_amount
Expected Normal Value
Observed Value

e.

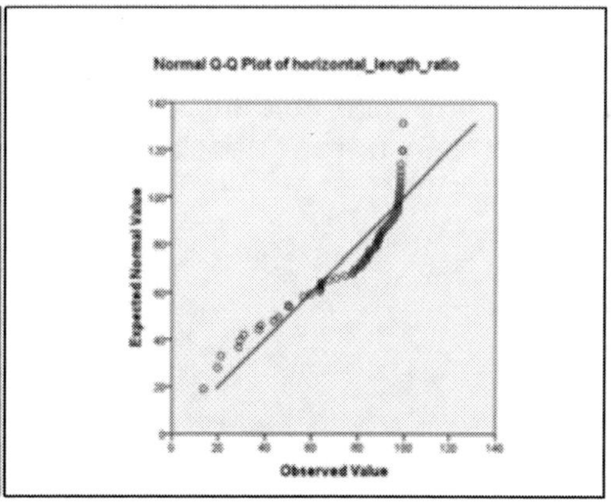
Normal Q-Q Plot of horizontal_length_ratio
Expected Normal Value
Observed Value

f.

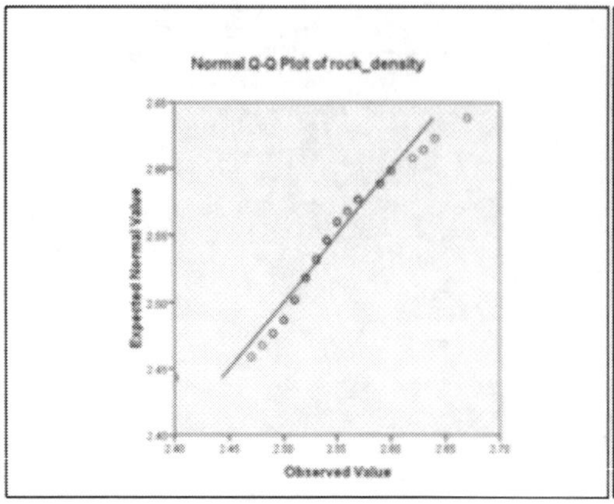

g.

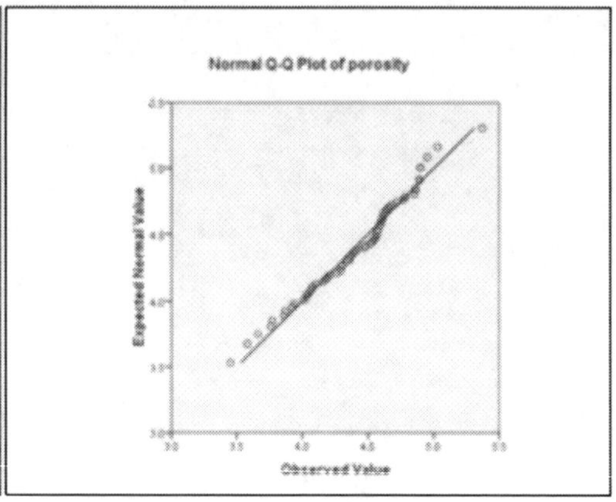

h.

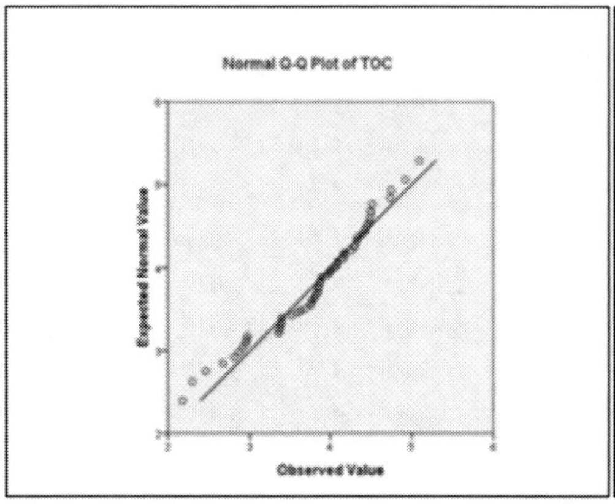

i.

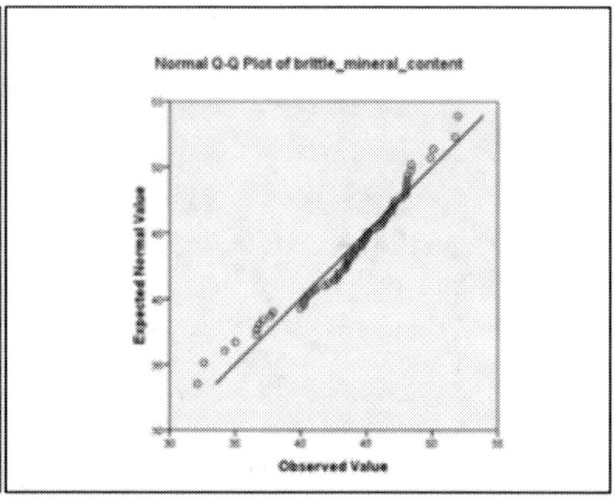

j.

Fig.4 The Q–Q (quantile-quantile) plots for the 10 variables of the 86 wells: a. AOF potential; b. number of fracturing stages; c. horizontal length of well; d. total liquid volume; e. total sand amount; f. horizontal length ratio through stage 1 and stage 3 of Long1 member; g. rock density; h. porosity; i. TOC; j. brittle mineral content of rock

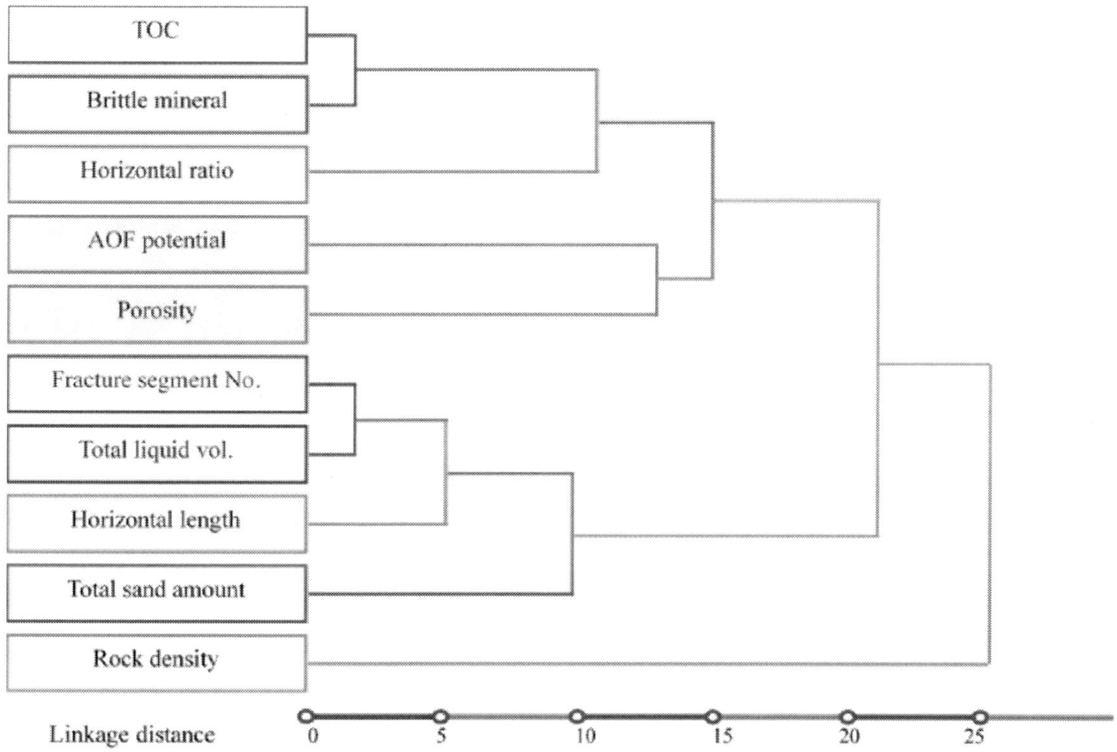

Fig.5 Hierarchical clustering analysis result of the sample data of 86 wells with linkage distance

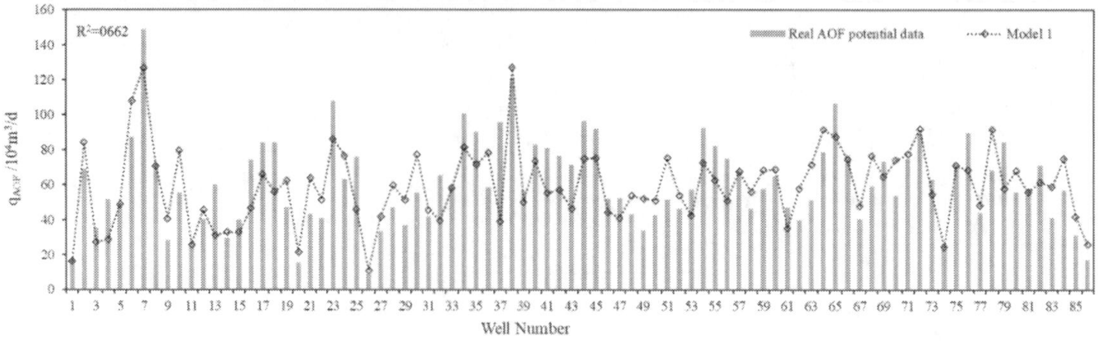

(1)

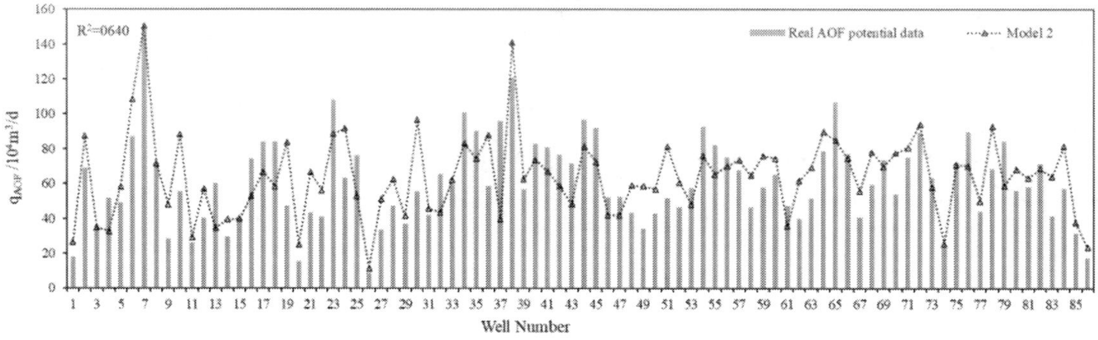

(2)

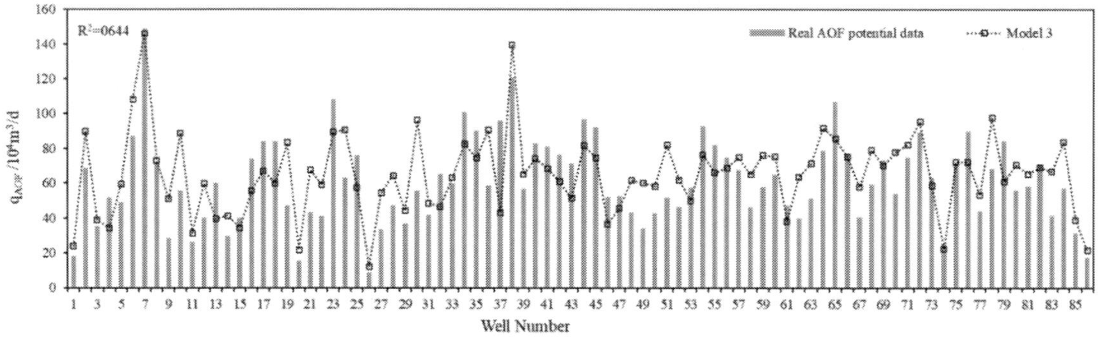

(3)

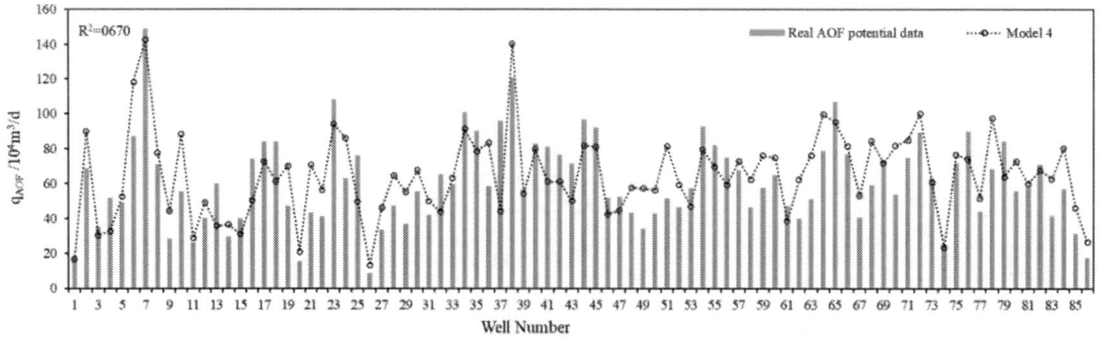

(4)

Fig.6 Comparison of actual and predicted AOF potential by the 4 models for the 86 wells with consideration of different key factors in Fuling Jiaoshiba shale gas field

Table 1 Correlation coefficients matrix of all variables in Jiaoshiba area

Factors	AOF potential	No. of fracturing stages	Horizontal length	Total liquid Volume	Total sand amount	Horizontal length ratio through stage 1 and stage 3	Rock density	Porosity	TOC	Brittle mineral content of rock
AOF potential	1	0.354	0.348	0.405	0.194	0.285	-0.373	0.246	0.433	0.438
No. of fracture stages	0.354	1	0.897	0.946	0.765	0.273	-0.203	-0.013	0.241	0.235
Horizontal length	0.348	0.897	1	0.863	0.633	0.331	-0.241	0.024	0.26	0.252
Total liquid Volume	0.405	0.946	0.863	1	0.736	0.302	-0.248	-0.01	0.274	0.282
Total sand amount	0.194	0.765	0.633	0.736	1	0.22	-0.002	-0.072	0.072	0.07
Ratio through stage 1 and stage 3	0.285	0.273	0.331	0.302	0.22	1	-0.492	0.287	0.568	0.578
Rock density	-0.373	-0.203	-0.241	-0.248	-0.002	-0.492	1	-0.289	-0.908	-0.917
Porosity	0.246	-0.013	0.024	-0.01	-0.072	0.287	-0.289	1	0.419	0.336
TOC	0.433	0.241	0.26	0.274	0.072	0.568	-0.908	0.419	1	0.982
Brittle mineral content of rock	0.438	0.235	0.252	0.282	0.07	0.578	-0.917	0.336	0.982	1

Table 2 Hierarchical clustering analysis result of factors in Jiaoshiba area

Parameter	Hierarchical clustering analysis result
AOF potential	1
Number of fracturing stages	2
Horizontal length	2
Total liquid volume	2
Total sand amount	2
Horizontal length ratio through stage 1 and stage 3	1
Rock density	3
Porosity	1
TOC	1
Brittle mineral content of rock	1

Table 3 Factors chosen in regression model according to hierarchical clustering analysis results

Regression model	Factors considered
1	a. Total liquid volume (X_1) b. Porosity (X_2) c. TOC (X_3)
2	a. Fracture segment No. (X_4) b. Porosity (X_2) c. TOC (X_3)
3	a. Fracture segment No. (X_4) b. Horizontal length ratio through stage 1 and stage 3 (X_5) c. TOC (X_3) d. Porosity (X_2)
4	a. Total liquid volume (X_1) b. Horizontal length ratio through stage 1 and stage 3 (X_5) c. Porosity (X_2) d. Brittle mineral content of rock (X_6)

601617 Solutions of C6F13- and C4F9-Based Polyfluoroacrylates in CO_2; Solubility, Viscosity, and Adsorption Onto Cement

Abstract:

High molecular weight poly(fluorinated acrylates) (PFAs) have drawn interest in the chemical and petroleum engineering fields due to their remarkable solubility in liquid and supercritical carbon dioxide (sCO_2) without the need for any co-solvent. This study does not include any examples of PFA based on monomers with the -C8F17 moiety because the ultimate hydrolysis degradation products of such polymers leads to the formation of PFOA, a compound deemed unsafe by the EPA. Instead, we have synthesized PFA based on fluoroacrylayte monomers containing the -C6F13 group (the functionality used in current commercial fluoroacrylate copolymers). These -C6F13-based polymers would degrade to form perfluorohexanoic acid (PFHxA) that, while significantly safer than PFOA, is also receiving more scrutiny from regulatory agencies in recent days. Therefore, we have also synthesized PFA based on monomers containing the -C4F9 functionality. The -C4F9-PFA ultimate degradation products include perfluorobutanoic acid (PFBA), which is significantly more environmentally benign than the -C6F13-based PFA. Therefore, one of the main objectives of this study was to determine if -C4F9-PFA retained the remarkable CO_2-solubility and CO_2-thickening capability as PFA based on the -C6F13 PFA and previously reported -C8F17-based PFA.

Many samples of PFA based on the -C6F13 and -C4F9-based fluoroacrylate monomers and varying ratios of initiator:monomer were bulk polymerized and then characterized for molecular weight and polydispersity. The -C4F9-PFA was shown to be nearly identical in CO_2-solubility as the -C6F16-PFA. Between 1-4 wt% PFA, the solubility of PFA in CO_2 was nearly independent of PFA molecular weight. Further, the cloud points loci for mixtures of CO_2-C4F9-PFA were nearly identical to the cloud point curves for the CO_2-C6F13-PFA at each temperature, with higher pressure required at higher temperature. This demonstrated that the -C4F9-PFA was just as CO2-soluble as the -C6F13-PFA. Further, both polymers were capable of thickening CO_2 in dilute concentration to a comparable degree (e.g. a 6-fold increase in viscosity with 4 wt% C4F9 PFA, 1.24×10^6 Da PFA in CO_2 at low shear rate).

PFA has previously been considered for use as a "CO_2 thickener" in enhanced oil recovery (EOR) operations. Unlike common chemical engineering applications where low viscosity is beneficial to reduce mass transfer resistance, low CO_2 viscosity is often problematic for EOR because it leads to viscous fingering (mobility control problems) and preferential loss of disproportionate amounts of CO_2 to high permeability "thief zones" that contain little recoverable oil (conformance control problems). However, in 2019 it was shown that PFA dissolved in CO_2 adsorbs so strongly onto sandstone or carbonate rock that the use of PFA in large-scale EOR projects to address mobility control would not be feasible, although near-wellbore conformance control applications (i.e. injecting the CO_2-PFA into isolated thief zones) may be viable.

In the current study, however, we are attempting to exploit the adsorption of PFA on solid surfaces for improved wellbore integrity. We are proposing to inject PFA-CO_2 solutions into small cracks within cement or between steel and cement (i.e. microannular cracks), thereby allowing a portion of the PFA to absorb and greatly reduce the crack permeability or completely seal the crack. Because PFA is a high molecular weight, amorphous, transparent, sticky, stretchy, oil-repellant, water-repellant polymer that is insoluble in water, oil or natural gas, it is a promising sealant for wellbore applications. Even though PFA is inherently expensive, this is a small volume application in which the cost of PFA would not exclude its consideration. Relative to other sealants (cement, resins, aqueous emulsions), PFA-CO_2 solutions are orders of magnitude less viscous, which would allow them deeper access into smaller cracks than other options. The PFA-CO_2 solutions would, however, be an inappropriate selection for the remediation of extremely large cracks or voids in cement.

In order to lay the groundwork for this proposed application, we have also determined the equilibrium adsorption of PFA onto Portland cement particles. This was accomplished by a direct measurement of the mass gain of a known quantity of sieved (22-30 mesh) Portland cement particles that was immersed in a PFA-CO2 fluid of specified original composition (e.g. 4 wt% PFA). The equilibrium concentration of PFA in the liquid CO_2 was determined via mass balance (original PFA placed in the cell – the amount of PFA found to adsorb onto the cement). This is, to the best of our knowledge, the first report of polymeric adsorption levels of solid surfaces immersed in CO_2-polymer solutions

611710 Competitive Sorption of CO2 in Shale Nanopores for Sequestration and Enhanced Gas Recovery Using Molecular Density Functional Theory

Abstract:

Enhanced gas recovery (EGR) by carbon dioxide (CO_2) injection to gas-bearing shale formations has been shown to be feasible and beneficial experimentally from pore scale to field scale. In addition to the improved recovery factor, CO_2 is trapped in the geological formation, which provides an opportunity to store CO_2. When supercritical CO_2 is injected, the preferential adsorption and dissolution of CO_2 in nanopores and organic matter enables the displacement of shale gas. This mechanism has been investigated by molecular simulation in graphite pores, but rarely for permeable pores by simulation or classical density functional theory (DFT). The compositional distribution of components from the organic matter source (kerogen) to nanoscale pores in the organic matter and to large fractures makes the estimation of fluid storage and the multicomponent compositional distributions in shale a challenging problem. We have introduced a density functional theory that accounts for molecular size and shape and kerogen maturity and swelling to predict the partitioning of components between the kerogen matrix and nano-pores in the kerogen [1].

In this work, a molecular thermodynamic model is presented that predicts both the adsorption of hydrocarbons in nanoscale pores and the dissolution of fluids in organic matter in equilibrium with the bulk kerogen phase. The swelling ratios of five kerogens of varying maturities and types in different solvents are well quantified using the new model. This kerogen matrix model is incorporated in interfacial statistical associating fluid theory (iSAFT) DFT to form a nanoporous kerogen composite model. The equilibrium partitioning of mixtures of fluids from the bulk phase to the nanoporous kerogen phase show enrichment of aromatic and cyclic molecules in kerogen, the preferential adsorption/absorption of high molecular weight molecules and carbon dioxide, and the significant amount of hydrocarbon storage both in pore space and kerogen matrix [1, 2].

Further, we apply molecular DFT to study the competitive sorption of CO_2 with shale gas in kerogen at various conditions. The DFT model is verified by grand canonical Monte Carlo (GCMC) simulation in graphite slit pores for pure and binary component systems at different temperatures, pressures, pore sizes and bulk compositions for methane/ethane/ CO_2. [2]. The model is utilized to predict the CO_2 adsorption selectivity in multicomponent systems which include two different shale gases. We show the selectivity of CO_2 decreases as temperature / pressure/ nanopore size/ shale gas average molecular weight increase. Subsequently, we model the CO_2/CH_4 adsorption in pores of more realistic conditions, in which we consider different levels of moisture in the pore and the gas dissolution in organic matrix. Interestingly, the presence of water inside a nanopore greatly reduces the CO2 adsorption but improves the CO_2 selectivity with respect to methane. The presence of organic matter kerogen also increases the CO_2 selectivity. In summary, a novel permeable pore DFT model enables prediction of partitioning and distribution of CO2 and hydrocarbons in shale related to enhanced gas recovery and CO_2 storage. [1, 2].

1. Jinlu Liu and Walter G. Chapman, "Thermodynamic Modeling of the Equilibrium Partitioning of Hydrocarbons in Nanoporous Kerogen Particles," Energy & Fuels, 33, 891-904 (2019).
2. Jinlu Liu, Shun Xi, and Walter G. Chapman, "Competitive Sorption of CO2 with Gas Mixtur in Nanoporous Shale for Enhanced Gas Recovery from Density Functional Theory," Langmuir, 35, 8144-8158 (2019).

609261 Spatial and Temporal Evolution of Natural Gas and Natural Gas Liquid Feedstocks in Shale Basins

Abstract:

Technological advancements in horizontal drilling and hydraulic fracturing have led to increasing production of light hydrocarbons from shale resources, including natural gas and natural gas liquids (NGLs). However, some basins that have insufficient gas gathering and pipeline networks (e.g. the Permian Basin in Texas) may flare significant quantities of gas. For example, in the Permian Basin, about 4.6% of gross gas production was estimated to be flared between May 2018 and March 2019 (Zhang et al., 2020). To mitigate flaring issues, hydrocarbon resources can be locally transformed to chemical or energy products. The National Science Foundation's Center for Innovative and Strategic Transformation of Alkane Resources (CISTAR) is developing modular processes to transform natural gas and NGLs to gasoline-like transportation fuels.

Determining the scale of operation of such modular chemical manufacturing requires a comprehensive understanding of the temporal and spatial availability of the hydrocarbon feedstocks. Preliminary work was performed in the Eagle Ford oil and gas producing region in south central Texas. With over 20,000 wells, the Eagle Ford produced 5.4 billion cubic feet natural gas per day and 1.2 million barrels of condensate and oil per day in 2019 (RRC, 2020). From the northwest to the southeast portion in the Eagle Ford Shale, well production varies from oil, to gas with condensate (wet gas), and to dry gas, as shown in Figure 1 (Gherabati et al, 2016). From oil to wet gas to dry gas regions, propane to ethane ratio varies from 1:1 to 1:3 (mass basis). In addition to varying gas compositions, the magnitude of available feedstocks also vary across the region. Ethane available at individual gas plants was estimated by assuming that produced gas from wells is sent to their nearest gas plant. Among 44 gas processing plants in the Eagle Ford, 60% of ethane production are represented by 43 plants with 200,000 – 750,000 kg of ethane per day, with one plant processing at ~3,000,000 kg/day, accounting for 40% of ethane production, shown in Figure 2.

Similar work has been performed in the Permian Basin in Texas. The Permian Basin is one of the largest oil-producing regions in the world and accounts for over 30% of oil production in the United States. Recent study (Zhang et al, 2020) shows that methane emissions in the Permian Basin are the highest among all the U.S. oil/gas producing regions. This presentation will provide detailed characterizations of spatial and temporal variability in light alkane production, and losses due to flaring for a case study region in the Permian Basin.

Collectively, the analyses in the Eagle Ford and the Permian Basin will provide two case studies of spatial and temporal characteristics of feedstock availability for modular light alkane manufacturing.

References

Gherabati, S. A., Browning, J., Male, F., Ikonnikova, S. A., & McDaid, G. (2016). The impact of pressure and fluid property variation on well performance of liquid-rich Eagle Ford shale. *Journal of Natural Gas Science and Engineering*, *33*, 1056-1068.

Railroad Commission of Texas (RRC) – Eagle Ford Shale Information. (2020). Available at: https://www.rrc.state.tx.us/oil-gas/major-oil-and-gas-formations/eagle-ford-shale-information/

Zhang, Y., Gautam, R., Pandey, S., Omara, M., Maasakkers, J. D., Sadavarte, P., ... & Zhang, R.

(2020). Quantifying methane emissions from the largest oil-producing basin in the United States from space. *Science Advances*, *6*(17), eaaz5120.

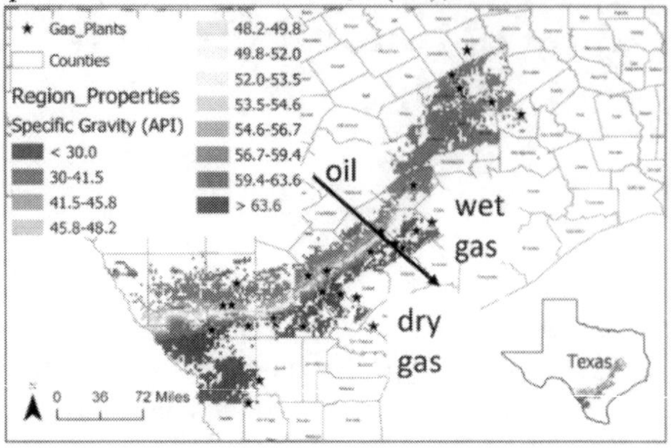

Figure 1. Spatial variabilities in well production in the Eagle Ford Shale.

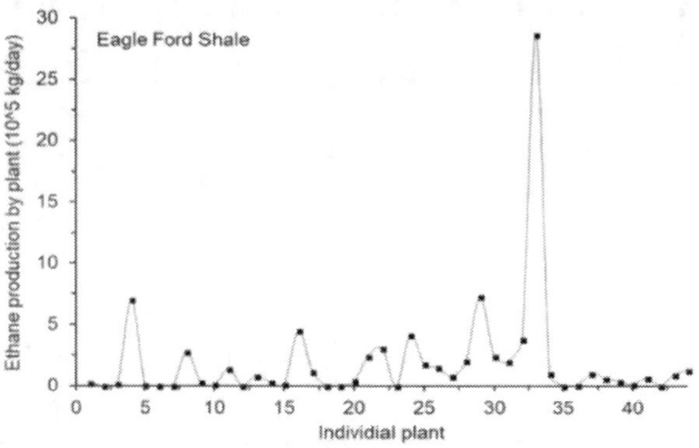

Figure 2. Ethane available at 44 gas processing plants in the Eagle Ford Shale

592706 A New Geology-Engineering Integration Solution for Shale Gas Exploration: A Case Study from Changning Block in China

Abstract:

Rich and quality shale gas reservoirs have been found in the Lower Silurian Longmaxi strata in the southern Sichuan Basin, which was identified as great development potential. However, shale gas reservoir is characterized by ultra-low permeability, which usually has no natural productivity and is called a typical "artificial gas reservoir". We must pay attention to reservoir quality, drilling quality and completion quality in order to obtain the maximum EUR and the highest oil recovery. To tackle this challenge, the integrated Geoengineering solution was deployed in Changning Shale Gas Block, which consists of geological modeling, geomechanics, fracturing simulation and gas reservoir simulation. Geoengineering integration technology is used on the Changning A platform for field applications. The results show that: (1) Carrying out integrated research on geological engineering can realize the organic combination of geology and engineering. The geological model provides technical support for engineering optimization. The engineering practice constantly corrects the geological model and improves the precision; (2) Through the integration of geological engineering, we can create "transparent" shale gas reservoir, which can improve the predictive accuracy of geological and engineering sweet spot mapping; (3) Hydraulic fractures can be fine characterization by using geological engineering integration technology, which is useful for quantitative description and analysis of key parameters such as geometrical shape, net pressure, proppant distribution and conductivity of hydraulic fractures; (4) the refined unstructured-grid (<10m) based 3D numerical simulation model was built for hydraulic fractured horizontal wells in shale gas reservoir, the history match for shale gas wells has been hugely improved: gas production rate was matched to more than 98%; the wellhead pressure was matched to more than 80%; the cumulative liquid production rate was matched to more than 90%. With the implementation of the Geoengineering study, the shale gas production in Changning area has been increasing rapidly in the past 2 years.

593227 New Technology of EOR of Ultra Deep and Complex Condensate Gas Reservoirs in Tarim Basin, China

Abstract:

A number of condensate gas reservoirs (both sandstone and carbonate) with depth of over 5000m have been developed in Tarim Basin, China. Those reservoirs are featured with thick reservoirs pay zones (over 100 meters) with ,strong reservoir heterogeneities in both rock properties and fluid properties. Gas channeling in the process of gas injection development and retrograde condensation in the process of depletion development have seriously affected the recovery of condensate. In order to enhance the condensate recovery in those condensate gas reservoirs, a series of new technologies have been developed during the last decade.

Laboratory experiment analysis and development practice discovered that three major mechanisms controlling the gas migration process in the reservoirs: displacement, gravity and diffusion for cyclic gas injection. It is proposed that gravity differentiation is the main factor affecting the development effect of cyclic gas injection, and the fluid distribution law of underground condensate gas reservoir is re-recognized. For the sandstone condensate gas reservoir developed with large well spacing (over 1000 meters), the displacement and diffusion will be replaced by gravity differentiation gradually in the process of cyclic gas injection, and the fluid will be redistributed vertically according to its density, which makes different phase behaviors. Similarly, the reservoir space of depleted carbonate condensate gas reservoir is mainly dissolution cave and fracture. Under the action of gravity, the reverse condensate oil usually accumulates at the bottom of fracture cave system. On this basis of this understanding, the technology of gravity assisted three-dimensional displacement gas injection for sandstone condensate gas reservoirs and horizontal well placement technology for carbonate condensate gas reservoirs were developed to optimize the well placement. The technologies have been successfully applied to the developed condensate gas field in Tarim Basin, and the recovery rate of condensate oil has been increased by over 19%.

Production capacity evaluation method of condensate gas reservoir considering the influence of stress sensitivity and condensate oil

DU Chunxiao, LIU Xiaohong, ZHANG Fengyi, XU Jing, GENG Zhigang

(Tianjin Branch of CNOOC Ltd., Tianjin 300459)

Abstract: Accurate evaluation of production capacity is an important guarantee for efficient development of condensate gas reservoir. For the deep condensate gas reservoir with low permeability and low pressure difference between the initial formation pressure and the dew point, the influence of reservoir stress sensitivity and condensate oil on the production capacity of the gas reservoir cannot be ignored. Nevertheless, at present, the widely used method to calculate the gas reservoir production capacity is based on single-phase gas, without considering the influence of condensate oil and stress sensitivity on reservoir permeability. On the basis of researching a large number of literatures, combined with laboratory experiments and numerical simulation, the production capacity evaluation model of condensate gas reservoir under the comprehensive influence of stress sensitivity and condensate oil is derived by using the seepage flow mechanics theory, and the research results are applied to the Bohai Z gas reservoir. The results show that the production capacity evaluation model can more accurately reflect the production status of this kind of gas reservoir in different development stages, and can be widely used in the productivity evaluation of similar condensate gas reservoir.

Key words: deep condensate gas reservoir; stress sensitivity; condensate oil; evaluation of production capacity; low permeability

Gas reservoir production capacity evaluation is an important research content of gas reservoir engineering research, which provides key guidance for the design of gas reservoir development plan. For condensate gas reservoir, the fluid flow law and phase change are more complex, and the reverse condensation phenomenon has a serious impact on gas well productivity. According to theoretical and experimental studies, many scholars have confirmed that low permeability reservoirs have strong stress sensitivity. At present, the widely used method to calculate the gas reservoir production capacity is based on single-phase gas, without considering the influence of condensate oil and stress sensitivity on reservoir permeability. Many scholars have carried out research on the influence of reservoir stress sensitivity or retrograde condensation on gas well productivity, but lack of comprehensive consideration of both. Therefore, a new productivity evaluation method of condensate gas reservoir considering stress sensitivity and retrograde condensation is derived by combining laboratory experiments and numerical simulation methods, and is applied to BZ gas reservoir in Bohai oilfield.

1. Influence of stress sensitivity on reservoir permeability

The relationship between rock permeability and effective confining pressure has been studied deeply by many scholars using a large number of experiments. The relationship between permeability and effective confining pressure can be divided into three types: power-law relationship, exponential relationship and polynomial relationship.

BZ gas reservoir in Bohai oilfield is a low permeability gas reservoir. The core stress sensitivity test results of BZ-1 well are shown in Table 1 and figure 1. It can be seen that the dimensionless permeability of core decreases with the increase of effective stress, and BZ gas reservoir has strong stress sensitivity.

Table 1 Data of core stress sensitivity test

Net pressure	Sample 1	Sample 2	Sample 3	Sample 4
MPa	K_n/K_i	K_n/K_i	K_n/K_i	K_n/K_i
3.5	0.748	0.792	0.774	0.781
5.0	0.637	0.670	0.642	0.643
7.0	0.476	0.521	0.503	0.524
9.0	0.391	0.394	0.402	0.414
11.0	0.313	0.329	0.308	0.369
15.0	0.208	0.238	0.239	0.248
20.0	0.152	0.170	0.180	0.170
25.0	0.117	0.121	0.141	0.131
30.0	0.095	0.102	0.122	0.112
35.0	0.081	0.087	0.097	0.083

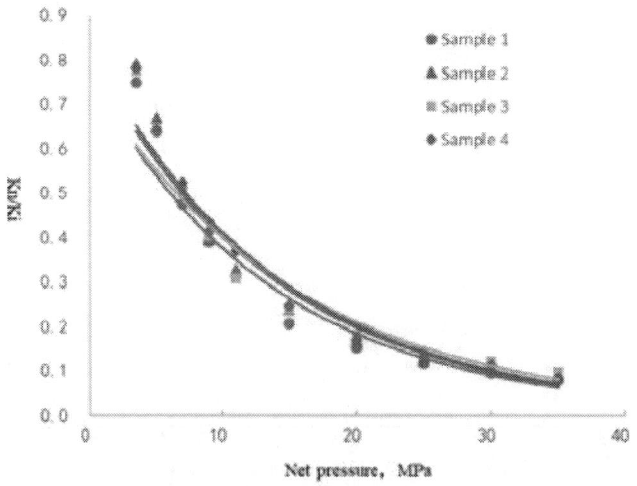

Fig 1 Stress sensitivity curves of the cores

By fitting the relationship between rock permeability and effective confining pressure, it is found that there is a good exponential relationship between them（Table 2）. It can be expressed as:

$$k_n = \mathrm{N}k_i e^{-\alpha(p_i-p)} \quad (1)$$

Table 2 Stress sensitivity index fitting of core permeability

Core number	Exponential fitting expression	R^2
1	$y = 0.7721e^{-0.072x}$	0.9553

2	$y = 0.8186e^{-0.071x}$	0.9595
3	$y = 0.7609e^{-0.064x}$	0.9489
4	$y = 0.8331e^{-0.07x}$	0.9727

According to the principle of seepage mechanics, the square distribution expression of stable seepage pressure of gas reservoir is as follows.

$$p^2 = p_e^2 - \frac{p_e^2 - p_{wf}^2}{\ln(\frac{r_e}{r_w})} ln\frac{r_e}{r} \qquad (2)$$

Formula (1) is deformed.

$$k_n = Nk_i e^{-\alpha\left\{p_i - \left[p_e^2 - \frac{p_e^2 - p_{wf}^2}{ln(\frac{r_e}{r_w})} ln\frac{r_e}{r}\right]^{\frac{1}{2}}\right\}} \qquad (3)$$

2 Influence of condensate on gas phase relative permeability

The influence of reverse condensate oil is studied by numerical simulation method. By using single well numerical simulation, the variation law of condensate oil saturation at different distances from well during gas field development is obtained.

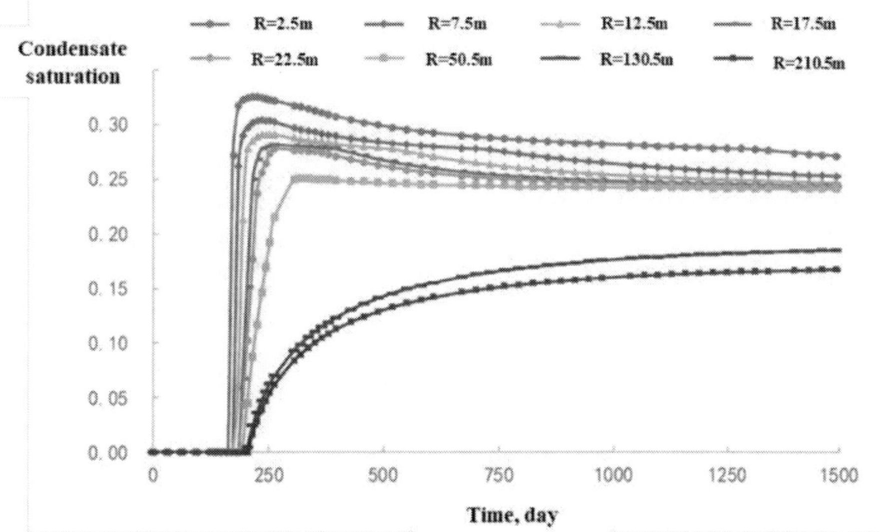

Fig.3 variation law of condensate oil saturation

It can be seen from Figure 3 that the condensate saturation decreases with the increase of radial distance, and tends to be stable after increasing with the extension of production time. The results show that the condensate saturation changes with time in the near well zone, and the distribution function of condensate saturation with radial distance $s_o(\mathrm{r})$ can be obtained at any time.

According to the relative permeability curve, the regression relationship between gas relative permeability and oil saturation can be obtained. The relative permeability curve of BZ gas reservoir in Bohai oilfield is shown in Fig. 4, the regression expression is:

$$k_{rg}(\mathrm{r}) = 4.7367s_o(r)^3 - 2.0497s_o(r)^2 - 2.1398s_o(r) + 1.0015 \qquad (4)$$

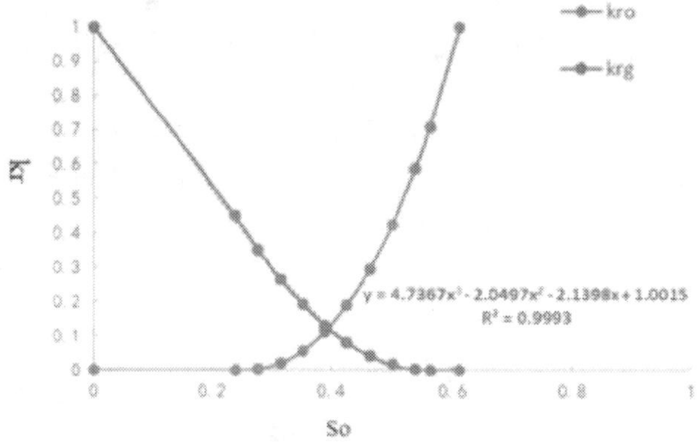

Fig.4 Relative permeability curve of Z gas reservoir

3 Improved gas well productivity evaluation model

The expression of effective permeability is:

$$k_g = k_{\mathrm{n}} \cdot k_{rg} \tag{5}$$

Replace formula (3) and formula (4) into formula (5):

$$k_g = Nk_i[4.7367s_o(r)^3 - 2.0497s_o(r)^2 - 2.1398s_o(r) + 1.0015]e^{-\alpha\{p_i-[p_e^2-\frac{p_e^2-p_{wf}^2}{\ln(\frac{r_e}{r_w})}ln\frac{r_e}{r}]^{\frac{1}{2}}\}} \tag{6}$$

The condensate saturation beyond a certain distance from the well is lower than the critical flow saturation, the distance is defined as the effective radial distance and expressed as r_o.

The whole stratum can be divided into two regions. The effective permeability can be expressed as:

$$k_g(\mathrm{r}, p_{wf}) = \begin{cases} Nk_a[4.7367s_o(r)^3 - 2.0497s_o(r)^2 - 2.1398s_o(r) + 1.0015]e^{-\alpha\{p_i-[p_e^2-\frac{p_e^2-p_{wf}^2}{\ln(\frac{r_e}{r_w})}ln\frac{r_e}{r}]^{\frac{1}{2}}\}}, & r_w < r < r_o \\ Nk_a e^{-\alpha\{p_i-[p_e^2-\frac{p_e^2-p_{wf}^2}{\ln(\frac{r_e}{r_w})}ln\frac{r_e}{r}]^{\frac{1}{2}}\}}, & r_o \le r \le r_e \end{cases} \tag{7}$$

① $r_w < r < r_o$, average permeability can be expressed as:

$$k_1(p_{wf}) = \frac{\int_{r_w}^{r_o} k_g(r, p_{wf}) \cdot 2\pi r dr}{\int_{r_w}^{r_o} 2\pi r dr} \tag{8}$$

② $r_o \le r \le r_e$, average permeability can be expressed as:

$$k_2(p_{wf}) = \frac{\int_{r_o}^{r_e} k_g(r, p_{wf}) \cdot 2\pi r dr}{\int_{r_o}^{r_e} 2\pi r dr} \tag{9}$$

According to the principle of seepage mechanics, the average gas permeability of the formation can be obtained:

$$k(p_{wf}) = \frac{k_1(p_{wf}) \cdot k_2(p_{wf}) \cdot \ln(\frac{r_e}{r_w})}{k_2(p_{wf}) \cdot \ln(\frac{r_o}{r_w}) + k_1(p_{wf}) \cdot \ln(\frac{r_e}{r_o})} \tag{10}$$

Based on the above research, the productivity evaluation formula of gas well considering stress sensitivity and condensate oil is obtained.

$$p_e^2 - p_{wf}^2 = \frac{1.273\times10^{-3}T\mu Z}{k(p_{wf})h}\ln\left(\frac{r_e}{r_w}\right) q_{sc} + \frac{2.162\times10^{-10}\gamma_g ZT}{r_w h^2 k(p_{wf})^{1.5}} q_{sc}^2 \quad (11)$$

4 Application

BZ is a rare condensate gas reservoir with complex reservoir and fluid characteristics. The main performance is low pressure difference between the initial formation pressure and the dew point (0.21-7.13MPa), the maximum amount of condensate is 40%, the reservoir permeability is 1.00~6.80mD, strong stress sensitivity.

The evaluation equation is applied to BZ gas field. The IPR curves of three cases are calculated, under the comprehensive influence of condensate oil and stress sensitivity, only considering the influence of condensate oil or stress sensitivity. The results are compared with those of the traditional binomial formula.

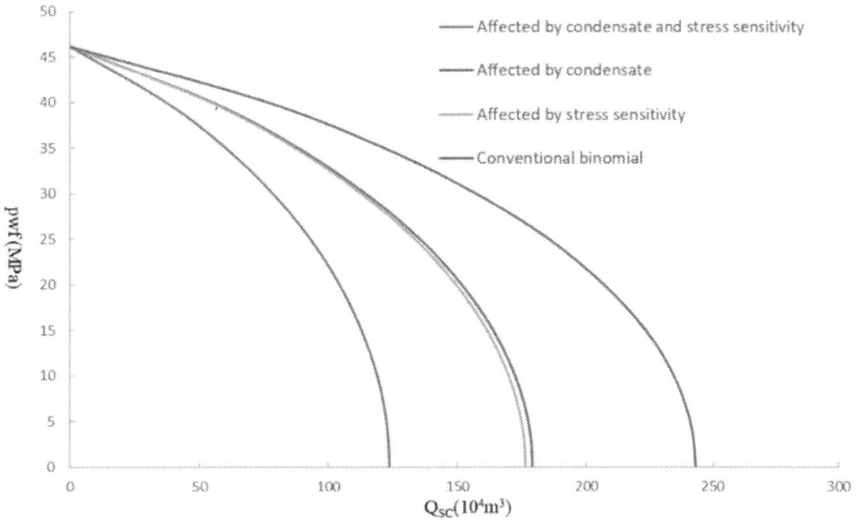

It can be seen that reservoir stress sensitivity and retrograde condensation have great influence on gas reservoir productivity. Under the combined effect of both, the productivity of gas reservoir is reduced by 50%. The IPR curves of gas wells under different formation pressures under the combined influence of stress sensitivity and retrograde condensation are further obtained.

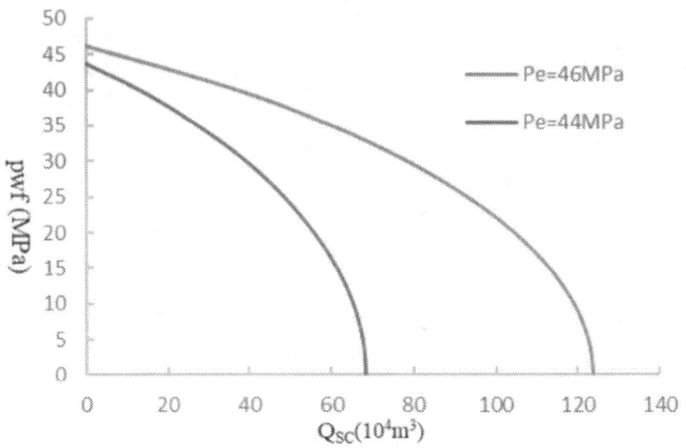

It can be seen that the formation pressure decreases, stress sensitivity and retrograde condensation increase, the productivity of gas wells also decreases obviously. The formation pressure drops by 2MPa, and the open flow rate of gas wells decreases by 57%. Therefore, in order to ensure the gas well productivity, it is necessary to adopt the development mode of gas injection and pressure maintaining, and the gas well production allocation should not be too high, in order to slow down the adverse impact of pressure drop.

6 Conclusion

(1) Stress sensitivity and condensate oil have great influence on the productivity of low permeability condensate gas reservoirs with low pressure difference between the initial formation pressure and the dew point. Most studies lack of comprehensive consideration of the two factors.

(2) Combined with laboratory experiment and numerical simulation, the formation permeability considering the effect of reservoir stress sensitivity and retrograde condensation is derived by using the principle of seepage mechanics, and then the evaluation method which can reflect the real productivity of gas reservoir is obtained.

(3) The research results show the necessity of gas injection to maintain reservoir pressure, and put forward suggestions for reasonable production of gas wells.

Symbol description

k_i-Original permeability, $10^{-3}\mu m^2$

k_n-Absolute permeability, $10^{-3}\mu m^2$

k_g-Effective permeability, $10^{-3}\mu m^2$

k_{rg}-Gas phase relative permeability, $10^{-3}\mu m^2$

p_i-Original formation pressure, MPa

p-Formation pressure, MPa

N-Stress sensitivity coefficient

α-Stress sensitivity index

p_e-Formation pressure at supply radius, MPa

p_{wf}-Bottom hole pressure, MPa

r_e-Supply radius, m

r_w-Wellbore radius, m

s_o-Condensate saturation near wellbore

r-Radial distance from well, m

T-Formation temperature, K

μ-Gas viscosity, mPa·s

Z-Gas deviation factor

h-Thickness of gas reservoir, m

β-Characteristic parameters of pore structure affecting turbulence and inertia resistance

γ_g-Relative density of natural gas

q_{sc}-Gas well production under standard conditions, m^3/d

References

[1] Fussel D.D. Single well Performance for gas condensate reservoir[J]. JPT， 1973， 25(12)： 860-876.

[2] Jones J.R. and Raghavan R. Interpretation of Pressure buildup reservoirs in gas condensate well[J]. SPEFE， 1989.

[3] FAN Xueping, XU Xiangrong. Experimental and mechanism research about permeability damage with the change of stress[J]. PETROLEUM EXPLORATION AND DEVELOPMENT, 2002

[4] XUE Zhan, LIU Pengcheng. A New Method of Correction of Gas Condensate Well Productivity Considering Capillary Number[J]. Journal of Chongqing University of science and Technology (NATURAL SCIENCE EDITION), 2014.

[5] LUO Ruilan, CHENG Linsong. A new method of determining relationship between permeability and effective overburden pressure for low-permeability reservoirs[J]. Journal of China University of Petroleum, 2007.

Numerical simulation study of a complex naturally-fractured gas reservoir with extra-high condensate in Bohai Oilfield

GAO Zhennan, SU Yanchun, GENG Zhigang, LIU Weilin, HUANG Baogang

(CNOOC China Limited Tianjin Branch, Tianjin 300459, China)

Abstract:

The BZ condensate gas field is the largest natural gas field in the history of the Bohai bay basin, with proven reserves of more than 100 billion cubic meters of natural gas and 100 million cubic meters of condensate oil. This gas field has unique features, such as deep burial, fracture development, strong stress sensitivity, small pressure difference between formation pressure and dew point pressure, high condensate content, etc. There is no similar condensate-gas field development cases in the world and proper reservoir numerical simulation study is crucial for efficiently developing such a reservoir.

This paper develops a numerical simulation flowchart for this kind of complex naturally fractured condensate gas reservoirs. First, based on logging, core and well testing data, the reservoir of BZ gas field is divided into three different types of deposits in the vertical direction: glutenite, semi-weathered zone and inner buried hill. The fluid flow among the three different types of deposits is calculated based on the equivalent fluid flow conduction between adjacent grids. Secondly, the correlation of permeability changes with stresses is established based on experimental data and implemented in the reservoir simulation based on the imbedded programming module in the simulator. Thirdly, through PVTi numerical simulation, EOS are tuned based on 7 pseudo-components. Finally, a dual-porosity compositional model is established, and the numerical simulation study is carried out to optimize the development plan. The developed flowchart will be a very useful tool for modeling naturally-fractured condensate gas reservoirs.

The simulation results indicated that: 1) considering the strong stress sensitivity, gas injection can effectively increase the permeability of the reservoir and control the condensate precipitation; 2) considering the structural characteristics of the reservoir, gas injection in the high position can form top-gas drive and expand the swept volume; 3) considering the influence of fracture, injection- production line perpendicular to fracture trend can effectively delay the breakthrough of injected gas and expand the injection capacity; 4) considering the characteristics of condensate, keeping formation pressure higher than dew point pressure can minimize the loss of condensate oil; 5) considering the mechanism of gas injection displacement, slug gas injection can form pressure pulse, expand the sweep volume and improve the recovery of condensate oil. Those findings can also be used to guide the development of other gas condensate reservoirs.

Key words: buried-hill fracture; condensate gas reservoir; stress sensitive characterization; reservoir numerical simulation; cyclic gas injection development

● 0 Introduction

BZ condensate gas field is the largest natural gas field in Bohai Sea, with proven natural gas reserves of over 100 billion cubic meters and proven condensate reserves of more than 100 million cubic meters. The gas field has the following geological characteristics: buried deep(-4400~-5500m), complex seepage (fracture development), stress sensitivity (permeability loss 89.2%), and the fluid characteristics are as follows: low pressure difference between formation pressure and dew point pressure (1.07MPa), ultra-high condensate (711g / m3), large amount of reverse condensate (32.3%). The efficient development of BZ condensate gas field is a world problem! It is found that there is no similar gas field development case at home and abroad, so it is impossible to accurately evaluate the development effect of this gas field. Therefore, it is urgent to carry out reservoir numerical simulation research to provide support for early development plan and subsequent efficient development[1-3].

● 1 Research method

According to the reservoir classification, the complex seepage system gas reservoir model is established. Based on the stress sensitive experiment, the variation law of seepage capacity is regressed. Based on the matching of fluid phase, the dual medium component model is established. And the development mode and gas injection development strategy are studied through reservoir numerical simulation.

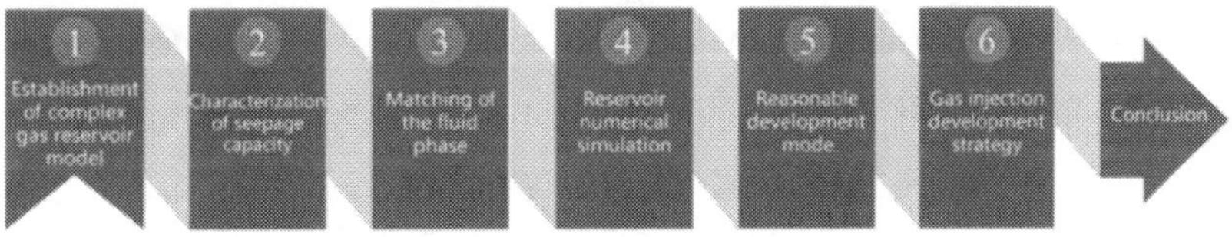

Fig 1. Technology Roadmap

● 2 Gas reservoir model

2.1 Reservoir zonation

Combined with the analysis of logging, core and testing data, BZ gas field has good reservoir connectivity, and it can be divided into sandy conglomerate zone, weathered zone and buried hill inner zone from top to bottom[4].

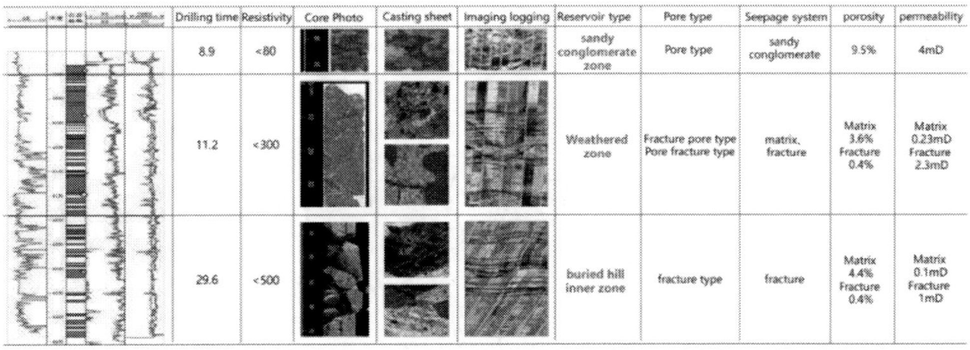

Fig 2. Vertical development characteristics of BZ condensate gas field reservoir

2.2 Establishment of complex gas reservoir model

Establish a structural model based on the understanding of reservoir zoning; and using lithofacies modeling and discrete fracture network modeling to generate matrix and fracture porosity, permeability and interporosity flow coefficient. By zeroing the matrix conductivity inside the buried hill, the coexistence of three types of complex seepage models in the BZ gas field can be realized.

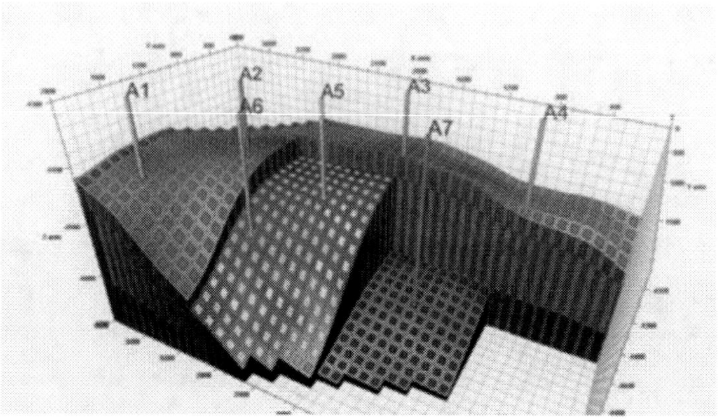

Fig 3. Gas reservoir model of complex seepage system

● 3 Stress sensitive characterization

3.1 Stress sensitivity analysis

According to the strong stress sensitivity of BZ gas field, the relationship between core permeability residual coefficient and net stress is fitted, and there is a good power function relationship between them; with the increase of net stress (reservoir pressure decreases), the rate of permeability loss gradually decreases[5-11]. In early stage, the net stress increased to 10MPa, Kr reduced by 60%; in mid stage, the net stress increased to 20MPa, Kr reduced by 20%; when reach the abandonment pressure(the net stress increased to 30MPa), the remaining permeability is about 10%.

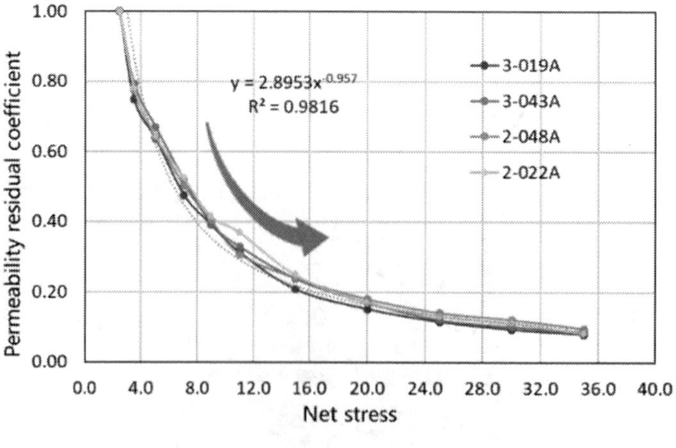

Fig 4. Permeability versus stress curve

$$K_r = 2.8953\ (P - P_0)^{-0.957}$$

K_r: Ratio of permeability to initial permeability, zero dimension;

P: Reservoir pressure at any time, MPa;

P_0: Initial formation pressure, MPa。

3.2 Stress sensitive characterization

According to the regression relationship, the characterization code of reservoir physical properties with time is compiled, and the sensitivity analysis is carried out by reservoir numerical simulation.

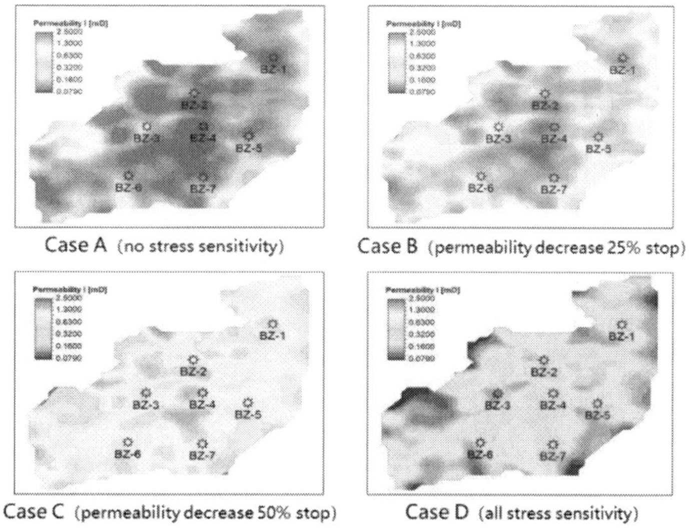

Fig 5. Comparison of permeability of different stress sensitive cases

4 Characterization of fluid characteristics

4.1 Fluid phase analysis

The BZ gas field is a condensate gas reservoir with high saturation and ultra-high condensate content through the study of fluid phase behavior characteristics, and the maximum condensate oil saturation is as high as 41%. When the pressure drops about 15MPa, the maximum saturation can be reached. Condensate oil precipitation seriously pollutes the reservoir, increases gas phase seepage resistance, and reduces gas well productivity and recovery[12-13].

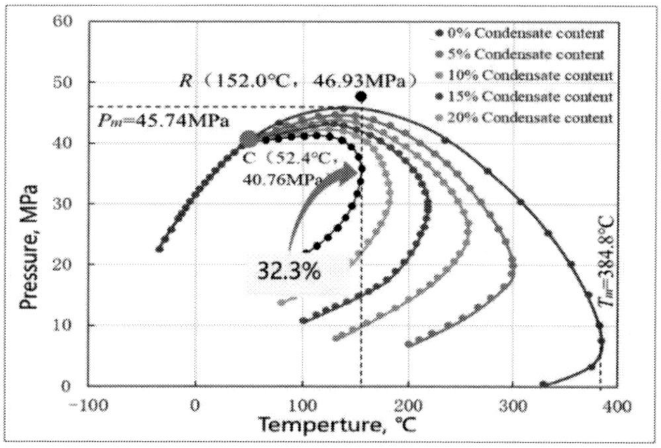

Fig 6. Phase diagram of fluid temperature and pressure in BZ gas field

4.2 Fluid phase matching

Considering the accuracy and speed of numerical simulation, 4 cases of component splitting combinations were set according to the principle of similarity combination. Combined with the calculation time and results of Numerical simulation of component model, the 7-component splitting plan is determined as the best plan.

11 components：C1、N2、C2、CO2、C3、C4、C5、 C6、C7-C10、C15+C22、C23+

7 components：C1+N2、C2+CO2、C3-C4、C5-C6、C7-C10、C15+C22、C23+

5 components：N2、CO2、C1、C2- C6、C7+

3 components：C1+N2、C2-C6+CO2、C7+

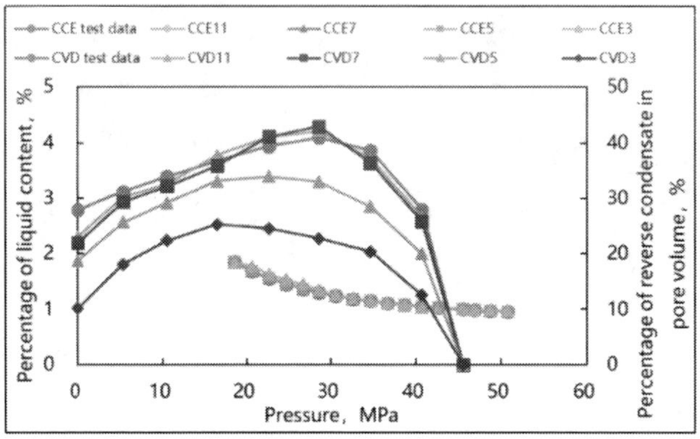

Fig 7. Comparison of experimental fitting accuracy

5 Study on gas reservoir numerical simulation

5.1 Optimization of development mode

As oil and gas fluids are produced, the formation pressure decreases, the physical properties become worse, the seepage ability decreases, and the recovery of gas field decreases. In depletion exploitation mode, the formation pressure drops rapidly, the condensate oil quickly reaches the maximum condensate velocity, and lost in the formation, so the condensate oil recovery rate is low. In gas injection development mode, the formation pressure is maintained through gas injection, and the retrograde condensate rate is controlled, and the condensate oil recovery is high. Considering reservoir stress sensitivity and retrograde condensate characteristics, from the perspective of improving oil and gas recovery and reservoir protection, it is recommended that the BZ condensate gas field development adopts first cycle gas injection and then depletion development[14-17].

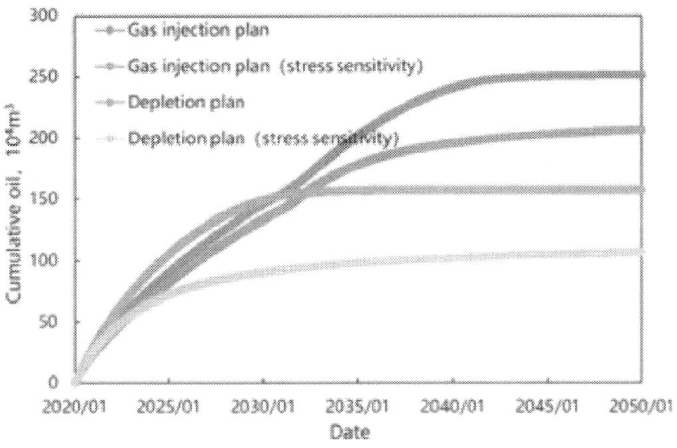

Fig 8. Comparison of different development mode

5.2 Cycle gas injection development strategy

Condensate gas reservoirs are usually developed with gas injection. The main purpose is to supplement formation energy, reduce condensate oil precipitation, and improve condensate oil recovery. When evaluating the effect of development, gas injection sweep volume and formation pressure maintenance level are two very important factors.

5.2.1 Optimization of injection and production

During gas injection development, the gas injection well position is directly related to the dry gas sweep volume. The structural and the fracture trend are two important factors.

A. Structural factor

The sandy conglomerate zone in structural high area of the BZ gas field has good physical properties. Priority is given to production in the early stage, and gas injection at the low area. In later stage, the condensate oil is enriched in the structural low area by gravity, and the top gas injection forms secondary gas and top drive. Vertical displacement, effectively expanding the gas sweep volume.

Tab 1. Development effect of structural factor

Plan	Cumulative gas 10^8m^3	Gas recovery %	Cumulative oil 10^4m^3	Oil recovery %
Inject in low area 10 years	36	53	181.3	26.6
Inject in low area 5 years,And in high area 5 years	36.6	53.3	192.4	28.3
Inject in high area 10 years	36.1	53.1	189.2	27.8

B. Fracture trend factor

When the injection-production line is parallel to the fracture trend, it is easy to form gas channeling, which reduce oil and gas recovery. When the injection-production line is vertical to the fracture trend, it can maximize the sweep volume, and effectively maintain formation energy and the seepage ability, and inhibit condensate oil to separate out.

Tab 2. Development effect of fracture trend factor

Plan	Cumulative gas 10^8m^3	Gas recovery %	Cumulative oil 10^4m^3	Oil recovery %
The injection-production line isparallel to the fracture	35.1	51.1	168.6	25.1
The injection-production line isperpendicular to the fracture	36.6	53.3	192.4	28.3
The injection-production line isoblique to the fracture	36	53	186.1	27.3

5.2.2 Reasonable pressure maintenance level

The most important thing in the development of condensate gas reservoirs is control the formation pressure to slow down the precipitation of condensate oil. The timing of gas injection and the rate of reinjection are two key factors.

A. Timing of gas injection

The timing of gas injection determines the degree of condensate oil loss, seepage capacity, and condensate precipitation rate during gas injection development. Since the difference between the formation pressure and the dew point pressure of the BZ gas field is very small, gas injection at the beginning of production can effectively delay the retrograde condensation time and reduce pollution, which is the optimal solution.

Tab 3. Development effect of timing of gas injection

Plan	Cumulative gas 10^8m^3	Gas recovery %	Cumulative oil 10^4m^3	Oil recovery %
Formation pressure 30MPa (4 years later)	26.4	48	127	23.4
Formation pressure 40MPa (2 years later)	26.3	47.9	132	24.4
Formation pressure 50MPa (0years later)	26.3	47.9	135.9	25.1

B. Re-injection rate

The higher the gas re-injection rate, the more stable the formation pressure remains. Before the formation pressure is higher than the dew point pressure, the 100% re-injection rate can always maintain the formation pressure higher than the dew point pressure, minimize the loss of condensate oil, and increase condensate oil recovery rate.

Tab 4. Development effect of Re-injection rate

Plan	Cumulative gas 10^8m^3	Gas recovery %	Cumulative oil 10^4m^3	Oil recovery %
100% re-injection rate	36.2	52.7	198.5	29.2
80% re-injection rate	36.4	53	196.1	28.8
50% re-injection rate	36.6	53.3	192.4	28.3

5.2.3 Optimization of gas injection mode

Considering the mechanism of gas injection and the demand for gas supply in the Beijing-Tianjin-Hebei region in winter, a comparative study of 3 plans: cyclic gas injection, slug gas injection, and Inject gas and soak. Circulating gas injection is prone to gas channeling, thereby reducing the condensate oil and gas recovery rate. The plan of inject gas and soak delays the gas channeling time on the basis of maintaining formation pressure, so the recovery rate is higher. In slug gas injection plan, the pressure pulse formed by the gas injection and gas production cycle can effectively expand the swept volume. And when injection stopped, the condensate in matrix will flow into the fracture due to the gravity. Therefore, the slug gas injection plan has the highest oil and gas recovery.

Tab 5. Development effect of Optimization of gas injection mode

Plan	Cumulative gas 10^8m^3	Gas recovery %	Cumulative oil 10^4m^3	Oil recovery %
Cyclic gas injection	35.3	50.2	188.4	26.8
Slug gas injection	37	53.9	198.4	29.1
Inject gas and soak	36.5	52.3	194.7	28.2

- **6 Conclusion**

(1) The BZ gas field are divided into sandy conglomerate zone, weathered zone and Buried hill inner zone. The gas reservoir model of complex seepage system is established by the zeroing matrix conductivity inside the buried hill.

(2) The BZ gas field has strong stress sensitivity, and Nearly 90% loss of seepage capacity under abandonment pressure. Therefore, it is necessary to carry out numerical simulation research through time-varying reservoir physical properties.

(3) BZ gas field is a gas field with high saturation and high condensate content. Through PVTi matching and numerical simulation sensitivity analysis, the 7-component splitting plan is determined as the best plan.

(4) The numerical simulation study of the BZ gas field shows that: firstly inject gas in low area of the structure and then in the high area, inject gas at the start of production, the injection rate is 1.0, and seasonal slug gas injection is the optimal plan. This research provides strong support for

ODP program preparation and subsequent efficient development.

Reference:

[1] Fang Na, Jiang Guanghong, Li Yunpeng, et al. Development characteristics and potential tapping direction of buried-hill fractured oil reservoirs[J]. Special Oil & Gas Reservoirs, 2017,24(3):90-94.

[2] Xu Chunqiang, Zhang Zhen, Zhang Xintao, et al. Tectonic evolution and hydrocarbon accumulation model of western sag in western Bohai sea[J]. Journal of Northeast Petroleum University, 2018, 42（1）: 68-76.

[3] Zhang Feifei. Characterization and evaluation of pore-throat structures in ultra-low permeability and tight sandstones and their impacts on petrophysical properties: a case study of the lower Cretaceous in northern Songliao Basin[J]. Journal of Northeast Petroleum University, 2018,42(5):54-62.

[4] Tong Kaijun, Zhao Chunming, Lv Zuobin, et al. Petroleum exploration and development[J]. Petroleum Exploration and Development, 2012,39(1):56-63.

[5] Lin Guangrong, Liu Xuegang, Lu Yan, et al. Impact of stress change on percolation pattern of two-phase in low permeability reservoir[J]. Special Oil & Gas Reservoirs, 2008,15(1):43-45.

[6] Yang Bin, Hou Dali, You Gang, et al. Research and application of stress sensitive experiment on low permeable reservior[J]. Reservoir Evaluation and Development, 2012,5(2):41-45.

[7] Cao nai, Lei Gang. Stress sensitivity of tight reservoirs during pressure loading and unloading process[J]. Petroleum Exploration and Development, 2019,46(1):132-138.

[8] Jiang Ruizhong, Huang Lei, Cui Yongzheng, et al. Transit pressure analysis of a horizontal well in fractured composite reservoirs with stress-sensitivity effect[J]. Journal of Northeast Petroleum University,2018,42（3）: 84-91.

[9] Fang Yang. Research on relationship between deliverability of gas reservoir and the stress-sensitiveness of the reservoir[J]. Well Testing,2009,18（6）:18-21.

[10] Fu Lanqing. The effect of fractures on stress sensitivity and seepage characteristics in sandstone reservoirs[J]. Journal of Yangtze University（Natural Science Edition）,2016,13（20）: 14-18.

[11] Zhang Rui, Ning Zhengfu, Zhang Haishan, et al. New insights and discussions on stress sensitivity of fractured tight reservoir[J]. Natural Gas Geoscience, 2016,27(5):918-923.

[12] Wang Lezhi, Qi Zhilin, Jiang Yiwei. Study on retrograde condensate characteristics of deep condensate gas reservoirs with low permeability [J]. Natural Gas Technology, 2007, 1（1）: 52-54.

[13] Xiao Lixian, Du Jianfen, Guo Ping, et al. Pollution caused by retrograde condensate in condensate gas reservoir[J]. Fault-block Oil & Gas Field, 2009, 16（4）: 102-104.

[14] Cheng Liang, Zhang Xiaodong, Wang Juan, et al. Study on periodic gas injection mechanism of condensate gas reservoirs with high to high-super content of condensate oil in central Tarim by numerical simulation[J]. Journal of Xi'an shiyou University (Natural Science Edition),2016,31(2): 57-64.

[15] Wang Jing, Liu Huiqing, Liu Renjing, et al. Numerical simulation for low-permeability and extra-low permeability reservoirs with considering starting pressure and stress sensitivity effects[J]. Chinese Journal of Rock Mechanics and Engineering, 2013,32(2):3317-3327.

[16] Li Yong, Zhang Jing, Wu Xueyong, et al. A new reservoir simulation approach for fractured gas-condensate reservoirs[J]. Petroleum Exploration and Development, 2010,37(5):592-595.

[17] Jiang Hai, Yang Zhaozhong, Hu Yuehua, et al. Optimization of the cyclic gas injection parameters in condensate gas reservoir[J]. Journal of Xi'an shiyou University (Natural Science Edition),2009,24（2）:54-56.

610412 Development of a Cost-Effective and Modular Closed Loop Process for Methanol Production in Wastewater Treatment Plants

Abstract:

Wastewater treatment plants (WWTPs) are under increasingly strict regulations to limit nitrogen and phosphorous in effluent discharge since they can have deleterious effects on aquatic species. In WWTPs, ammonia in wastewater is converted to nitrites, then nitrates and finally nitrogen in biological nitrification-denitrification processes. Methanol is commonly used as an external organic carbon source for denitrification since it is readily available, relatively cheap, and residual methanol can be removed from the treated effluent by aeration. Denitrification using methanol exhibits high potential removal efficiency, high process stability and reliability, relatively easy process control, low land area requirement, and moderate cost [1]. However, high competing demand for methanol from utility chemicals manufacturing can result in ever increasing cost of methanol for denitrification. Additionally, safety concerns related to methanol storage and handling facilities might preclude the purchase and transport of commercial methanol to WWTPs, highlighted by the ban of methanol use for denitrification in New York City in the late 2000s.

The presence of anaerobic digesters in WWTPs offers an opportunity to use biogas as a feedstock for on-site methanol production. Biogas, a mixture of 50-70% methane, 30-40% carbon dioxide, along with sulfur and trace elements, is used only in about half of WWTPs in the United States with digesters, for digester heating and combined heat and power [2]; in the rest of the cases, biogas is flared. On-site conversion of biogas to methanol has the potential to mitigate risks from cost and supply volatility by reducing the dependence on external carbon sources, address safety concerns and reduce the overall carbon footprint of the WWTPs [3]. However, traditional methanol synthesis processes from steam-methane reforming-based syngas ($CO+H_2$) are not viable since they cannot be scaled-down economically for use at WWTPs. This provides an opportunity for the development of a cost-effective and modular closed-loop process for on-site production of methanol at WWTPs.

RTI has developed and demonstrated a cost-competitive, internal combustion engine-based, impurity-tolerant syngas generator at near-deployment scales. RTI's pilot plant processes 50,000-60,000 standard cubic feet per day (SCFD) natural gas to produce syngas in a partial oxidation process. We have also integrated methanol production, with a 15% exhaust slipstream producing 1 barrel/day (BPD) of crude methanol. Instead of conventional process equipment that rely on economies of scale, the engine-based process uses economies of mass manufacturing to lower costs. The production of crude (75vol%) methanol alleviates safety concerns of handling, storage, and use. This paper will present technoeconomic evaluation of integrating the engine-based modular methanol synthesis technology in WWTPs for cost-effective on-site production.

A detailed biogas-to-methanol conversion process, including biogas cleanup, syngas production, conditioning and compression, and methanol synthesis sections, was simulated using Aspen HYSYS chemical process simulation software. Data from the operation of the engine and methanol synthesis pilot plant was used for reconciliation and validation of model performance. Several plant configurations were studied to optimize the biogas to methanol conversion process. Heat and mass balances, utility requirements, auxiliary power loads and equipment specifications were generated to calculate operating cost estimates. ASPEN Process Economic Analyzer (APEA) and vendor quotes were used to generate capital cost estimates. WWTP flow capacities determine the amount of external carbon needed and the biogas potential. Our analysis shows the engine-based methanol production process can meet methanol demand (~9 BPD) for WWTPs as small as 5 million gallons per day (MGD) capacities. Methanol costs are expected to be at par with commercial prices [4] from large-scale production at a 20 MGD scale, and much lower for larger WWTPs, up to 20% lower for a 30 MGD WWTP as an example. Our analysis indicates that the engine-based methanol production system is ideal for medium-sized WWTPs from 10-50 MGDs, with short payback periods and high returns on investment.

References

[1] H. H. Kung, MethanolProduction and Use Chemical Industries ; V. 57, 1994.

[2] United States Environmental Protection Agency, "Digesters at Water Resource Recovery Facilities," [Online]. Available:https://www.epa.gov/anaerobic-digestion/types-anaerobic-digesters#WRRFdigesters. [Accessed April 2020].

[3] S. K. C. Yu-Chen Su, "Bioaugmented methanol production using ammonia oxidizing bacteria in a continuous flow process," *Bioresource Technology*, pp. 101-107, 2019.

[4] Methanol Institute, "Methanol Price and Supply/Demand," March 2020. [Online]. Available: https://www.methanol.org/methanol-price-supply-demand/.

599047 Determining Materials Specifications Based on Technoeconomic and Life Cycle Assessment: A Case on Oxygen Career Material for Chemical Looping Steam Methane Reforming

Abstract:

Development of new process is one of the keys to achieve low-carbon production in the chemicals industry. Many new processes being studied rely on use of novel materials such as catalysts and chemicals careers. Since there are numerous possibilities of materials that can be used, it is important that a target-based development of these materials and processes are practiced for efficient process development with short lead time.

In this study, to delineate the feasible materials and their performance under operating conditions and a reactor of choice, process systems modeling, followed by assessment of economic and life cycle greenhouse gases emission is performed in an integrated manner. A novel chemical looping process for steam methane reforming (CL-SMR) as a part of methanol production is used as an example to demonstrate this approach. SMR is an endothermal reaction to produce CO and H2, the raw materials for commodity chemicals (ex. methanol). Conventional SMR is controlled at 700 - 900 °C to achieve a desired conversion, which not only leads to high energy consumption for supplying heat but also incurs expensive reformer equipment. CL- SMR is an attractive alternative, dividing the reaction system into two reactors where metal and its oxides is circulated among the reactors to carry oxygen atoms using two fluidized bed reactors connected with each other. In the steam reactor, H2O is reduced to yield hydrogen while oxidizing the metal into an oxide form, while in the methane reactor, CH4 is oxidized to produce carbon oxides (CO, CO2) and H2. In this way, the overall conversion is no longer bound by their original equilibria, allowing the operation at a significantly lower temperature while maintaining a sufficient conversion.

Using the process simulation combined with economic calculation and assessment of life cycle GHG emission, we present the target region for the career materials on a space defined by oxygen storage capacity and oxygen release rate in the methane reactor.

By presenting the simulation results, target temperatures and conversion, as well as the oxygen storage ccapacity and oxygen release rate were discussed.

610626 Minimization of Capital and Operating Costs in Cooling Water Distribution Systems

Abstract:

Cooling water is the fluid most commonly used in heat exchangers for cooling processes. The most widely used concept for cooling water semi closed loops is the one that a pumping system supplies water from the cooling tower basin through a header to branches where heat exchangers are located, and after heat absorption in these exchangers, the water returns from these branches through another header to the cooling tower top to reject the absorbed heat. The use of this kind of concept results in a reasonably simple design procedure for the cooling water system, but it can yield high electrical energy demands. As good engineering practice, the pressure drops in heat exchangers must not exceed 0.5 to 1.0 bar, but there are cases that this pressure drop may be higher than the recommended values. Cooling water consumers distant, above 500 m, from the cooling tower also impose high pressure drops in the control valves of the branches closest to the main pumping system. To avoid a short circuit, branches with heat exchangers with low pressure drops or near the cooling tower may have control valves that impose additional pressure drops and, therefore, mechanical energy losses. As a result, the pumping system must compensate these energy losses by increasing the consumed electrical energy.

Alternatives are analyzed to reduce the system total electrical energy consumed. Among these alternatives are the use of auxiliary (booster) pumps in the branches with the highest head losses, the installation of secondary headers, and the use of separate cooling towers. On the other hand, these alternatives result in an increase of capital costs. The mathematical modeling of the pumping system allows the estimate of system mechanical energy demands as a function of the presence of booster bumps and secondary headers. The synthesis of the cooling water distribution system must minimize the overall costs.

The optimal synthesis of the cooling water system shall determine the values for the decision variables of the system model that minimizes the capital and operating costs. Among the model decision variables are: capacity of the cooling water main pumping system; number of pumps in the cooling water main pumping system; capacity and location of the cooling water booster pumps; layout of the cooling water supply and return headers; required minimum pressure drop through control valves. An existing cooling water distribution system is selected as study case, so the optimal synthesis solutions can be compared to the existing one.

A sensitivity analysis is performed varying the design parameters such as cooling water total and individuals demands, cooling water consumers' locations, heat exchangers pressure drop and cooling tower elevation. Evidently, these project parameters have a direct impact in the cooling water distribution system and the resulting cost. As a result, a methodology is established so it can be applied to the detailed design of the cooling water distribution system.

609711 Economic Evaluation of Wastewater Treatment Plant Construction in Small Communities

Abstract:

Wastewater collection and treatment systems are an integral part of communities. In the life of a town, its investment in wastewater collection and treatment systems is hugely important for attracting industry, protecting the environment, and conforming to state law. Through this work we seek to understand the optimal size of a town for construction of a wastewater collection and treatment system.

We utilize data covering wastewater treatment plants in small towns across Oregon to estimate parameters utilized in a series of costing models. Using several capital cost estimating techniques, we identify the optimal size of town for capital investment in wastewater collection and plant systems.

We then compare early, late, and optimal construction timing with status quo (septic) setups to gain a better understanding of the consequences of ill-timed investment. Through the cost models, we analyze a number of different configurations for wastewater treatment plants including utilizing different machinery and perform sensitivity analysis on major parameters to gain an understanding of the uncertainties in our model.

Lastly, we utilize our model setup to observe several real cities amid wastewater collection and plant investment controversy to see the consequences in practice of regulatory and investment decisions. We quantify some of the projected costs associated with those communities and compare those to costs under ideal timeliness of investment.

611335 Invited Talk: Understanding Radiative Transport in Flowing Particles Via Monte Carlo Ray Tracing and Data-Driven Techniques

Abstract:

Particle-laden flows comprising assemblies of individual grains/agglomerates or fluidized gas-solid flows have widespread applications in particle-based concentrated solar power thermal energy storage systems[1–4]. At high temperatures, the contribution of radiation to the overall energy transport intensifies. However, quantitative predictions and mechanistic insights to evaluate the coupled impacts of the flow regimes, underlying morphology, and optical properties on radiative transport, and therefore the overall heat transfer behavior are lacking in the literature. To fully resolve radiative energy transport in particle-laden multiphase flows, the challenge, in part, is to compute absorption, emission, and scattering of radiative energy as a function of dynamically evolving particle positions. In this talk, I will highlight two ongoing research activities to address these challenges. First, we investigate the influences of spatial correlations of a packed-bed of particles on radiative intensity transport . Spatial distributions of the particles were statistically quantified and radiation-particle interactions were calculated by performing collision based Monte-Carlo ray tracing simulations. Results from this study indicated that the transmittance is strongly influenced by the spatial correlation of particles in the medium. Next, we leverage results obtained from the fully resolved ray tracing simulations to inform the development of data-driven correlations for the predictions of view factors using least-square regression and Gaussian kernel regression techniques. Three distinct regimes were identified for the dependence of the particle-particle view factors on the dimensionless distance between particles—a monotonic decay regime for small distances, a shading-dominated regime for medium distances, and a large-distance regime, where the view factors have negligibly small values. Results from this study reveal the potential for data-driven techniques to be applied to deduce correlations that are computationally more efficient to model radiative transport and overall heat-transfer, especially for flowing particle systems.

References

1. Ho, C. K. *et al.* On-sun testing of an advanced falling particle receiver system. in 030022 (2016). doi:10.1063/1.4949074
2. Ma, Z. & Martinek, J. Fluidized-Bed Heat Transfer Modeling for the Development of Particle/Supercritical-CO 2 Heat Exchanger. in *ASME 2017 11th International Conference on Energy Sustainability* V001T05A002 (ASME, 2017). doi:10.1115/ES2017-3098
3. Ge, W. *et al.* Discrete simulation of granular and particle-fluid flows: from fundamental study to engineering application. *Rev. Chem. Eng.* **0**, (2017).
4. Mehos, M. *et al. Concentrating Solar Power Gen3 Demonstration Roadmap.* (2017).

609626 Through-the-Glass Thermal Imaging of Solar Cavity Receivers

Abstract:

Solar cavity receivers lie at the heart of many high-temperature solar thermal processes. Measurement of the temperature distribution inside the cavity is of critical important to determining the efficiency and successful operation of the receiver. Traditional contact thermometry approaches, e.g. using point-measurement thermocouples, can be costly and challenging at high temperatures, and cannot provide a high-resolution map of the temperature distribution inside a solar cavity receiver. Non-contact approaches, e.g. infrared thermography, allow for convenient spatial mapping without the need to physically touch the high temperature process. However, the application of infrared thermography to solar cavity receivers has been severely limited by the fact that many receivers utilize a glass front window which is opaque to thermal radiation, rendering infrared thermography of the cavity's inner surfaces ineffective.

In this work, we present the design, construction and successful demonstration of a thermal camera capable of seeing through glass. This "through-the-glass" thermal camera utilizes a high-resolution grayscale CMOS optical sensor equipped with a narrow bandpass filter with a center wavelength of 772 nm (which is near the transmission peak of many optical glasses), to effectively probe the monochromatic spectral radiance of any high temperature surface. Being sensitive to short-wave photons, our device can effectively see through glass and similar optical glazings, allowing direct contactless measurement of the inner surfaces of a cavity receiver. By utilizing a standard optical CMOS sensor, the device achieves much higher resolution at a much lower cost than can traditional thermal cameras operating in the mid-infrared. An additional advantage of the short-wave approach is that the device is also impervious to any radiatively participating media present inside the cavity, making it a very flexible measurement technique.

Using a calibrated photometry sphere with a known spectral radiance, we experimentally determined the device sensitivity (grayscale value per photon per second), which allows the spectral radiance and ultimately temperature to be determined from a simple grayscale image. A thermal blackbody operating at above 1000 °C was used as a secondary calibration and a thermal validation of the photometric calibration. We then successfully used our device to map the temperature distribution inside a solar cavity receiver operating at temperatures well in excess of 1000 °C. The results serve as an experimental proof-of-concept for through-the-glass contactless thermal measurement of solar cavity receivers. We believe this low-cost and flexible technique will become a staple in characterization of high-temperature solar receivers, and may well find utility in many other high temperature industrial applications.

611303 Robust and Spectrally Selective Aerogels for Next-Generation Concentrated Solar Thermal Conversion

Abstract:

Although silica aerogels have led to improvements in solar collection efficiency at low to intermediate temperatures, they are plagued by low efficiency and performance degradation at temperatures necessary for next-generation concentrated solar thermal (CST) plants (>700 °C). The low efficiency of silica aerogels at high temperatures is a result of large thermal re-radiative losses through their infrared transparency window between 3 – 5 microns. Furthermore, previous work shows that exposure to such temperatures decreases the spectral selectivity (i.e. increased infrared transparency) and increases thermal conductivity of silica aerogels resulting in even lower receiver efficiency. Here, we investigate multicomponent aerogels which exhibit increased spectral selectivity and mitigate performance degradation at 700 °C. Specifically, we stabilize the aerogels at elevated temperatures by applying an ultrathin coating of solar-transparent refractory oxides via atomic layer deposition (ALD). Interestingly, our multicomponent aerogels absorb strongly within the aerogel's infrared transparency window, leading to >50% reduction in thermal losses at 700 °C without sacrificing high solar transmission. We characterize the effects of the ALD coating on the optical and thermal properties of the aerogels using UV-Vis-NIR, ellipsometry, FTIR, and hot-disk techniques. Additionally, we report on our methods to develop a uniform and conformal ALD coating by modeling the diffusion and reactant requirements of ALD on aerogels. Lastly, we compare the experimentally determined receiver efficiencies with our simulations of coupled radiative and conductive heat transfer. Our results reveal that ultrathin, conformal ALD coatings not only slow down the rate of efficiency degradation, but significantly improve the performance of aerogels in high temperature CST solar receivers.

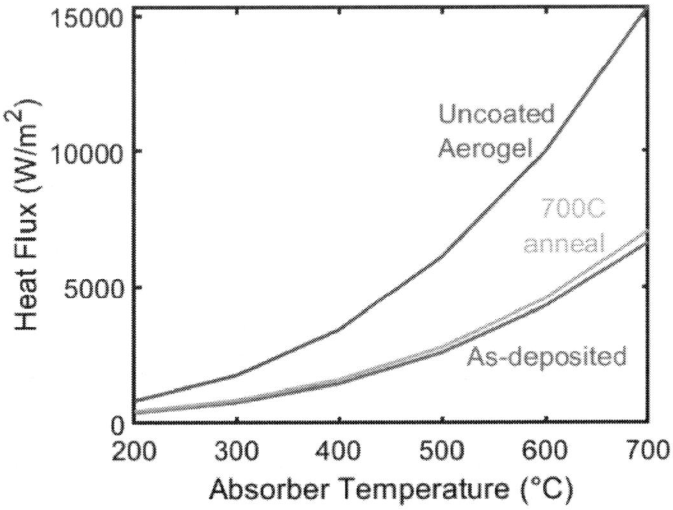

Modeling of a Small-scale Solar Central Receiver with Pressurized Water for Heat Collection and Storage – 612g

[a] **Ahmed Gamil**, [a] **Peiwen Li** *, [b] **Roger Angel**

[a] Department of Aerospace and Mechanical Engineering

[b] Department of Astronomy and Optical Sciences

University of Arizona, Tucson 85721, United States

*Email: peiwen@arizona.edu

ABSTRACT

This paper presents a new modeling and algorithm to analyze and optimize a Central Receiver System (CRS) for solar thermal energy collection and storage. The analysis includes associated modeling of the heliostat field and the heat transfer at the receiver to couple the energy in the two subsystems. Transient optical performance of the heliostat field and predicted heat flux at the receiver is provided. The modeling analyzed a case which is a medium temperature solar concentrating system with pressurized water as heat transfer fluid and thermal storage medium. A tubular external solar receiver is considered and the modeling determines the geometry to maintain a best thermal efficiency. The transient optical performance and heat flux mapping are conducted using Monte Carlo Ray Tracing (MCRT) technique. The new receiver modeling and approach of optimization design was verified through analysis of two reference solar thermal power plants with molten salt heat transfer fluid (i.e. SolarTwo and GemaSolar Plants) with a satisfactory agreement to the reported design parameters. The computations and algorithm was implemented in a use-friendly interface via Visual Basic. The modeling and algorithm provides an efficient tool for solar thermal power plant design in industry.

Keywords: Solar Power Tower (SPT), Central Receiver System (CRS), Heliostat Field.

INTRODUCTION

The ever growing energy demand and the requirement of environmental protection makes it contingent to adopt more and more alternative energy and reduce the use of fossil fuels and the associated CO_2 emissions. Solar energy is the most readily available renewable energy. Compared to solar photovoltaic technology, Concentrating Solar Power (CSP) has the advantage of being able to store large quantity of thermal energy for extended power generation in days or weeks.

CSP systems are typically classified by how the sunlight is concentrated for thermal energy. The four widely known systems are illustrated in Fig. 1, which include Fresnel Reflector (FR), Solar Power Tower (SPT), Solar Dish (SD), and Parabolic Trough (PT). The concentration ratios of these technologies largely decide the maximum temperature of the heat transfer fluid that is used to take and carry the thermal energy to thermal-to-electrical energy conversion system and thus determines the energy conversion efficiency.

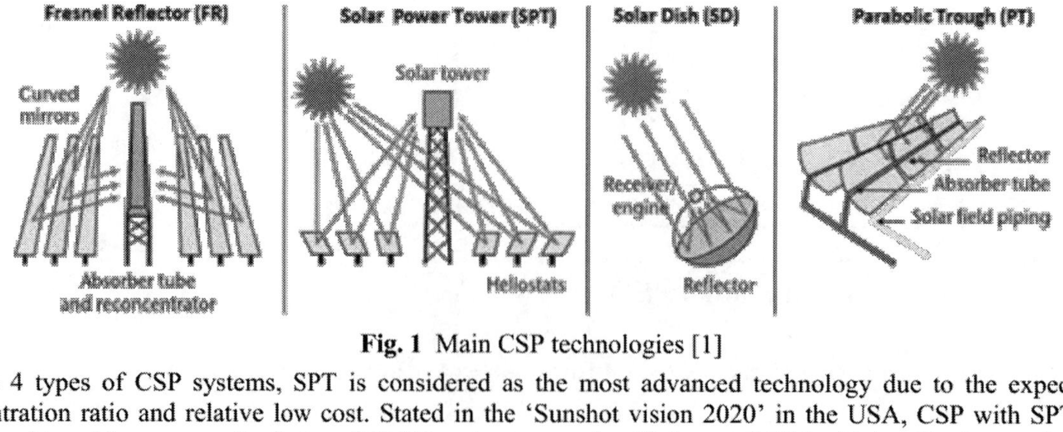

Fig. 1 Main CSP technologies [1]

Of the 4 types of CSP systems, SPT is considered as the most advanced technology due to the expected high concentration ratio and relative low cost. Stated in the 'Sunshot vision 2020' in the USA, CSP with SPT will be developed to achieve a capacity factor of 69.1% and power cycle efficiency of 55% with levelized cost of electricity

(LCOE) of USD 0.05/kWh [2]. SPT is designated to receive concentrated sunlight and transfer the heat to a fluid called Heat Transfer Fluid (HTF). The HTF can then deliver heat for various purpose of use, such as, power generation, industrial heating, desalination, and enhanced oil recovery and refineries in oil and gas industry. A SPT-based CSP system consists of heliostat field, a tower receiver, storage system, power conversion system, and control system. The heliostat field and the receiver system are major subsystems that determine the overall efficiency and the cost of the system [3], [4]. This fact highlights the importance of a correct estimation of the receiver thermal efficiency to avoid oversizing the heliostat field and thereby caused unnecessary cost of the systems.

The modeling studies about central receivers has been done for molten salt as the HTF in cavity receivers in large scale applications. Design of the cavity receivers has been an important topic. There is no sufficient work on research and optimum design for cylindrical receivers especially using water as HTF [5], [6].

The performance of a SPT-based CSP system heavily depends on the efficiency of the receiver because it is used to collect the heat received from the solar field to the working fluid. In addition, the receiver works at a medium to high temperature environment and is subjected to different heat losses. The energy input for the receiver is stemmed from the heliostat field, however it is equally important to model the heat loss and receiver performance due to different effects such as the tower diameter and height, as well as the field size.

The current research focuses on the analysis of thermal performance of a cylindrical external receiver, which has pressurized water as HTF. An algorithm to couple the energy analysis and optimization of receiver and heliostat field has been established and implemented in a Visual Basic computer code. As the results of the analysis, the provided parameters include: the optimal geometry of the receiver (height, diameter, tube diameter, number of panels, tubes and headers), the operational parameters, such as flow velocity, surface temperature, heat flux, pumping power, and field geometry (height, size, and number of heliostats), and efficiencies of both heliostat field and receiver.

ALGORITHM FOR OVERALL SYSTEM ANALYSIS AND OPTIMIZATION

Fig. 2 shows the flow of energy from solar radiation to thermal energy being available to the load. First, most of the incident sunlight reflected from the heliostat field is absorbed by the receiver surface (assuming that mirrors are perfect and the tracking errors are minimal). In this process, some energy is dispersed to the surroundings by means of optical loss, and heat loss occurs at the surface of the receiver due to convection and radiation to the ambient air and surroundings. The energies in the process are represented by terms: Q_{DNI} , the Direct Normal Irradiance (W/m^2), Q_{solar}, the intercepted solar power (W) by heliostat field , Q_{th_rec}, the incident thermal power to receiver (W), Q_{htf}, the thermal power absorbed by HTF (W) in the receiver, Q_{Load} , the desired heating load (W). The proposed model and algorithm will analyze the energy balance and in the meantime to optimize the designs of the heliostat field and the receiver for the best of energy efficiency of the entire system.

In the algorithm the computation for optimization of the entire system starts from a prescribed power demand (thermal or electrical) by the load. A thermal efficiency of the receiver is assumed to decide the total thermal energy that needs to get by the receiver, which leads to the determination of the solar field later. The heat transfer fluid temperatures at the inlet and outlet of the solar receiver are prescribed, which leads to the determination of the flow rate of the HTF, the number of tubes in the solar receiver, the lengths of the tubes, and thus the dimension of the receiver. The analysis of the heat transfer in the tubes and the heat loss at the receiver allows the determination of the thermal efficiency of the receiver. If the new determined thermal efficiency of the receiver converges to the same value as assumed at the beginning of the analysis, the design for the receiver is complete; otherwise the thermal efficiency of the receiver needs to be updated for a new round of computation. It is noted that because of the update of the thermal efficiency of the receiver, the total thermal energy that needs to get by the receiver is also updated until a convergence is achieved.

For the solar receiver to get the required amount of thermal energy determined in the above stage of analysis, the heliostat field has to be designed. The optimal design of the heliostat field will consider optical loss, blockage of incident and reflected sunlight. The parameters for the solar receiver, including the total thermal energy, elevation of the receiver from the ground, the height and diameter of the receiver are the input for the determination of the solar heliostat field. Multiple cases of the layout of the heliostat fields were trail-and-error optimized to determine the needed number of heliostat and their layout pattern.

The above discussion covers two parts of optimization analyses, receiver design and heliostat field design, which are coupled with each other because of the energy demand. The coupling and the detailed computation in each part of optimization are illustrated in Fig. 3.

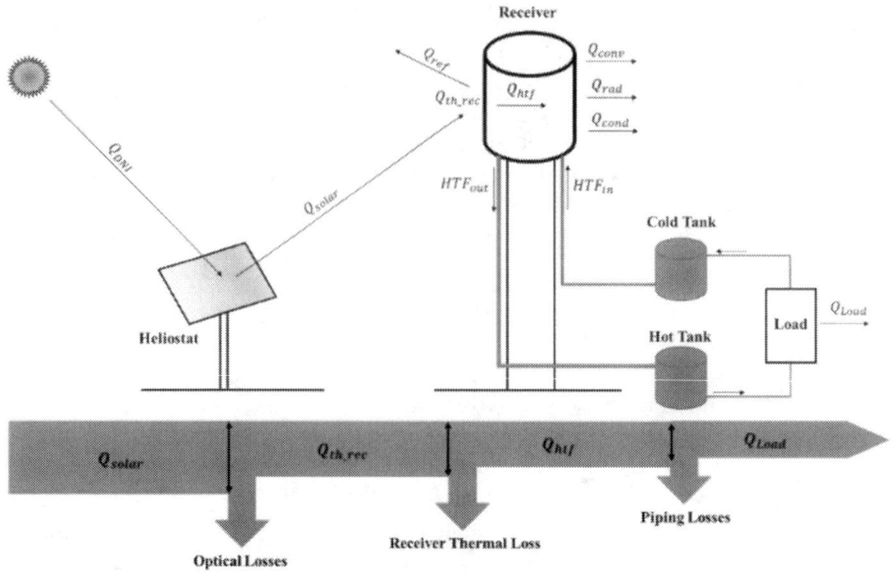

Fig. 2 Illustration of energy flow of the solar power plant

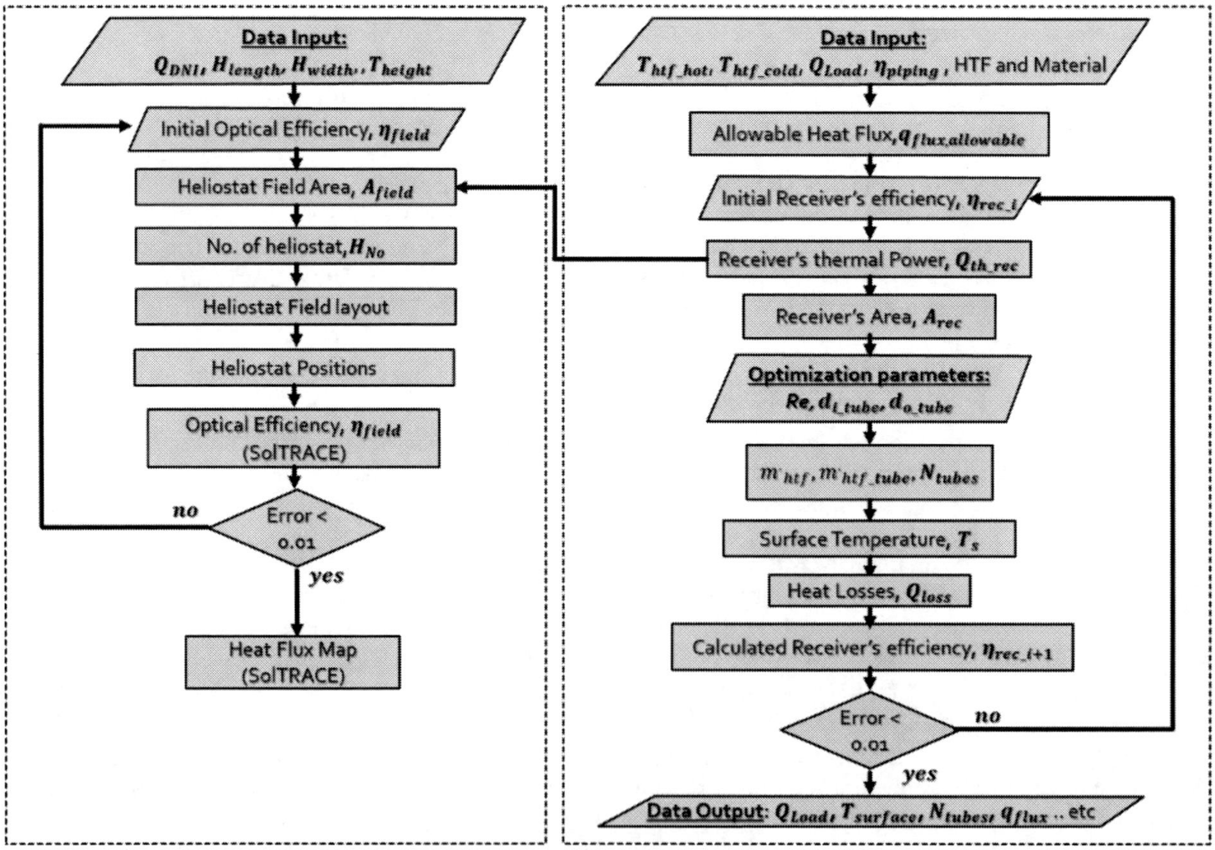

Fig. 3 The flow chart of integrated optimization computation for solar receiver and heliostat field

The following discussion will elaborate the steps of computations for receiver design, heliostat field design, and the coupling of the two designs.

1) An initial assumption to the receiver's thermal efficiency η_{rec_i} is made in order to decide the needed incident thermal power Q_{th_rec} which will provide the required load of Q_{Load} after counting any losses. The energy for the load can be given as the electrical power which derives the need of thermal energy, or directly thermal power, depends

on the application of the collected thermal energy. With the known thermal energy load to be taken by the heat transfer fluid, the flow and heat transfer in the receiver can be analyzed, which will allow the calculation of the receiver's thermal efficiency which is compared to the initial assumption.

2) Analysis and calculation of the flow and heat transfer in the receiver to meet the thermal energy load demand. To accomplish this, properties of heat transfer fluid (HTF) and materials for the heat transfer tubes are needed. The HTF inlet and outlet temperatures, $T_{htf,cold}$ and $T_{htf,hot}$ are pre-determined from the view point of utilization of the thermal energy. Therefore, the mass flow rate of HTF going to the entire receiver can be calculated based on the thermal power load, Q_{Load}, and the specific heat of the HTF, $C_{p,htf}$. It is noted that the fluid inlet and outlet temperatures at the receiver are decided and checked based on the power plant requirement as well as the limit of the maximum temperature of the fluid and the pipe materials. The average fluid temperature in receiver tubes is decided as $T_{htf,avg} = (T_{htf,hot} + T_{htf,cold})/2$, which is used in the heat transfer analysis. A large difference of inlet and outlet temperatures will lead to less flow rate going to the receiver, or vice versa, a small difference will lead to a large flow rate.

3) The allowable maximum heat flux on the receiver is checked based on the type of HTF and the maximum allowed temperature of the heat transfer tube material. Upon the determination of a safe heat flux, the geometry and dimensions of the receiver, in respect of height to diameter aspect ratio AR, as well as the overall receiver surface area can be obtained. It is noted that the receiver surface area is kept small to reduce thermal losses. Correspondingly, the flow Reynolds number and Nusselt number of the turbulent flow in heat transfer tubes are determined. The heat transfer tube thickness is determined considering the pressure of the fluid inside the tube.

4) Determine the number of heat transfer tubes in the receiver that the total mass flow being distributed. The cross-sectional area required for the fluid flow is calculated using the total mass flow rate, and thus the number of tubes in each panel and number of panels can be calculated by selecting the number of flow paths.

5) From the obtained receiver surface temperature (i.e. outer wall temperature of tubes), the heat loss from the surface of the receiver to the ambient air is calculated, and consequently the receiver thermal efficiency is obtained. If the new obtained receiver thermal efficiency agrees with the initial assumption, the iterative computation is converged. Otherwise, more iterative computations with updated receiver thermal efficiency is conducted until convergence (relative change of receiver thermal efficiency less then 0.01) is achieved.

6) Make an assumption for the initial optical heliostat field efficiency η_{field_i} that can feed the above determined thermal power (in steps 1 to 5) to the receiver. Find the size of the heliostat field based on the design DNI point. This requires determination of the number of total heliostats based on the input dimensions of the heliostat (i.e. length and width) and the elevation of the tower as well as the receiver height and diameter.

7) Configure the biomimetic heliostat field, and determine the individual heliostat position based on the angular and radial position of the Fibonacci's spiral formula. The field layout and determination of the optical efficiency will be calculated using MCRT via SolTRACE. Check and update for the optical efficiency in the calculation until the obtained optical efficiency converges at relative variation less than 0.01.

8) Several different values of the elevation of solar receiving tower will be simulated to find the best solar tower design with the highest field efficiency.

9) With all the design parameters determined, the transient local heat flux maps on the receiver will be examined to ensure that there is no extreme hot spot exceeding the limit of heat flux.

MATHEMATICAL MODELING FOR SUBSYSTEMS AND COMPONENTS

Receiver Incident Thermal Power

The receiver incident thermal power from the solar field, Q_{th_rec}, is determined based on the required load and the extra energy storage for extra period of time. The receiver incident thermal power is determined by:

$$Q_{th\_rec} = \frac{Q_{Load}}{(1-f_{pumping})\times\eta_{rec}} SM \tag{1}$$

where Q_{Load} is the desired heating load power (W); f_{puming} is the fraction of pumping power versus the total of the load power; η_{rec} is the receiver's thermal efficiency; and SM is the Solar Multiple. Solar multiply is a factor to count the X hours of extra energy storage when the operation time period with sunlight Y hours, which thus gives:

$$SM = \frac{X+Y}{Y} \tag{2}$$

For the present study, $Q_{th\_rec} = 1.752\ MW_{th}$ as a direct input to the model.

In the analysis, the current analysis, the HTF is liquid water, with the inlet temperature of water set as $T_{htf,cold}$= 80 ˚C @ 500 kPa, and the outlet temperature of water being set at $T_{htf,hot}$ = 150 ˚C @ 500 kPa. The pressure of 500 kPa makes the water at 150 ˚C compressed and being slightly subcooled. The HTF properties are based on the average fluid temperature, $T_{htf,avg} = \frac{T_{htf,hot}+T_{htf,cold}}{2}$, which is 115˚C in this case. The equations for water properties used in this work are adopted from [7].

Sizing the Solar Receiver

The receiver's diameter is determined through choosing the diameter of the solar receiving tubes and calculation of the number of tubes for a suitable flow rate of the heat transfer fluid. The receiver height is then determined based on a proper heat flux for a proper total surface area of the receiver. The proper heat flux for tubes is determined to avoid thermal failures and fatigue. Based on the HTF mass flow-rate, one can find the number of panels, and the number of tubes in each panel for the receiver.

Allowable Heat Flux

The allowable heat flux for the solar receiver is due to the limitation of the materials of the tube. The receiver sizing is then associated to the allowable flux, $q_{flux,allowable}$ in order to increase the lifetime of the receiver. The allowable heat flux, $q_{flux,allowable}$ can be calculated as:

$$q_{flux,allowable} = \frac{q_{flux,limit}}{R_{peak,avg}} \tag{3}$$

where $q_{flux,limit}$ is the material limited heat flux (W/m^2), $R_{peak,avg}$ is the limit-to-average heat flux ratio.

For instance the present study assumes 316 stainless steel as a receiver's material which has a limited heat flux of 600 kW/m^2 in terms of the thermal capacity of water [8], [9]. According to Stine and Harrigan [10], the limit-to-average flux ratio, $R_{peak,avg}$ can be between 2 to 3. In present study, a conservative value of $R_{peak,avg} = 2.3$ is chosen, and the allowable heat flux, $q_{flux,allowable}$ is 260 kW/m^2. The value of the flux on the receiver might have fluctuations during the day and the year, but an instantaneous heat flux shall never be over 600 kW/m^2.

Mass Flow Rate

The mass flow rate of the HTF going to receiver is calculated based on the determined thermal power, Q_{htf}. The total mass flow rate, $\dot{m}_{htf}$ is calculated by:

$$\dot{m}_{htf,total} = \frac{Q_{htf}}{c_{p,htf}(T_{htf,hot}-T_{htf,cold})} \tag{4}$$

where $\dot{m}_{htf,total}$ is the total HTF mass flow rate in the receiver (Kg/s), Q_{htf} (W) is the power absorbed by HTF, $c_{p,htf}$ is the specific heat (kJ/kg.K) of HTF, $T_{htf,hot}$ (˚C) and $T_{htf,cold}$ (˚C) is the HTF outlet and inlet temperatures, respectively. The power absorbed by HTF, Q_{htf} is calculated in the form of:

$$Q_{htf} = Q_{th,rec} \times \eta_{rec} \tag{5}$$

where η_{rec} is the receiver thermal efficiency.

The mass flow rate per tube is related to the selected flow Reynolds number (or flow regime) and the tube inner diameter, which gives,

$$\dot{m}_{htf,tube} = \left(\frac{1}{4}\pi\, di_{tube}{}^{2}\right).\rho_{htf}.v_{htf,tube} \tag{6}$$

where the velocity of HTF $v_{htf,tube}$ is determined as:

$$v_{htf,tube} = \frac{Re_{htf} \times \mu_{htf}}{\rho_{htf} \times di_{tube}} \tag{7}$$

Here, Re_{htf} is the Reynolds number, $\mu_{htf,tube}$ is the dynamic viscosity (Pa s) and $\rho_{htf,total}$ the density (kg/m^3) of the HTF, di_{tube} (m) is the inner diameter of the tube.

Determination of Receiver Geometrical Dimensions

The receiver outer geometry is important in defining the receiver's efficiency because it can affect the total heat losses associated to the surface area, A_{rec} , which is given by height, H_{rec}, and diameter D_{rec} of the receiver. For better performance, the aspect ratio (AR, height-diameter ratio) of a cylindrical external receiver is selected, which lies between 1 to 2 [11], [12] and should be determined for maximum thermal efficiency. The opted aspect ratio in this study is 1.5.

Using the allowable heat flux, the area of the receiver, A_{rec} (in m^2) can be calculated by:

$$A_{rec} = \frac{Q_{th,rec}}{q_{flux,allowable}} \tag{8}$$

The height and diameter of the receiver can be determined based on the area of the receiver, A_{rec} , which gives:

$$\pi D_{rec} H_{rec} = A_{rec} \tag{9}$$

The receiver height can be expressed as a function of the receiver diameter multiplying the height-diameter aspect ratio in Eq. (10). Bring Eq. (10) into Eq. (9), the height of the receiver is calculated.

$$H_{rec} = AR \cdot D_{rec} \tag{10}$$

The receiver inner and outer diameters are related to number of flow tubes that HTF is distributed at the manifold, as well as how many flow paths are adopted. One flow path is made in one panel of tubes that shares an inlet manifold and an outlet manifold. Since D_{rec} is determined from Eq. (10), the number of flow turns for each tube can be determined.

The tube outer diameter usually varies between 20 mm to 45 mm [13]. The flow velocity of HTF is given in Eq. (7) as already discussed. The flow tube thickness depends on the pressure of the fluid in the tube. For instance, direct water/steam receivers needs thicker tubes to withstand the higher pressure. In case of molten salt receivers, it can be as thin as 1 mm [14].

Wagner [15] suggested that the total flow may be distributed to one or two manifolds to form one or two flow branches (N_b). This will lead to the determination of the number of tubes controlled by one manifold, as well as number of tube cluster/panel. When viewing a horizontal cross-section of the receiver, the total number of tubes, $N_{tube/rec}$, on a receiver can be determined.

$$N_{tube/rec} = \frac{\pi \cdot D_{rec}}{do_{tube}} \tag{11}$$

where D_{rec}, is the outer tube diameter (m), do_{tube} is the outer tube diameter (m). The number of tubes per panel, $N_{tube/panel}$, can be calculated with the total mass-flow rate and mass flow rate in each tube in the form of

$$N_{tube/panel} = \frac{\dot{m}_{htf,total}/N_b}{\dot{m}_{htf,tube}} \tag{12}$$

where $\dot{m}_{htf,total}$ is the total HTF mass flow rate in the receiver (kg/s), $\dot{m}_{htf,tube}$ is the mass flow rate per tube (kg/s).

Once the number of tubes per panel is calculated, the total number of receiver panels, N_{panels} can be calculated as:

$$N_{panels} = \frac{N_{tube/rec}}{N_{tube/panel}} \tag{13}$$

It should be noted that the number of tube panels is even if two branches of flow are adopted.

Receiver Surface Temperature

The receiver's surface temperature, T_s depends on the amount of incident thermal power on the receiver and the temperature of the HTF. It is determined by calculating the resistance due to conduction and convection by the following equation, assuming that the surface temperature along the receiver tubes is uniform.

$$T_s = Q_{htf} \times R_{total} + T_{htf,avg} \tag{14}$$

where Q_{htf} is the power absorbed by HTF (W), $T_{htf,avg}$ is the average temperature (K) of HTF, R_{total} is the total thermal resistance, (K/W).

The total thermal resistance, R_{total} for a tube is the sum of the conduction resistance, $R_{cond,tube}$ between the outer and inner surface of the tube and the convection resistance, $R_{conv,htf}$ between the inner wall of the tube and the HTF. Therefore, the total thermal resistance (shown in Fig. 4), R_{total} is calculated by:

$$R_{total} = R_{cond,tube} + R_{conv,htf} \tag{15}$$

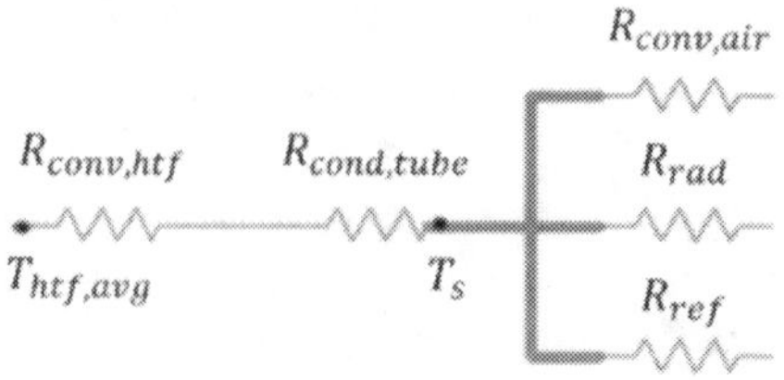

Fig. 4 Thermal resistance circuit diagram for a solar receiving tube

The resistance due to conduction through the tube walls, $R_{cond,tube}$ is given as,

$$R_{cond,tube} = \frac{ln(r_{o,tube}/r_{i,tube})}{2\pi \cdot H_{rec} \cdot k_{tube} \cdot N_{tube/rec}} \tag{16}$$

where $r_{o,tube}$ (m), and $r_{i,tube}$ (m) are the outer and inner radius of the receiver's tube, respectively. $N_{tube/rec}$ is the total number of tubes in the receiver, k_{tube} is the thermal conductivity (W/m·K) of tube material.

The resistance due to internal convection between tube wall and the HTF, $R_{conv,htf}$ can be given by:

$$R_{conv,htf} = \frac{1}{h_{inner} \cdot \pi \cdot r_{i,tube} \cdot H_{rec} \cdot N_{tube/rec}} \tag{17}$$

The internal convective heat transfer coefficient between tube wall and the HTF, h_{inner} is determined as:

$$h_{inner} = \frac{Nu_{htf} \cdot k_{htf}}{d_{i,tube}} \tag{18}$$

where k_{htf} is the thermal conductivity, W/m·K of HTF.

The correlation by Dittus-Boelter [16] gives the Nusselt number as:

$$Nu_{htf} = 0.023\, Re_{htf}{}^{0.8} . Pr_{htf}{}^{0.4} \tag{19}$$

which is applicable at conditions of $0.7 < Pr < 120$ and $10^4 < Re < 1.2 \times 10^5$.

Receiver Thermal Efficiency

The thermal efficiency of solar receiver is determined as a function of the heat loss and the total thermal energy that reaches to the receiver surface:

$$\eta_{rec} = 1 - \frac{Q_{loss}}{Q_{th\_rec}} \tag{20}$$

The heat loss model includes heat loss due to reflection, external convection and radiation. The heat conduction between the receiver body and panels is ignored. The total heat loss from the receiver, Q_{loss} is calculated by:

$$Q_{loss} = Q_{conv} + Q_{rad} + Q_{ref} \tag{21}$$

where Q_{conv} is the heat loss due to external convection (W), Q_{rad} is the heat loss due to long-wave thermal radiation (W), Q_{ref} is the heat loss due to reflection (W).

Energy Loss due to Convection

The convective heat loss from the receiver surface to ambient air is calculated that:

$$Q_{conv} = h_{mix} . A_{rec} . (T_s - T_{amb}) \tag{22}$$

The heat transfer coefficient h_{mix} is determined by the following correlation [17]:

$$h_{mix} = (h_{nat}{}^m + h_{for}{}^m)^{1/m} \tag{23}$$

where h_{nat} is the heat transfer coefficient due to natural convection (W/m^2.K), h_{for} is the heat transfer coefficient due to forced convection (W/m^2 · K) in wind. The value m lies between 3 and 4 [17], where a value close to 3 suits better for vertical surfaces and larger values are suitable for horizontal surfaces. For the present study, a value of 3.2 is opted for m in Eq. (23).

Natural convection

The natural convection heat transfer coefficient h_{nat} in Eq. (23) can be calculated with the following equation:

$$h_{nat} = \frac{Nu_{nat} \cdot K_{air}}{H_{rec}} \quad (24)$$

where Nu_{nat} is the Nusselt number of natural convection, K_{air} is the thermal conductivity of air (W/m.K). In natural convection analysis, Siebers and Kraabel [18] correlation can be used to determine Nu_{nat} .

$$Nu_{nat} = 0.098 \,.\, Gr^{1/3} . (\frac{T_s}{T_{amb}})^{-0.14} \quad (25)$$

Forced convection

The forced convection loss occurs due to the influence of wind, which is determined as:

$$h_{for} = \frac{Nu_{for} . k_{air}}{D_{rec}} \quad (26)$$

where Nu_{for} is the Nusselt number for forced convection (W/m^2.K).

Siebers and Kraabel correlation for Nusselt number is again used due to its validity in wide range of Reynolds number. The dry air thermos-physical properties is calculated at the film temperature $T_{film} = \frac{T_s+T_{amb}}{2}$ [19].

Energy Loss due To Radiation

The radiation heat loss Q_{rad} occurs due to infrared radiation emitted from the receiver's surface, which is considered based on an assumption that the surface is a grey body.

$$Q_{rad} = \sigma \,.\, \epsilon_{eff} \,.\, A_{rec} \,.\, ({T_s}^4 - {T_{amb}}^4) \quad (27)$$

where σ is the Stefan-Boltzmann constant (5.670×10^8 W/m^2.K^4), ϵ_{eff} is the effective emissivity of the receiver, A_{rec} is the receiver's surface area (m^2), T_s is the surface temperature (K), T_{amb} is the ambient temperature (K). Taking into account the tubes curvature, the effective emittance is given by the following equation [20]:

$$\epsilon_{eff} = \frac{\epsilon_{rec}}{\epsilon+(1-\epsilon) \cdot \frac{2}{\pi}} \quad (28)$$

Energy Loss due to Reflection

Some of the incoming solar energy is reflected due to the receiver's surface reflectivity. The following equation gives the reflective heat loss, Q_{rad} from the receiver surface,

$$Q_{ref} = (1 - \alpha_{eff}) . Q_{th\_rec} \quad (29)$$

where α_{eff} is the effective absorptivity due to the curvature of receiver tubes and the paint on the tube surface. Similar to effective emittance, curvature of tubes is considered, which thus determined the effective absorptivity in the form of [20]:

$$\alpha_{eff} = \frac{\alpha_{rec}}{\alpha+(1-\alpha) \cdot \frac{2}{\pi}} \quad (30)$$

HELIOSTAT FIELD ANALYSIS AND DETERMINATION

Field Size

In order for the receiver to get enough heat for the HTF, an equivalent amount of energy should be supplied by the heliostat field after accounting for optical losses. The size of the heliostat field, A_{field} needs to be determined based on the receiver thermal power, Q_{th_rec} obtained from the receiver's model and defined by below equation:

$$A_{field} = \frac{Q_{th\_rec}}{Q_{DNI} \times \eta_{optical}} \quad (31)$$

where A_{field} is the area of the heliostat field (m^2), $\eta_{optical}$ is the field optical efficiency. The total number of heliostats, N_{helio} is defined to be:

$$N_{helio} = \frac{A_{field}}{L_{helio} \times W_{helio}} \quad (32)$$

where L_{helio} is the heliostat length (m), W_{helio} is the heliostat width (m). With a given heliostat dimensions (2m x 2m), a model is developed to determine the number of heliostat and the patterns of distribution, and consequently the optical efficiency is found based on the given elevation of solar tower. The peak design point of sunlight energy density is 1000 W/m^2.

Heliostat Field Layout

Due to its simplicity and reduced number of heliostats (less area) with higher performance compared to the conventional radial staggered layout [21], a biomimetic layout is opted in this study as shown in Fig. 5. In this arrangement, the position of a heliostat k is defined based on angular position, θ_k and radial position, r_k as following:

$$\theta_k = 2\pi\ \varphi^{-2} * k \tag{33}$$

$$r_k = a * k^b \tag{34}$$

where φ is the golden angle between the heliostat ($\varphi = 137.5°$), k is the count number of a heliostat, a is the linear Fermat's spiral parameter, b is the power Fermat's spiral parameter. The range of values of a and b are 2- 8 and 0.45-0.7, respectively. For the present study, the optimized value of "a" is 3; while "b" is 0.6.

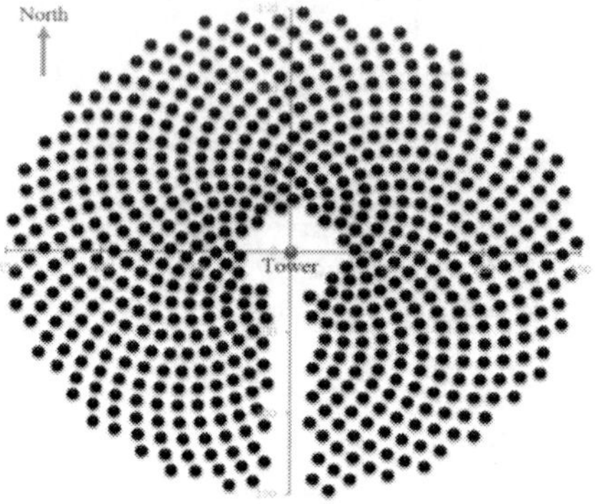

Fig. 5 Proposed biomimetic layout of heliostat field

Heliostat Positions

Two angles are used to indicate the relative position of a heliostat against the solar power tower: the facing angle, $\emptyset$ and target angle, λ. Facing angle is defined as the rotation angle about the heliostat's zenith from the north direction pointing towards the target (central tower) on the ground level as illustrated in Fig. 6. The Target angle is defined as the angle between the heliostat horizontal level and the line connecting heliostat and the tower [22].

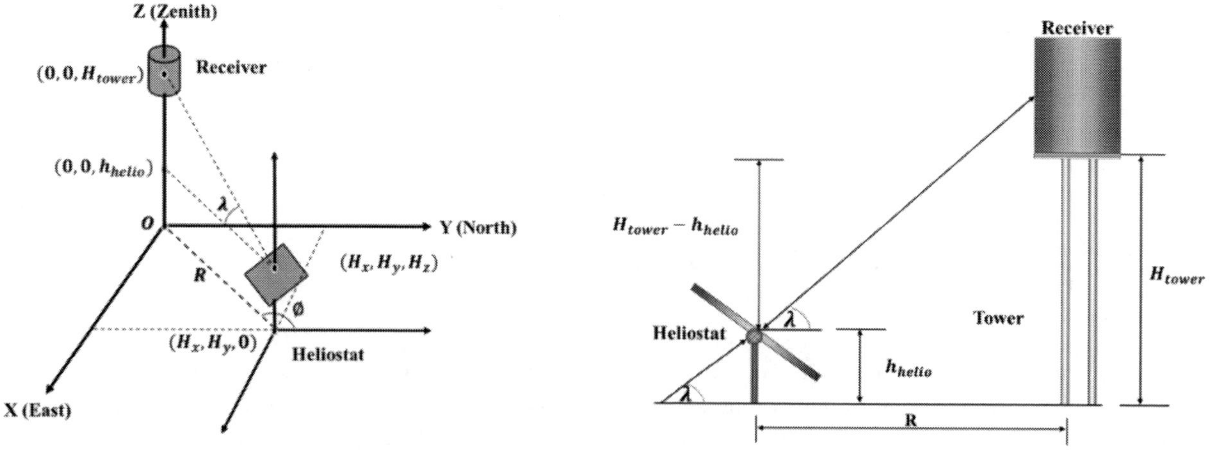

Fig. 6 Cartesian system and the relative positions of tower and heliostat

In this analysis, the facing angle will be 0° if the heliostat is placed due south of the tower and 180° if it is placed due north. Using the heliostat height, h_{helio} the target angle is derived as:

$$\lambda = tan^{-1}(\frac{H_{tower} - h_{helio}}{R}) \tag{35}$$

Field Optical Performance and Heat Flux

Regardless of heliostat geometry, the heliostat field is subject to number of losses that result in a reflected image on the receiver less than the amount of solar energy originally incident on the surface of the heliostat. In heliostat fields, the net field efficiency (field optical efficiency) is computed according to the following equation:

$$\eta_{optical} = \eta_{cosine} \times \eta_{ref.} \times \eta_{blocking\_shading} \times \eta_{tarcking} \times \eta_{spill} \times \eta_{atten.} \quad (36)$$

where η_{cosine} is the cosine efficiency, η_{ref} the reflection efficiency, $\eta_{blocking\_shading}$ the blocking and shadowing efficiency, $\eta_{tarcking}$ the tracking efficiency, η_{spill} the spillage efficiency, η_{atten} the attenuation efficiency.

Due to the transient nature of losses and heat flux on receiver, they were numerically evaluated using Monte Carlo Ray Tracing (MCRT) technique via SolTRACE [23], a general purpose tool of ray tracing developed at the National Renewable Energy Laboratory (NREL).

The optimized heliostat field has a minimum and maximum land radius of 25 m, 125 m respectively, the plant location is Tucson, Arizona. The field consists of 617 heliostats, each has an aperture area 4 m^2. According to the allowable heat flux calculated in the model, the optimized receiver's surface area is found to be 6.8 m^2.

USER INTERFACE (UI) DEVELOPMENT FOR COMPUTATIONAL CODE

The receiver model is developed in Visual Basic. It consists of: optimization parameters, design inputs and outputs. Design inputs are fixed inputs while the optimization parameters represent variable inputs as shown in Table 1. Heat losses were assessed in a separate tab but were considered as outputs of the model.

Table 1 Receiver Model: Optimization parameters, Inputs and Outputs

Optimization Parameters			
Properties	**Range**	**Value**	**Unit**
Heat Load		1.752	MW_{th}
Initial Receiver Thermal Efficiency	0.7 – 0.9	**0.876**	
Reynolds Number	$\geq 1.00 \times 10^4$	6.00×10^4	
Receiver Height	1 – 4	1.8	m
Receiver Diameter	1 – 4	1.2	m
Inner tube diameter	20 – 40	22	mm
Outer tube diameter	20 – 40	30	mm
Tube thickness	1 – 10	4	mm
Receiver Height/Diameter (Aspect Ratio)	1 – 2	1.5	
Inputs			
Properties	**Value**		**Unit**
HTF hot temp	150		°C
HTF cold temp	80		°C
System Pressure	5		bar
Receiver's Material	316 stainless steel		
HTF type	Water		
Outputs			
Properties	**Values**		**Unit**
Receiver Area	6.8		m^2
Power Absorbed by HTF	1.535		MW
HTF mass flow rate	5.176		kg/s
HTF mass flow rate / tube	0.252		kg/s
HTF Velocity	0.687		m/s
No. of tubes per receiver	126		
No. of tubes per panel	21		

No. of Panels	6	
Receiver Allowable Heat Flux	258	kW/m2
Total pressure drop	3.80E+05 (3.8bar)	Pa
Pump power	2.716	kW

The heliostat field model iteratively requires the size of the heliostat field, A_{field} and consequently the receiver thermal power, Q_{th_rec} is called up from the receiver model. Table 2 shows the main parameters for the heliostat field model.

Table 2 Heliostat Field Model: Optimization parameters, Inputs and Outputs

Input		
Properties	**Values**	**Unit**
Heliostat Length	2	m
Heliostat Width	2	m
Heliostat Height	1	m
Design Point (Direct Normal Irradiance)	1000	kW/m^2
Tower Height	40	m
Initial Optical Field Efficiency	70%	
Heliostat Field Area	2468	m^2
Outputs		
Properties	**Values**	**Unit**
No. of Heliostats	617	
Golden Angle	137.5	Degree
Linear Fermat's spiral parameter "a"	3	
Power Fermat's spiral parameter "b"	0.6	
Min. Heliostat Field Radius	25	m
Maximum Heliostat Field Radius	125	m
Optical Field Efficiency	56.1%	
Average Heat Flux (Mean)	210	kW/m^2

VALIDATION AND RESULTS

Receiver Model Validation

The receiver model methodology is validated using two selected reference plants with molten salt as HTF because water receivers data are not available in literature, in addition the opted plants are commercially well-established plants with adequate available data. The molten salts properties were specified as inputs for the model. These plants are the 10 MW SOLARTWO power plant in California and the 20 MW GEMASOLAR power plant in Spain. In order to mimic the real plant behavior at design points, the inputs to the models are: receiver's thermal rating, receiver's height and diameter, tube inner diameter and thickness, while the validated outputs are: number of receiver's panels, number of tubes per panel, number of total tubes, mass flow rate, and receiver's thermal efficiency.

The two solar plant characteristics are summarized in Table 3 [25] [36].

Table 3 Reference plants characteristics

Design Parameters	Solar Two Plant	Developed Model	Gemasolar Plant	Developed Model
Electric power	10 MW_e	10 MW_e	19.9 MW_e	19.9 MW_e
Storage capacity	3 hours	3 hours	15 hours	15 hours
Receiver's thermal power	43 MW_{th}	43.2 MW_{th}	120 MW_{th}	120 MW_{th}
Receiver's fluid	Molten Salt	Molten Salt	Molten Salt	Molten Salt
Receiver's area	99.25 m^2	99.3 m^2	270 m^2	270 m^2

Inputs				
Receiver's height	6.2 m	6.2 m	10.6 m	10.6 m
Receiver's diameter	5.1 m	5.1 m	8.1 m	8.1 m
Tube diameter	23 mm	23 mm	25 mm	25 mm
Tube thickness	1.2 mm	1.2	1.65 mm	1.65 mm
Outputs				
No. of receiver's panel	24	24	16	16
No. of tubes per panel	32	32	64	64
Total No. of tubes	-	696	1024	1017
Mass Flow rate	82.5 kg/s	84.8 kg/s	-	244.5 kg/s
Receiver's thermal efficiency	84 – 87% [25], [26]	83.8%	84%	83.7%

Comparing the above values with the reference plants, it can be concluded that the developed receiver model outputs matches the published receivers' data.

Results and Discussions

This section examines and discusses the tower height effects on both field efficiency and average incident heat flux via MCRT using the weather data of the proposed location. In addition, fluid velocity impact is evaluated in terms of Reynolds number where it has to be greater than or equal to 10^4 in the turbulent region. The analysis and simulation carried out in this study used conditions mentioned in Table 1 and Table 2.

Effect of Tower Height on Optical Efficiency and Incident Flux

The effect of tower height on the optical efficiency is shown in Fig. 7(a). It shows that the field optical efficiency is improved when tower height is increased to reach its peak at 40 m then it declines with increasing the height. Therefore, 40 meters was the opted tower height for the proposed biomimetic field. Fig. 7(b) shows the tower height effect on the transient average incident heat flux simulated using SolTRACE, we can inferred that the highest transient average incident heat flux reaches 210.5 kW/m^2 at 40 m height and it begins to drop as the tower height increase which agrees with the designed allowable flux. Noting that another analysis is carried out for transient peak incident heat flux as shown in Fig. 7(c) in order to ensure that the heating flux on the receiver has not exceed an instantaneous 600 kW/m^2 (limiting value for 316 stainless steel with water system) as stated previously.

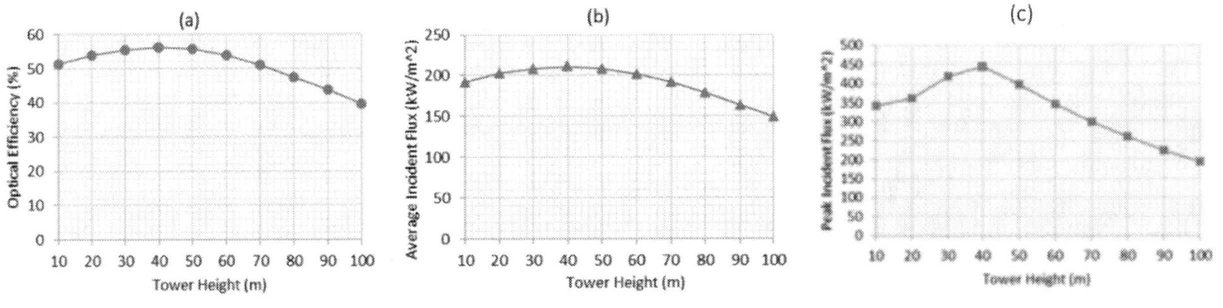

Fig. 7 The tower height effect on (a) field optical efficiency (b) average incident flux (c) peak incident flux

Effect of Fluid Velocity on Receiver's Efficiency and Surface Temperature

In this case, the fluid velocity is not directly analyzed, alternatively the Reynolds number is used as an optimization parameter and the effect is depicted in Fig. 8. The inner tube diameter is fixed at 22 mm with 4 mm thickness; it can be observed that the receiver's efficiency increases from 86.3% at $Re = 10 \times 10^3$ to reach 87.7% at $Re = 100 \times 10^3$. The receiver's efficiency is held constant for the range $60 \times 10^3 \leq Re \leq 95 \times 10^3$. The surface temperature decreases from 286°C to 166°C for the same interval. It is clear that at higher Reynolds No. (aka high fluid velocity) the receiver's efficiency is improved due to decreased convective losses, and reduced surface temperature at turbulent flow regimes. At the same time, this increase in receiver's efficiency moderately increases the feed pump power as shown in Fig. 9. The pump power increases from 2.668 kW at $Re = 10 \times 10^3$ to 2.745 kW at $Re = 100 \times 10^3$.

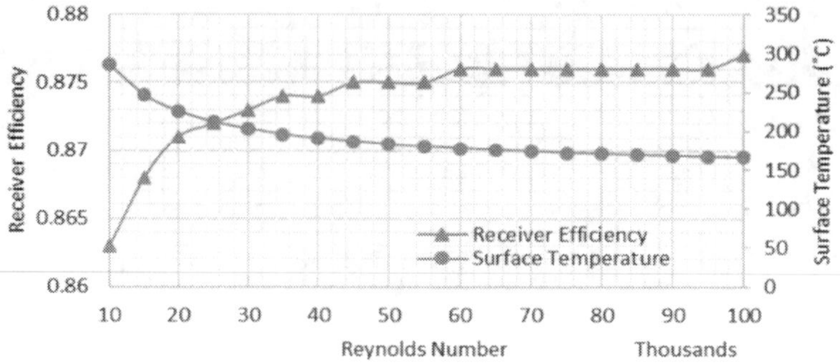

Fig. 8 Fluid velocity effect on receiver's efficiency and surface temperature

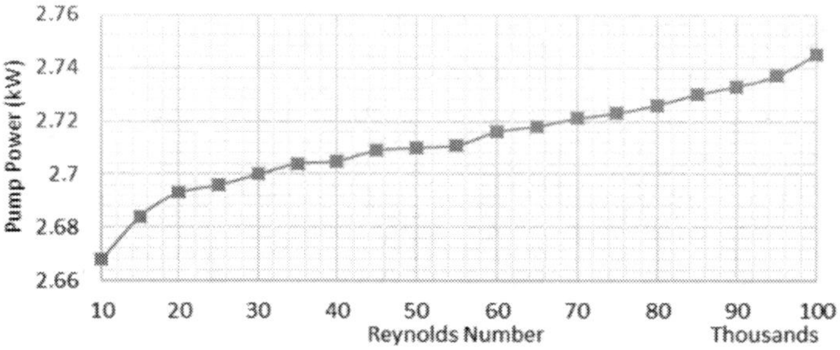

Fig. 9 Fluid velocity effect on feed pump power

Effect of Fluid Velocity on Panels and Tubes Number

The receiver's outer geometry is affected by the flow regimes in tubes, taking $Re = 10 \times 10^3$ for the fully turbulent condition as a start-up. From Fig. 10, at $Re = 10 \times 10^3$ the total number of tubes per receiver is 63 tube with 2 panels while at $Re = 30 \times 10^3$ the panels required become 4 panel each has 32 tubes, also at $Re = 50 \times 10^3$ the number of panels increases to be 6. It is clear that, increasing the Reynolds Number by 20,000 at the pure turbulent regime increases the number of panels by 2 and less tubes in each panel to maintain the overall number to receivers' tubes (i.e. 126). Because of the two-flow path branches, the number of panels must be rounded to the next even number due to the reason that receivers with only one flow path have higher pressure drop and therefore higher power consumption of the feed pumps. However, this analysis opted $Re = 60 \times 10^3$ for best receiver efficiency and accordingly the geometry shall be 6 panels, each has 21 tubes.

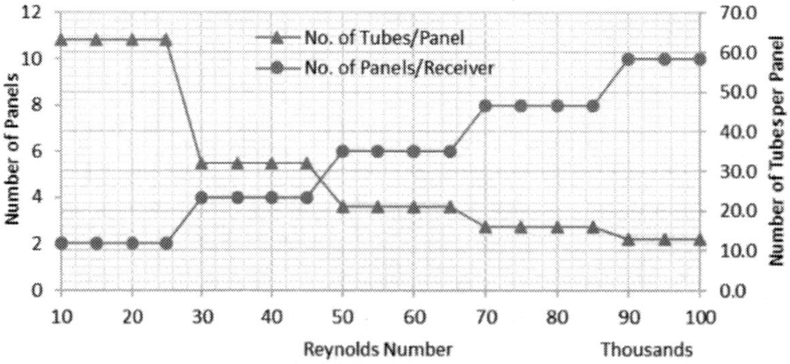

Fig. 10 Fluid velocity effect on panels and tube number

Flux maps

The heat flux calculation were carried out in SolTRACE simulation as a MCRT, the transient behavior is observed for all year long but as a representative day 21st June at 1.00 pm (Tucson, Arizona) is shown in Fig. 11. The highest instantaneous heat flux on receiver is 465 kW/m^2 while the average heat flux is 182 kW/m^2. The simulated flux maps indicate that the modelled receiver ensured acceptable level of heat flux that does not exceed the peak heat flux of 600 kW/m^2, consequently the lifetime of the receiver is optimum for the given design conditions.

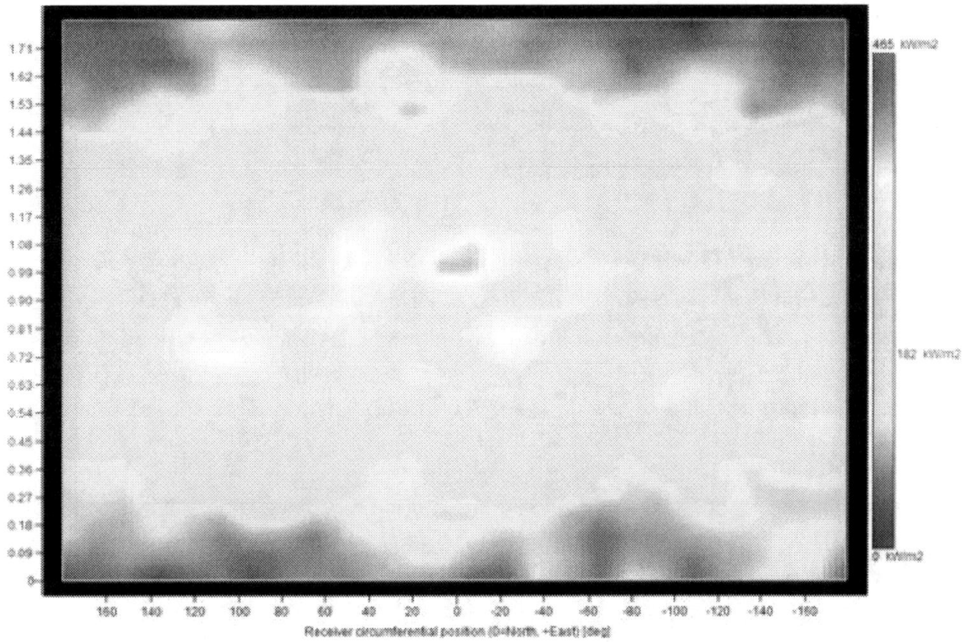

Fig. 11 Heat Flux on Receiver on June 21, at 1:00 pm

CONCLUSION

Computational Fluid Dynamics (CFD) modeling to simulate the entire SPT system is expensive and unfeasible therefore this research develops a simple method for designing a small-scale central receiver system using water as HTF. The receiver model was coupled with a heliostat field model in Visual Basic platform due to their performance impact and cost on overall the plant. The model has evaluated the thermal efficiency based on the energy balance approach taking into account the critical design factors such as allowable heat flux, fluid mean temperature and water properties at specified system pressure. The model had demonstrated an excellent match for two well-established commercial molten salt plants (i.e. SolarTwo and GemaSolar).

The results show that the highest transient average incident heat flux reaches 210.5 kW/m^2 at 40 m height and it begins to drop as the tower height increases, which agrees with the designed allowable flux; while the transient peak incident heat flux did not exceed the critical flux for 316 stainless steel with water system.

Besides, the receiver's efficiency is impacted by the fluid velocity, therefore at higher Reynolds No., the receiver's efficiency is improved due to decreased convective losses, and reduced surface temperature at turbulent flow regimes. At the same time, this increase in receiver's efficiency moderately increases the feed pump power. On the other hand, the receiver's outer geometry is affected by the fluid velocity where increasing the Reynolds No. by 20,000 at the turbulent regime increases the number of panels by 2 with less tubes in each panel to maintain the overall number of receivers' tubes (i.e. 126). However, this analysis opted $Re = 60 \times 10^3$ because it yields an efficiency of 87.6% and accordingly the geometry shall be 6 panels, each has 21 tubes.

Finally, the heat flux maps showed that the modelled receiver ensured acceptable level of heat flux that doesn't exceed the peak heat flux of 600 kW/m^2 therefore the lifetime of the receiver is at optimum.

REFERENCES

[1] IEA, "Technology Roadmap: Solar Thermal Electricity," 2014.

[2] SunShot U.S. Department of Energy, "SunShot Vision Study," *U.S. Dep. Energy*, no. February, pp. 69–96, 2012.

[3] G. Kolb and S. Jones, "Heliostat cost reduction study," *SAND2007- …*, no. June, 2007.

[4] IRENA, "Renewable Energy Technologies: Cost Analysis Series Power Sector," 2012.

[5] M. A. Qaisrani, J. Wei, J. Fang, Y. Jin, Z. Wan, and M. Khalid, "Heat losses and thermal stresses of an external cylindrical water/steam solar tower receiver," *Appl. Therm. Eng.*, vol. 163, no. September, p. 114241, 2019.

[6] M. A. Qaisrani *et al.*, "Thermal losses evaluation of an external rectangular receiver in a windy environment," *Sol. Energy*, vol. 184, no. October 2018, pp. 281–291, 2019.

[7] C. O. Popiel and J. Wojtkowiak, "Simple formulas for thermophysical properties of liquid water for heat transfer calculations (from 0°c to 150°c)," *Heat Transfer Engineering*, vol. 19, no. 3. pp. 87–101, 1998.

[8] A. B. Zavoico, "Solar Power Tower - Design Basis Document," no. July, p. 148, 2001.

[9] Z. Liao, X. Li, C. Xu, C. Chang, and Z. Wang, "Allowable flux density on a solar central receiver," *Renew. Energy*, vol. 62, pp. 747–753, 2014.

[10] W. B. Stine and R. W. Harrigan, *Solar Energy Fundamentals and Design with Computer Applications*. New York: Wiley-Interscience (April 1985), 1985.

[11] P. K. Falcone, *A Handbook for Solar Central Receiver Design*. Sandia National laboratories Livermore, 1987.

[12] L. L. Vant-Hull, "The role of 'allowable flux density' in the design and operation of molten-salt solar central receivers," *J. Sol. Energy Eng. Trans. ASME*, vol. 124, no. 2, pp. 165–169, 2002.

[13] J. M. Lata, M. Rodr guez, and M. A. de Lara, "High Flux Central Receivers of Molten Salts for the New Generation of Commercial Stand-Alone Solar Power Plants," *J. Sol. Energy Eng.*, vol. 130, no. 2, p. 021002, 2008.

[14] M. R. Rodríguez-Sánchez, "On the design of solar external receivers." pp. 1–256, 2015.

[15] M. M. J. M. Wagner, "Simulation and predictive performance modeling of utility-scale central receiver system power plants," 2008.

[16] H. Benoit, L. Spreafico, D. Gauthier, and G. Flamant, "Review of heat transfer fluids in tube-receivers used in concentrating solar thermal systems: Properties and heat transfer coefficients," *Renew. Sustain. Energy Rev.*, vol. 55, pp. 298–315, 2016.

[17] Y. Cengel, *Heat transfer a practical approach*. Mcgrow Hill, 2003.

[18] D. L. Siebers and J. S. Kraabel, "Estimating Convective Energy Losses From Solar Central Receivers," pp. 1–69, 1984.

[19] J. R. C. and M. M. Y. F.J. McQuillan, "Properties of Dry Air at One Atmosphere," *Microelectron. Heat Transf. Lab*, no. June, 1984.

[20] C.-J. Winter, R. L. Sizmann, and L. L. Vant-Hull, *Solar Power Plants: Fundametals, Technology, Systems, Economics*, 2nd ed. Springer, 2011.

[21] C. C. J. Noone, "Optimization of central receiver concentrated solar thermal: site selection, heliostat layout & canting," MASSACHUSETTS INSTITUTE OF TECHNOLOGY, 2011.

[22] A. Gamil, S. Ul-haq Gilani, and H. Al-kayiem, "Simulation of incident solar power input to fixed target of central receiver system in Malaysia," in *2013 IEEE Conference on Sustainable Utilization and Development in Engineering and Technology (CSUDET)*, 2013, pp. 92–97.

[23] T. Wendelin, "SOLTRACE: a new optical modeling tool for concentrating solar optics, Solar Engineering," in *ASME 2003 International Solar Energy Conference*, 2003, no. March 2003, pp. 253–260.

[24] M. J. Wagner and T. Wendelin, "SolarPILOT: A power tower solar field layout and characterization tool," *Sol. Energy*, vol. 171, pp. 185–196, 2018.

[25] M. R. Rodriguez-Sanchez, A. Sanchez-Gonzalez, and D. Santana, "Revised receiver efficiency of molten-salt power towers," *Renew. Sustain. Energy Rev.*, vol. 52, pp. 1331–1339, 2015.

[26] J. E. Pacheco, "Final Test and Evaluation Results from the Solar Two Project," *Contract*, no. January, 2002.

613422 Keynote: Solar fuel processing: Research, technology development, and scaling

Abstract:

Solar radiation is the most abundant renewable energy source available but it is distributed and intermittent, thereby necessitating its storage via conversion to a fuel or chemical commodity (e.g. hydrogen or carbohydrates) for practical use. Solar thermo-chemical and photo-electro-chemical approaches (and combinations thereof) provide viable, non-biological routes for the direct synthesis of solar fuels and chemical commodities. Both approaches involve complex interactions between electromagnetic wave propagation, multi-mode heat transfer, multiphase flow, charge transfer, and chemical reactions. The design of practical and scalable reactors and their scaling into solar fuel processing systems is therefore challenging, but considerable advancements have been demonstrated over the last decades. I will describe research and technology development of solar fuel processing devices and systems, and compares the various approaches in terms of efficiency, selectivity, production rate, and stability. One specific example I will discuss, is LRESE's concentrated photoelectrochemical approach, starting from the underlying idea, the design implementation [1], laboratory-scale demonstration [2] with world-record power density [3], and ongoing scaling into an operational solar fuel processing system utilizing our 7m-diameter solar concentrating dish. Remaining challenges to be addressed at a fundamental research level will be discussed in an outlook.

[1] S. Tembhurne, S. Haussener, Integrated Photo-Electrochemical Solar Fuel Generators under Concentrated Irradiation- Part II: Thermal Management a Crucial Design Consideration, *Journal of The Electrochemical Society*, 163: H999-H1007, 2016.

[2] S. Tembhurne, F. Nandjou, S. Haussener, A thermally synergistic photo-electrochemical hydrogen generator operating under concentrated solar irradiation, *Nature Energy*, doi: 10.1038/s41560-019-0373-7, 2019.

[3] http://specdc.epfl.ch

602630 Invited Talk: Solar Water Splitting: A Decoupled Photoelectrochemical Benchtop Demonstration System

Abstract:

Photoelectrochemical (PEC) water splitting is an elegant route for direct conversion of solar energy to green hydrogen. Theoretically, photovoltaic-photoelectrochemical (PV-PEC) tandem cells with ideal photoelectrodes could surpass the efficiency of PV-electrolysis systems that use the same PV cells, owing to the enhanced solar energy conversion efficiency of multi-junction tandem cells.

Due to the diffuse nature of natural sunlight, an immense area must be covered by PEC cells to produce hydrogen at a commercial scale. However, in the current architecture, PEC cells are unsuitable for application at such a large scale. Namely, the current architecture is that of a single, hermetically sealed cell, that produces hydrogen and oxygen simultaneously in the same cell. Accordingly, a PEC hydrogen production plant would require hermetic sealing of the entire solar array, construction and maintenance of an immense hydrogen piping infrastructure, and constant monitoring in accordance with hydrogen safety guidelines. The safety hazards associated with conventional electrolyzers are also exacerbated in this case due to the low-current densities and intermittent power supply characteristic of PEC cells.

To overcome these challenges, we proposed a disruptive approach of decoupling of the hydrogen and oxygen evolution reactions using solid redox mediating electrodes based on nickel hydroxide[1]. This strategy enables oxygen evolution alone to take place at the PEC cell, while hydrogen evolution takes place in a separate electrolytic cell that does not require exposure to light. This eliminates the need to hermetically seal PEC cell. Furthermore, there is no need for a membrane or hydrogen collection from the solar (PEC) site, as hydrogen can be generated in a separate compact cell.

Thus far, only a handful of studies have proposed scalable PEC-PV tandem devices, all of which are based on the conventional coupled architecture. In this work, we present a first of its kind prototype of a benchtop-scale decoupled and membrane-free PEC device based on a 100 cm^2 hematite photoanode stack and commercial Si PV modules.[2] Nickel hydroxide electrodes are used as hydroxide-mediating auxiliary electrodes, enabling cell separation over > 10 m distance. We address the issues of material selection, stability and light harvesting, as well as designing, building and optimizing the system for continuous hydrogen production. Finally, we demonstrate successful operation of our prototype device under solar simulated sunlight and under natural sunlight.

[1] Landman, Avigail et al. "Photoelectrochemical water splitting in separate oxygen and hydrogen cells." *Nature Materials* 16.6 (2017): 646-651.

[2] Landman, Avigail and Halabi, Rawan, et al. "Decoupled photoelectrochemical water splitting system for centralized hydrogen production." *Joule* (2020)

609236 Dynamic Process Simulation and Control of a High Performance Thermally-Integrated Concentrator Photovoltaic and Electrolysis Co-Generation Device

Abstract:

Solar hydrogen producing devices utilising photonic energy conversion encompass a large spectrum of device types [1], and in this work we investigate the performance of a compact thermally integrated and directly electrically coupled photo-voltaic (PV) and electrolyser (EC) operating at high solar concentrations (~800 suns). Previously, a lab scale prototype (30 W scale) using concentrated light (~470 suns) has experimentally demonstrated the synergetic effect of the thermal integration of the water splitting reaction and the PV cell due to the improved kinetics of the electrochemical reaction at higher temperatures and the reduced recombination in the PV cell at lower temperatures [2]. In this work, a larger scale similarly integrated device is simulated in order to study the operability and real world performance and provide control guidelines for the pilot kW scale integrated solar hydrogen device demonstration (38.5 m^2 solar parabolic dish) underway at EPFL. A lumped mixed differential-algebraic model was constructed from the energetic, mass and momentum continuity equations for the PV, EC, solar dish, water pump and piping. This was numerically solved using gPROMS ModelBuilder for representative parameters extracted from experimental data and literature. A high efficiency solar parabolic dish (86 % optical efficiency), a triple-junction (InGaP/InGaAs/Ge) concentrator solar module (30-36 % electrical efficiency at maximum power point) and a proton exchange membrane (PEM) electrolyser (50-70 % electrical efficiency based on Gibbs energy) was considered.

Firstly, the steady-state behaviour of the model was investigated. Energetic system efficiency to hydrogen (based on the Gibbs energy) was found to be between 16-21 % depending on configuration and operational point. A parametric study varying the water mass flowrate and the Direct Normal Irradiance (DNI) discovered disparate behaviour at different ratios of EC to PV cells, where a pseudo thermal runaway situation is identified at operating voltages higher than maximum power point voltages and for low flowrates. Whilst we demonstrate that the PV temperature remains bounded (i.e. mathematically stable), it will quickly exceed the maximum allowable value and therefore, the control strategy must be adapted accordingly (via PV voltage and current monitoring).

Secondly, the system dynamics is explored through system response to step changes in the major disturbance variable (DNI) and controlled variable (water flowrate). Non-linear system dynamics are observed as the operating point of the PV-EC non-linearly depends on system temperatures which have differing time constants. The fast dynamics (temperature time constants ~5-30 seconds) observed leads to fast start-up times, suitable for the operation under varying meteorological conditions.

Thirdly, control strategies were identified and evaluated using the developed model. The two major requirements to ensure safe and efficient operation are: 1) PV temperature should not exceed 100 °C 2) EC power must be maintained within a large operational window typical of PEM electrolysers (10-110 %). In order to meet the first requirement, whilst simultaneously stabilising the thermal output of the system both feedback and feedforward control for the water flowrate are investigated. The impact of time delay leads us to recommend a model-based feedforward-feedback control, which maximises the amount of useful heat than can be extracted and stabilises the system temperatures over the diurnal solar irradiance, whilst remaining responsive to large changes in DNI. Furthermore, the system response to major deviations from normal operation, such as pump failure or electrical disconnection, is quantified and this reiterates the need for sub-second monitoring and control.

Control strategies for the daily, seasonal and yearly variations in solar irradiance and ambient temperature is investigated. If the EC stack power falls below a critical value, then the stack must temporarily be supplied with external electricity or shut-down completely, both of which possess energetic penalties to the overall system efficiency. We implement a discrete state-based process control routine (with conditional transitions between states based on monitored measurements such as DNI, electrolyser power and time elapsed) to demonstrate continued operation on intermittently cloudy days, whilst minimising the external power required.

[1] Holmes-Gentle, I., Alhersh, F., Bedoya-Lora, F., & Hellgardt, K. (2018). Photoelectrochemical Reaction Engineering for Solar Fuels Production. *Photoelectrochemical Solar Cells*, 1-41.

[2] Tembhurne, S., Nandjou, F., & Haussener, S. (2019). A thermally synergistic photo-electrochemical hydrogen generator operating under concentrated solar irradiation. *Nature Energy*, *4*(5), 399.

602047 CO_2 Gasification of Sewage Sludge, Plastic Waste, and Municipal Solid Waste in an Internally-Circulated Fluidized Bed Under High-Flux Solar Irradiation

Abstract:

Solar thermal gasification of carbonaceous feedstocks is a promising approach to produce high-quality syngas. Previous work in solar gasification demonstrated merits of fluidized beds such as high residence time of particles and highly-effective solid–gas heat and mass transport. However, large temperature gradients between absorbing surfaces and particles may arise in solar thermal fluidized beds. Internally circulating fluidized beds, i.e. clapboard and draft tube configurations, are attractive for eliminating the stagnation zones and increasing the solid–solid and solid–gas interaction. The most favorable merit is the capability of mitigating the hot spot on the absorber surface by enhancing the interaction and heat transfer of solid particles and the wall. In this work, a novel indirectly irradiated solar gasifier featuring a clapboard-based internally circulating fluidized bed was developed and tested under high-flux irradiation provided by a 28 kW_e solar simulator. The solar gasifier is composed of an internally-circulating fluidized bed and Al_2O_3-SiO_2-insulated cavity. The internally-circulating fluidized bed consists of a 70-mm wide, 100-mm high and 10-mm thick clapboard, a 5-mm thick square bed of 70-mm inner diameter, and a gas distributor at the bottom of the bed. The biomass particles pass through the gap between the clapboard and the gas distributor to form the internal circulation. A SiC absorber plate located at the center of the cavity is attached to the outer surface of the square bed. The biomass feedstock is delivered through a semi-continuous auto feeder. In the present work, dried sewage sludge, plastic waste, and municipal solid waste from Singapore are used as the example carbonaceous feedstocks. The effects of the superficial gas velocity and the CO_2/C ratio on the carbon conversion ratio, the solar upgrade factor, and solar-to-chemical efficiency are investigated.

614005 Finding Our Way to Drive Sustainability Forward Through a Global Pandemic

Abstract:

Ford commitment to sustainability does not disappear during a pandemic, resource scarcity, low oil price, or with many other challenges. The presentation will outline how the growth of the automotive industry and R & D efforts encourage the use of cost effective, lightweight, sustainable and advanced materials for automotive applications. Pioneering developments of sustainable and advanced materials by Ford scientists including nanocomposites, aerogels, natural fiber reinforced composites, polymeric and soft materials made from renewable feed stocks, 3-D printed plastic parts, bio-inspired and patterned functional materials as well as plastic parts made from recycled carbon dioxide will be featured.

614006 Creating New Bioproducts Targeting "Zero Waste" Production

Abstract:

This presentation outlines USDA strategies for converting agricultural coproducts into commercially-viable products with the ultimate goal of achieving "zero waste" agriculture. The Western U.S., especially the Central Valley of California, is a world leader in the production of an array of specialty crops, including wine grapes, olives, artichokes, avocados, almonds, walnuts, pistachios, and, most recently, hemp. These crops contribute billions of dollars to the rural economy. The processing of each of these crops generates their own specific coproducts and waste streams, which in turn can generate billions more in revenue or result in significant costs. This presentation will discuss successful development of several ag-derived biomass feedstocks into commercial products, outlining specific processing methods that meet the cost-points required to meet market demands. Discussion will include description of torrefaction of biomass whereby biomass sources were heat-treatment to 200 C-300 C in limited oxygen, to create fillers for modifying the properties of traditional plastics. Char-like ground torrefied additives were melt-blended into recycled plastics and bioplastics as a replacement for carbon black. In another example, we describe converting organic acids and/or methane from agricultural biomass into polyhydroxyalkanoate (PHA), a commodity bioplastic with significant market potential. As will be discussed, the drive toward "zero waste" agriculture can be applied universally to generate value-added bioproducts from a wide array of agriculturally-derived biomass sources.

614007 Can Bioplastics Alleviate Plastic Waste Issues?

Abstract:

Plastic wastes on land and in oceans have become major societal issues. Articles in print, television, and social media about plastics waste issues and bans on plastic items are on the rise everywhere in the world. Most serious, is the issue relating to microplastics found in oceans and living systems. Against this backdrop, this lecture reviews the role of BioPlastics in providing solutions for plastic waste issues. There are many misperceptions about bioplastics, biobased, biodegradable, compostable, and the value proposition they offer for addressing the plastic waste issues. In this context, this @EnvSciTech viewpoint summarizes necessary requirements for assessing and reporting plastic biodegradation as solution for plastic waste https://pubs.acs.org/doi/10.1021/acs.est.9b04513. The presentation reviews the science, the definitions, standards, and misleading claims that abound in the marketplace.

614008 Circular Economy and Sustainable Materials

Abstract:

Plastic waste and their adverse environmental impacts have become daily front news. Since 1950, we have accumulated more than 7.5 billion tons of plastic wastes in the globe. These wastes quite often end-up in in ocean. According to United Nations, each year we create around 300 million metric ton of plastics wastes that is equivalent to the weight of our entire human population. Similarly, billion tons of agro-residues and forestry wastes are unused or burnt very often-creating environmental pollution.

Agro-residues and forest wastes including waste wood and sawdust on reinforcement with waste plastics can create a range of sustainable hybrid biocomposites. Valorization of waste plastics in biocomposites uses is a strategy towards lessening green house gas emissions. Giving new birth to waste plastics supports circular economy and this is a strategy towards improved sustainability. The sustainable biocomposites can find applications in consumer products, rigid packaging, and green housing structures as well as lightweight automotive parts.

611744 Lignin-Based Carbon Fibers from Hybrid Poplar and Corn Stover: Renewable Aqueous Solvents for Lignin Fractionation and Solvation

Abstract:

Because of the growing interest in cellulosic bioethanol, lignin from agriculture residue sources has drawn attention as a renewable and inexpensive material for value-added applications. Similar to Kraft lignins, the heterogeneity and impurities of these biorefinery byproduct lignins present significant challenges that negatively impact their economic viability as precursors. The ALPHA process was developed as a means of both controlling the molecular weight and reducing the metallic impurities of technical lignins through the use of aqueous, renewable solvents. Previous work has demonstrated that higher molecular weight cuts of Kraft lignin, purified via ALPHA, produce carbon fibers with the best tensile strength and Young's modulus of any lignin to date. This same approach of lignin purification is being applied to lignins isolated via alkaline pulping from hybrid poplar wood chips and corn stover. The resultant ALPHA-processed lignin fractions are being prepared into spinning dopes -- using the same ALPHA solvents as for the fractionation/purification step – and then dry-spun into fibers. The effect of lignin molecular weight and purity on spinning, stabilization, and fiber properties will be reported, and results compared against a commercial softwood Kraft lignin.

610204 Partial Substitution of Phenol in Phenol-Formaldehyde (PF) Resin By Plant Biomass for Wood Adhesives

Abstract:

The modified phenol-formaldehyde resin was synthesized by substituting the significant component of the resin by plant biomass. This research is bifurcated into three parts. The first part of this study was performed to understand the structure of plant biomass used by various characterization techniques like FTIR, ^{1}H-NMR, ^{13}C-NMR, ^{31}P-NMR, Py-GCMS, GPC and solubility. In the second part of the research, PF (novolac) resins were produced by polycondensation reaction between phenol and formaldehyde under an acidic condition with formaldehyde (F) to phenol (P) ratio less than. The bio-PF resin was synthesized by replacing 50wt% of phenol by kraft lignin, lignocellulosic bio-oil and lignin bio-oil to produce 50% PF resin of lignin, 50% PF resin of lignocellulosic bio-oil and 50% PF resin of lignin bio-oil respectively. These PF resins show the linear or partially cross-linked structure, and the addition of curing agent makes them formed a fully cross-linked network. The behaviour of the physical mixture of HMTA curing agent with PF resins was studied using DSC (differential scanning calorimetry) analysis. The result obtained from DSC analysis was for bonding wood laminate in a pressing machine. In the later part of the research, the wood laminate was bonded using each prepared PF resins, and their physical mixture with 9% HMTA curing agent and the properties like tensile shear strength wet and dry condition, water absorption and thickness swelling were measured.

611539 Hairy Cellulose Nanocrystals: An Emerging Class of Nanocelluloses for Advanced Industrial Applications

Abstract:

For decades, cellulose nanocrystals (CNCs) have been produced by hydrolyzing the amorphous regions of hierarchical cellulose fiber structures. These colloidal particles are made up of highly ordered arrays of cellulose chains, impeding the physicochemical modifications of inner crystalline layers, which in turn result in limited/compromised dispersion stability, functionalizability and charge, response to external fields, and transport. Nanoengineering cellulose fibrils via controlled oxidation partially disintegrates the amorphous cellulose chains attached to the crystalline body, which yields Janus-like nanoparticles with a needle-shaped crystalline body (similar to CNCs) sandwiched between two highly functionalized disordered cellulose regions (hairs). This newly emerged class of nanocelluloses are called hairy cellulose nanocrystals (HCNC). The protruding soft brushes from HCNC poles impart significant modifications to the colloidal properties of HCNCs, enhancing their functionality, charge density, stability, and self-assembly. In this presentation, we will review our recent progress in engineering HCNCs for advanced applications, such as water treatment, scale inhibition, biomimetic mineralization, rheology modification, nanocomposites, hydrogels, and so forth. We will show how HCNCs enable several technologies that would otherwise be impossible to develop using conventional CNCs. The outcome of our research may leverage universal biomass-based sustainable, green solutions for long-lasting industrial challenges in water, healthcare, food, and energy sectors.

605570 Novel Double-Shell Lignin Nanocapsules Are a Stable Vehicle for Fungicide Encapsulation and Release

Abstract:

This paper describes the fabrication and characterization of a newly developed lignin nanocapsule, its performance towards encapsulating a fungicide, and effectiveness at controlling its release. Lignin nanoparticles are biologically and environmentally compatible, the amphiphilic properties of lignin, adsorption capacity, as well as its natural cross-linking tendency make it suitable for use as a vessel for stable cargo delivery. Here we report a sustainable approach for producing hollow, double-shell capsules from hardwood kraft lignin. The recovered lignin was dissolved without further modification in tetrahydrofuran and inserted dropwise into a water/ethanol solution containing sacrificial surfactant templates. Monodispersed, hollow, double-shell lignin capsules were produced via a two-step self-assembly. The water/ethanol solution acted as a non-solvent while simultaneously providing strategic sites for controlled-size production. The internal shells of the nanoparticles were cross-linked with biodegradable maleic anhydride to bolster the structural stability. The average hydrodynamic diameter of the particles was 202.2 ±66 nm, and these were stable in water for a period of four months. Following synthesis, the nanoparticles were loaded with propiconazole at a drug loading of 16%. This system could represent an effective wood preservative.

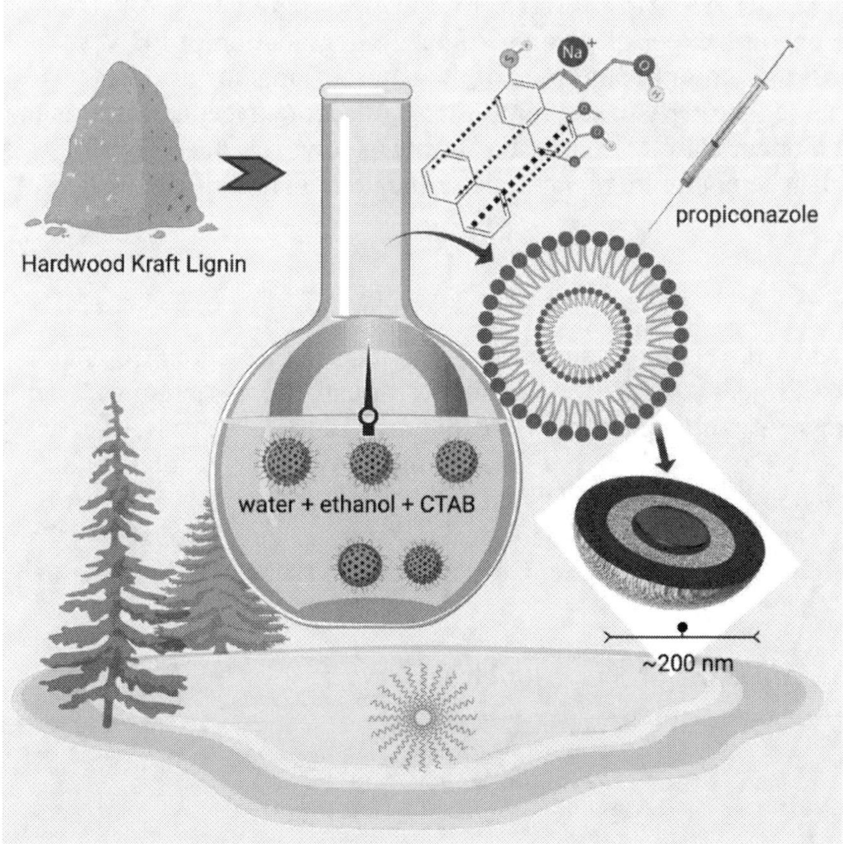

614981 The Effect of Natural Fillers on the Marine Biodegradation Behaviour of Poly(3-hydroxybutyrate-co-3-hydroxyvalerate) (PHBV)

Abstract:

Each year various plastics are being produced worldwide and disposed of improperly, fueling significant environmental concern around the accumulation of plastics in the environment and the world's oceans. As a result, the plastic debris floating/buried in the oceans are steadily increasing every year and can be found in most large water bodies across the globe. The rapid growth of biodegradable and bio-based plastics is being observed and driven by consumers demanding environmentally friendly plastic products. However, all the known bio-based plastics are not able to degrade in marine environment. One class of biodegradable plastics i.e. polyhydroxyalkanoate (PHA) have shown the most promising potential for marine biodegradability to date. However, as per our literature survey, there is no information available which can confirm the 100% degradability of PHA composites in natural marine environments. Poly(3-hydroxybutyrate-co-3-hydroxyvalerate) (PHBV) biocomposites with Miscanthus (Misc) fibres and distillers' dried grains with solubles (DDGS) were studied to ascertain the marine biodegradable performance of natural fibres versus proteinaceous fillers.

We simulated a marine environment (as per ASTM D7991-15) to study the degradation of PHA-based biocomposites. The results from the ongoing experiment suggests the % degradation of PHAs in a marine environment improved significantly after the addition of the natural fibres in various proportions. This evidence can conclude that PHA-based composites with proteinaceous fillers is the more effective in improving marine biodegradability than natural fibres. However, the effects of various experimental conditions such as temperature on natural marine degradation of plastics is still unclear.

Acknowledgements

This research is financially supported by the Natural Sciences and Engineering Research Council (NSERC), Canada Discovery Grants (Project # 400320); the Ontario Ministry of Agriculture, Food and Rural Affairs (OMAFRA) – University of Guelph, the Bioeconomy Industrial Uses Research Program Theme (Project # 030177); the Ontario Ministry of Economic Development, Job Creation and Trade ORF-RE09-078 (Project # 053970 and 054345); and the Agriculture and Agri-Food Canada (AAFC), Maple Leaf Food, Canada and Bank of Montreal (BMO), Canada through Bioindustrial Innovation Canada (BIC) Bioproducts AgSci Cluster Program (Project # 054015, 054449 and 800148).

595838 Hydrothermal Liquefaction for Co-Valorization of Plastics and Biopolymers

Abstract:

Hydrothermal liquefaction (HTL) takes place in water at near- and supercritical conditions (T_c = 374 °C, P_c = 22 MPa). HTL has been studied to valorize components of municipal solid waste (MSW), namely synthetic and biopolymers. Biopolymers such as those in food waste and wood are broken down to produce energy-dense oil under subcritical conditions (280-374°C) while synthetic polymers generally decompose under supercritical conditions (374-500°C). HTL of plastics (polypropylene (PPE), polyethylene terephthalate (PET), polystyrene (PS), polycarbonate (PBC)) can give higher yields of energy dense (33 – 91 wt% oil yield) than HTL of biopolymers (10 – 50 %). The heating value of the oil produced from some plastics is higher than that from biopolymers and comparable to petroleum. The oil, solid, and aqueous products obtained contain moieties similar to monomeric units of plastics that could be recycled.

Separation of biopolymers from plastics for recycling has been a challenge due to contamination, source management and enormity of waste produced. This limitation is extended to most chemical and mechanical recycling techniques. HTL offers a unique platform to valorize together most components of municipal solid waste to obtain useful products while bypassing the need for separation. The products obtained are separated into 4 phases – non-polar liquid, polar liquid, solid, and gas. HTL can be hence be viewed as an intensified step that combines reaction and separation.

Co-liquefaction experiments were performed with mixtures of nine model compounds (5 biopolymers, 4 plastics) to mimic the composition of MSW. Experiments were conducted from 300 – 425 °C with a batch holding time of 30 min. Synergistic interactions were calculated by comparing the oil yield of the mixture with the oil yields from HTL of the biopolymer and plastics individually. Note that synergistic interactions are observed among plastic, biopolymers in their respective mixtures. The percentage synergy in the nine-component mixture exceeds the synergies obtained among the separate mixtures of plastic alone and biopolymer alone. This indicates the presence of positive interactions between plastics and biopolymers that increase the oil yield and hence the energy recovered from the process. A high energy recovery of 45 percent was obtained at the lowest operated temperature (300 °C). The largest percentage synergy (105 %) is also observed at the same temperature. The presence of these synergistic behaviors at a low operating temperature aids at obtaining higher energy recoveries at lower operating temperatures which is economically and environmentally beneficial.

To further pinpoint the binary interactions between biopolymers and plastics that cause such synergy – more experiments were conducted by manipulating the mixture composition. We will also report on a mathematical model that would predict the energy recovered from isothermal operation of HTL for any given mixture of synthetic polymers and biopolymers. This would further aid in enabling feedstock engineering to increase the net energy recovered to more than 45% i.e. adding waste feedstocks to optimize yield obtained from available waste feedstock.

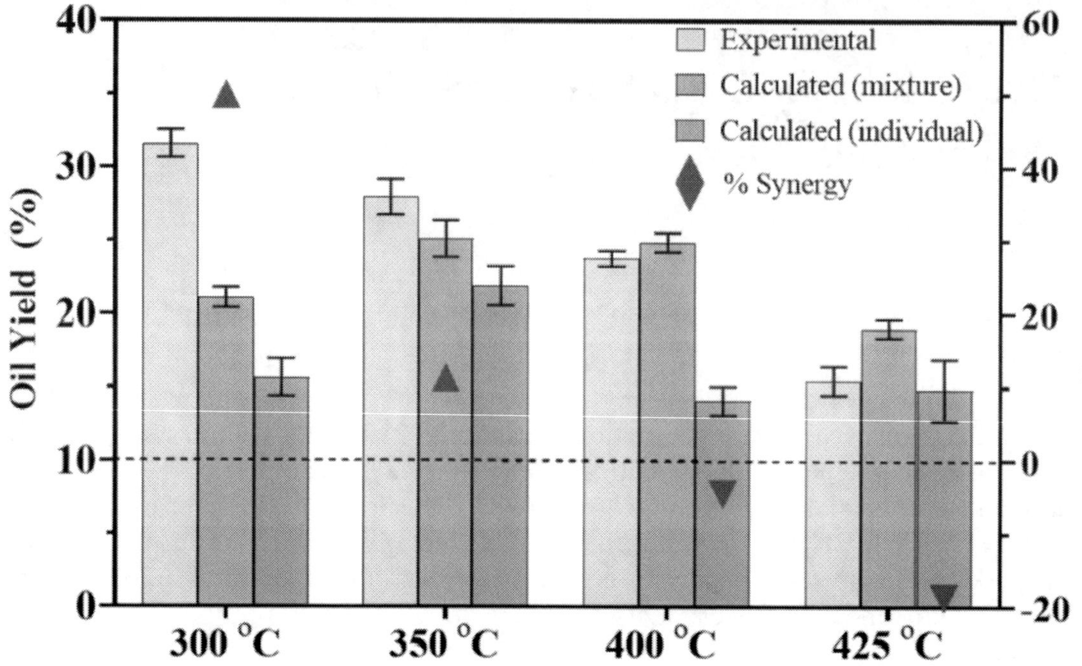

Fig: Oil yield from HTL treatment of MSW mixture based on model compounds (cellulose, starch, lignin, soy protein and lipid, PPE, PBC, PS, PET) at different temperatures with calculated (individual) weight averaged yield based on experiments on model compounds, calculated (mixture) weight averaged yield based on experiments on plastic and biopolymer mixture and % synergy

613277 Polystyrene Sulfonic Acid As a Recoverable Catalyst for the Hydrolysis of PET

Abstract:

Poly (4-styrenesulfonic acid) (PSSA) was applied as a recoverable acid catalyst to hydrolyze polyethylene terephthalate (PET) to terephthalic acid (TPA) and ethylene glycol. The impact of proton concentration and temperature on depolymerization of PET were compared for PSSA and H_2SO_4. The PSSA system exhibited more rapid hydrolysis kinetics than the sulfuric acid at all conditions studied in this work. The hydrophobic backbone of PSSA improved surface wetting of the PET as measured by contact angle. The increase in reaction rate was attributed to higher localized concentration of protons at the surface of the PET by the PSSA solutions. The PSSA maintained catalytic activity after use and recovery for five cycles at 150 °C.

613276 Conversion of Waste Solids to Monomers and Intermediates

Abstract:

In this study a process design and assessment for a waste plastics recycling plant to produce monomers and refinery-like liquids for use as intermediate chemicals is presented. The catalyst is comprised of a promoted and modified ZSM-5 zeolite catalyst spray dried into an attrition resistant, fluidizable, spherical shape to be used in a circulating fluid bed reactor. A feed stock of mixed, co-mingled solid waste composed of plastics, polyols, polyethers, polyurethanes, paper, rubber, tires, auto shredder residue, composites, agricultural waste, MRF bales (including rejected and residuals) and the like is added to a fluidized bed reactor in a melted state and the fluidized catalyst bed is operated in the range of 550 – 600° C at 1.5 – 3 atm, typically. The catalyst regeneration will take place at 600 – 750° C. The mass balances are in the range of 100 $\pm$ 10% and the analytical measurements of the reaction products are conducted by a duel function gas chromatograph equipped with a flame ionization detector and thermal conductivity detector. The light products wt% yields, CO, CO_2, H_2, C_1-C_3, sum to below 20 wt% yield and the wt% liquids falls above 80%.

601264 Valorization of Plastic Waste Using Hot Pressurized Water

Abstract:

Plastic waste can be processed in hot pressurized water (hydrothermal liquefaction) to recover energy (oil) and chemicals (monomers). Most waste plastics (76 wt%) are landfilled, with just 9 wt% being mechanically recycled and 15 wt% being burned for energy recovery. The need for separation of plastics, costs, and environmental impact has greatly curbed the growth of mechanical recycling. Direct energy recovery processes such as combustion release pollutants such as oxides of sulfur and nitrogen. Pyrolysis is a potentially viable process for valorizing some plastics (e.g., polyolefins), but it is not well suited for others (e.g., polyesters). If done at scale, pyrolysis will likely require feedstock separation, both of different types of plastics and of plastics from non-plastics. This difficulty might be overcome by processing plastics in hot pressurized water.

Hydrothermal liquefaction (HTL) takes place in water at near- and supercritical conditions (T_c = 374 °C). Water can act as a solvent, reactant, or catalyst for an array of acid- or base-catalyzed organic chemical reactions. These favorable conditions aid in depolymerization reactions at relatively low temperatures. In this study, we focused on the energy recovery in the oil phase and monomer recovery in oil, aqueous, and solid phases from HTL of polyesters - polyethylene terephthalate (PET), polycarbonates - poly bisphenol-A carbonate (PBC) and polyolefins - polypropylene (PPE), polystyrene (PS) at 300 – 450 °C and 30 – 60 mins batch holding times.

Energy-dense oils are produced from not just polyolefins but polycarbonates and polyester as well. PPE and PET produce considerable oil under supercritical conditions (425 °C), while PS and PBC produce considerable oil even under subcritical conditions (350 °C). Plastics other than polystyrene gave the highest energy recoveries at 425 °C, 30 mins – PPE (29 %), PBC (66 %), PET (35 %). This means that different plastics can be processed using HTL under the same process conditions to achieve optimal energy recoveries. Hence, HTL can recover energy efficiently in the oil phase from plastics without the need for any separation.

The water in the HTL process acts as a reactant and several products obtained from PBC and PET (as detected from GC/MS) likely arose from hydrolysis. Also, PPE (a hydrocarbon) mainly produces long-chain alcohols. Hence, water molecules interact with PPE or PPE-derived intermediates as the PPE decomposes. Correlations between GC/MS, oil yield, and CHNS analysis show that dehydration reactions (occurring at 400 – 425 °C) cause an increase in heating values in oils from HTL of PBC and PET, in turn, resulting in high energy recoveries at these temperatures. At higher temperatures (450 °C), aromatization becomes the major pathway and leads to a reduction in heating value (oil phase).

The suitability of HTL for chemical recycling of plastics depends on the individual plastic in view. We found high concentrations of Bisphenol A in the oil phase from HTL of PBC at 300 °C. Note that no other plastic decomposes at 300 °C. Hence, differential HTL processes can be used to recover selective monomers from a mixture of plastics. PET is another material from which recovery of chemicals is possible. Benzoic acid (derivative of terephthalic acid) was the major product in the oil phase from HTL of PET at higher temperatures (450 °C). Thus, there may be opportunities to recover chemicals, along with oil, from the product streams from HTL of plastics. We will also report on HTL processing of nylon 6, nylon 6,6, polylactic acid, and polysulfone under similar conditions.

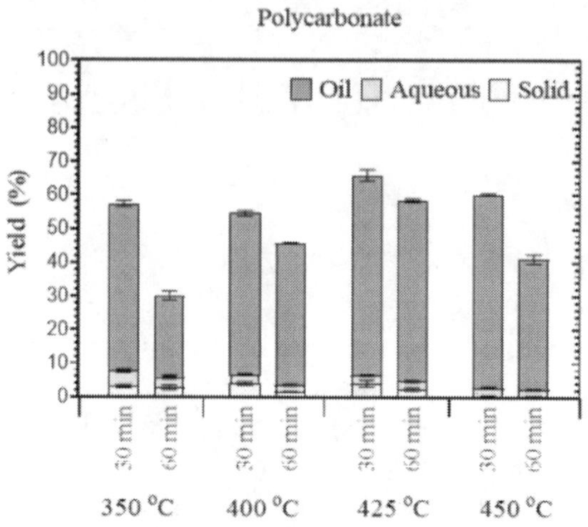

Top 3 compounds for PBC at various temperatures as detected from GC-MS									
300 °C		350 °C		400 °C		425 °C		450 °C	
Bisphenol-A	72	Benzene, 1-methoxy-2,4-dimethyl-	40	Benzene, 1-methoxy-2,4-dimethyl-	39	Benzene, 1-methoxy-2,4-dimethyl-	56	Benzene, 1-methoxy-2,4-dimethyl-	48
Phenol, 4-(1-methyl-1-phenylethyl)-	21	Phenol	17	2-Hepten-4-yne, 2-methyl-	19	Phenol, 4-(1,1-dimethylpropyl)-	13	Phenol, 3,4-dimethyl-	11
Phenol	2	Phenol, 4-(1,1-dimethylpropyl)-	15	Phenol	11	Bisphenol-A	12	Phenol, 4-(1-methylpropyl)-	6

Fig: Variation of product fraction yields (wt.%) from HTL of polycarbonate with temperature and time and top 3 concentrated products detected by GC/MS with corresponding percentage peak area at different temperatures with batch time of 60 mins.

600896 Iron Oxide Supported on Silicalite-1 As a Multifunctional Material for Biomass Chemical Looping Gasification and Syngas Upgrading

Abstract:

Iron oxide supported on silicatite-1 was synthesized as a novel oxygen carrier for biomass chemical looping gasification (BCLG) with additional functionalities of its reduced form as a catalyst for catalytic decomposition of tar and an adsorbent for capturing CO_2 in syngas. The temperature-programmed-reduction/oxidation profiles showed that the iron oxide supported on silicatite-1 had lower reduction and oxidation temperatures than the iron oxide supported on conventional silica. It was found the iron oxide supported on silicatite-1 was resistant to the formation of strong metal-support interaction and irreversible silicate at a high temperature and exhibited high thermal/chemical stability in a cyclic process. The experimental results showed that the novel oxygen carrier can gasify biomass into high value compounds of H_2 and CO at a moderate temperature. After reduction at 800°C, the iron supported on siliclaite-1 showed higher catalytic activity in tar decomposition than on a conventional silica support, due to the higher surface area and better metal dispersion. As an adsorbent for CO_2 capture, the reduced oxygen carrier with 15% Fe had a 1.01 mmol/g CO_2 adsorption capacity at 1 atmosphere pressure, which is comparable to other CO_2 adsorbents reported in literature. Therefore, the iron oxide supported on silicatite-1 is an effective looping material as an oxygen carrier for biomass gasification, a catalyst for tar decomposition and an adsorbent for CO_2 capturing.

EVALUATION OF THE IRON-BASED MATERIAL WITH Al_2O_3/ZrO_2 ADDITION AS OXYGEN CARRIER APPLIED IN CHEMICAL LOOPING PROCESS

C.T. Kao*, J.Y. Cheng, C.H. Shen and H.P. Wan

Green Energy and Environment Research Laboratories, Industrial Technology Research Institute, Taiwan

Abstract

Chemical looping process is one of the promising technologies for carbon capture. Oxygen carrier plays an important role in the chemical looping process. By the reaction of oxidation and reduction, iron could carry or release the oxygen by its phase change. In the present study, iron-based composite with alumina and zirconia addition were used for oxygen carriers in chemical looping. Material properties including relative density, microstructure, reactivity and mechanical ones were measured. Feasibility of this composite for chemical looping could therefore be evaluated.

The result indicated that all of the specimens formed a layer of Fe_2O_3 in the surface. Different ionic diffusion rate made this phase separation occurred. Moreover, the reactivity was then decreased. In the case of the addition of ZrO_2, an extra oxygen vacancy will produce due to the substitutional reaction between Zr and Fe. Although the specimen which only composited with ZrO_2 and Fe_2O_3 showed the best reactivity, the consuming of Fe_2O_3 might lead it not suitable to be the material for chemical looping.

Keywords: chemical looping, oxygen carrier, zirconia, microstructure, redox reaction

1. Introduction

There was more than 70% electricity generated by fossil fuel in Taiwan during 2010 to 2018. Moreover, over 90,000 kton CO2 are produced each year. According to the statics from IEA 2016 (International energy the global average CO_2 emission per capita was 4.35 tCO_2. While the CO_2 emission in Taiwan was 10.98 tCO_2 per capita, which was twice as much as global average.

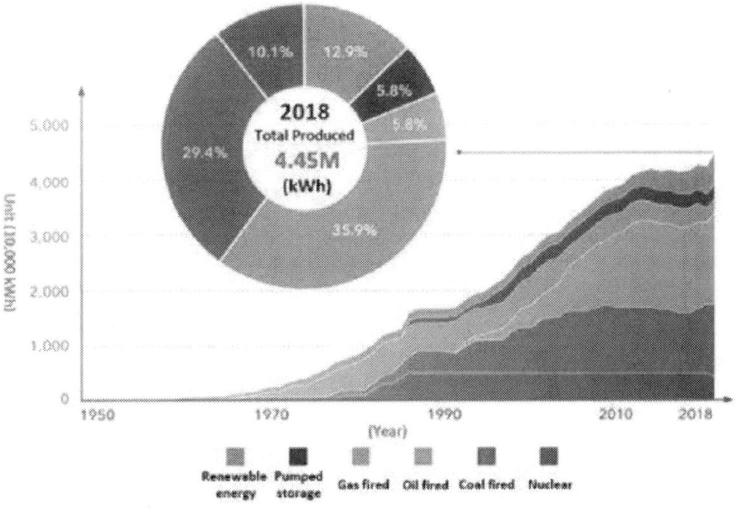

Fig.1 Resource for electricity generation in Taiwan

Among the types of fossil fuel using for electricity generation in Taiwan, coal occupied more than 60% of CO_2 emission. The problem of the CO_2 emission in Taiwan should indeed be solved. The greenhouse gas reduction and management act came into force in 2015. This act laid down GHG emission reduction targets in 2050 for Taiwan should reduce by 50% from 2005 levels.

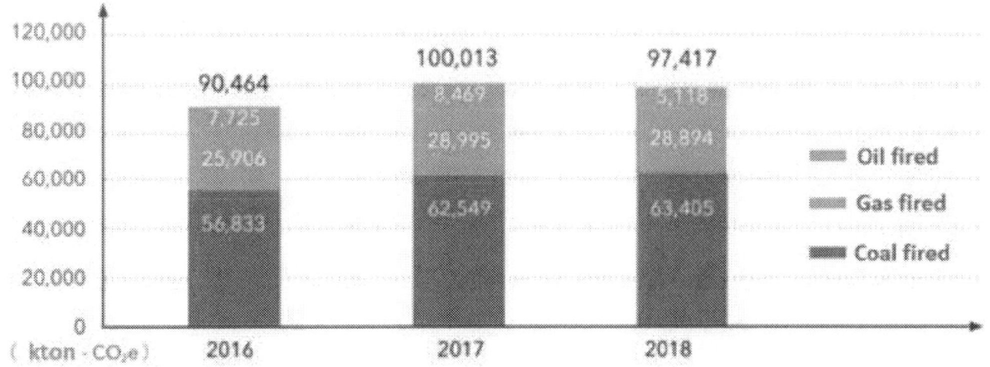

Fig.2 Amount of CO_2 emissions produced from different type of fossil fuel

Chemical looping is one of the most promising routes to solve the CO2 emission fundamentally. It doesn't need extra facility to separate CO2 and other gas.

The concept of chemical looping was shown in Fig. 4. Metal oxide (Me_xO_y) which is called oxygen carrier (O.C.) plays an important role for the chemical looping. The major function for oxygen carrier was to transport oxygen between oxidation and reduction reactor.

Fossil fuel will pass through the reduction reactor and then react with metal oxide (Me_xO_y). Only CO2 and H2O are produced during this reactor. Afterward, H2O could be removed by cooling process. High purity of CO2 was collected and could storage directly. Afterward, the oxygen depleted particle (Me_xO_z) will be transported to oxidation reactor and then react with air. The oxygen could be recovered again. Meanwhile, heat will also generate during this process.

Chemical looping combustion(CLC) was one of the applications in chemical looping process. The major purpose of CLC technology is a low-carbon alternative to the fossil fuel power generation. Fossil fuel such as coal, natural gas could be reacted with the oxygen carrier and then produce the heat and then generate the electricity.

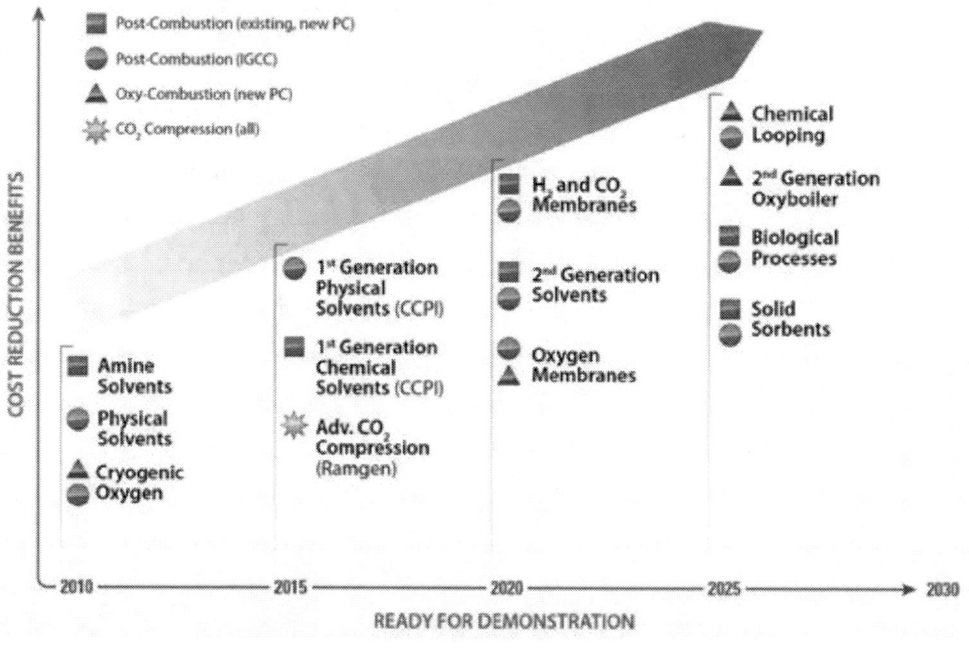

Fig.3 Progress and the future trend for CCS

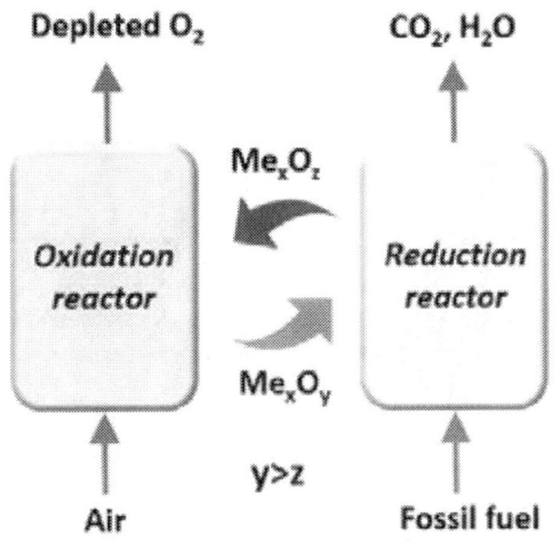

Fig.4 Concept for chemical looping process

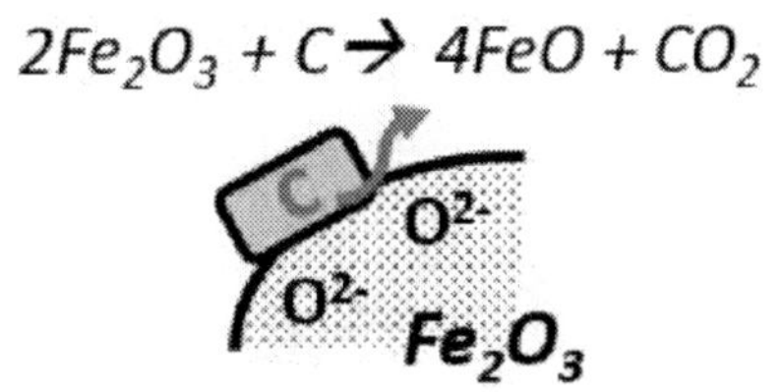

Fig.5 Concept for chemical looping combustion (CLC) process

The progress of chemical looping system developed in ITRI was exhibited in Fig. 6. Both lab. scale testing system including TGA (Thermogravimetric analyzer) and fixed bed system which combined with GC (Gas chromatograph) were used for the performance measurement. Moreover, a 30 kW_t pilot scale system was also established by our group.

For the initial stage (2013-2018), chemical was used as the material for oxygen carrier in chemical looping. Recently, we tried to search the material which was cheaper and environment friendly as the oxygen carrier.

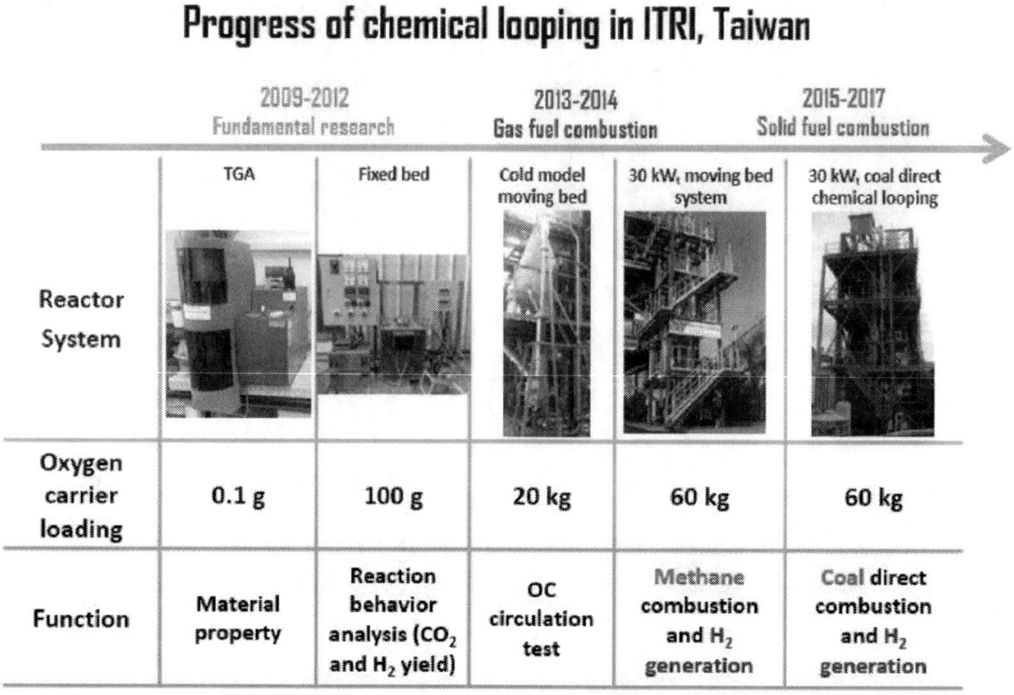

Fig.6 Progress of chemical looping system developed in ITRI

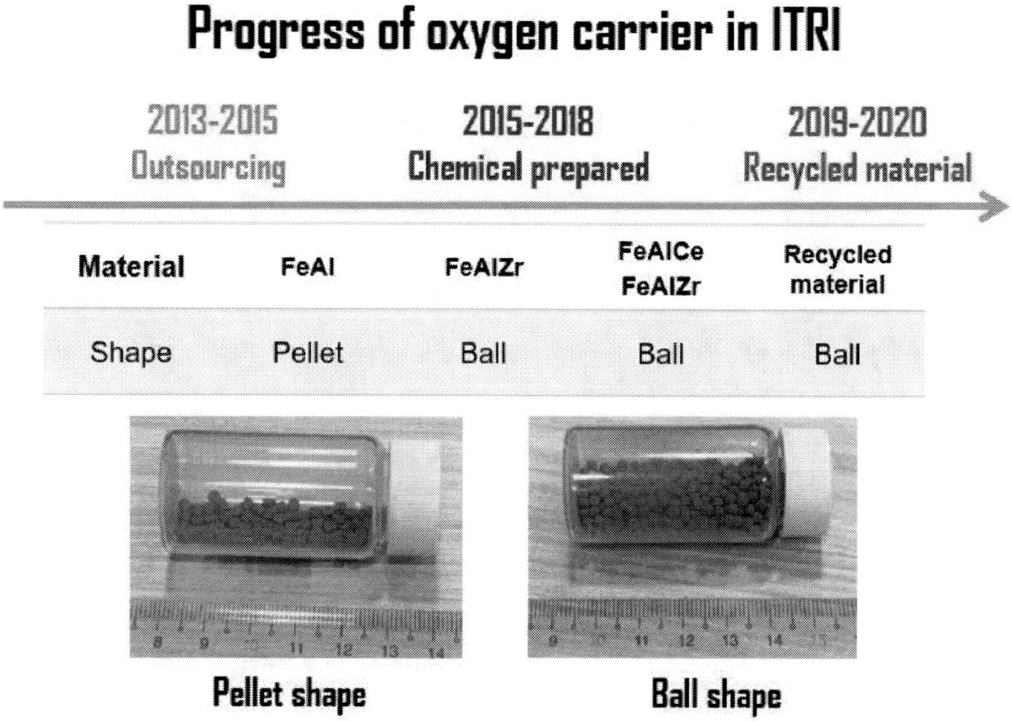

Fig.7 Progress of oxygen carrier developed in ITRI

2. Experiment procedure

The preparation procedure for the composite of Al_2O_3/ZrO_2 addition to iron-based showed in Fig. 8. Starting powders including Fe_2O_3, Al_2O_3, ZrO_2 and pore former were mixing in dilute water by mechanical method for 15 h. Drying, granulation and sieving process were then conducted. Particle size around 270 to 500 um was chosen for the following pressing process. Pellet shape with 3 mm in diameter was pressed by die-pressing machine. De-binder and sintering process were then conducted in order to remove the pore former and sinter the specimens.

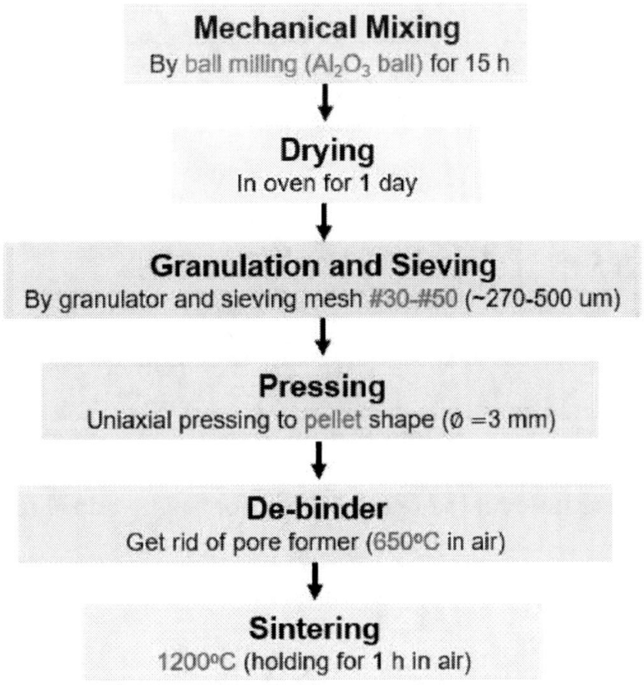

Fig. 8 Flow chart of the specimens prepared in present study

Image of the as-sintered specimens could be found in Fig. 9. For the reactivity test, 50 redox reaction cycle was performed for all of the as-sintered specimens. The parameter of reactivity test for each cycle was shown in Fig. 10. The definition of reactivity could be found in the following equation.

$$X_{OC} = \frac{w_O - w_{(t)}}{w_O - w_R}$$

W_0 is the weight of the composite being fully oxidized
W_R is the weight of the composite being fully reduced
$W_{(t)}$ is the weight of the composite at specific time point

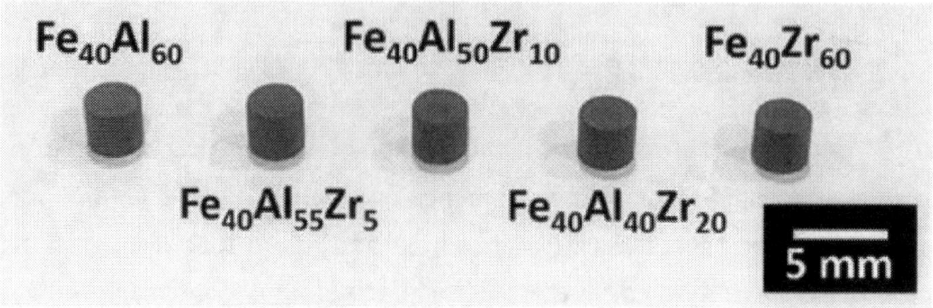

Fig. 9 Image of the as-sintered specimens

Atmosphere	Composition and purity	Flow rate	Holding time
Oxidation	Air 99.9%	200 cc/min	10 min
Purge gas	N_2 99.9%	200 cc/min	5 min
Reduction	H_2 20% with N_2 compensation	200 cc/min	40 min
Purge gas	N_2 99.9%	200 cc/min	5 min
Ramp 10°C/min from room temperature to 900°C			

Fig. 10 Parameter of the reactivity test for single redox reaction cycle

Image for the specimens reacted with 50 redox reaction cycle was shown in Fig. 11. There was a red layer attached in the surface of specimens. $Fe_{40}Zr_{60}$ was broken after reactivity test.

Fig. 11 Image of the as-sintered specimens reacted with 50 redox reaction cycle

One of the bottlenecks for the chemical looping was its cost. A thousand MW_t chemical looping hydrogen generation system (CLHG) was applied for simulation the economic analysis in by Aspen plus in ITRI. The result showed that nearly 40% of the cost for CLHG was derived from the consuming of oxygen carrier material. For both moving bed or fluidized bed reactor, the particle of oxygen carrier should withstand such as impact or abrasion during circulation. Afterward, the particle will loss or break, and need to supplement again.

To elevate the reactivity and durability for oxygen carrier was the major purpose. Both Al_2O_3 and ZrO_2 were applied as the inner support material in the iron-based material. The feasibility of Al_2O_3/ZrO_2 addition in iron-based for chemical looping was then revealed.

3. Result and discussion

The reactivity of the Al_2O_3/ZrO_2 addition in iron-based composition was studied in present study. Microstructure variance was introduced to explain how it affect the reactivity for as-sintered specimens.

Microstructure observation:

The microstructure of the as-sintered were shown in Fig. 12. All of the specimens exhibited porous structure which was attributed by the elimination of pore former during de-bindering process. High porosity could provide more area to oxygen carrier and fuel gas to react.

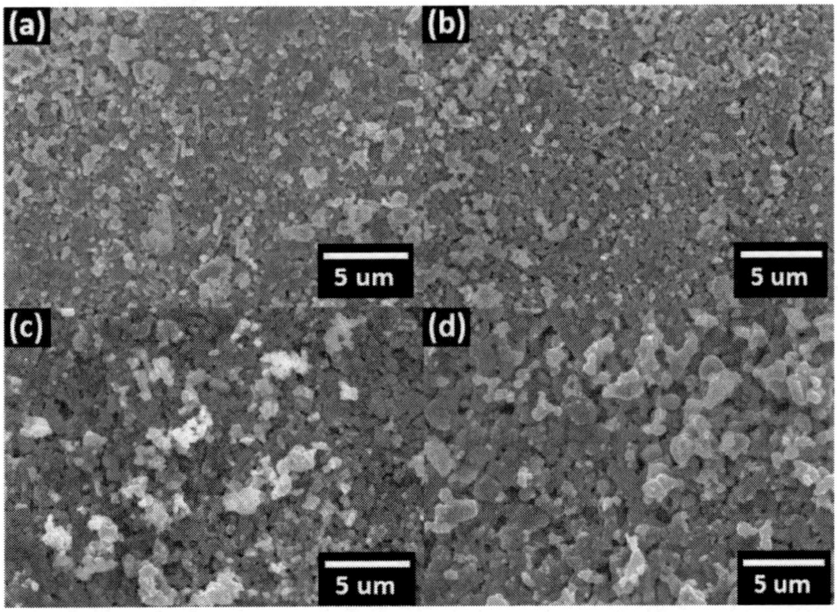

Fig. 12 Microstructure of the as-sintered specimens (a)$Fe_{40}Al_{60}$; (b)$Fe_{40}Al_{50}Zr_{10}$; (c)$Fe_{40}Al_{40}Zr_{20}$ and (d)$Fe_{40}Zr_{60}$.

The specimens after 50 cycle redox reaction was mounted into the epoxy in order to observe its cross-section view. The corresponded microstructure was shown in Fig. 13. And the brighter particle was identified as the Fe_2O_3. Along with the addition of

ZrO_2, the thickness of Fe_2O_3 layer formed in the surface was also increased. Moreover, this layer will detach from the surface in the specimen of $Fe_{40}Zr_{60}$.

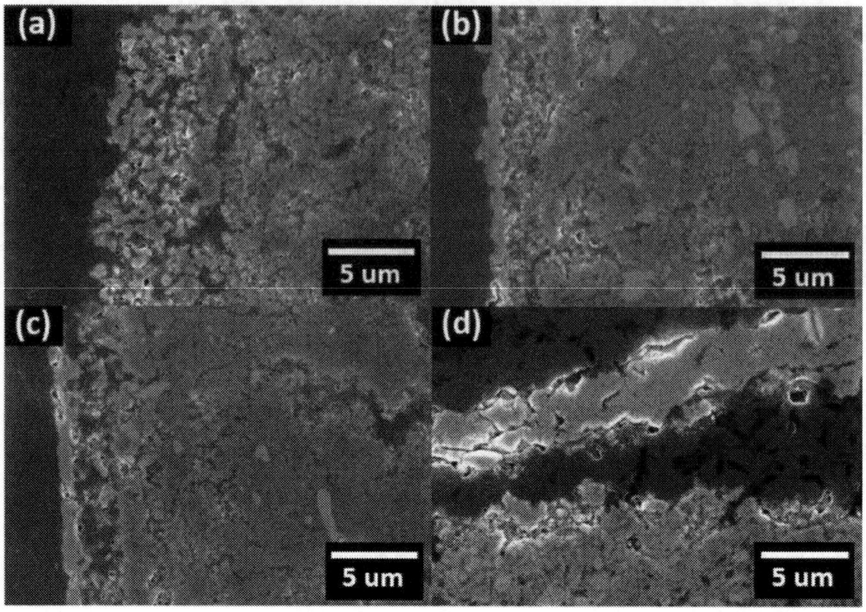

Fig. 13 Microstructure of the as-sintered specimens reacted with 50 redox reaction cycle (a)$Fe_{40}Al_{60}$; (b)$Fe_{40}Al_{50}Zr_{10}$; (c)$Fe_{40}Al_{40}Zr_{20}$ and (d)$Fe_{40}Zr_{60}$.

Layer formation mechanism:

The formation of Fe_2O_3 layer could be regard as the phase separation. By Ellingham diagram, only Fe_2O_3 has phase change during redox reaction. Fe_2O_3 will migrate to the surface in order to react and then segregate in the surface.

Defect in this composite could explain the result which exhibited in Fig. 12. Oxygen vacancy is deem as the most critical defect which could accelerate the ion diffusion in ceramic material. It provides the lower energy gap and easier route for ion diffusion in ceramic material.

For the addition of Al_2O_3 into Fe_2O_3. There was no any extrinsic oxygen vacancy produced.

$$Fe_2O_3 \xrightarrow{Al_2O_3} Fe_{Al}^{\times} + O_O^{\times}$$

However, the addition of ZrO_2 into Fe_2O_3 will produce extra oxygen vacancy. The diffusion rate in this composite was then accelerated.

$$\frac{1}{2}Fe_2O_3 \xrightarrow{ZrO_2} Fe_{Zr}' + \frac{1}{2}V_O^{\cdot\cdot} + O_O^{\times}$$

Therefore, the thickness of the Fe_2O_3 layer for ZrO_2 addition was thicker than that in Al_2O_3 addition.

The reactivity for 50 cycle redox test was shown in Fig. 14. $Fe_{40}Zr_{60}$ showed the best reactivity with around 70%. Other as-sintered specimens showed the similar reactivity with around 45%.

The Fe_2O_3 layer formed in surface might play as the barrier which inhibited the reaction inside the specimen. And the reaction area was thus only occurred in surface.

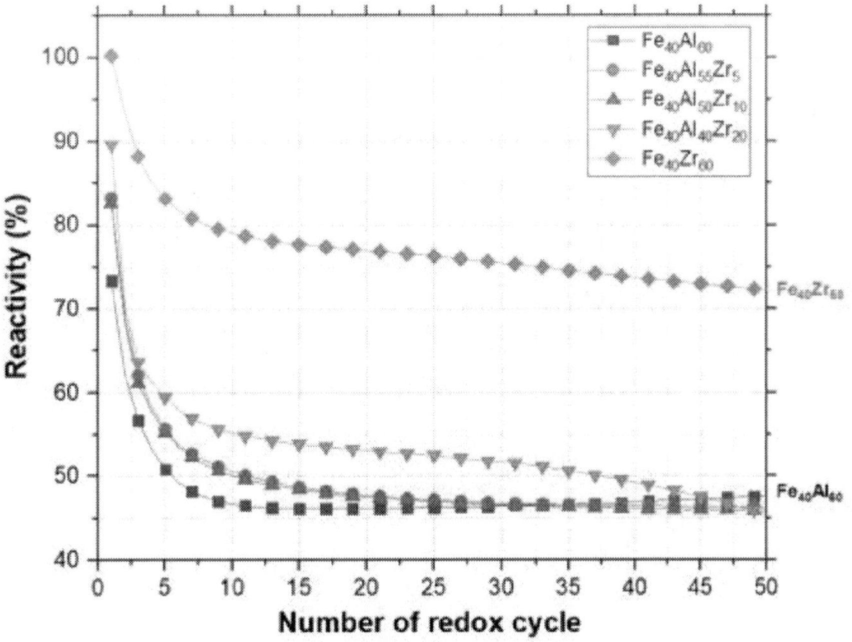

Fig. 14 Reactivity of the as-sintered specimens reacted with 50 redox reaction cycle.

In case of $Fe_{40}Zr_{60}$, because of the Fe_2O_3 layer was too thick to detach, Fig. 13(d), the inner material could thus expose and react with fuel gas again. The highest reactivity for $Fe_{40}Zr_{60}$ to all of specimens was therefore exhibited.

Although the specimen of $Fe_{40}Zr_{60}$ showed the best reactivity, the expected reactivity of $Fe_{40}Zr_{60}$ which over 50th redox cycles might be gradually decreased due to the consuming of Fe_2O_3. Fe_2O_3 layer will formed and then detach to the mother body repeated.

The addition of ZrO_2 in Fe_2O_3 showed the negative effect to the mechanical properties, Fig. 15. The major reason could be caused by the phase transformation from zirconia and it decreased the relative density.

Composition	Relative density (%)	Crash strength (kgf)	Attrition (%)
$Fe_{40}Al_{60}$	66.42 ± 1.96	57.17 ± 11.75	0.75
$Fe_{40}Al_{50}Zr_{10}$	62.43 ± 5.49	40.93 ± 7.60	1.22
$Fe_{40}Al_{40}Zr_{20}$	65.39 ± 4.61	54.45 ± 7.86	0.70
$Fe_{40}Zr_{60}$	54.45 ± 5.81	27.81 ± 5.42	2.03

Fig. 15 Mechanical property of crash strength and attrition.

4. Conclusion

The feasibility of the addition of Al_2O_3 and ZrO_2 in iron-based oxygen carrier was revealed in the present study. The corresponded microstructure before and after 50 redox reaction cycle were also exhibited.

For the reactivity test conducted in present study, all of the as-sintered specimens showed a drastic decrease within the first ten redox cycle. The Fe_2O_3 layer formed in the surface played as the barrier to inhibit the deeply reaction to the inner layer material.

Due to the substitution reaction between Zr and Fe, additional oxygen vacancy was produced and then promoted the ion diffusion. The thickness of the Fe_2O_3 layer which formed in the surface was thus increased.

$Fe_{40}Zr_{60}$ showed the best reactivity among the as-sintered specimens in present study. However, the worst mechanical property and the consuming of Fe_2O_3 during redox reaction make it be excluded the candidates for being the material for chemical looping.

Acknowledgements

This project was supported by Bureau of Energy, Ministry of Economic Affairs, project number 108-E0216.

610444 Feasibility Analysis of Chemical Looping Ammonia Synthesis from Stranded Natural Gas.

Abstract:

Stranded Natural Gas (SNG) flaring is intermittent flaring of natural gas at a volume or location which makes it uneconomical to recover by traditional methods (compression to liquid natural gas, or combustion for electricity). SNG flaring releases a substantial amount of greenhouse gas emissions, prompting public and private entities to focus efforts on developing technologies to convert SNG to a transportable value-added product.[1] Ammonia production from SNG is a practical solution, considering it is industrially produced from natural gas today through the Haber Bosch process (1).

N_2+H_2NH_3, *T=450°C, P=200atm*

While the high temperature and pressure of the Haber Bosch process is needed to push the reaction equilibrium forward at an industrially acceptable rate, it makes the process uneconomical at small scales [2]. Recently attention has been given to chemical looping for ammonia synthesis (CLAS), because of its ability to produce ammonia at atmospheric pressures and favorable small-scale economics [2]. A literature review identified previously proposed chemical loops for ammonia synthesis with varying levels of technical success. Therefore, the objective of this work is to provide a techno-economic assessment of ammonia production from SNG using chemical looping synthesis.

Chemical looping is capable of circumventing the high pressures of the Haber Bosch process by separating the reaction into individual steps mediated by a material carrier. Generally, in CLAS a material carrier is used to fix nitrogen, then a hydrogen source is used to reduce the carrier and evolve ammonia leaving the material carrier to complete the chemical loop. Recently, a chemical loop using alkali metals as the material carrier was demonstrated to produce ammonia at atmospheric pressures and 100 [3]:

$xN_2 + 4AH_x\ 2xA_{2/x}NH + xH_2$

$2xA_{2/x}NH + 2xH_2xNH_3 + 2AH_x$

(A denotes an alkali metal and x is the valence of A)

A different chemical loop was proposed in [4] using water as the hydrogen source mediated by a calcium carrier. Although, the hydrolysis of calcium nitride is thermodynamically favored to produce ammonia it also produces a stable calcium oxide. This results in an energy intensive process to regenerate calcium and complete the chemical loop:

$3Ca + N_2\ Ca_3N_2$

$Ca_3N_2 + 3H_2O\ 2NH_3 + 3CaO$

$CaO + R\ Ca + RO$

(R denotes suitable reducing agent)

Here, to evaluate previously proposed chemical loops for mild condition synthesis of ammonia, focus is given to their thermodynamic feasibility. Thermodynamic feasibility is determined as a negative Gibbs free energy of reaction at a temperature in which the constituents of the reaction do not phase change. Along with thermodynamic feasibility, process safety, reaction yield and other criteria are used to create a decision tree to rationally evaluate chemical loops presented in this work. For each material carrier and chemical loop combination, the temperature that minimizes the Gibbs free energy of reaction at a constant pressure of 1 atm was found, assuming ideal states. Aspen plus was used to simulate the equilibrium reactions and to assess economic viability of the process using its economic analyzer. From this work it was discovered that modifying a previously proposed unsuccessful chemical loop, creates a new feasible chemical loop, shown in Fig.1. This new chemical loop along with other techno-economically feasible chemical loops are studied and presented in this work. Further studies are recommended to study the kinetic feasibility of the proposed chemical loops for industrial applications.

Acknowledgements

This work was partially sponsored by the United Technologies Corporation Institute for Advanced Systems Engineering (UTC-IASE) of the University of Connecticut. Any opinions expressed herein are those of the authors and do not represent those of the sponsor.

References

[1] Natural Gas Flaring and Venting: State and Federal Regulatory Overview, Trends, and Impacts, 2019.

[2] R. Michalsky, B.J. Parman, V. Amanor-Boadu, P.H. Pfromm, Solar thermochemical production of ammonia from water, air and sunlight: Thermodynamic and economic analyses, Energy. 42 (2012) 251–260. doi:10.1016/j.energy.2012.03.062.

[3] W. Gao, J. Guo, P. Wang, Q. Wang, F. Chang, Q. Pei, W. Zhang, L. Liu, P. Chen, Production of ammonia via a chemical looping process based on metal imides as nitrogen carriers, Nat. Energy. 3 (2018) 1067–1075. doi:10.1038/s41560-018-0268-z.

[4] R. Michalsky, Thermochemical production of ammonia using sunligth, air, water and biomass, Kansas State University, 2012.

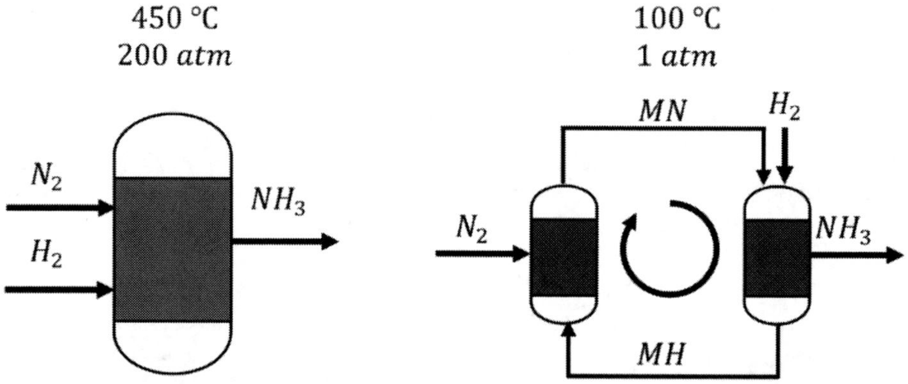

Figure 1. Ammonia production via the Haber Bosch process (left) compared to chemical looping process (right). Where M is the metal carrier, MN is the metal nitride and MH is the metal hydride.

596792 Oxygen Vacancy Healing for Desorption of Stored Ammonia from a Metal/Zeolite Nanocomposite

Abstract:

Recently, the US Department of Energy (DOE) has set targets for the storage and delivery of hydrogen for light passenger vehicles which require a sorbent material with reversible storage, low cost, fast kinetics, and stability in ambient conditions. In this regard, gaseous ammonia has gained major attention for its potential as a high density hydrogen carrier. Storage of ammonia in porous materials could address many of the DOE target metrics and offer additional benefits including hazard mitigation, minimal pressurization and cooling, and facile material handling.

Zeolites offer highly tunable pore chemistries and geometries, which allows precise control of the chemical environment of the adsorbate. The intrinsic acidity of zeolites promotes strong gas-solid interactions but simultaneously hinders the effective release of the stored gas due to chemisorption. To overcome this strong binding energy, temperature or pressure swing is typically employed to remove adsorbed ammonia.

In the current work, the partial reduction of a nickel-oxide/zeolite nanocomposite led to a remarkable heat release upon exposure to air, which was attributed to the exothermic healing of oxygen vacancies in the NiO_{1-x} lattice. This heating phenomenon was shown to be repeatable over 10 at least cycles. Temperatures as high at 694 °C were measured in the catalyst bed on exposure to air. By applying the redox properties of the nickel-oxide/zeolite nanocomposite, significant desorption of ammonia was promoted. The removal efficiency was found to be 38%, an improvement in comparison to vacuum swing desorption (30%). Future outlooks are proposed to improve the total ammonia uptake and to improve the cycling process.

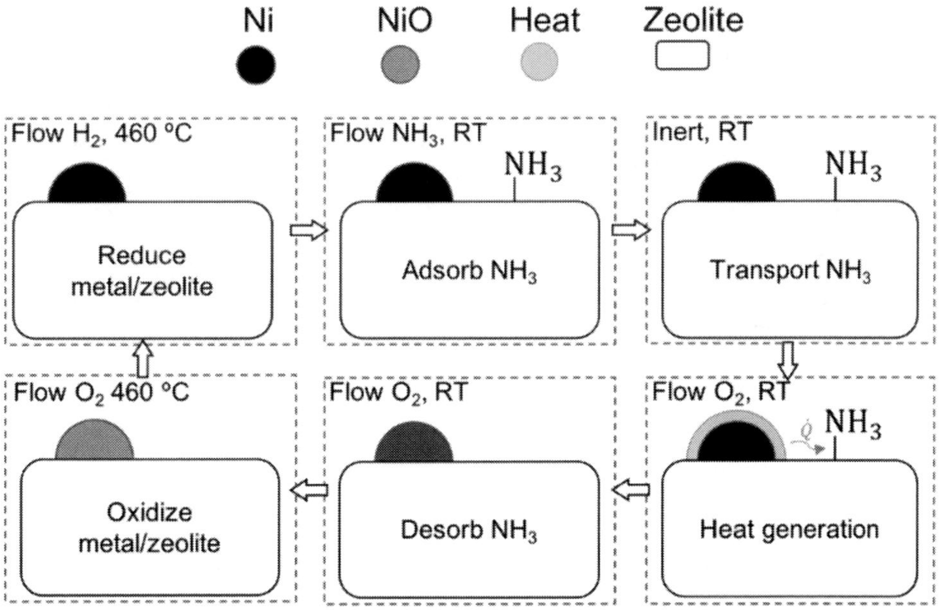

Scheme 1. Ammonia storage on metal/zeolite sorbents using the redox properties of the reducible metal sites

601361 A Molten Carbonate Shell Modified Perovskite Redox Catalyst for Anaerobic Oxidative Dehydrogenation of Light Alkanes

Abstract:

Acceptor doped, redox-active perovskite oxides such as $La_{0.8}Sr_{0.2}FeO_3$ (LSF), $La_{0.8}Sr_{0.2}MnO_3$ (LSM) and $La_{0.8}Sr_{0.2}Co_{0.2}Fe_{0.8}O_3$ (LSCF) are active for light alkane oxidation to CO_x. However, they show poor selectivity to olefins. This study investigates molten alkali metal salts as effective "promoters" to modify the surface of perovskite oxide for chemical looping–oxidative dehydrogenation (CL-ODH) of light alkanes including ethane, propane and butane. Li_2CO_3 coated LSF (LSF@Li_2CO_3) is an efficient redox catalyst for CL-ODH of ethane [1]. Unlike previously reported redox catalysts which are typically endothermic during the lattice oxygen assisted ethane ODH step, the labile oxygen species in LSF@Li_2CO_3 facilitate exothermic operation in both redox steps as confirmed by TGA-DSC measurements (-69.5 kJ/mol [O] during the ODH step and -40 kJ/mol [O] during the re-oxidation step). Up to 92.2% ethylene selectivity and 63.6% ethane conversion were obtained. TEM, XPS, XRD characterizations and DSC measurement indicate that the redox catalyst is composed of a layer of molten Li_2CO_3 covering the solid LSF substrate. $^{18}O_2$-exchange experiments and electrochemical impedance spectroscopy indicate that the molten Li_2CO_3 layer facilitates oxygen shuttling from LSF bulk to the molten carbonate layer surface while blocking the non-selective sites for ethane oxidation. Further investigations of the potential reaction pathways indicates that peroxide species (O_2^{2-}) are the most likely active species for CL-ODH. TGA measurements, *in-situ* XRD and Mössbauer spectroscopy indicate that Fe^{4+} species reduction to Fe^{3+} is responsible for the formation of the active peroxide, which are subsequently transported to the outer surface of the molten Li_2CO_3 layer for the ODH reaction (**Fig. 1a**). The formation of active peroxide via Fe^{4+} to Fe^{3+} transition is further supported by DFT calculation. With other molten alkali metal salt promoters, the perovskite oxide can also be modified for CL-ODH of propane and butane. More than 50% of C_{2+} olefin yield (ethylene + propylene) at 600 °C for CL-ODH of propane was obtained. In a similar manner, >40% C_4 olefin yield at 500 °C for CL-ODH of butane was obtained (**Fig. 1b**). With high olefin selectivity, high olefin yield and favorable heat of reactions, the core-shell redox catalyst has excellent potential to be effective for intensified light alkane conversion. The mechanistic findings also provide a generalized approach for designing CL-ODH redox catalysts.

Reference

[1] Gao et al., *Science Advanced*, Accepted in 2020, **In Press**

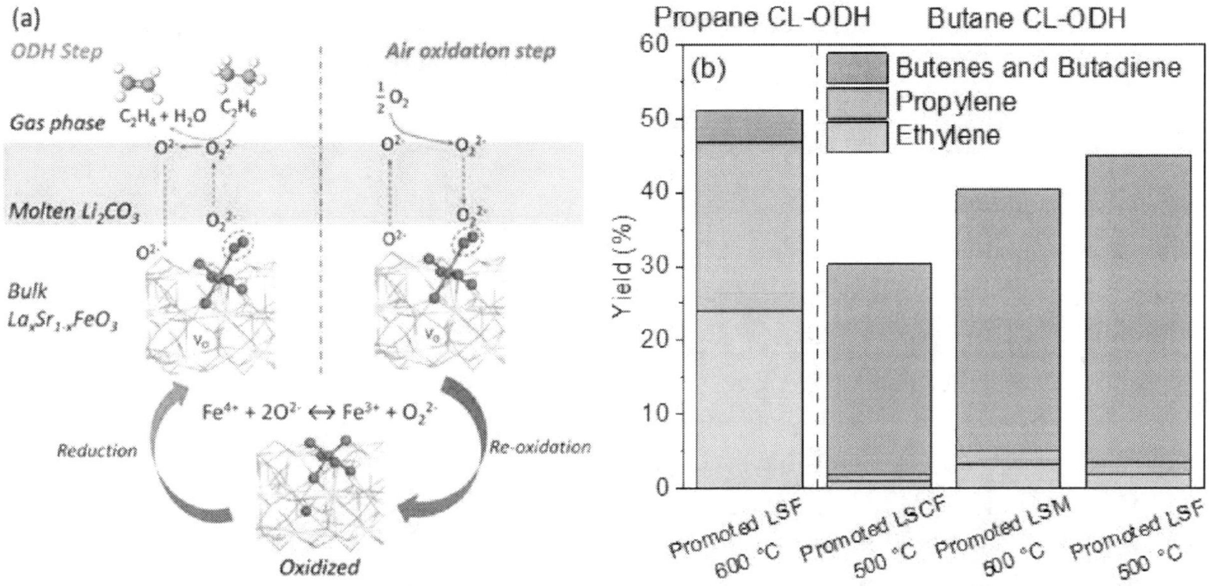

Figure 1. (a) Mechanistic drawing of ethane CL-ODH on Li_2CO_3 promoted LSF; (b) Butane and propane CL-ODH on perovskite oxides promoted with Li-containing salts.

609815 Solar Driven Iron Oxide Based Thermochemical Methane Reforming and Water Splitting for H_2 and Syngas Production

Abstract:

Due to the continuous increase in the population of world and drastic depletion of the fossil fuel reservoirs, it is highly essential to invest towards renewable energy technologies such as solar energy (storage, conversion and utilization). A two-step solar thermochemical H_2O and/or CO_2 splitting process which utilizes metal oxide based redox reactions is one of the promising ways of producing solar fuels such as solar H_2 or renewable precursors for fuels such as solar syngas (a mixture of H_2 and CO). This process includes the endothermic reduction of metal oxides at elevated temperatures by releasing O_2 and oxidation of the reduced metal oxide by H_2O, by CO_2, or by the mixture of the two producing H_2, CO or syngas. One of the major limitations associated with the MO based cycles is the higher thermal reduction temperature. To solve this issue associated with the higher reduction temperatures, methane can be used as the reducing agent. In this investigation, thermodynamic analysis of the two-step iron oxide based solar thermochemical methane reforming and water-splitting process was investigated in two sections with the help of commercial thermodynamic software's such as HSC Chemistry and FactSage. In the first section, the thermodynamic equilibrium compositions were determined as a function of temperature and other operating parameters at different experimental conditions. In the second section, the second law thermodynamic analysis of iron oxide based solar thermochemical methane reforming and water-splitting process was performed. Once the process flow diagram was set, effects of operating parameters on solar absorption efficiency of the solar reactor (η_{abs}), solar energy input to the solar reactor (Q_{solar}), rediation heat losses from the solar reactor (Q_{re-rad}), net energy absorbed in the solar reactor ($Q_{reactor-net}$), rates of entropy produced in the solar reactor ($I_{rr,reactor}$) and cooling unit ($I_{rr,cooling}$) were calculated and plotted. At the end, the solar-to-fuel conversion efficiency ($\eta_{solar-to-fuel}$) for this process was estimated and compared with the other thermochemical fuel production cycles.

598068 Kinetic Studies on Chemical Looping – Oxidative Dehydrogenation of Ethane Using Surface Modified Mixed Oxides

Abstract:

Ethylene is commercially produced from ethane via high-temperature pyrolysis in the presence of diluting steam a.k.a. steam cracking. The process requires high temperature (>1100 K) and a large steam generation load, making it energy intensive with an energy consumption of 17-21 GJ/tonne ethylene. Alternatively, Chemical Looping – Oxidative Dehydrogenation (CL-ODH) has the potential to substantially reduce the energy demand and to increase the single-pass ethane conversion. The ability of a redox catalyst for selective hydrogen combustion (SHC), i.e. selectively oxidizing hydrogen co-product from ethane dehydrogenation, represents an effective strategy for CL-ODH because it can shift the reaction equilibrium and facilitate autothermal operation.

In this work, redox kinetics of unpromoted and Na_2WO_4-promoted $CaMnO_3$ based redox catalysts were investigated in the context of CL-ODH. In the case of unpromoted $CaMn_{0.9}Fe_{0.1}O_3$, the reduction kinetics follow reaction-order models. The activation energy for C_2H_4 reduction is approximately three times greater than that for H_2 reduction. The greater dependence on active lattice oxygen concentration and the larger energy barrier under C_2H_4 reduction result in an order-of-magnitude decrease in the reduction rate. Despite of its favorable SHC properties, $CaMn_{0.9}Fe_{0.1}O_3$ redox catalyst leads to significant CO_2 formation in CL-ODH of ethane. To further improve the redox catalyst's selectivity towards hydrogen combustion, a Na_2WO_4 promoter was impregnated on $CaMn_{0.9}Fe_{0.1}O_3$. The reduction of Na_2WO_4-promoted $CaMn_{0.9}Fe_{0.1}O_3$ follows an Avrami-Erofe'ev nucleation and nuclei growth model. The addition of Na_2WO_4 more significantly suppressed C_2H_4 combustion relative to H_2 oxidation. As a result, the reduction rate of Na_2WO_4-promoted $CaMn_{0.9}Fe_{0.1}O_3$ under H_2 was three orders of magnitude greater than that under C_2H_4, demonstrating its excellent SHC properties.

In addition, a number of other mixed oxide redox catalyst particles with different alkali metal salt promoters were also investigated, showing promising performance in terms of long-term stability and single-pass ethylene yield at different operating temperature ranges. Both the kinetic studies and long-term experiments validate the feasibility and attractiveness of the CL-ODH approach.

601024 A High-Throughput Method for Analyzing Struvite Formation and Morphology: Exploring the Effect of Peptide Additives

Abstract:

Precipitation of struvite ($MgNH_4PO_4*6H_2O$), a slow-release fertilizer, provides a means of recycling phosphate from wastewater streams. Through the implementation of high throughput methods, many different precipitation parameters can be investigated at once. In this work, these reactions occur in small volumes (300 µL or less) in a 96 well plate over the course of 45 minutes. A high throughput platform allows for a minimal use of chemicals and a fast way to compare many reaction conditions. The formation of struvite was investigated by fitting absorbance at 600 nm over time to a first order model with induction time. The impact of peptide addition was assessed through this platform, determining the peptide causes significant changes in the yield parameter and formation constant – allowing for control of precipitation. The impact of struvite seed dosing was also investigated, highlighting the importance of system optimization when peptide is present. Composition of struvite was confirmed through Fourier-transform infrared spectroscopy (FTIR) while morphology and crystal size were analyzed through optical microscopy. Crystals were elongated when precipitated with the peptide present. Finally, the utility of the high-throughput platform was demonstrated with a 2^5 full factorial design to capture the effects and interactions of: magnesium dose, mixing time, seed dose, pH, and temperature.

A Techno-Economic Analysis of Greenhouse Waste and Water Recycling Using Anaerobic Digestion: A Case Study in Food-Energy-Water Nexus

Abstract

The US DOE estimates that about 111 million tons of agricultural crop residues are produced every year(Turhollow, Perlack et al. 2014). The total world production is estimated at nearly 1 billion tons/year(Cherubin, Oliveira et al. 2018). Sustainable agriculture would require us to recover and recycle as much of the materials and energy contained in these wastes. In this paper, we will examine the techno-economics of crop residue recycling using anaerobic digestion for greenhouses in Almeria, Spain which supplies nearly 20% of produce for Spain (About 3 million tonnes/year)(Sayadi-Gmada, Rodriguez-Pleguezuelo et al. 2019). Much of the cultivation is done on a year around basis. Large amount of waste material is produced after every season, which is typically hauled away to landfills where it decays producing unwanted greenhouse gases polluting the atmosphere. The total waste is estimated to contain around 1.35 million tonnes of crop residues and 0.5 million tonnes of rejected produce(Egea, Torrente et al. 2018).

Anaerobic digestion (AD) of the waste represents a potential route to create an integrated and sustainable agricultural model (Kalra and Panwar 1986, Corneli, Dragoni et al. 2016, Sukhesh and Rao 2018). As Shown in attached figure, AD produces mainly 4 product streams all of which has value associated with it. The biogas produced can be used to generate heat and power both of which are needed commodities for greenhouse operation. The waste water produced is rich in nutrients and therefore can supply much needed irrigation water thus reducing the stress on local aquifers and desalination plants. The solid digestate produced can be converted to compost manure after treatment and thus forms a source of organic fertilizer for the next round of crops. The main issue facing the application of this technology is one of economics. Implementation of such a waste recycling system can provide a sustainable solution from the point of view of Food-Water-Energy Nexus as applied to greenhouse agriculture (Haltas, Suckling et al. 2017, Hoolohan, Soutar et al. 2019, Falconer, Haltas et al. 2020).

For the case study, we selected a 100 hectare greenhouse cooperative and carried out an assessment of the economics of constructing and operating an AD system from the crop residues produced. Based on prior studies, the total crop residues was estimated at 4900 tonnes/year (tpy) containing about 3400 tpy of Volatile Solids (VS) and about 500 tpy of water. Additional water needed for dilution in the AD was estimated at 20,000 tpy. The biogas production from the digester was estimated at 1400 tpy containing 660 KW of energy which is roughly 1/3 of the energy content in the waste biomass. The biogas can be used to produce nearly 200 KW of electricity and 460 KW of heat energy both of which are needed in the greenhouse operation and therefore directly recyclable. The wet digestate is sent to a filtration system where much of the water is recovered but is enriched with nutrients. Therefore after some pretreatment, this water can be recycled back for greenhouse irrigation. Likewise the solid digestate is rich in minerals and after treatment, becomes valuable as a compost fertilizer which is also a needed commodity for the greenhouse cultivation.

In this paper, we will present the results of a preliminary economic analysis of the process. We will take a look at the economic feasibility of the process as well as the sensitivity to variables such as waste disposal tipping fee, capital expenditures required and the cost of energy recovered from the AD system.

References

Cherubin, M. R., D. M. D. Oliveira, B. J. Feigl, L. G. Pimentel, I. P. Lisboa, M. R. Gmach, L. L. Varanda, M. C. Morais, L. S. Satiro, G. V. Popin, S. R. de Paiva, A. K. B. dos Santos, A. L. S. de Vasconcelos, P. L. A. de Melo, C. E. P. Cerri and C. C. Cerri (2018). "Crop residue harvest for bioenergy production and its implications on soil functioning and plant growth: A review." Scientia Agricola **75**(3): 255-272.

Corneli, E., F. Dragoni, A. Adessi, R. De Philippis, E. Bonari and G. Ragaglini (2016). "Energy conversion of biomass crops and agroindustrial residues by combined biohydrogen/biomethane system and anaerobic digestion." Bioresource Technology **211**: 509-518.

Egea, F. J., R. G. Torrente and A. Aguilar (2018). "An efficient agro-industrial complex in Almeria (Spain): Towards an integrated and sustainable bioeconomy model." New Biotechnology **40**: 103-112.

Falconer, R. E., I. Haltas, L. Varga, P. J. Forbes, M. Abdel-Aal and N. Panayotov (2020). "Anaerobic Digestion of food waste: Eliciting sustainable water-energy-food nexus practices with Agent Based Modelling and visual analytics." Journal of Cleaner Production **255**.

Haltas, I., J. Suckling, I. Soutar, A. Druckmann and L. Varga (2017). "Anaerobic digestion: a prime solution for water, energy and food nexus challenges." Proceedings of 1st International Conference on Sustainable Energy and Resource Use in Food Chains (Icfes 2017), Including Symposium on Heat Recovery and Efficient Conversion and Utilisation of Waste Heat **123**: 22-29.

Hoolohan, C., I. Soutar, J. Suckling, A. Druckman, A. Larkin and C. McLachlan (2019). "Stepping-up innovations in the water-energy-food nexus: A case study of anaerobic digestion in the UK." Geographical Journal **185**(4): 391-405.

Kalra, M. S. and J. S. Panwar (1986). "Anaerobic-Digestion of Rice Crop Residues." Agricultural Wastes **17**(4): 263-269.

Sayadi-Gmada, S., C. R. Rodriguez-Pleguezuelo, F. Rojas-Serrano, C. Parra-Lopez, S. Parra-Gomez, M. D. Garcia-Garcia, R. Garcia-Collado, M. B. Lorbach-Kelle and T. Manrique-Gordillo (2019). "Inorganic Waste Management in Greenhouse Agriculture in Almeria (SE Spain): Towards a Circular System in Intensive Horticultural Production." Sustainability **11**(14).

Sukhesh, M. J. and P. V. Rao (2018). "Anaerobic digestion of crop residues: Technological developments and environmental impact in the Indian context." Biocatalysis and Agricultural Biotechnology **16**: 513-528.

Turhollow, A., R. Perlack, L. Eaton, M. Langholtz, C. Brandt, M. Downing, L. Wright, K. Skog, C. Hellwinckel, B. Stokes and P. Lebow (2014). "The updated billion-ton resource assessment." Biomass & Bioenergy **70**: 149-164.

607360 Engineering a Purple Non-Sulfur Bacterium to Expand Symbiotic Nitrogen Fixation

Abstract:

The production of nearly 50% of the world's food relies on ammonia fertilizer. While this source of bioavailable nitrogen has led to a marked increase in the number of people that can be fed, the impact on the environment is striking. Ammonia production requires more energy and releases more carbon dioxide than any other industrial reaction, and up to 90% of the product is used to manufacture fertilizers. Furthermore, soil microbes quickly convert ammonia to nitrate, which is easily washed into our water supply. Elevated levels of nitrates can be toxic to humans and animals. Sustainable innovations are required to maintain nutritional security while reducing the environmental damage. When nitrogen-stressed, legumes form an endosymbiotic relationship with bacteria that can fix the critical nutrient. Unraveling the mechanisms responsible for this symbiosis could lead to the expansion of the relationship to cereal crops and a decrease in agriculture's reliance on nitrogen fertilizers. Using the more basic Nod factor-independent relationship between the legume *Aeschynomene evenia* (*A. evenia*) and the photosynthetic bacterium *Bradyrhizobium* sp. ORS278 as a template, the long-term goal is to engineer the closely-related, non-endosymbiotic *Rhodopseudomonas palustris* CGA009 (*R. palustris*) to nodulate the legume while resolving the prerequisites for triggering the symbiotic signal transduction pathway employed by both rhizobacteria and arbuscular mycorrhizal fungi.

Determining *R. palustris*' response to *A. evenia*'s Nodule-specific Cysteine-Rich (NCR) peptides is the first step towards this goal and the focus of this work. NCR peptides are expressed in the nodules of some legumes, such as *A. evenia*, and cause their symbiotic partner to differentiate into nitrogen-producing bacteroids. It has been shown that the NCR peptides from some legumes induce signs of differentiation in their symbiotic partner in vitro, but kill other bacteria. Since *R. palustris* harbors the proteins that have been shown to be essential for the *Bradyrhizobium* sp. ORS278's differentiation in *A. evenia*, the hypothesis is that *R. palustris* and *Bradyrhizobium* sp. ORS278 will respond similarly to the in vitro treatment with the legume's NCR peptides. Changes in both bacteria's ploidy, morphology, membrane permeability, and gene expression, all signs of differentiation, will be presented, in addition to work towards consistently creating knockout strains in the polyploid *R. palustris* using CRISPR.

610606 A Microalgae-Methanotroph Coculture Platform for Fuels and Chemical Production from Wastewater

Abstract:

Wastewater resource recovery facilities are major energy consumers in a community, as well as major contributors for greenhouse gas (GHG) emission. Although anaerobic digestion (AD) is widely employed in wastewater treatment to reduce the amount of solid organic waste and the sludge produced, the use of the produced biogas is mostly limited to heating and electricity generation, while the nutrient rich digestate still requires further treatment. In this work, we propose a waste-to-value (W2V) platform based on a microalgae-methanotroph coculture, which can convert AD-generated biogas into value-added products, while simultaneously removing nutrients from digestate. The coculture platform explores the synergistic interactions within a microalgae-methanotroph coculture to achieve significantly improved productivity of microbial biomass and enhanced nutrient recovery performance.

Using *Chlorella sorokiniana – Methylococcus capsulatus* (Bath) as the model coculture, we demonstrate that the coculture offers a highly promising platform for W2V technologies, which can efficiently recover energy (from CH_4) and carbon (from both CH_4 and CO_2) to produce microbial biomass, while removing nutrients from wastewater to produce treated clean water. Specifically, the coculture demonstrated robust growth on minimally processed AD effluent (via gravitational setting), and showed better growth performance than autoclaved AD effluent. In addition, the nutrient recovery performance of the coculture was compared with microalgae single culture, and the coculture demonstrated significantly improved nutrient removal performance, a direct result of enhanced coculture growth. Next, we showed that the coculture could achieve zero GHG emission without external supply of oxygen, as well as complete removal of inorganic nitrogen and phosphorus from AD effluent. Finally, the potential applications of the wastewater-derived coculture biomass are discussed.

607714 Auxin Degradation By *Variovorax* Maintains Stereotypic Plant Root Development within the Complex Plant Microbiome

Abstract:

Plant-microbe and microbe-microbe interactions within the rhizosphere contribute both positively and negatively to plant physiology, health, and productivity. To test hypotheses about these interactions and define the bacterial genes responsible, we are using a synthetic community of 185 fully genome sequenced bacteria isolated from *Arabidopsis* roots. Through a large screening of *Arabidopsis* plants across various stress gradients (pH, temperature, phosphate, and salt), bacterial colonization patterns robust to these stresses were identified. Then through analysis of subset communities, the genus *Variovorax* was identified as being required to maintain stereotypic root development in these mixed bacterial communities. This ability was traced to *Variovorax* reverting the phenotypic effects of many root growth inhibition (RGI)-inducing bacteria by consuming indole-3-acetic acid (IAA), an auxin plant hormone produced by both plants and bacteria. Using a combination of comparative genomics, RNA-Seq, and screening of a gain-of-function plasmid library in *E. coli*, a single locus conserved in all sequenced *Variovorax* genomes, containing IAA degradation genes, was identified. Using bacterial genetics, this locus was deleted, confirming its role in stereotypic root development. This locus was also expressed in a closely related bacterium from the genus *Acidovorax*, conferring this strain with the ability to degrade IAA and partially revert IAA-induced plant phenotypes. Sub-cloning of genes from this locus have identified the essential genes for IAA degradation. Further, we are investigating the regulation of this locus through a MarR transcriptional regulator. Isothermal titration calorimetry has demonstrated this MarR binds to IAA and related auxins, as well as a promoter element upstream of the degradation genes. Crystal structures of the MarR regulator identified the IAA binding residues and mutation analysis confirms their role in IAA binding. Understanding the *Variovorax* genes required for auxin degradation, and their regulation through the MarR regulator, will enable more carful engineering of this locus in *Variovorax* or related bacteria. The ability to control auxin homeostasis and promote stereotypic root development through this mechanism is of interest for inclusion in biofertilizers and microbial probiotics in agricultural applications.

Reference

Finkel OM†, Salas-González I†, Castrillo G†, Conway JM†, Law TF, Teixeira PJPL, Wilson ED, Fitzpatrick CR, Jones CD, Dangl JL. 2020. A single bacterial genus maintains root development in a complex microbiome. bioRxiv:645655. Doi: 10.1101/645655 († These authors contributed equally to this work)

610492 Catalytic Pyrolysis of Waste Automobile Tires to Low Sulfur Fuels

- Download/View Extended Abstract

Abstract:

Energy production and utilization is one of the main drivers of human civilization and until we master how to produce and distribute it efficiently and sustainably, the human civilization will always be threaten for survival. Currently, there are millions of tons of organic materials, which are classified as waste because we do not know how to efficiently utilize them to benefit humankind. A classic example is the so-called waste automobile tires, which are ubiquitous in every country. Disposal of these tires are increasingly becoming a challenge because of limited landfills and fire problems. Improperly disposed waste tires can potentially become the sources of diseases such as malaria fever and zika virus because they become breeding grounds for mosquitoes that carry the disease. The traditional disposal methods such as land filling, road applications, playground applications, direct pyrolysis to produce high sulfur heating fuels, and combustion as tire-derived fuels are becoming inadequate because of pollution problems associated with them and the low cost petroleum crude oil prices currently prevailing. In the USA about 300 million tires are disposed of annually either in landfills, combustion, and other disposal methods. Tire pyrolysis is extensively practiced in several countries especially, China and India where they are used to produce tire pyrolysis oils for heating. These oils contain 1-1.35 wt% sulfur, which make them unsuitable for transportation fuels where the sulfur content is typically below 0.5 %. The sulfur content of these oils can be reduced through various sweetening or desulfurization processes, but these unit operations increase the cost of fuel and thus noncompetitive relative to petroleum derived heating fuels. At the USTAR Bioenergy Center we have developed a sulfur-tolerant catalyst for catalytic pyrolysis that effectively removes over 60 wt% of the sulfur in the pyrolysis oil and also removes the chlorine in the butyl rubber liner. Thus, the pyrolysis oil can be easily upgraded to higher value fuel. In general, most waste tire pyrolysis oils reported in literature contain 0.80 – 1.35 wt% sulfur for atmospheric pyrolysis oil compared to 0.38 wt% that we measured for the oils produced from the tire crumbs used for these studies. The oxygen content of the oil was less than 1%, which meant that the oil will need minimal hydrogenation to remove the residual oxygen. The hydrodeoxygenation of the tire pyrolysis oils also converted the residual sulfur to hydrogen sulfide and hence making the fuel sulfur free. The higher heating value (HHV) of the pyrolysis oil was 42.7 Mg/kg and after hydrotreatment increased to 45.7 MJ/kg. The ASTM D86 distillation of the pyrolysis showed 95% diesel fraction while after hydrotreating the diesel fraction was 55% and gasoline fraction was 35%. Thus, the catalytic tire pyrolysis oil would be suitable a marine diesel fuel because of its low sulfur content and high energy content.

599918 Probabilistic Analysis of Converting Marine-Borne Plastics into Usable Fuels

Abstract:

Oceanic plastics represent a major global environmental pollution problem. Every year between 4.8 and 12.7 million tons of plastic enter the ocean and as a result a potentially dangerously high amount of plastic has already accumulated in the oceans. This plastic is highly concentrated in the naturally occurring gyre's in the ocean which could aide in any clean-up efforts, but the question remains what should be done with the plastic once it is collected? We have proposed a process that would involve a ship collecting plastic in the ocean and converting it into fuel in an onboard hydrothermal liquefaction reactor to fuel itself. Before starting experimental work, we wanted to determine whether the afore mentioned process could in fact make more energy than it consumes. To do this in the presence of uncertainty a probabilistic exergy analysis was carried out on a theoretical system for three different and common plastics, polystyrene, polyethylene and polypropylene, as well as for a mixed plastic stream. The theoretical process consisted of a plastic shredder, reverse osmosis system, reactor, separator and an ultraviolet & hydrogen peroxide oxidation system. The exergy model accounted for the energy requirements of the equipment, heating up the reactants to the reaction temperature and the chemical exergy of the streams before and after reaction. This analysis allowed us to determine the break even point (when the net exergy was equal to zero) as well as the point at which there was a 50% chance of having net exergy production and create probability curves as seen in Figure 1. The analysis showed that with less than 10% by volume plastic in the feed (ocean) enough energy could be created to run the equipment and fuel a boat large enough to hold it. The work also allowed for the creation of a decision-making tool that allows one to assess the risk associated with a project.

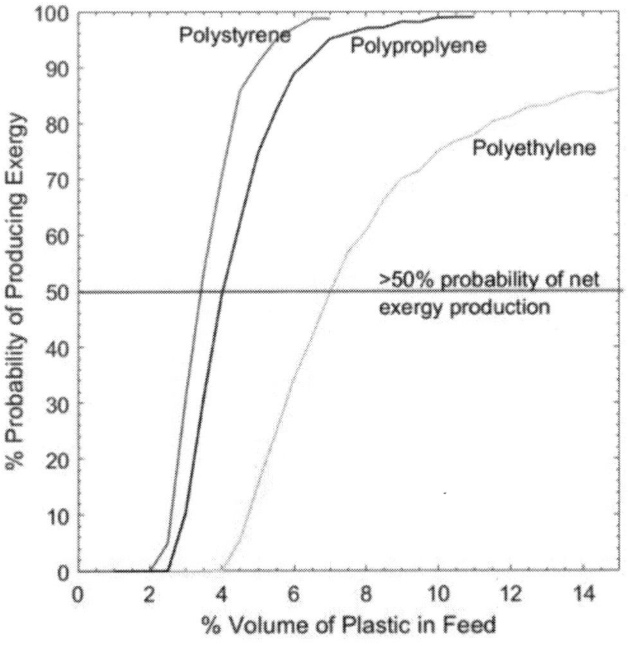

Figure 1 Exergy analysis of polystyrene, polypropylene and polyethylene including the energy required for the ship's engine.

611928 Struvite Recovery from Manure By Way of Hydrothermal Carbonization

Abstract:

Dairies and other animal feed lots have grown in size in recent years. Traditional practices of manure disposal were focused on land application, but this is problematic for large operations. The manure is either applied heavily in nearby fields, leading to overapplication and deleterious effects on environment, or the cost of hauling to spread on wider spaced fields is prohibitive. At the same time, there is an opportunity to recover valuable nutrients in manure, primarily phosphorous but also nitrogen, into a valuable form and reduce the costs associated with manure management.

Hydrothermal carbonization has been established as a promising treatment technology for wet biomass. A recent study has shown its potential in solubilization of nutrient content of cow manure into the hydrothermal aqueous product specially when its combined with a strong organic acid such as citric acid. In the current study dairy manure was treated by HTC process at 170C for 10 minutes with different concentration of citric acid. The resulting liquid phase was then used as a supernatant for struvite crystallization at pH range of 9-10. We show experimentally effects of HTC reaction conditions on recovery of phosphorous in struvite. Different ratios of N:P and Mg:P were studied to obtain the optimum precipitation condition. The effect of different concentrations of the organic acid on the crystallization mechanism and the quality of the final struvite were investigated.

603049 Conversion of Recycled Drywall into Nitrogen Efficient Fertilizers

Abstract:

New, cost-effective and sustainable methods are needed to decrease the reactive N-losses into the environment from mineral fertilizers. For this purpose, we demonstrate a successful mechanochemical synthesis of urea cocrystals containing secondary nutrients calcium (Ca) and sulfur (S) using widely abundant drywall gypsum waste. Resulting $CaSO_4 \times 4$urea cocrystal possesses unique reactive properties. In particular, it possesses a higher urea decomposition temperature of 195 ^oC as it foregoes the melting of urea at 137 ^oC as inferred from thermogravimetric measurements. A combination of Dynamic Vapor Sorption (DVS) and *in situ* Raman spectroscopy measurements suggested a much lower $CaSO_4 \times 4$urea propensity to adsorb and react with water as relative humidity than urea alone. *Ab initio* thermodynamics framework was devised and showed a large apparent thermodynamic barrier of crystalline $CaSO_4 \times 4$urea to transform into urea and the resulting reactive gases, NH_3 and CO_2. Detailed N and S measurements suggested their enhanced availability when using cocrystals especially during the early stages of the experiments. The synthesis process described using waste drywall gypsum to form $CaSO_4 \cdot 4$urea cocrystal mechanochemically thus can provide a new value proposition for this abundant waste and contributing to sustainable use as it can decrease the use of various enzyme inhibiting synthetic organophosphorus compounds.

604232 Synthesis and Characterization of Improved Energy Materials from Spent Lithium-Ion Battery Cathodes

Abstract:

Lithium-ion batteries (LiBs) are widely used as electrochemical power sources in energy applications because of their high energy and power density. With the global market for electric vehicles estimated to reach $150 billion by 2025, an efficient and sustainable metal recovery process has become essential to minimize pollution created by disposing of LiBs in landfills. In this work, a closed-loop, sustainable process for recovery of Li and Co from $LiCoO_2$ (the most common cathode of LiBs) is presented using oxalate chemistry. Oxalic acid has the unique property of separating Li and Co directly by precipitating cobalt oxalate dihydrate ($CoC_2O_4{\cdot}2H_2O$) during the digestion while lithium oxalate ($Li_2C_2O_4$) remains in solution. The addition of hydrogen peroxide as an external reducing agent during the metal extraction process reduces the amount of oxalic acid required for digestion and results in the precipitation of cobalt oxalate dihydrate having a unique micro-rod structure. The rod-like structure is maintained after calcination to form unique cobalt oxide materials (Co_3O_4 and CoO) that can be used in synthesizing high-performance $LiCoO_2$ cathode materials. This presentation will describe the experimental and mechanistic details of the synthesis and characterization of the improved energy materials along with their applications.

602611 Upgrading Biochar Via Co-Pyrolysis of Agricultural Biomass and Spent Tires/Sulfur Wastes for Environmental Remediation

Abstract:

This study was carried out to use 5 biochars (woochip (WC) biochar, raw sulfur/rice straw (RS) biochar, modified sulfur/RS biochar, spent tire/RS biochar, and CO_2-activated biochar) as sorbents for explosive (2,4-dinitrotoluene (DNT)), halogenated phenol (2,4-dichlorophenol (DCP)), toxic metals (Pb and Ba) and toxic anions (Cr and Se). RS biochar was also evaluated as control material. Spent tire/RS biochar was the most effective sorbent for DNT, showing the maximum sorption capacities of 23.4 mg/g. Statistical analysis indicates that BET surface area and hydrophobic may be involved in the sorption of DNT to biochar. Compared to RS biochar, WC biochar samples effectively remove DCP considering maximum sorption capacity and properties of sorbents and sorbates, it appears that the sorption capacity of biochar for DCP is related to the pH, pK_a of DCP to biochar. Raw sulfur/RS biochar and CO_2-activated RS biochar had the highest sorption capacities for Pb, showing 126.6 and 132.2 mg/g, respectively. In the case of Ba, CO_2-activated RS biochar was the most effective (23.98 mg/g). Correlation analysis between biochar properties and maximum sorption capacities suggests that surface functional groups and CEC may affect the sorption of cationic heavy metals to biochar. Microtox® bioassay and TCLP tests did not show any toxic effects of the synthesized biochar. Our results suggest that co-pyrolysis of spent tires with biomass wastes may be a viable option to dispose spent tires by decreasing mobility and toxicity of metals as well as increasing sorption capacity of biochar for some contaminants.

597288 Synthesis of Nitrogen-Containing Carbon Dots As Fe (III) Probe after Extracting Humic Acid from Compost of Sludge and Leaf

Abstract:

Sewage sludge is generated as a by-product during wastewater treatment and needs to be efficient and environmentally friendly disposal. In the past decades, the excess sludge in municipal wastewater treatment plants was mainly treated environmentally-unfriendly such as landfill, incineration. Therefore, recycling sludge into high value-added products, such as humic acid, arising great attention. Composting is an ideal aerobic transformation process for biowaste stabilization. Especially, humic acid contains conjugated double bonds and aromatic rings, the similar structure with natural biomass which were successfully used to hydrothermally synthesize carbon dots. Therefore, it's promising to utilize humic acid as a precursor for hydrothermal oxidation to synthesize N-CDs.

In this work, blue fluorescent N-CDs was synthesized for the first time by the use of humic acid. The schematic illustration on the synthesis of fluorescent N-CDs is in Figure 1. Firstly, the humic acid was extracted from the compost of sludge and leaf for recycling solid waste. Then hydrothermal treatment of humic acid using hydrogen peroxide as an oxidant, blue fluorescent N-CDs was synthesized within short time. A series of characterization methods including TEM, XPS, FTIR and Raman analyses were used to conform the successful synthesis of N-CDs. As shown in Figure 2, the N-CDs has an average diameter of 1.88 nm and exhibits a well-defined spherical shape, and the distances between the lattice fringes are further characterized to be 0.218 nm and 0.355 nm, corresponding to the (010) in-plane lattice spacing and inter-layer distance of graphite[1]. Furthermore, Raman analysis of N-CDs revealed the existence of D and G bands at ~ 1390 cm−1 and ~ 1600 cm−1, respectively, confirming their architectures belonging to the carbon nanostructure family. Notably, the observed higher intensity of the G band than the D band verified the high quality of the synthesized products. Moreover, XPS measurements suggested that the synthesized N-CDs contains C, O and N elements with atom ratio of C:O:N = 67.56: 27.61: 4.82.

Nowadays, a large amount of iron ions-polluted wastewater is discharged into nature due to the widespread application and imperfect post-processing system in industrialized countries, leading to an increased accumulation of Fe^{3+} in the water system. Therefore, it's an urgent task to probe and monitor the Fe^{3+} in the water system. Among the various method developed to probe Fe^{3+}, fluorescence detection is a favourable method due to high sensitivity, selectivity and fast detection. The aqueous solution of the obtained N-CDs with excitation-dependent properties showed a strong blue emission with 8.8% quantum yield (QY) under the excitation of 340 nm, which indicates the N-CDs a promising sensing probes for Fe^{3+} in water. With the addition of Fe^{3+}, the fluorescence intensity of the N-CDs gradually decreased. This research indicates an extremely sensitive method for detecting Fe^{3+} at a lower 1.9 nM level by using the synthesized N-CDs as fluorescent probes without a complicated procedure. The detectability is lower than the concentration of Fe^{3+} in the specified drinking water standard of 5.36mM[2]. Furthermore, it is lower than the limit concentration of Fe^{3+} in reverse osmosis influent, which assure the N-CDs can be used as a sensor to prevent the reverse osmose membrane from being contaminated by excess Fe^{3+}. On the other hand, the practical feasibility of the synthesized N-CDs for probing Fe^{3+} in the actual sample of water has also been investigated. The reduction rate of fluorescence intensity of the N-CDs had a good linear relationship with the added water volume of the Youth Lake when the amount of lake water added was changed from 0 to 300 μL. The concentration of Fe^{3+} in the lake water was calculated according to the standard curve to be 20.4 mM, which is consistent with the result detected by ICP-MS, confirming the accuracy of the N-CDs method for detecting Fe^{3+} in actual environment.

Moreover, the mechanism of the N-CDs sensing technique was investigated in detail. It can be explained by the fluorescence static quenching effect and a ground-state complex formed in this process. The formation of the ground-state complex ensures the specificity of the N-CDs for the detection of Fe^{3+}, that is, the N-CDs will have no quenching effect with other metal ions.

Acknowledgments

The authors are grateful to the financial support of National Natural Science Foundation of China (Nos. 21776203, 21576187 and 51478308).

References

[1]Tang, L , et al. Energy-level structure of nitrogen-doped graphene quantum dots. Journal of Materials Chemistry C 2013, 32: p. 4908-4915.

[2]Organization, World Health. "Guidelines for Drinking-water Quality 4th Ed". (2011).

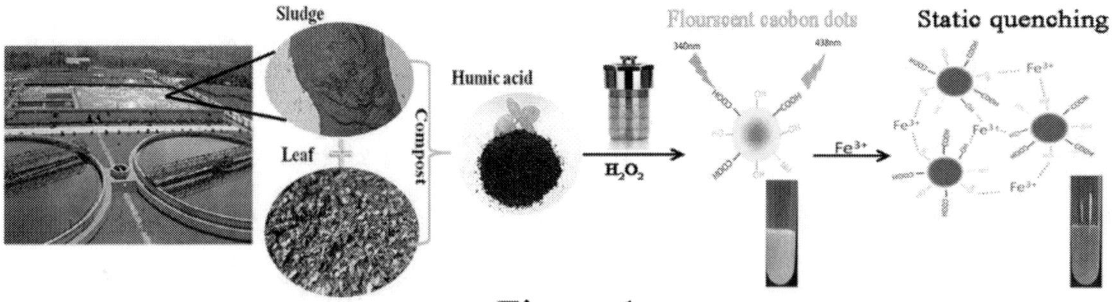

Figure 1

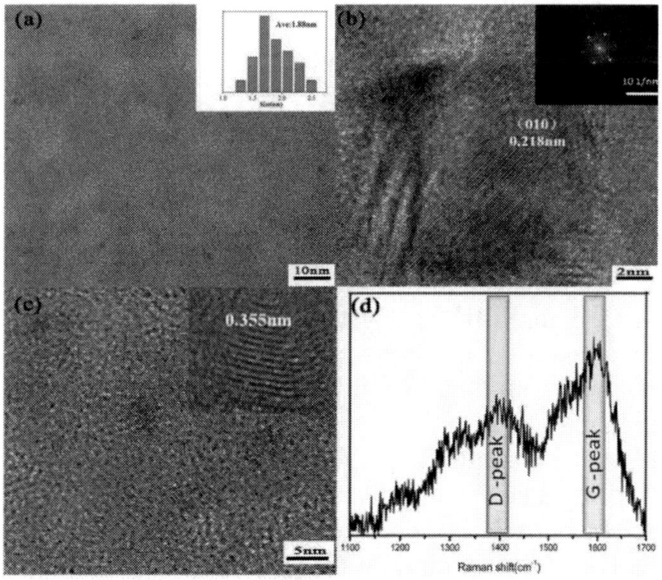

Figure 2

608441 Structural Characterization and Molecular Representation of Refined Unburned Carbon Recovered from Fly Ash

Abstract:

In order to improve and efficient utilization of fly ash derived from coal-based power plants, a structural study of refined unburned carbon (RUC) recovered from fly ash was carried out with multiple analytical tools, including X-ray diffraction (XRD), Raman spectroscopy, solid-state ^{13}C nuclear magnetic resonance (^{13}C-NMR) spectroscopy, Fourier transform infrared (FT-IR) spectroscopy, X-ray photoelectron spectroscopy (XPS), Laser desorption ionization time-of-flight mass spectroscopy (LD-TOF-MS), and high-resolution transmission electron microscopy (HRTEM). These analytical data was used to construct a molecular representation of GUC by using a built-in Amorphous Builder module in Materials Studio.

In the present study, GUC was recovered by a two-step process including refining of fly ash by demineralization with HF/HCl/HF process and then graphitization and crystallization by heat treatment up to 2500 °C.

The crystalline dimensions such as interlayer spacing (d_{002}), stacking height (L_a), and average lateral size (L_c) of GUC was calculated from XRD spectrum and found to be 0.338, 19.60, and 13.15 nm respectively. Moreover, the increase in peak intensity of D and G band was clearly observed during the heat treatment for carbon graphitization. The aromaticity (f_a) of RUC was found in the range of 0.45-0.51 by the multiple techniques. The relative intensity ratio of D and G band of GUC was 0.47.

The –OH, aliphatic and aromatic C–H, C=O, C-C, and C-O functional groups was clearly observed in FTIR and XPS spectrums. The quantitative analysis of carbon and oxygen functionalities such as –OH, aliphatic and aromatic C–H, C=O, C-C, C-O groups was found to be 12.1, 38.7, 10.4, 28.5, and 10.3% respectively from C1s and O1s XPS profile. The molar content of different carbon types such as aliphatic, aromatic, and carbonyl carbons of GUC was 37.5, 57.1, and 5.1% calculated from ^{13}C-NMR spectrum. Four aromatic rings per clusters with 3-4 substitute side chains on each aromatic cluster was calculated from ^{13}C-NMR spectrum. The aromatic layers was clearly observed in the HRTEM image in the form of fringes with a consistent crystalline dimension from XRD.

250 aromatic fringes was selected from the HRTEM image, and the heteroatoms and aliphatic side chains obtained from different analytical tools was used to construct 250 individual RUC fragments. A random cross-linking of individual fragments with different combinations of aromatic, aliphatic, and oxygen linkages till observing a consistent distribution of molecular weight observed in LD-TOF MS. A final molecular model was presented with 52 unique individual molecules with a molecular formula $C_{7754}H_{4576}O_{109}N_{48}S_{32}$. The structural and molecular properties obtained from analytical tools are consistent with the constructed model. This is the first molecular representation of unburned carbon recovered from fly ash ever proposed, which may lead to further understand the unburned carbon from fly ash and improve their utilization.

References:

1. Jaiswal, Y., Pal, S.L., 2020. Structural Characterization of Indian Vitrinite-Rich Bituminous Karharbari Coal. ACS Omega 5, 6336–6347. https://doi.org/10.1021/acsomega.9b03674.
2. Niekerk, D. Van, Mathews, J.P., 2010. Molecular representations of Permian-aged vitrinite-rich and inertinite-rich South African coals. Fuel 89, 73–82. https://doi.org/10.1016/j.fuel.2009.07.020.
3. Sharma, A., Kyotani, T., Tomita, A., 2001. Quantitative evaluation of structural transformations in raw coals on heat-treatment using HRTEM technique. Fuel 80, 1467–1473. https://doi.org/10.1016/S0016-2361(01)00018-7.

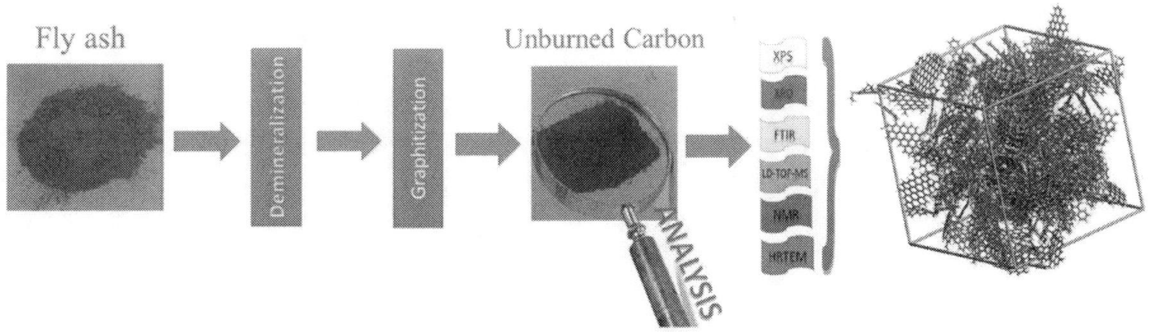

597490 Electrochemical Control of the CO_2 Capture Capacity of Amine Sorbents Using Cu (II)/Cu (I) Redox Chemistry

Abstract:

The most mature carbon capture technology, amine scrubbing, has been beneficial in reducing emissions from fossil fuel combustion. However, the inefficiencies of its thermal regeneration cycle lead to prohibitively high energy consumption. Since the high energy demand largely comes from the thermal capture/release cycle, alternative approaches to cycling are of interest.

One alternative is to use more efficient electrochemical cycles to switch between high capacity CO_2 capture states and low capacity sorbent regeneration states. In this study, the copper (II)/copper (I) redox couple is used to control the CO_2 capacity of the model sorbent 1,2-dimehtylimidazole in water. By alternating between oxidizing and reducing potentials, the solution changes between its 4-coordinate Cu (II) state and its 2-coordinate Cu (I) state. As a result, the redox reaction is coupled with the release or uptake of two sorbent molecules per copper, leading to an increase or decrease in CO_2 capture capacity, respectively.

Experimental results validate this approach, showing that for a given partial pressure of CO_2, the reduced Cu (I) state has a higher CO_2 capacity than the oxidized Cu (II) state. Energy consumption of a full oxidation-reduction cycle is measured using a bench-scale electrochemical cell. The aqueous inorganic carbon concentration agrees with predicted results based on chemical equilibrium calculations, allowing further computational investigation of the effects of solution composition, sorbent equilibrium constants, and inlet CO_2. Through these simulations, we have gained insights into the underlying mechanism, allowing identification of what characteristics the ideal sorbent molecule should have for this application.

604762 Epoxide-Modified Amine-Impregnated CO_2 Adsorbent with Lower Regeneration Energy Requirement and Enhanced Durability

Abstract:

Amino polymer-based adsorbents have been extensively investigated for post-combustion CO_2 separation for its low energy penalty as compared to amine-based absorption. However, most studies have focused on improving the CO_2 capacity, and the regeneration process is still required to separate high purity CO2 (>95%), minimize energy requirements, and enhance the chemical and mechanical durability of the sorbent such as sorbent degradation in other flue gas constituents in a temperature window of the separation process.

In this study, the adsorption/desorption cyclic performances were investigated using 1,2-epoxybutane (EB)-modified polyethylenimine (PEI)-impregnated mesocellular silica foam (MCF) in a thermogravimetric analyzer (TGA) and a fixed-bed reactor. The sorbent could demonstrated a stable working CO_2 capacity of ~2-3 mmol CO_2/g sorbent (50 °C, 15% CO_2 + in N_2 for adsorption and 120 °C, 100% CO_2 for desorption) over a 100 adsorption-desorption cyclic test. The desorption under pure CO_2 condition enabled the sorbent to be regenerated by temperature swing adsorption (TSA) without any dilution caused by sweeping gas. Additional modification could reduce degradation by SO_2 present at ~30-100 ppm level in typical coal combustion flue gas. This approach using modified PEI-based sorbent can offer a CO_2 adsorption capacity close to MEA absorption at a reduced desorption energy requirement of ~2 GJ/tCO_2 without a need for wastewater treatment. The detailed synthesis, characterization and performance results will be presented.

598733 Integrated Computational Screen and Design of Polymers for Post-Combustion Carbon Capture

Abstract:

Developing polymer membranes for cost-effective post-combustion carbon capture is challenging. Recent process modeling shows that for cost-effective post-combustion carbon capture using membrane separation technology, a polymer should have CO_2 permeance above 2000 GPU (10^{-6} cm^3(STP)/(cm^2 s cm Hg)) and CO_2/N_2 permeance selectivity above 25 [1] . Note that 1 GPU of permeance corresponds to 1 barrer (1×10^{-10} cm^3 $(STP)\cdot$ cm/cm^2 s cm Hg) of permeability if the selective polymer membrane thickness is assumed to be 1 mm (10^{-6} m). A survey of the MSA polymer gas separation database [2] shows that only four sets of experimental data out of a total of ~ 1500 data sets exhibit CO_2 permeability above 2000 barrer and CO_2/N_2 selectivity above 25, and those polymers are glass at room temperature and as a result the physical aging of the glass significantly affects their performance.

To address the above challenges of screening and designing rubber polymers for carbon capture, we adopted an integrated approach, which includes database surveys, chem-informatics and machine learning, and molecular modeling. Specifically, we have web-scraped the on-line CROW [3] polymer database. This database includes important polymer physical properties, such as glass transition temperature (*Tg*), polymer density, and van-der Waals volume for the repeat unit, from which the polymer free volume fraction was estimated. There are 238 polymers whose physical properties were collected; 111 polymers are rubbers at room temperature. Furthermore, CO_2 and N_2 permeability values in many polymers obtained from the CROW database were estimated using the polymer Genome platform and machine-learning models [4] developed by Ramprasad and coworkers. Preliminary calculations show that out of the 111 rubbers, the polydimethylsiloxane (PDMS) polymers exhibit a CO_2 permeability of ~4000 barrer, which is reasonably close to the experimental data and our molecular simulation results. All other rubbers calculated so far exhibit CO_2 permeability much less than 1000 barer. Note that both our molecular simulations and the literature experimental data show that the PDMS rubber polymer exhibits a CO_2/N_2 permeability selectivity of 7-9, much less than 25.

To further functionalize PDMS rubber to increase CO_2/N_2 permeability selectivity while maintaining its high CO_2 permeability, we utilized the computational database, which we recently developed to characterize CO_2 interactions with about 202 different functional groups. We also utilized the simplified molecular-input line-entry system (SMILES) and the RDKit software package implanted in Python to generate molecular structures for the above 111 rubbers to quickly look through the branch-structures of the polymers to functionalize the PDMS polymer. Finally, molecular simulation results of CO_2 and N_2 solubility[5,6], diffusivity, permeability, and permeability selectivity will be presented for the above functionalized PDMS polymers.

References:

1. Zoelle, R. Newby, "**Performance and Cost Sensitivities for Post-Combustion Membrane**

Systems", *NETL CO_2 capture Technology Project Review Meeting*, Pittsburgh, PA, August 13-17, 2018.
2. Membrane Society of Australasia (MSA) Polymer Gas Separation Membrane Database, "https://membrane-australasia.org/msa-activities/polymer-gas-separation-membrane-database/", accessed and downloaded locally on Aug 23, 2018.
3. CROW Polymer Science, http://www.polymerdatabase.com/, web-scraped on August 15 2019.
4. Polymer Genome, "https://www.polymergenome.org/"
5. Wei Shi, Robert L. Thompson, Megan K. Macala, Kevin Resnik, Janice A. Steckel, Nicholas S. Siefert, and David P. Hopkinson, "**Molecular Simulations of CO_2 and H_2 Solubility, CO_2 Diffusivity, and Solvent Viscosity at 298 K for 27 Commercially Available Physical Solvents**", Invited paper to the *Chem. Eng. Data*, DOI: 10.1021/acs.jced.8b01228, 2019
6. Wei Shi, E. J. Maginn, "Continuous Fractional Component Monte Carlo: An Adaptive Biasing Method for Open System Atomistic Simulations", *Chem. Theory Comput.* **2007**, *3*, 1451-1463.

607066 Process Modeling and Techno-Economic Analysis of a Diamine-Appended Metal–Organic Framework for CO_2 Capture

Abstract:

Aqueous solvents are the current leading commercial technologies for post-combustion CO_2 capture, but suffer from several drawbacks that include degradation, large energy penalties, and low working capacities. There has been considerable research directed toward investigating viable alternatives to these aqueous solvents in the area of solid sorbents. Most notably, a novel class of functionalized metal-organic frameworks (MOF's) have been shown to exhibit promising characteristics for post-combustion capture from a coal flue gas source[1,2,3]. The MOF studied in this work, Mg_2(dobpdc) (dobpdc^{4-} = 4,4′-dioxidobiphenyl-3,3′-dicarboxylate), contains one-dimensional hexagonal channels with unsaturated Mg^{2+} sites and is functionalized by binding the diamine 2,2-dimethyl-1,3-diaminopropane (dmpn) to these sites which then chemisorb CO_2 in cooperative mechanism to form carbamate and carbamic acid species. This MOF exhibits a step-shaped isotherm under low partial pressures of CO_2 which leads to higher working capacities under similar temperature and pressure intervals when compared to traditional sorbents. Along with high working capacities, this functionalized MOF also shows excellent long-term stability and maintains performance under humid conditions, both desirable attributes for flue gas CO_2 capture.[1]

To make this new class of MOF's economically feasible it is crucial that the type of contactor used minimizes the capital and operating costs of the system. For this system, high adsorption heat coupled with a large sensitivity of the adsorption capacity to temperature result in thermal management being key to realizing the full potential of the MOF's. Therefore, a contactor that can reject the heat generated during adsorption while maintaining efficient mass transfer is needed. In this work, four types of contactor models are analyzed: axial flow fixed bed, radial flow fixed bed, moving bed, and rotary packed bed. These dynamic, pressure-driven models are developed using first principles and consist of mass, heat, and momentum conservation equations and account for internal and external mass transfer resistances, heat transfer between gas and solid phases, and an embedded heat exchanger to supply heat or aid in heat removal of the system. The step-shaped isotherms that this MOF exhibits can also present a challenge when attempting to model the adsorption equilibrium. To achieve this, a new isotherm model is presented that can accurately predict adsorption equilibrium across the experimental temperature and pressure ranges. Kinetic models are also presented that were developed using available thermogravimetric analysis experimental data and fixed bed breakthrough experimental data.

These contactor models were then used to develop commercial scale CO_2 capture process models using flue gas from a 550 MWe net coal-fired power plant. A cost model is also presented that evaluates the capital costs of the reactors and compressors, if needed, and the operating costs of steam, cooling water, and electricity. Sensitivity of each contactor process to important operating variables are also investigated with a final comparison of each process as well as a traditional MEA capture system. Our works shows the critical impact of the contactor types on the commercialization potential of these next-generation MOFs.

References:

[1] – Milner, P.J., Siegelman, R.L., Forse, A.C., Gonzalez, M.I., Runcevski, T., Martell, J.D., Reimer, J.A., Long, J.R. A Diaminopropane-Appended Metal-Organic Framework Enabling Efficient CO_2 Capture from Coal Flue Gas via a Mixed Adsorption Mechanism. *Journal of the American Chemical Society*. 2017; 139 (38), 13541-13553

[2] Mcdonald, T.M., et al., Cooperative insertion of CO_2 in diamine-appended metal-organic frameworks. *Nature*. 2015; 519, 303-308

[3] Forse, A.C., Milner, P.J., Lee, J., Redfearn, H.N., Oktawiec, J., Siegelman, R.L., Martell, J.D., Dinakar, B., Porter-Zasada, L.B., Gonzalez, M.I., Neaton, J.B., Long, J.R., Reimer, J.A., Elucidating CO_2 Chemisorption in Diamine-Appended Metal-Organic Frameworks. *J. Am. Chem. Soc.* 2018; 140(51), 18016-18031

605585 Economic Optimization and Sensitivity Analysis of Ionic Liquid-Based Carbon Capture Plants

Abstract:

Carbon capture and sequestration (CCS) is an essential means to mitigate carbon emissions from power plants [1]. Current industrial-scale carbon capture systems rely on amine-based absorbents due to their high reactivity with CO_2. However, there are several disadvantages owing to the intrinsic properties of amine-based absorbents. Ionic liquids (ILs) have been proposed as potential absorbents for next-generation carbon capture processes. They are non-flammable, non-volatile, thermally stable, and non-corrosive relative to conventional amine solvents. In addition, ILs have a lower heat of absorption; consequently, energy requirements in the solvent regeneration step are lower compared to amine-based technologies. Aprotic heterocyclic anion ILs (AHAs) are particularly promising due to their ability to chemically absorb CO_2 without a significant increase in viscosity [2].

Although ILs have potential advantages at the molecular level, the process-level economics of IL-based carbon capture are not well studied. Developing process modeling and optimization tools for IL-based carbon capture plants is essential to elucidate their economic feasibility and scalability. However, robust simulation and optimization of the entire process flowsheet are challenging, particularly when rigorous models are used for key unit operations (e.g., absorber, stripper).

In this work, we present a novel framework for solving these flowsheet optimization problems, using a pseudo-transient modeling approach developed by our group [3,4]. The model is tailored to IL-based CCS: a Langmuir-type model that describes both chemical and physical uptake of CO_2 is used to describe the equilibrium absorption of CO_2 by the IL [5]. Further, a rigorous rate-based model for the absorption column is employed, that accounts for the effect of reversible chemical reactions on mass transfer. Detailed first-principles models of the other units in the plant flowsheet are used.

Economic optimization of the entire process flowsheet is performed. Sensitivity studies are then performed to elucidate the impact of the IL chemical absorption enthalpy and investigate the impact of other IL properties (viscosity, heat capacity, and molar volume) on process performance. We also present a comparison to amine-based processes, using recently published rigorous flowsheet optimization results for a modern amine-based CCS process [6].

References:

[1] J.-L. Fan. M. Xu, F. Li, L. Yang, X. Zhang, Carbon capture and storage (CCS) retrofit potential of coal-fired power plants in China: The technology lock-in and cost optimization perspective. Appl. Energy, 229 (2018), 326-334.

[2] S. Seo, M. Quiroz-Guzman, M.A. DeSilva, T. B. Lee, Y. Huang, B.F. Goodrich, W.F. Schneider, J.F. Brennecke, Chemically tunable ionic liquids with aprotic heterocyclic anion (AHA) for CO_2 capture. J. Phys. Chem. B, 118 (2014), 5740-5751.

[3] R.C. Pattison, M. Baldea, Equation-oriented flowsheet simulation and optimization using pseudo-transient models, AIChE J., 60 (2014), 4014-4123.

[4] R.C Pattison, C. Tsay, M. Baldea, Pseudo-transient models for multiscale, multiresolution

simulation and optimization of intensified reaction/separation/recycle processes: Framework and a dimethyl ether production case study, Comput. Chem. Eng., 105 (2017), 161-172.

[5] B. Hong, L. D. Simoni, J. E. Bennett, J. F. Brennecke, M. A. Stadtherr, Simultaneous process and material design for aprotic *n*-heterocyclic anion ionic liquids in postcombustion CO_2 capture, Ind. Eng. Chem. Res., 55 (2016), 8432-8849.

[6] C. Tsay, R.C. Pattison, Y. Zhang, G.T. Rochelle, M. Baldea, Rate-based modeling and economic optimization of next-generation amine scrubbing carbon capture processes, Applied Energy, 252 (2019), 113379.

Acknowledgement – This work was supported by the University of Texas Energy Institute under the Fueling a Sustainable Energy Transition program. Additional support was provided by the U.S. Department of Energy under Award Number DE-FE0026465. Portions of this work were performed at the University of Notre Dame (B. H. and M. A. S.).

610689 Optimal Temperature Profiles within a CCS Absorber

Abstract:

It is well known that absorber intercooling may be used to reduce column height and reboiler heat duty (Freguia et al. 2003, Rezazadeh et al. 2017, Plaza et al. 2010, Le Moullec et al. 2014, Walters et al. 2016). However, to date most research has focused on relatively simple intercooling geometries, and little work has been done to establish upper limits on the improvements more advanced intercooling techniques could provide.

In this work, we develop and validate a numerically efficient and flexible model of an absorption column containing monoethanolamine (MEA), similar to that developed by Saimpert et al. (2013). By coupling this model with a differential evolution optimization algorithm, we calculate the optimal temperature profile along the column's length which minimizes the column height required for 90% capture of CO_2. The optimal temperature profile represents a tradeoff at each point in space between increasing the temperature to increase reaction kinetics and decreasing the temperature to increase the driving force for CO_2 consumption. In general, the optimal profile is not isothermal (see Figure 1), however for some simple cases analytical expressions for the optimal profile may be derived.

By comparing these optimized columns to the adiabatic case, it is possible to put an upper limit on the reduction in column height intercooling can provide. In general, this is a strong function of the liquid flowrate: at flowrates close to the minimum, the temperature bulge tends to be large and near the top of the column (Kvamsdal and Rochelle 2008), and under these circumstances the adiabatic column height is substantially larger than the minimum possible height (see Figure 1). For example, for 90% capture from a CO_2 stream containing 10mol% CO_2, when the temperature bulge is near the top the adiabatic case is typically 10-20% larger than the minimum column height. On the other hand, at larger liquid flowrates (typically around $1.5L_{min}$ - $2L_{min}$) the temperature bulge tends to shrink in size and move towards the bottom of the column (Kvamsdal and Rochelle 2008) and under these circumstances the difference between the adiabatic and minimum possible column height tends to be marginal – typically less than 5%. The maximum possible improvement is also dependent on the inlet CO_2 concentration: for gaseous streams containing 20mol% CO_2, the adiabatic column may be more than 30% taller than column with optimized temperature profile, while for capture from a 5mol% CO_2 stream the maximum improvement intercooling could provide is correspondingly smaller.

The model was also used to investigate a range of heat integration strategies within the column. It was shown that under a wide range of circumstances the column height could be reduced to close to the minimal value simply by optimally rearranging heat within the column – moving heat away from regions where increasing the driving force is more important, and towards regions where higher temperatures are beneficial to improve reaction kinetics. The transfer of heat was consistent with both the first and second laws of thermodynamics. Such sophisticated heat integration may be possible through the use of active packings, which incorporate a heat exchange fluid within the column packing itself. For a number of cases of interest, it was also shown that traditional in-and-out intercooling (Rezazadeh et al. 2017) was able to reduce the column height to close to the minimal value, though approaches which incorporate heat redistribution within the column itself would have the benefit of reduced cooling loads.

Prepared by LLNL under Contract DE-AC52-07NA27344. This work was supported via the U.S. Department of Energy, Office of Fossil Energy, grant number FWP-FEW0225

References

Freguia, S. and Rochelle, G.T., 2003. Modeling of CO2 capture by aqueous monoethanolamine. *AIChE Journal*, *49*(7), pp.1676-1686.

Kvamsdal, H.M. and Rochelle, G.T., 2008. Effects of the temperature bulge in CO2 absorption from flue gas by aqueous monoethanolamine. *Industrial & Engineering Chemistry Research*, *47*(3), pp.867-875.

Le Moullec, Y., Neveux, T., Al Azki, A., Chikukwa, A. and Hoff, K.A., 2014. Process modifications for solvent-based post-combustion CO2 capture. *International Journal of Greenhouse Gas Control*, *31*, pp.96-112.

Plaza, J.M., Chen, E. and Rochelle, G.T., 2010. Absorber intercooling in CO2 absorption by piperazine-promoted potassium carbonate. *AIChE journal*, *56*(4), pp.905-914.

Rezazadeh, F., Gale, W.F., Rochelle, G.T. and Sachde, D., 2017. Effectiveness of absorber intercooling for CO2 absorption from natural gas fired flue gases using monoethanolamine solvent. *International Journal of Greenhouse Gas Control*, *58*, pp.246-255.

Saimpert, M., Puxty, G., Qureshi, S., Wardhaugh, L. and Cousins, A., 2013. A new rate based absorber and desorber modelling tool. *Chemical engineering science*, *96*, pp.10-25.

Walters, M.S., Edgar, T.F., Rochelle, G.T., 2016. Dynamic modeling and control of an intercooled absorber for post-combustion CO2 capture. Chemical Engineering and Processing: Process Intensification 107, 1–10.

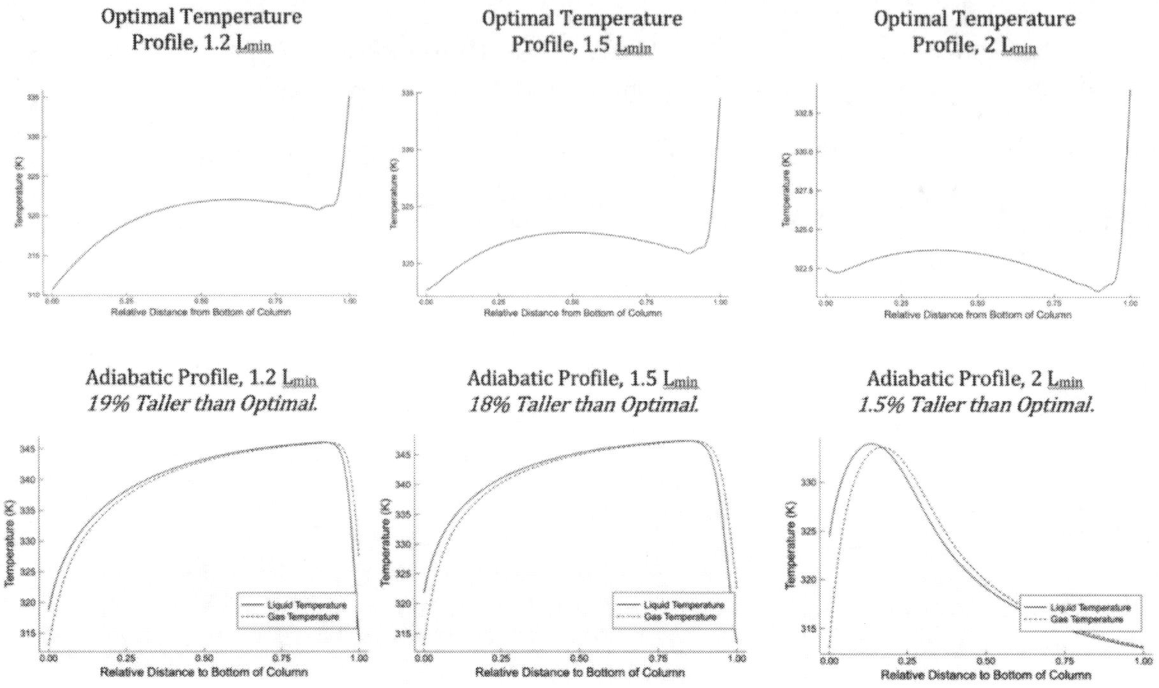

Figure 1. Comparison of optimal and adiabatic temperature profiles at a range of liquid flowrates. Top row: optimal temperature profiles which minimize absorber height. Bottom row: adiabatic temperature profiles. Conditions: 90% Capture from gas stream containing 10% CO_2 and 5% H_2O using 30wt% MEA with $\theta_{lean} = 0.25$. Inlet gas and liquid temperatures: 40C.

607447 Optimization of the Amines-CO2 Capture Process By a Nonequilibrium Rate-Based Modeling Approach to Determine Operating Policies

Abstract:

Amine based chemical absorption has stood as the most mature and studied technology among Carbon Dioxide capture processes, featuring major advantages such as large separation efficiencies and simplicity to attach itself to existent industrial facilities. Nevertheless, its operation requires vast amounts of energy, especially at the solvent regeneration unit, making the process expensive and poorly sustainable, thus stopping its practical implementation. To tackle this problem we propose a deterministic nonlinear parameter optimization of a typical CO2-amines process, aiming to reduce the overall energy consumption while meeting CO2 capture targets as well. For such purpose, we deploy a rigorous first principles nonequilibrium rate-based and chemically reacting steady-state model of the aforementioned system.

The mathematical model (consisting of approximately 32 800 equations and variables) was implemented in the GAMS optimization environment. Critical output parameters were validated against experimental data from a pilot plant, which consists of two columns interconnected (an absorber and a desorber). Moreover, relevant profiles were successfully compared against results from simulations in Aspen Plus. Once validated, the model was used to perform several sensitivity analyses on interesting input parameters in order to study the response of the system's separation degree and energy demand. Afterwards, a series of deterministic nonlinear programming (NLP) problems were proposed, where three different objective functions were studied in order to obtain optimal operational policies which enable both CO2 sequestration and low energy expenditure.

Up to our best knowledge, optimization studies with this degree of modeling complexity (i.e. rate-based models along with NLP) have not been reported in the open literature. Results reflect the importance of optimizing both columns simultaneously instead of modeling them separately, since a complete model like the one at hand takes into account complex interactions between variables all across the process, which may be counterintuitive. By applying the optimal operating policies suggested by the model, an energy reduction of up to 25% was achieved when compared against conventional industrial data from plants operating under heuristic policies. Furthermore, this work emphasizes the significance of performing optimization studies in advance of practical implementations of emerging technologies. As can be seen, the problem is challenging and timely due to global warming concerns.

603070 Planning and Optimization of a Regional Hydrogen Supply Chain: An ASEAN Case Study

Abstract:

The world is seeking scalable and affordable low carbon and renewable energy sources to reduce atmospheric CO_2 concentration levels while meeting rising global energy demand. Hydrogen is receiving significant attention worldwide as a versatile energy carrier and feedstock that can achieve deep decarbonization across energy and industrial sectors. Researchers have proposed various approaches to evaluate hydrogen supply chains. For example, Almansoori and Shah[1] developed a multi-period hydrogen supply chain model, based on the hydrogen demand of fuel cell vehicles in the Great Britain from 2005 to 2034. Reuß et al.[2] carried out a techno-economic comparison between different transport and storage infrastructure options for scenarios with 25%, 50% and 75% fleet penetration of fuel cell electrical vehicles in Germany. Seo et al.[3] developed a novel hydrogen supply chain model to compare centralized storage and decentralized storage modes. However, most previous work has focused on country-specific supply chains, such as for the UK, Germany, or South Korea, and considered the transport sector as the only hydrogen demand, neglecting that hydrogen can also be utilized in power and industrial sectors. There is no study on regional hydrogen supply chains with multiple H_2 demand sectors. A regional supply chain can capture the benefits of integrating energy-abundant areas with high energy-consuming areas.

In this research, a multi-period regional hydrogen supply chain model is developed for a better understanding of different pathways for a hydrogen economy in Singapore and the ASEAN countries through 2050. A comprehensive superstructure including hydrogen production, storage, delivery, and utilization is proposed. Both renewable and non-renewable hydrogen production technologies with applicable carbon capture and sequestration are included, reflecting the diversity of energy sources available in the region for hydrogen production. To compare different options for an ASEAN hydrogen supply chain through 2050, we carry out techno-economic analyses of hydrogen production taking into account location-dependent attributes, such as cost and carbon emissions. The variation of hydrogen demand over a long-term planning horizon is considered, as well as uncertainty of energy source prices and technological progress, to find the competitive production technologies in each year. Since the volumetric energy density of hydrogen is low, hydrogen must be transported and stored at high pressure or in liquid form, or via liquid organic hydrogen carriers or ammonia. Consequently, different transmission-storage-distribution pathways are also considered and evaluated to address the trade-off between the energy density, energy consumption, and facility cost. Based on the superstructure, a mixed-integer non-linear programming (MINLP) model is formulated to optimize both the total cost and carbon emissions of the hydrogen supply chain. The most effective pathways for producing, transporting, and storing hydrogen are being identified. Several what-if scenarios are explored to derive insight on relative emissions and costs associated with various hydrogen pathways for the ASEAN region.

1. Almansoori, A., Shah, N., 2009. Design and operation of a future hydrogen supply chain: Multi-period model. *International Journal of Hydrogen Energy*, 34(19): 7883-7897.

2. Reuß, M., Grube, T., Robinius, M., Stolten, D., 2019. A hydrogen supply chain with spatial resolution: Comparative analysis of infrastructure technologies in Germany. *Applied Energy*, 247: 438-453.

3. Seo, S.-K., Yun, D.-Y., Lee, C.-J., 2020. Design and optimization of a hydrogen supply chain using a centralized storage model. *Applied Energy*, 262.

611098 Predicting Reaction Mechanism of Electrochemical CO_2 Reduction Using Machine Learning

Abstract:

Understanding the mechanism of an electrochemical reaction could, in principle, be one of the important factors in the accelerated discovery of highly efficient electrocatalysts. Cyclic voltammetry (CV) is one of the techniques providing important mechanistic insight into an electrochemical reaction by the virtue of the change in current density when a reactant or an intermediate species is adsorbed/desorbed on the electrode surface. However, interpreting a CV can be influenced by subjectivity from the experience of the electrochemist. Moreover, quantifiable confidence in such interpretations is wanting. Machine Learning (ML) algorithms can be applied in such cases to circumvent the heuristics in CV interpretations and user subjectivity to present robust, quantifiable confidence in predicting electrochemical reaction mechanisms. One of the most attractive electrochemical reactions in recent years has been the electrochemical CO_2 reduction (CO2R) to value-added chemicals due to its potentially high impact on the industries, society, and the environment. This work shows the implementation of an ML model (Neural networks) to predict the reaction mechanism of CO2R by learning from different data sets of cyclic voltammograms. The simulated and experimental cyclic voltammograms are used to train the neural network to identify the dependence on the parameters like CV scan rates and uncompensated resistance. For each individual CV, the voltammogram serves as an input to the model. The expected output predicts the likelihood of the occurrence of either simple reversible electron transfer reactions or irreversible product formation after an electron transfer reaction. This ML model confirms that the predictions can be made even by simple models as a substitute for the conventional subjective approach.

610876 Revealing the Mechanism of Intriguing Predominant Urea Formation from Thermal Degradation of CO_2-Loaded Aqueous Ethylenediamine

Abstract:

Amine scrubbing using aqueous amine solutions remains to be the most efficient near-term solution for post-combustion CO_2 capture from flue gas at coal-fired power plants.[1,2] Monoethanolamine (MEA) is one of the most extensively tested alkanolamines, but its commercial application tends to be limited because of degradation and corrosion issues and enormous energy requirement for solvent regeneration.[3–6] About (~10 %) of total capital cost of CO_2 capture is attributed to makeup of amine solvent loss.[7–10] Moreover, some degradation products are considered to be dangerous if exposed to the environment.

Ethylenediamine (EDA) has been reported to have several advantages over MEA for its higher CO_2 capacity, faster CO_2 absorption rate, easier producible implementation from renewable resources, and slightly lower degradation rates at moderate CO_2 loading.[11–16] EDA and MEA have a similar molecular structure as a primary amine; the former contains two primary amine groups and the latter has one primary amine and one hydroxyl group, while the end groups are separated by two -CH_2- groups in both molecules. Despite their structural similarity, the major thermal degradation products of CO_2-loaded EDA and MEA solutions have been found to be distinctly different.[11,17]

Thermal degradation of amines is generally understood to occur primarily in the stripper due to high temperature (~ 100 - 120 °C) and is known to increase with CO_2 loading.[18] Previous experimental studies proposed that thermal degradation of primary and secondary amines would be initiated by polymerization of carbamates, which can ultimately degrade to urea and imidazolidinones.[19,20] The MEA degradation is thought to proceed via 2-oxazolidinone (OZD) formation that may undergo further reaction with amines to form more stable products.[21] Our previous study predicted isocyanate formation from dehydration of carbamic acid to be highly probable, and also showed that cyclization of intermediate isocyanate to OZD would be kinetically more facile than reaction with MEA to form urea in aqueous MEA solution. The formation of isocyanate, albeit very small quantity, has also been experimentally reported.[22,23] In contrast to the case of MEA, the major thermal degradation products of EDA have been identified to be 1,3-Bis(2-aminoethyl)urea and 2-imidazolidone from mass spectrometry and high performance liquid chromatography measurements, while urea formation appears to be preferred.[11,17]

In this talk, we will present molecular mechanisms underlying thermal degradation of EDA in CO_2-loaded solution, especially the underlying reason for the preferential formation of urea, rather than cyclic compounds, in contrast to the case of MEA. The thermodynamic favorability of previously proposed reaction pathways, including the relative stabilities of intermediates and products, were evaluated using quantum mechanical calculations with an implicit solvent model. The free-energy barriers for key elementary reactions involved in the EDA degradation were determined using *ab initio* molecular dynamics simulations combined with metadynamics. Besides the simulation results, further analysis of the interactions between short-lived intermediates and surrounding solvent molecules provide an explanation as to why EDA and MEA show different degradation behaviors. This work highlights the importance of kinetic effects associated with the local structure and dynamics of solvent molecules around the intermediates that may significantly influence the degradation process of amine solvents.

References

[1] G. T. Rochelle, *Science*, 2009, **325**, 1652–4.

[2] P. Luis, *Desalination*, 2016, **380**, 93–99.

[3] M. R. M. Abu-Zahra, L. H. J. Schneiders, J. P. M. Niederer, P. H. M. Feron and G. F. Versteeg, *Int. J. Greenh. Gas Control*, 2007, **1**, 37–46.

[4] M. R. M. Abu-Zahra, J. P. M. Niederer, P. H. M. Feron and G. F. Versteeg, *Int. J. Greenh. Gas Control*, 2007, **1**, 135–142.

[5] D. M. D'Alessandro, B. Smit and J. R. Long, *Angew. Chem. Int. Ed. Engl.*, 2010, **49**, 6058–82.

[6] H. M. Stowe and G. S. Hwang, *Ind. Eng. Chem. Res.*, 2017, **56**, 6887–6899.

[7] A. B. Rao and E. S. Rubin, *Environ. Sci. Technol.*, 2002, **36**, 4467–4475.

[8] A. J. Reynolds, T. V. Verheyen, S. B. Adeloju, A. L. Chaffee and E. Meuleman, *Int. J. Greenh. Gas Control*, 2015.

[9] G. S. Goff and G. T. Rochelle, *Ind. Eng. Chem. Res.*, 2006, **45**, 2513–2521.

[10] S. A. Mazari, B. Si Ali, B. M. Jan, I. M. Saeed and S. Nizamuddin, *Int. J. Greenh. Gas Control*, 2015.

[11] S. Zhou, X. Chen, T. Nguyen, A. K. Voice and G. T. Rochelle, *ChemSusChem*, 2010, **3**, 913–918.

[12] M. Rabensteiner, G. Kinger, M. Koller, G. Gronald and C. Hochenauer, *Int. J. Greenh. Gas Control*, 2014, **27**, 1–14.

[13] S. Kadiwala, A. V. Rayer and A. Henni, *Chem. Eng. J.*, 2012, **179**, 262–271.

[14] J. Li, A. Henni and P. Tontiwachwuthikul, *Ind. Eng. Chem. Res.*, 2007, **46**, 4426–4434.

[15] Z. Wang, M. Fang, Y. Pan, S. Yan and Z. Luo, *Chem. Eng. Sci.*, 2013.

[16] A. P. Salvi, P. D. Vaidya and E. Y. Kenig, *Can. J. Chem. Eng.*, 2014, **92**, 2021–2028.

[17] J. Thompson, H. Richburg and K. Liu, in *Energy Procedia*, 2017.

[18] D. Hatchell, O. Namjoshi, K. Fischer and G. T. Rochelle, *Energy Procedia*, 2014, **63**, 1558–1568.

[19] H. Lepaumier, D. Picq and P. L. Carrette, in *Energy Procedia*, 2009.

[20] H. Lepaumier, E. F. Da Silva, A. Einbu, A. Grimstvedt, J. N. Knudsen, K. Zahlsen and H. F. Svendsen, in *Energy Procedia*, 2011.

[21] J. Davis and G. Rochelle, *Energy Procedia*, 2009, **1**, 327–333.

[22] T. Sakakura, J. C. Choi and H. Yasuda, *Chem. Rev.*, 2007, **107**, 2365–2387.

[23] T. E. Waldman and W. D. McGhee, *J. Chem. Soc. Chem. Commun.*, 1994, 957.

611409 Techno-Economic and Sustainability Analysis of the Production of Acetic and from Methane and Carbon Dioxide

Abstract:

Methane and carbon dioxide together represent 90% of greenhouse gases, and one way of diminishing their emissions is by converting them into added value chemicals. Acetic acid is one of such chemicals, and efforts at the laboratory scale have been carried out to prove it can be directly and indirectly synthesized from CH4 and CO2. However, there are very few studies that evaluate the possibility of implementing any of these processes at a larger scale. In this thesis, techno-economic and sustainability analyses were used to assess the feasibility of upscaling two pathways to produce acetic acid: (1) by direct conversion of methane and carbon dioxide using a plasma-assisted reactor and (2) by indirect catalytic conversion, using syngas (obtained from dry methane reforming) as an intermediate. ASPEN Plus was used for simulation, and with literature values and plant cost estimation techniques, the capital and operational expenses were estimated and compared to those of the commercial route.

The indirect and direct pathways attained return on investment (ROI) values of 1.5% and 5.5%, respectively, and unless credit is given to the reduction in the carbon footprint, neither process can economically compete with the commercial route, which attained an ROI of 19%. In terms of sustainability, the indirect and direct processes captured 0.7 and 2.5 times more carbon dioxide compared to the emitted amount by the commercial pathway. For the direct pathway it was found through a sensitivity analysis the carbon credit value can be reduce down to a value of $50/tonne to render the process economically viable.

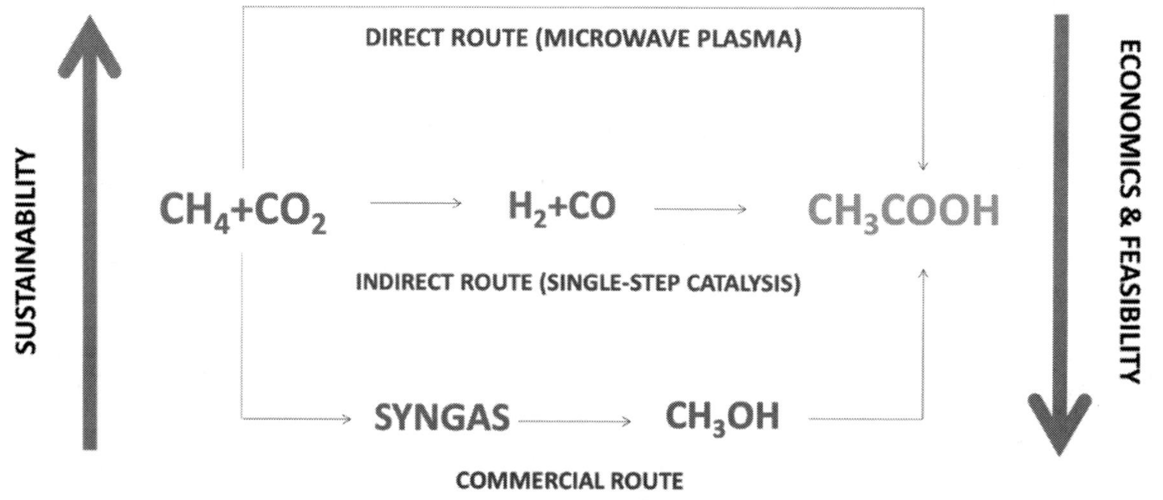

THE CARBONSAFE INITIATIVE: LAUNCHING THE CONVERGENCE OF CARBON CAPTURE AND STORAGE PROJECTS BY 2025

Mary Sullivan[1], Kanwal Mahajan[2], Darin Damiani[3], and Tomas Mora[4]

[1]National Energy Technology Laboratory, Pittsburgh, PA 15236-0940, U.S.A

[2]National Energy Technology Laboratory, Morgantown, WV 26507-0880, U.S.A

[3]U.S. Department of Energy, Washington, DC 20585, U.S.A.

[4]KeyLogic Systems, Inc., Morgantown, WV 26505, U.S.A.

Abstract

The Carbon Storage Program implemented by the U.S. Department of Energy (DOE) Office of Fossil Energy and managed by the National Energy Technology Laboratory (NETL) is helping to develop technologies that safely and permanently store carbon dioxide (CO_2) without impacting human health and the environment or hindering economic growth. Building upon almost two decades of knowledge and experience gained from the field lab testing projects and Regional Carbon Sequestration Partnership (RCSP) Initiative efforts, the Program initiated the Carbon Storage Assurance Facility Enterprise (CarbonSAFE) Initiative. The CarbonSAFE Initiative addresses key gaps on the critical path toward carbon capture, utilization and storage (CCUS) deployment by reducing technical risk, uncertainty, and cost of a geologic storage complex with capacity to store 50+ million metric tons of CO_2 from industrial sources over a 30-year period. The CarbonSAFE Initiative is taking a phased approach: (I) Integrated CCS Prefeasibility, (II) Storage Complex Feasibility, and (III) Site Characterization, Permitting and Carbon Capture Assessment. Subject to the availability of funding, a fourth phase would include construction and completion of wells and infrastructure leading to authorization to begin injection operations. A total of 13 projects were completed under Phase I, which provided high-level evaluations of potential CCS scenarios in regions throughout the United States. Six projects were awarded under Phase II to confirm the adequacy of storage complexes through initial site characterization. Phase III projects commenced in 2020 and will involve the entire integrated process, including a capture assessment, detailed site characterization, and preparing an application for a Class VI Underground Injection Control (UIC) permit to construct. These efforts are designed to help launch the industry by 2025 and contribute to increasing confidence that geologic storage of captured CO_2 can be a safe enterprise on a broad scale.

Keywords: U.S. Carbon Storage DOE Program; National Energy Technology Laboratory (NETL); carbon capture, utilization and storage (CCUS) research; geologic storage technologies; Regional Carbon Sequestration Partnerships (RCSPs); field projects; accomplishments; CarbonSAFE

1. United States Leadership in Commercial Deployment of CCUS

One of the great challenges facing the world today is the need to reduce global greenhouse gas (GHG) emissions and thereby mitigate the potential impacts of climate change without provoking undue detriment to global economies. In the United States (U.S.), energy-related CO_2 emissions have declined over the last decade in part due to a trend in fuel-switching from coal-based to natural gas-based sources in power generation, as well as increased production of renewable electricity, and improved energy efficiencies of vehicles. However, the U.S. Energy Information Administration (EIA) projects that U.S. energy-related emissions will remain essentially unchanged for the next 30 years unless carbon reduction programs and policies are adopted [1, 2]. Numerous international organizations, from the Intergovernmental Panel on Climate Change to the International Energy Agency (IEA), recognize that CCUS is an essential clean energy technology if a low-carbon economy is to be achieved at the lowest possible cost. Geologic Carbon storage includes the capturing of carbon dioxide (CO_2) from a stationary source; compressing it to a liquid-like state; transporting it via pipeline, ship, truck, or other means; and injecting it underground into deep reservoirs safely and securely. Although not part of CarbonSAFE, CO_2 recycling to produce economically valuable products and services such as improved materials, processes, fuels, and chemicals can be added to the CCUS value chain.

The United States is ideally situated for commercial CCUS projects because of its abundant carbon storage resources in proximity to many industrial CO_2 sources such as coal- and gas-fired power plants, gas-processing plants, ethanol plants, cement manufacturers, hydrogen production facilities and a host of others.

The United States also has over 50 years of CO_2-enhanced oil recovery experience and more than 5,000 miles of CO_2 pipeline infrastructure. Moreover, enhancements to federal and state policies are providing the needed incentives for advancing commercial CCUS projects. According to the IEA, 45Q tax credits could trigger new capital investments of up to $1 billion (B) for CCUS technologies over the next 6 years, resulting in an additional 10 to 30 million (M) metric tons (MT) of CO_2 capture capacity [3]. In the U.S power sector alone, 45Q credits could incentivize retrofit technologies in coal and natural gas power plants and capture of 49 MMT of CO_2/year (yr) by 2030 [4]. To date, there are over 30 carbon capture projects in various stages of project development in the U.S. [5, 6, 7]. With respect to regulations, recent U.S. Environmental Protection Agency (EPA) approval of primacy for North Dakota and Wyoming for Underground Injection Control (UIC) Class VI wells (used for geologic saline storage of CO_2) has provided developers an opportunity to work more closely with state jurisdictions, which might help facilitate the process and expedite permitting in these two states.

Over the past two decades, the U.S. Department of Energy (DOE) Office of Fossil Energy (FE) and its partners have reduced the cost of carbon capture from power and industrial sources and advanced the integration of CCUS systems in storage complexes with different geologic settings. One example of such an integrated system was NRG Energy's, Inc. Petra Nova project, the world's largest coal-fired post-combustion CO_2 capture system. The captured CO_2 was sent for use in enhanced oil recovery (EOR) in a partially depleted oil field in Jackson County, Texas. Lessons learned from Petra Nova and other industrial projects, such as the Air Products and Chemicals, Inc. hydrogen production facility in Port Arthur, TX and the Archer Daniels Midland Company ethanol facility in Decatur, IL, will result in lower costs for future projects. The next large-scale coal-fired post combustion CO_2 capture and compression project is expected to reduce costs by nearly 30% compared to Petra Nova's cost [8].

The knowledge gaps impeding growth in commercial CCUS deployment are both technical and non-technical. FE's Carbon Storage Program is working to close these remaining gaps by supporting a portfolio of research and development (R&D) projects and initiatives that investigate the technical gaps and aid in addressing the non-technical aspects of CCUS, such as the regulatory process, finance and markets, and public perception. Perhaps one of the best-known and most successful endeavours is the Regional Carbon Sequestration Partnerships (RCSPs). Over the last 17 years the RCSPs have characterized the CO_2 storage resource potential in their respective regions, completed pilot and large-scale CO_2 injection projects that cumulatively stored more than 11 MMT of CO_2, demonstrated the effectiveness of various technologies to verify storage containment in multiple reservoir types, and established a broad national network among companies and professionals supporting the deployment of commercial-scale CO_2 storage projects throughout the United States. Perhaps most importantly, the RCSPs have helped to validate storage security by virtue of its injection projects showing no evidence of impacts to human health or the environment. Through these successes the RCSP Initiative has laid the foundation for the CCUS industry, with key findings and recommendations captured in a series of Best Practice Manuals [9] made available to the public.

2. Convergence of DOE-FE's Carbon Capture and Carbon Storage programs

At present, the current state of CCUS technology is sufficient to develop and execute a CCUS project. However, cost reduction and performance optimization are among the biggest challenges to an industry launch by 2025. Increasing the number of CCUS projects at commercial scale enables development of the following:

- Cost-effective and accurate approaches for appraising site-specific storage resources and supporting investment-level decisions.
- Fully optimized and cost-effective long-term monitoring technologies and tools for commercial-scale onshore storage sites.
- Information and approaches for more targeted data acquisition and understanding of risk(s) in support of the permitting process.

DOE-FE's Carbon Storage Program is addressing these knowledge gaps primarily through its Carbon Storage Assurance Facility Enterprise (CarbonSAFE) Initiative, as described in more detail below. In parallel, DOE-FE's Carbon Capture Program is advancing R&D post-combustion carbon capture technologies that have the potential to reduce capture cost to $30/MT (at 95% purity). Both DOE-FE programs are planned to converge in 2025 with the intent to deploy integrated CCUS projects that would link commercial-scale storage complexes with state-of-the-art capture facilities.

3. CarbonSAFE Initiative

The Carbon Storage Program envisioned CarbonSAFE as the natural next step following the success of the Regional Carbon Sequestration Partnership (RCSP) efforts [10, 11], which provided the technical and non-technical foundations for scaling up CCUS operations to commercial scale. RCSP successes have already been demonstrated by contributions to major demonstration projects (Archer-Daniels-Midland [ADM], Air Products, Petra Nova), which cumulatively have stored more than 11 MMT of captured CO_2. The CarbonSAFE Initiative was created to ensure carbon storage complexes with capacities of 50+ MMT of CO_2 will be ready for integrated CCUS system deployment starting in 2025. The vision of the CarbonSAFE Initiative is to provide lessons learned and examples of the development of a CCUS storage complex from the feasibility study up until the point of injection. The progress made by this effort will instil greater confidence that commercial-scale CCUS projects are technically feasible and can be integrated in an economically feasible manner. The CarbonSAFE Initiative was first envisioned in 2016 as consisting of four phases:

- Phase I: Integrated CCS Prefeasibility – 13 projects awarded in 2016
- Phase II: Storage Complex Feasibility – 6 projects awarded: 3 in 2016, 3 in 2018
- Phase III: Site Characterization and CO_2 Capture Assessment – 5 projects awarded in 2020
- Phase IV: Permitting and Construction – potential future phase, dependent on appropriations

Thirteen prefeasibility studies were conducted in ten states and two offshore locations in the lower 48 states. Prefeasibility projects conducted economic feasibility analyses for targeted sites under varying scenarios. In addition, extensive regional data was collected, analyzed, and modeled; multiple storage sites and infrastructure were evaluated; and business plans for a variety of CCUS scenarios were developed. Currently, the program is implementing the Storage Complex Feasibility and Site Characterization and CO_2 Capture Assessment efforts. Research conducted within the six Storage Complex Feasibility projects focused on one or more specific reservoirs within the defined storage complex, encompassing (a) data collection; (b) geologic analysis; (c) identification of contractual and regulatory requirements; (d) subsurface modeling to support geologic characterization, risk assessment, and monitoring; and (e) public outreach.

The Site Characterization, Permitting and CO_2 Capture Assessment phase (Phase III) began in 2020. These projects will converge with Front-End Engineering Design (FEED) studies awarded by DOE's National Energy Technology Laboratory (NETL) Carbon Capture Program for commercial-scale post-combustion capture technologies implemented in the power sector.* The key objectives of this phase are to (a) complete a detailed site characterization of a commercial-scale CO_2 storage site (50+ MMT of captured CO_2 within a 30 yr period); (b) apply for and obtain UIC Class VI permit(s) to construct injection well(s); (c) complete a CO_2 capture assessment; (d) perform all work required to support a National Environmental Policy Act determination for each proposed project; and (e) address all other permitting processes and requirements, liability relief, and finance agreements in support of the business model for eventual commercial operations. Once the UIC Class VI permit applications to construct are submitted, additional activities will include working with the regulators to satisfy their requirements until construction permits are granted. Figure 1 shows locations of the CarbonSAFE Phase III projects with FEED studies across the U.S. The characterization projects in Phase III will converge with five of nine FEED studies awarded by DOE's NETL Carbon Capture Program for commercial-scale carbon capture systems on coal and natural gas power plants.

Subject to availability of funding, the final phase of CarbonSAFE, Construction of the Storage Complex, will involve the drilling and completion of the injection and monitoring wells and development of risk and mitigation plans to obtain authorization by the regulator to begin injection operations. Issued final reports for all project phases are located at the DOE's Office of Scientific and Technical Information site (https://www.osti.gov/search). Project-specific data produced by the CarbonSAFE effort are available on the Energy Data Exchange (edx.netl.doe.gov).

---

* The Funding Opportunity Announcements (FOAs) for the CarbonSAFE Phase III projects and Carbon Capture Program FEED studies were run independently with no dependence between them in terms of project selection. CO_2 capture assessments are being completed under CarbonSAFE for sources that did not receive FEED funding from the Carbon Capture Program.

Figure 1. Map of the contiguous U.S. showing locations of CarbonSAFE Phase III projects and Carbon Capture Program FEED studies

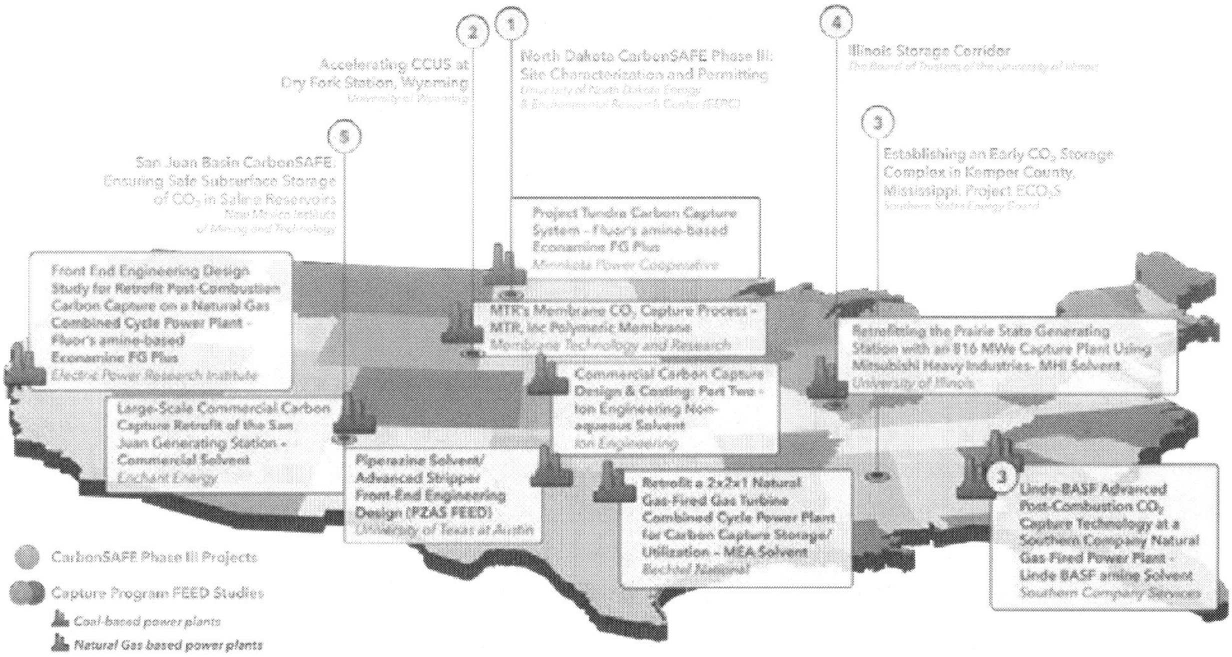

3.1 Integrated CCS prefeasibility projects

While not all prefeasibility projects were able to receive funding to continue their investigations into the next phase, a significant amount of useful and actionable research was produced. Extensive regional data was collected, analyzed, and modeled; multiple storage sites and infrastructure were evaluated; and business plans for a variety of CCUS scenarios were developed. A summary of each project's sources and storage estimates were provided in a previous publication [12]. Further characterization of the subsurface is needed at each site to verify the general initial findings. Common themes identified for making an integrated CCUS project successful from the prefeasibility projects include the following:

- Collaboration with oil and gas companies and strong host site support is essential.
- Uncertainty in federal policy results in lack of commitment from industry and the public for CCUS projects.
- Capture costs will need to decrease to make integrated CCUS projects economical.

- Reservoir heterogeneity requires improved models and data to increase plume behavior accuracy.
- Pressure management may be necessary to meet target rates and volumes in applicable geologic settings.
- Delineating the area-of-review for a geologic storage project is a critical component of permitting and the technical risk assessment.
- Regional state-of-stress and geomechanical characterization are important for site selection.

3.2 Financial and economic analysis

The prefeasibility projects performed a regional storage resource assessment to explore economic strategies for capturing, transporting, and storing large-scale anthropogenic sources in potential 50 MMT storage sites. Revenue from 45Q tax credit incentives for dedicated and associated storage scenarios and the potential sale of CO_2 to EOR fields were considered to offset the costs of these projects. Associated storage revenues were calculated at $20-33/MT of CO_2 sold for EOR at an oil market price range of $55-60/barrel (bbl). Table 1 includes the total cost of several fully integrated systems for CO_2 capture, compression, and transportation from sources to injection sites in the Onshore U.S. Prefeasibility projects with primary storage sites in the U.S. Offshore and Coastal regions in the Gulf of Mexico, Coastal Louisiana, and Offshore Washington are not summarized in this section.

Table 1. Total cost and revenue (shown in parentheses) for prefeasibility projects in onshore U.S.

Key	Project Name	Source(s) by Location	Econ. Cases	Revenue Type	Total Cost ($/MT)
1	Integrated Mid-Continent Stacked Carbon Storage Hub	Coal-fired power plant in Lincoln County, NE	3	Dedicated storage	31 to 67
		Ethanol plant corridor in eastern NE	3	Oil+Net CO_2 stored for EOR	(1.72) to 57
2	Nebraska Integrated Carbon Capture and Storage Prefeasibility	Coal-fired power plant in Lincoln County, NE	1	Dedicated storage	20
			1	CO_2 sold for EOR	32
3	Integrated Carbon Capture and Storage for Kansas (ICKan)	Two coal-fired power plants in Finney and Pottawatomie County, KS	6	Dedicated storage	2 to 27
			6	CO_2 sold for EOR	14 to 39

		Ethanol plant corridor in northeast KS			
4	Integrated CCS Prefeasibility Study at Dry Fork Station	Coal-fired power plant in Campbell County,WY	1	CO_2 sold for EOR and dedicated storage	(0.14)
5	CarbonSAFE Illinois East Basin[1]	Chemical plant in Vigo County IL	1	CO_2 sold for EOR and dedicated storage	-
6	CAB-CS: Central Appalachian Basin CarbonSAFE Integrated Prefeasibility Project	Coal-fired power plant, a Gas-fired power plant, and Chemical plant in eastern Ohio. Potential sources: steel mill and gas-fired power plant in eastern and southern Ohio	6	dedicated storage	9 to 66
			6	CO_2 sold for EOR and dedicated storage	9 to 68
			6	CO_2 sold for EOR	1 to 67
7	Northern Michigan Basin CarbonSAFE Integrated Prefeasibility Project	Coal-fired power plant in Otsego Co., MI. Potential sources: steel mill and gas-fired power plant in northern and central MI	4	Dedicated storage	21 to 116
			4	CO_2 sold for EOR and dedicated storage	17 to 114
			4	CO_2 sold for EOR	12 to 108
8	California CO_2 Storage Assurance Facility Enterprise	Three integrated energy corridors in Southern San Joaquin Valley, CA	1	CO_2 sold for EOR and dedicated storage	(40)
9	CarbonSAFE Rocky Mountain Phase I: Ensuring Safe Subsurface Storage in the Intermountain West	Two coal-fired power plants in Emery County, Utah.	4	Dedicated storage [NA]	50 to 51
			9	CO_2 sold for EOR [NA]	17 to 27
10	Integrated Commercial CCS Prefeasibility Study at Rock Springs Uplifts, Wyoming[2]	Coal-fired power plant in Sweetwater Co., WY.	1	CO_2 sold for EOR and Dedicated storage	-

[1]Capital investment was estimated at $58.7 M. Annual operating costs were estimated at $12.7 M to supply 1.7 MMT/yr of CO_2.
[2]Capital investment was estimated at $787 M. Annual operating costs were estimated at $54 M to supply 2 MMT/yr of CO_2.
[NA] Economic analysis did not include 45Q tax credit incentives.

The economic analysis for each prefeasibility study can be categorized into three types of scenarios: coal-fired power plant point-to-point transport (Scenario 1), multi-source point-to-point transport (Scenario 2), and long-range energy corridor transport (Scenario 3). The scenarios are meant to offer potential commercial synergies for reducing the unit cost of CO_2 storage and the risk of investment. One representative prefeasibility study for each scenario is summarized in the following sections.

3.2.1 Scenario 1: Coal-fired power plant point-to-point transport

The University of Utah performed a prefeasibility study in central Utah (Table 1, Key 9). The PacifiCorp's Hunter Power Plant and Huntington Power Plant are located in Emery county near a 7,000-foot deep brine-bearing sedimentary rock formation. A commercial amine-based capture technology (Mitsubishi Heavy Industries (MHI) KM-CR Process® with KS-1™ solvent) was evaluated to estimate the capture and compression costs. Approximately 1.84 MMT and 7.2 MMT of CO_2 per yr can be captured from each plant at $50 and $61/MT, respectively. The proposed locations of two storage sites named Buzzards Bench and Drunkards Wash have estimated storage capacities of 201 MMT and 380 MMT of CO_2, respectively.

Table 2 lists the point-to-point economic scenarios for each power plant. Cost analysis supports the primary core scenario consisting of 65% capture (equivalent to 1.84 MMT of CO_2 per year) from each coal-fired power plant and transporting and injecting it into a sandstone formation at the Buzzards Bench or Drunkards Wash area. Simulation studies suggest that one well is sufficient to store 50 MMT of CO_2 over 30 years in each storage complex. The second scenario involves selling CO_2 to an EOR field located in the Uinta Basin in Utah and a natural source of CO_2 in the McElmo Dome in southwestern Colorado. Findings suggest that larger-scale infrastructure deployments to oil fields in the Uinta Basin for EOR applications could result in a cost savings of up for 67%/MT of CO_2. None of the scenarios considered splitting tax credit incentives from storing CO_2 in both dedicated and associated storage sites. Detailed discussion of the results of this study are provided by McPherson [13].

Table 1. Economic scenarios without 45Q tax credit incentives

Case	Source(s) and Sink(s)	Pipeline (Miles)	Total Cost ($M/yr)	Capture ($M/yr)	Trans. ($M/yr)	Storage ($M/yr)	Total Unit[1] ($/MT)

1	Hunter – Buzzards	22	86	78	0.8	6.6	50.40
2	Hunter – Drunkards	11	87	78	2.4	6.7	51.37
3	Huntington – Buzzards	9	86	78	1.3	6.6	50.67
4	Huntington – Drunkards	7	87	78	1.8	6.7	50.98
5	Hunter – Uinta EOR	100 >	29	78	1.5	(51)	16.87
6	Hunter – McElmo Dome	300 >	42	78	14.5	(51)	24.57

Note: Numbers in parentheses indicate negative costs or profit derived from 45Q tax credit.
[1] Total Unit cost includes storage and monitoring costs for 30 years.

3.2.2 Scenario 2: Multi-source point-to-point transport

Battelle performed a prefeasibility study in northern Michigan (Table 1, Key 7). Multiple CO_2 sources, rather than a single source, were considered in the analysis. These include: 1) the existing Wolverine Power Cooperative-owned natural gas-fired power station in Otsego County that will be retrofitted to a combined cycle facility with CO_2 capture supplemented by CO_2 from the DCP Midstream-owned natural-gas separation facility, also located in Otsego County; 2) a potential new source (PNS) assumed to be located close to the saline storage site(s) and the EOR fields that is a conventional natural gas combined cycle (NGCC) facility with capture; 3) a second variation of the PNS that incorporates the emerging natural gas-fired technology being developed and built by NET Power, LLC, based on the Allam Cycle; and 4) Project TIM, a proposed steel manufacturing facility by New Steel, Inc. in Shiawassee County. The saline storage sites, Saint Peter Sandstone (SPSS) Site 2 and SPSS Site 7, were down selected for detailed economic analysis. Site 2 is located in Montmorency County and Site 7 is located in Kalkaska and Missaukee counties in Michigan. Simulation studies suggest that Site 2 requires six injection wells to inject 50 MMT of CO_2 and has a projected 50 MMT CO_2 plume area of 18,200 acres. Site 7 requires only three injection wells and has a projected plume area of 14,000 acres.

Table 4 lists the multi-source point-to-point economic scenarios evaluated. Interest costs during construction and escalation were included for each of the storage, pipeline, and capture facilities during the construction period prior to the commencement of operations in 2025. Economic analysis of Net Power, an emerging gas-power technology, was not included. The three types of scenarios include: 50 MMT of CO_2 stored at Site 2 for each of the identified sources, 50 MMT of CO_2 supplied to the EOR fields (i.e., carbonate reefs), and a hybrid case in which the CO_2 is divided between saline storage (50%) and CO_2-EOR (50%). In the hybrid case, only one group of EOR reefs is needed to store 25 MMT CO_2. The availability of EOR (ongoing and expanded) as an option for stacked storage with saline storage is an attractive feature for the Northern Michigan Basin area, as this can help defray some of the capital and operating costs for saline storage, increasing the feasibility of the project. Results show that either a 20% reduction in the cost of capture or a $30/MT of CO_2 sale price is necessary for a new natural gas-fired combined cycle with 100% CO_2-EOR storage to achieve a breakeven net revenue requirement. Core Energy, a key member of the Battelle Carbon Storage – Northern Michigan Basin project team, operates in EOR fields and could benefit from this project. Detailed discussion of the results of this study are provided by Kelly [14] in Table 4.

Table 2. Economic scenarios with 45Q tax credit incentives

Case	Source(s) and Sink(s)	Pipeline (Miles)	Total Costs ($M)	Capture ($M)	Trans. ($M)	Storage ($M)	Total Unit Cost[1] ($/MT)
1	Wolverine – SPSS-2 site	21	1,056	691	40	325	54-74
2	Wolverine – EOR and SPSS-2 site	74	1,017	691	131	195	55-73
3	Wolverine – Niagaran Reefs	105	881	691	190	-	47-60
4	New NGCC – SPSS-2 site	17	1,016	658	33	325	44-58
5	New NGCC – EOR and SPSS-2 site	46	937	658	84	195	43-56
6	New NGCC – Niagaran Reefs	83	808	658	150	-	38-48
7	Project Tim – SPSS-2 site	154	1,450	848	277	325	89-116
8	Project Tim – EOR and SPSS-2 site	183	1,370	848	327	195	89-114
9	Project Tim – Niagaran Reefs	225	1,250	848	402	-	85-118

[1]Cost components and anticipated amount of revenue for a scenario on a 30-year levelized basis, expressed in 2018 $/tonne (the project-start date) on a low or high cost of capital basis.

3.2.3 Scenario 3: Large-scale energy corridor transport

The University of Kansas, along with the Kansas Geological Survey, conducted a preliminary feasibility study of two coal-fired power plants in Kansas connecting to large storage corridors in Nebraska (Table 1, Key 3). The Westar Energy's Jeffrey Energy Center (JEC) and S.E. Holcomb station were selected to evaluate multiple capture levels using a commercially available Linde-BASF system as the basis for the capture technology. The estimated capture costs for JEC (2.5 MMT/yr) and S.E. Holcomb Station (1.7 MMT/yr) were \$45–67/MT and \$46–\$72/MT, respectively. In addition, about 34 ethanol plants within the region could supply CO_2 to EOR and storage sites located in southwest Kansas and the Permian basin in Texas. Capture and compression costs from these ethanol plants were estimated at \$19/MT and represent a yearly production 10.9 MMT of CO_2 from 3.6 B gallons of ethanol.

Saline aquifers that are overlain by oil reservoirs were studied in the state of Kansas. As a primary site option, the Patterson site, located in southwest Kansas, has an estimated storage capacity of 60 MMT. Previous studies in the region estimated that four of these fields in the NHSC have an estimated associated storage potential for EOR of 1.8-2.3 MMT/yr [15]. To date, 5.5 M barrels (bbls) of oil have been produced from the combined Patterson and Heinitz fields before and after unitization, and an additional 1.2 Mbbls of oil are expected from the waterflood (6.7 Mbbls ultimate recovery). As a secondary site option, the Davis Ranch and John Creek fields, located in the Forest City Basin in Wabaunsee county, have an estimated storage capacity of 24 MMT.

Table 5 lists the scenarios presented by the University of Kansas. All cost estimates include both capital costs and operating expenses over a 2-yr construction period followed by 20 years of operation, and a 10% cost of capital. The most feasible option, as outlined in Case 4, consists of delivering 4.3 MMT/yr of CO_2 from ethanol plants in Kansas and Nebraska to several EOR fields located in southwest Kansas. The total project cost can be further reduced by supplying 2-2.5 MMT to dedicated storage and selling 1.8-2.3 MMT of CO_2 to EOR in the nearby EOR field. Profits from the sale of CO_2 for EOR could offset transportation. Detailed discussion of the results of this study are provided by Holubnyak [16].

Table 3. Economic scenarios with 45Q tax incentives

Case	Source(s) and	Pipeline	45 Q	Pipeline	Capture	Total

	Sink(s)	(miles)	Revenue Type	Costs ($/MT)	Costs[1] ($/MT)	Cost ($/MT)
1	Holcomb – Patterson Site	25	EOR	(23.3)	46-72	23
			Saline	(35.1)	46-72	11
2	JEC – Davis Ranch/John Creek	50	EOR	(22.9)	45-67	22
			Saline	(34.8)	45-67	10
3	34 ethanol plants – Permian Basin	1,546	EOR	9.0	10	19
			Saline	0.8	7	8
4	15 ethanol plants – SW Kansas	737	EOR	5.0	9	14
			Saline	(2.5)	5	2

Note: Numbers in parentheses indicate negative costs or profit derived from 45Q tax credit. [1]The derived costs for scenarios 1 and 2 used lowest capture costs from pre-FEED studies

3.3 CarbonSAFE storage complex feasibility projects

CarbonSAFE feasibility projects focused on the next phase of advancing carbon storage toward commercial scale. Work conducted under the feasibility projects included activities such as completing at least one characterization well, updating of static earth models, updating of dynamic and geomechanical modeling, 3D seismic analysis, core and log analysis updates and well testing updates. Initial characterization results demonstrated the likely viability of multiple 50+ MMT geologic storage complexes in deep saline formations. A risk assessment exercise was also conducted to identify and assess technical and nontechnical risks that could prevent potential candidate storage complexes within the study area from serving as commercial storage sites. The assessment identified and evaluated six technical risk categories: 1) CO_2 injectivity, 2) storage capacity, 3) lateral migration of CO_2, 4) lateral pressure propagation, 5) vertical migration of CO_2 or formation brine, and 6) induced seismicity. The risk assessment results can be used to complete a site characterization plan (e.g., future site characterization, modeling and simulation, monitoring activities) and management strategies for the project. Three feasibility projects that have not yet advanced to the characterization phase are briefly described below.

3.3.1 Wabash CarbonSAFE

The Wabash CarbonSAFE project, led by the University of Illinois at Urbana-Champaign, determined the feasibility of commercial-scale CO_2 storage in the Mt. Simon sandstone formation located beneath the future Wabash Valley Resources

(WVR) ammonia production plant. Targeted to begin operation in 2021, the ammonia plant is under construction to repurpose and expand the former integrated gasification combined cycle plant located near Terre Haute, Indiana. Approximately 1.65 MMT of CO_2/yr will be produced as a by-product, which can then be captured through a process using Rectisol.

The selected storage site, located on the eastern side of the Illinois Basin, is on the WVR plant site. The primary target reservoir is the Mt. Simon sandstone formation. This formation is estimated to have a storage capacity of 12-172 billion metric tons of CO_2 in the Illinois Basin [17]. Secondary storage targets include units in the Knox Group, and the St. Peter Sandstone. A stratigraphic well was drilled in the WVR facility for the acquisition of 265 (ft) feet of core and to obtain geologic data to update models and confirm reservoir properties. Core samples were collected from the Maquoketa Group (secondary seal), the Eau Claire Formation (primary seal), and the Upper Mt. Simon Sandstone. An extensive suite of geophysical logs was completed to evaluate CO_2 storage potential, hydrocarbon potential, and general lithological and petrophysical characteristics of the sedimentary column. In addition, 20 kilometers (km) of 2D seismic data is being processed to assist with subsurface geologic characterization. Future vertical seismic profiles (VSPs) can be performed to improve delineation of geologic structures. Final reservoir simulation, based on the stratigraphic well's core and log analyses and drill stem test data, would be performed to assess the CO_2 injection potential of the Mt. Simon sandstone.

3.3.2 CarbonSAFE Illinois Macon County

The CarbonSAFE Illinois Macon County project, led by the Illinois State Geological Survey, assessed the feasibility of developing a CO_2 storage complex that can store more than 50 MMT of CO_2 in the central Illinois Basin. The closest potential CO_2 sources are the Abbot Power Plant located in Champaign County, Illinois; City Water, Light, and Power located in Springfield County, Illinois; and ADM located in Decatur County, Illinois. These three sources emit a total of approximately 3.0 MMT/yr of CO_2. A fourth, more distant potential source is the 1,600-megawatt (MW) coal-fired Prairie States Energy Campus, which began operations in 2012 and emits greater than 10 MMT of CO_2/yr.

The original project site in Macon County was switched to the Podolsky Oil Company's field in Christian County, Illinois to accommodate the larger potential sources of CO_2. The target reservoir was the Lower Mt. Simon sandstone formation.

Initial characterization work included drilling the T.R. McMillen #2 (TRM) stratigraphic well for the acquisition of 148 ft of core and to obtain geologic data to update models and confirm reservoir properties. The well was drilled through the targeted storage interval into the Precambrian basement to a total depth of 6,490 ft. Preliminary results indicated that the Mt. Simon sandstone at the location of the TRM #2 well was suitable for commercial-scale CO_2 storage. Initial modeling showed that 30 years of injection in three wells with 80 acre-spacing projected a 190 MMT CO_2 plume extension of 2.7 miles with little interaction with the adjacent Illinois Basin–Decatur Project (IBDP). A 30 km 2D seismic survey and a VSP were performed to obtain data needed to interpret the subsurface structure of the site. Seismic data is being processed and analyzed. Additionally, stacked storage in oil-bearing carbonate oil reservoirs in the Mt. Auburn trend of the Sangamon Arch in the Illinois Basin are being studied for CO_2-EOR potential.

Previous characterization efforts at the IBDP site suggested that induced microseismicity migrated from sediment into Precambrian basement. At the site, the microseismicity was monitored using deep borehole sensors. About 1400 events were recorded, with moment magnitudes between 1.1 and -1.7. [18, 19]. Induced seismicity risk was assessed at the Macon site through the characterization of geomechanical properties of rocks and fault analysis, and the stress state field measurements in the basement. Well log cross sections displaying the geomechanical properties at the Macon and IBDP sites were prepared to illustrate the differences between the sites. Laboratory tests performed on the cores from the TRM well were also performed on the basement rock. In situ stress direction characterization at the Macon site was completed using indicators observed in a Formation Micro-Imager log and well tests (e.g., Diagnostic Fracture Injection Test). Future characterization work would include the integration of these data and well site studies with improved geomechanical models and dynamic reservoir models.

3.3.3 Integrated Midcontinent Stacked Carbon Storage Hub

The Integrated Midcontinent Stacked Carbon Storage Hub project, led by Battelle Memorial Institute, is a combination of three prefeasibility projects located in Kansas and Nebraska. The intent of the project is to develop a source corridor that can run southwest from eastern and central Nebraska toward Red Willow County, Nebraska. Detailed discussion of the Phase I results of this study are provided by Duguid [20]. The feasible pipeline routes connecting sources along the source corridor to potential saline sinks (i.e., dedicated storage) or CO_2-EOR sinks (i.e., associated storage) in the storage corridor account for environmentally and culturally sensitive areas. The team identified 17 technically and economically feasible CO_2-EOR oilfields and three saline reservoir fields in the storage corridor. The designated oilfields have estimated storage of 577.4 MMT of CO_2 with the potential to produce 181.9 Mbbls of oil via EOR, equivalent to an estimated gross revenue for stacked storage and EOR of $30.9 B.

Three saline sinks were evaluated: Madrid field in southwest Nebraska, Patterson-Heinitz-Hartland field (PHH) in southwestern Kansas, and the Sleepy Hollow Field (SHF) in southwest-central Nebraska. One stratigraphic well was drilled at the SHF site to obtain geologic data through well logging and 110 ft of core for analysis. Two additional stratigraphic wells were drilled at the PHH site to collect geologic data through well logging and obtaining 778 ft of core for analysis. The effort also included the acquisition of 26 square miles (mi^2) of new 3D seismic data to interpret the subsurface structure in the PHH site. Results from seismic data processing revealed two major reverse faults that offset the reservoir and seal intervals and constitute an uplifted block in the Patterson area. In addition, three- and four-way structural closures were identified that can assist with trapping CO_2 in the Arbuckle-Osage reservoirs. Updated models indicated that the original 50 MMT target injection amount was not viable in the SHF site. Numerical simulation for the PHH site indicated that three wells would be required to store 50 MMT of CO_2 in deep stacked saline Paleozoic formations over 30 years. Future recommended characterization work would include fault analysis at the PHH site and conducting a 3D seismic survey and drilling a characterization well at the Madrid site.

3.4 CarbonSAFE site characterization and CO_2 capture assessment projects

Three of the five Site Characterization and CO_2 Capture Assessment projects build upon previous phases that indicated the commercial viability of storage complexes located in North Dakota, Mississippi, and Wyoming. The two remaining projects, located in New Mexico and Illinois, demonstrated the viability of their respective sites through previous work conducted, unrelated to the CarbonSAFE effort. The site characterization efforts conducted by these five projects, and the association of these projects with capture assessment projects, are discussed below.

3.4.1 Illinois Storage Corridor (Phase III)

The Illinois Storage Corridor is a region with proven geologic storage performance and numerous industrial carbon sources. Numerous DOE-FE funded projects in the Illinois Basin are already addressing the challenges posed by CCUS deployment at scale through the field validation of storage complexes and new technologies. These projects include two pilot carbon capture technology projects [21] two commercial-scale CCUS projects led by RCSPs [22], and Wabash and Macon CarbonSAFE phase II projects. As part of the CarbonSAFE phase III work, the Board of Trustees of the University of Illinois (Champaign, IL) will determine the feasibility of developing a storage complex near two facilities: One Earth Energy ethanol facility in Ford County, Illinois and Prairie State Energy Campus (PSEC) facility, a 1,200-MW coal-fired electricity generation plant in St. Clair County, Illinois. Annually, these two sources have the potential to capture 0.45 and 6 MMT of CO_2, respectively. A Carbon Capture Program funded FEED study at PSEC will evaluate the use of the commercial MHI's KM CDR Process™ CO_2 capture technology, and a capture assessment at the One Earth Energy ethanol facility will be completed under CarbonSAFE. Future characterization work at each site will include the drilling and logging of one characterization well, coring new samples, and acquiring 26 miles of 2D seismic data and 4-6 mi^2 of 3D seismic data.

3.4.2 San Juan Basin CarbonSAFE Phase III: Ensuring safe subsurface storage of CO_2 in saline reservoirs

New Mexico Institute of Mining and Technology (Socorro, NM) is performing a comprehensive site characterization study of a mature oil and gas field in the San Juan Basin, northwest New Mexico. The field, operated by Hilcorp Energy, has great stacked storage potential in alternating sequences of deep saline formations and oil-bearing reservoirs. Saline storage targets include the Salt Wash Member, Bluff sandstone, and Slick Rock Member.

The selected site is approximately 20 miles from the San Juan Generating Station (SJGS), an 847-MW coal-fired electricity generation plant. Annually, this plant has the potential to capture 6-7 MMT of CO_2. The CCUS project will consist of supplying 2 MMT of CO_2/yr to the storage complex and transporting the remaining CO_2 via the Cortez pipeline network to EOR fields in the Permian Basin in Texas. A Carbon Capture Program funded FEED study at SJGS will be completed using MHI's commercial amine-based solvent CO_2 capture technology. Ongoing and upcoming efforts at the storage complex site include the acquisition and evaluation of existing field data (e.g., 3D seismic, logging and coring samples), the drilling and logging of one characterization well, and coring of new samples.

3.4.3 Establishing an early CO_2 storage complex in Kemper County, Mississippi: Project ECO_2S

Southern States Energy Board (Peachtree Corner, GA) will fully characterize the storage complex facility in Kemper County, Mississippi. Previous characterization efforts recognized Kemper County as a world class geologic setting capable of securely storing over 1,200 MMT of CO_2 over a 30,000-acre area. Detailed discussion of the Phase II results of this study are provided by Reistenberg [23]. Three high-quality sandstone reservoirs include the lower Cretaceous Paluxy Formation, Washita-Fredericksburg interval, and the Upper Cretaceous Lower Tuscaloosa group. The initial proposed CCUS scenario comprised capture from the Kemper County Energy Facility with minimal pipelines required to transport the CO_2 to the storage complex. Due to operational changes at the facility, the project now includes two additional sources totaling 22.5 MMT of CO_2/yr for the development of a regional hub. A Carbon Capture Program funded FEED study will be completed using Linde-BASF's OASE blue Advanced Amine solvent-based capture technology at Southern Company's Plant Daniel in Moss Point, Mississippi. Additional CO_2 capture assessments of Plant Ratcliffe in Kemper County, Mississippi and Plant Miller in Jefferson County, Alabama will be completed under this project. Ongoing and upcoming efforts at the storage complex site include the drilling and logging of three characterization wells, coring new samples, and acquisition of 4-6 mi^2 of 3D seismic data.

3.4.4 North Dakota CarbonSAFE Phase III: Site characterization and permitting

The University of North Dakota Energy & Environmental Research Center (Grand Forks, ND) will fully characterize a storage complex near Minnkota's Milton R. Young Station (MRYS) one of two storage sites considered in earlier feasibility efforts. Detailed discussion of the Phase II results of this study are provided by

Peck [24]. The site is capable of storing over 3 MMT of CO_2/yr from the plant's 455-MW Unit 2. The project will investigate three storage reservoirs for further study: Broom Creek, Inyan Kara, and Deadwood sand reservoirs. Previous work suggests that Broom Creek formation is capable of securely storing over 563 MMT of CO_2. Coordinating with this CarbonSAFE Initiative is the Project Tundra initiative - an integrated CO_2 capture and storage project led by Minnkota Power Cooperative Targeted to begin operation in 2022-2023, the MRYS Unit 2 will be using a commercial amine-based Econamine FG Plus℠ capture technology from Fluor Corporation. A FEED study of this technology is being studied under the Carbon Capture program. Ongoing effort at the MRYS site will facilitate the completion of all necessary regulatory requirements for two Class VI well permits. Future characterization work will include the drilling and logging of two characterization wells, coring new samples, and 11 mi^2 of 3D seismic acquisition.

3.4.5 Wyoming CarbonSAFE: Accelerating CCUS commercialization and deployment at Dry Fork Power Station and Wyoming Integrated Test Center

The University of Wyoming (Laramie, WY) is finalizing the characterization of the storage complex in Campbell County, Wyoming. Detailed discussion of the Phase I results of this study are provided by Quillinan [25]. The selected storage site is centrally located in the Powder River Basin, which produces more than 40% of the coal mined in the United States [26], and is accessible to existing CO_2 pipeline infrastructure and potential CO_2-EOR fields. The project will investigate four storage reservoirs for further study: the Minnelusa, Lower Sundance, Lakota/Fall River Group, and the Muddy Sandstone. Initial analysis suggests a P50 storage volume of 3.4 MMT/mi^2 for the combined reservoirs. The project would capture CO_2 from Basin Electric's Dry Fork Station (DFS), one of the newest 385-MW commercial-scale coal-fired power plants in the United States. The power station is also home to the Wyoming Integrated Test Center (https://www.wyomingitc.org/), which is used to study the capture, storage, and management of carbon emissions from DFS. A Carbon Capture Program funded FEED study at the DFS will be completed for the commercial Polaris™ membrane CO_2 capture technology from Membrane Technology and Research, Inc. Future characterization work will include analysis of previous phase II acquisitions (legacy 2D seismic, newly acquired 3D seismic, and coring new samples), re-entering one well to gauge for monitoring, constructing a second well for data collection and injection tests, and a new VSP seismic data acquisition.

4. Summary

CCUS is becoming widely accepted as a viable and necessary option for fossil-based energy sources, such as coal- or gas-fired power plants and other industrial sources to lower their carbon dioxide (CO_2) emissions. Over the past two decades DOE research efforts have provided supporting evidence that CO_2 can be stored in geologic formations safely and securely across multiple regions of the U.S. Efforts now must shift focus on the justification of the business case for widespread CCUS deployment. Research will continue to focus on technologies and methods to reduce cost, reduce operational risk and increase efficiencies to make commercial-scale CCUS economical.

The CarbonSAFE Initiative is providing this understanding of how to safely, efficiently and cost-effectively characterize and permit project site(s) at the commercial-scale throughout the U.S. Phase I and Phase II projects have investigated potential integrated CCUS project scenarios and identified both technical and non-technical research gaps. Each Phase III CarbonSAFE project now works to fill those gaps. Projects will need to consider and negotiate various aspects and business arrangements as part of the process to evaluate and secure CO_2 source(s), develop a plan for pore space ownership, and complete a UIC Class VI permit application(s) to construct the injection wells. To achieve wide adoption of CCUS, it is imperative that we build upon these learnings to test, mature and prove CCUS technologies at commercial scale.

5. Disclaimer

This report was prepared as an account of work sponsored by an agency of the United States Government. Neither the United States Government nor any agency thereof, nor any of their employees, makes any warranty, express or implied, or assumes any legal liability or responsibility for the accuracy, completeness, or usefulness of any information, apparatus, product, or process disclosed, or represents that its use would not infringe privately owned rights. Reference herein to any specific commercial product, process, or service by trade name, trademark, manufacturer, or otherwise does not necessarily constitute or imply its endorsement, recommendation, or favoring by the United States Government or any agency thereof. The views and opinions of authors expressed herein do not necessarily state or reflect those of the United States Government or any agency thereof.

6. Acknowledgement

KeyLogic's contributions to this work were funded by the National Energy Technology Laboratory under the Mission Execution and Strategic Analysis contract (DE-FE0025912) for support services.

7. References

[1] National Petroleum Council, "Meeting the Dual Challenge: A Roadmap to At-Scale Deployment of Carbon Capture, Use, and Storage, Chapter One: The Role of CCUS in the Future Energy Mix," 2019. [Online]. Available: https://dualchallenge.npc.org/files/CCUS-Chap_1-042320.pdf. [Accessed November 2020].

[2] EIA, "U.S. Energy-Related CO2 Emissions in AEO2019 Reference Case, 1990 to 2050 (figure 1-10)," EIA, 2019.

[3] S. Bennett and T. Stanley, "US budget bill may help carbon capture get back on track," 12 March 2018. [Online]. Available: https://www.iea.org/commentaries/us-budget-bill-may-help-carbon-capture-get-back-on-track. [Accessed November 2020].

[4] D. Nagabhushan and J. Thompson, "Carbon capture and storage in the US power sector - The impact of 45Q Federal Tax Credits," Clean Air Task Force, The impact of 45Q Federal 2019. [Online]. Available: https://www.catf.us/wp-content/uploads/2019/02/CATF_CCS_United_States_Power_Sector.pdf. [Accessed November 2020].

[5] Carbon Capture Coalition, "Carbon Capture Jobs and Project," June 2020.

[Online]. Available: https://carboncapturecoalition.org/wp-content/uploads/2020/06/Carbon-Capture-Jobs-and-Projects-1.pdf. [Accessed November 2020].

[6] Clean Air Task Force, "Potential Carbon Capture Projects Database," undated. [Online]. Available: https://docs.google.com/spreadsheets/d/115hsADg3ymy3lKBy4PBQRXz_MBknptqlRtlfuv79XV8/edit#gid=1540463113. [Accessed November 2020].

[7] Clean Air Task Force, "Interactive Map of CCUS Projects in Development in the U.S.," 27 July 2020. [Online]. Available: https://www.catf.us/2020/07/ccus-interactive-map/. [Accessed November 2020].

[8] H. Tanakaa, T. Tsujiuchi, T. Kamijo, S. Kishimoto, Y.-J. Lin, S. Kawasaki, Y. Nakagami and T. Nojo, "Advanced KM CDR Process using New Solvent," October 2018. [Online]. Available: https://www.cfaenm.org/wp-content/uploads/2019/03/GHGT14_manuscript_20180913Clean-version.pdf. [Accessed November 2020].

[9] NETL, "BEST PRACTICES MANUALS," undated. [Online]. Available: https://netl.doe.gov/coal/carbon-storage/strategic-program-support/best-practices-manuals. [Accessed November 2020].

[10] T. Rodosta, M. Ackiewicz and A. Albenze, "Status Update and Results from the U.S. Department of Energy Regional Carbon Sequestration Partnership Initiative, Proceedings GHGT-12," *Energy Procedia,* vol. 63, pp. 6039-6052, 2014.

[11] T. Rodosta, W. Aljoe, G. Bromhal and D. Damiani, "U.S. DOE Regional Carbon Sequestration Partnership Initiative: New Insights and Lessons Learned, Proceedings GHGT-13," *Energy Procedia,* vol. 114, pp. 5580-5592, 2017.

[12] M. Sullivan, T. Rodosta, K. Mahajan and D. Damiani, "An overview of the Department of Energy's CarbonSAFE Initiative: Moving CCUS toward commercialization," *Separations: Materials, Devices and Processes,* 6 November 2019.

[13] B. McPherson, M. Cather, R. Middleton, T. Chidsey, J. Heath, M. Saunders and S.-Y. Lee, "CarbonSAFE Rocky Mountain Phase I: Ensuring Safe Subsurface Storage of Carbon Dioxide in the Intermountain West," 31 August 2018. [Online]. Available: https://www.osti.gov/biblio/1559990. [Accessed November 2020].

[14] M. Kelley, A. Haagsma, P. Champagne, J. Glier, N. Gupta, M. H. Yugulis, J. Hawkins, J. Main, A. Pasumarti, J. Sminchak, S. Weber, B. Mannes, R.

Pardini, K. Sanders and W. Harrison, "Integrated Pre-Feasibility Assessment for a Northern Michigan Basin CarbonSAFE CO 2 Storage Complex," 2 September 2018. [Online]. Available: https://doi.org/10.2172/1469190. [Accessed November 2020].

[15] M. K. Dubois, E. T. Williams, J. C. Youle and D. E. Hedke, "Potential for CO2 storage and enhanced oil recovery in four southwest Kansas oil fields, an extended abstract; in W. L. Watney, PI, Final Report for DOE Award Number: DE-FE0002056," DOE, 2015.

[16] Y. Holubnyak, M. Dubois, T. Bidgoli, D. Wreath, L. Watney, S. Stover, D. Newell, F. Fazelalavi, A. Hollenbach, J. Jennings, C. Steincamp, J. Schremmer, B. Jordan and B. Crabtree, "Integrated CCS for Kansas (ICKan) Final Technical Report," December 31, 2018.

[17] United States Department of Energy, "Carbon Storage Atlas: 5th edition," DOE, Washington, D.C., 2015.

[18] B. Goertz-Allmann, S. Gibbons, V. Oye, R. Bauer and R. Will, "Characterization of induced seismicity patterns derived from internal structure in event clusters," *Journal of Geophysical Research: Solid Earth,* vol. 122, p. 3875–3894, 2017.

[19] R. Bauer, R. Will, S. Greenberg and S. Whittaker, "Chapter 19: Illinois Basin-Decatur project," in *Geophysics and Geosequestration*, T. M. Davis, M. Landrø and M. Wilson, Eds., Cambridge University Press, 2019, pp. 339-369.

[20] D. H. Bacon, D. Blankenau, D. Divine, A. Duguid, I. Fukai, J. Glier, J. Hawkins, M. Jimenez, R. Joeckel, S. McDonald, R. Middleton, R. Peterson and M. Scharenberg, "Integrated mid-continent stacked carbon storage hub, Phase I Final Report," 12 October 2018. [Online]. Available: https://doi.org/10.2172/1478726. [Accessed November 2020].

[21] NETL, "Large Pilot Testing of Linde-BASF Advanced Post-Combustion Carbon Dioxide Capture Technology at a Coal-Fired Power Plant," 2018. [Online]. Available: https://www.netl.doe.gov/projects/project-information.aspx?k=FE0031581. [Accessed November 2020].

[22] S. McDonald, "Illinois Industrial Carbon Capture & Storage Project: Eliminating CO2 Emissions from the Production of Bio Fuels - A 'Green' Carbon Process," 11 July 2017. [Online]. Available: https://www.energy.gov/sites/prod/files/2017/10/f38/mcdonald_bioeconomy_2017.pdf. [Accessed November 2020].

[23] D. Riestenberg, R. Esposito and K. Gray, "Establishing an Early CO2 Storage Complex in Kemper County, Mississippi: Project ECO2S (FE0029465)," 28 August 2019. [Online]. Available: https://www.osti.gov/biblio/1617036.

[Accessed November 2020].

[24] W. Peck, K. Glazewski, D. Nakles, N. Azzolina, C. Crocker, R. Klapperich, B. Oster, D. Daly, A. Livers-Douglas, S. Butler, S. Smith, B. Botnen, I. Feole, J. He, N. Dotzenrod, A. Salazar, S. Patil and J. Crossland, "North Dakota Integrated Carbon Storage Complex Feasibility Study," 24 March 2020. [Online]. Available: https://doi.org/10.2172/1606011. [Accessed November 2020].

[25] S. Quillinan and K. Coddington, "Integrated Commercial Carbon Capture and Storage (CCS) Prefeasibility Study at Dry Fork Station, Wyoming," 30 May 2019. [Online]. Available: https://www.osti.gov/biblio/1557442/. [Accessed November 2020].

[26] U.S. Department of the Interior, U.S. Geological Survey, "Assessment of Coal Geology, Resources, and Reserve Base in the Powder River Basin, Wyoming and Montana," February 2013. [Online]. Available: https://pubs.usgs.gov/fs/2012/3143/fs-2012-3143. [Accessed November 2020].

ANATOMICAL FRACTION SEGMENTATION IN THE BIOMASS BALES

Rahul Reddy Kancharla [1,2], William A. Smith [1], Elisa H. Barney Smith [2], and Jordan L. Klinger [1]
[1]Idaho National Laboratory, Idaho Falls, Idaho, USA
[2]Boise State University, Boise, Idaho, USA

Abstract

According to the Bioenergy Technologies Office (BETO), creating a robust next-generation domestic bioenergy industry is an essential pathway for providing sustainable renewable energy alternatives. Using non-food feedstocks, like corn-stover and forest residue, in the biorefineries doesn't affect the food supply chain. In the commercial-scale bioenergy operations, a significant development in the technological advancements is required to determine the biomass feedstock quality at the preprocessing stage. The penetrating ability of the x-rays helps study the big biomass bales, but the feedstock heterogeneity—physical size, shape, and chemical composition—poses a significant challenge during milling, conveyance, feeding, and biofuel conversion processes. The inherent complexity introduced during harvesting and bailing makes the reconstruction and interpretation of baled biomass materials from x-ray data time consuming, laborious, and expensive. The presence of similar low-dense materials showed a small contrast difference in the x-ray images, which makes the characterization based on the x-ray attenuation values not promising. This paper focuses on using the shape and texture properties extracted with image processing techniques to characterize the different tissue samples in the biomass bales.

1 Introduction

Recently, renewable energy resources have attracted a lot of attention as countries worldwide are trying to find sustainable energy alternatives while reducing the impact on the climate. Among the resources, renewable biomass represents an abundant carbon-neutral energy source, which has the potential to meet the expanding energy needs in the transportation sector while limiting greenhouse gas emissions. According to the 2016 Billion-Ton report [1], there is a future potential of producing one billion tons of biomass resources (composed of agricultural, forestry, waste, and algal materials) available in the United States on an annual basis. It can generate 85 billion gallons of biofuels that can potentially displace 30% of U.S. petroleum consumption. Now, the Bioenergy Technologies Office's (BETO's) future looks to "drop-in" replacements such as bio-butanol or other bio-derived intermediates that can be used to replace petroleum intermediates within existing/modified refineries.

Current research is more focused on using lignocellulosic biomass materials as a feedstock at the biorefineries. Lignocellulosic materials refer to the biomass originating from plants that cannot be used for food, which recycles unused plant matter and preserves food resources. The use of lignocellulosic biomass presents a crucial step in moving toward more sustainable renewable energy systems.

Lignocellulosic biomass constituents are made of complex polymer structures that need to be broken into simple sugars before converting them to biofuels. In order to convert these complex polymers biologically, pretreatment is required. Much research has been done on understanding each constituents' behavior and optimizing the pretreatment procedures to achieve high yield.

The main constituents of lignocellulosic biomass are cellulose, hemicellulose, and lignin. During pretreatment, the cellulose and hemicellulose are broken down to glucan and xylan, then converted into biofuels. Lignin is considered to be the most recalcitrant component of lignocellulosic biomass. Acid soluble lignin fragments formed during pretreatment can cause irreversible cellulose loss during enzymatic saccharification. The remaining solid lignin restricts enzymatic hydrolysis by physically impeding cellulase's accessibility to cellulose and un-productively binds cellulose enzymes [2]. Different plant parts are composed of different carbohydrates, have different permeabilities, crystallinities, etc., and therefore do not behave the same way physically or chemically.

And other factor that affects the feedstock quality is the ash content in the lignocellulosic biomass. Ash content during the conversion process can cause problems like slagging, fouling, and corrosion in the thermochemical conversion processes and can cause displacement of fermentable carbohydrates and potential buffering capacity during pretreatment in the biochemical conversion, which increase operational costs for both conversion pathways [3]. This could increase the costs of pretreatment, handling and disposal which can affect the end-product price. A review conducted by Kenney et al. in [4] estimates that a 5% increase in soil-derived ash would increase a 227 $MLyear^{-1}$ conversion facility's costs by \$1.15 $Myear^{-1}$, which translates to a 1% increase in the minimum ethanol selling price of \$0.57 L^{-1}.

Moisture content impacts process performance. High moisture content in storage promotes biological degradation of fermentable sugars, which represents a loss of valuable biomass components after harvesting. Moisture content determination at the biorefinery can be an indication of the efficacy if the off-site storage conditions. Higher moisture content bales (> 25%) during the grinding reduces throughput which increases the overall production cost, which in turn affects the price of the final product.

Biorefineries' economic sustainability relies mainly on the available biomass cost and quality, which should meet the cellulosic biofuel cost and also get some profits. A cost-effective conversion process to produce biofuels from lignocellulosic biomass material relies not just on the material quality, but also on the biorefinery's ability to measure and adapt their process to the biomass's feedstock quality. For that, the biomass-to-energy conversion cycle at the biorefinery should include some sensing systems. These sensing technologies need to provide accurate, repeatable, and timely biomass indices through the biomass-to-energy chain. The application of appropriate sensing technology will be critical to developing a successfully integrated biomass management system [5]. The goal of this research is to extract the fundamental knowledge of the volumetric content of the material composition, ash content and moisture content early in the conversion process, which can help to optimize the conversion process. This work mainly lays a foundation to such work.

2 Background

2.1 Skeletonization

Skeletons are shape descriptors with a broad spectrum of applications in shape matching, recognition, animation, retrieval, and compression [6]. Skeletons provide medial axis representation of an object. Blum laid a foundation for the skeletonization algorithms [7]. Figure 1a, b shows an example image and its skeleton representation.

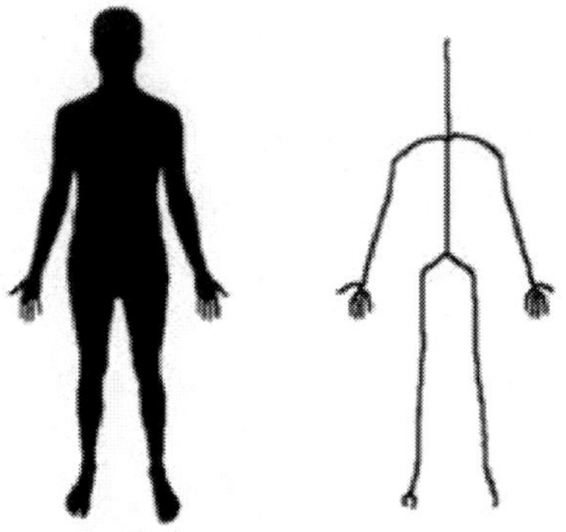

(a) (b)

Figure 1: (a) Example image (b) Skeleton

2.2 Graph theory

A graph represents a set of elements and a set of pairwise relationships between those elements. Here the elements are called nodes or vertices and the relationships are called edges. $G = (V, E)$ represents the graph of an object with vertices or nodes V and edge links E. Figure 2 shows an example of a simple graph. Any set of data points can be represented as a graph. In this work, skeletons are represented as graphs, in order to extract the shape information.

Figure 2: Example of a simple graph with four vertices represented by dots, and its four-edge links connecting them.

3 Materials and Methods

3.1 Image Acquisition

The CT scans used in this research for the corn stover bales are taken using the North Star Imaging (NSI) industrial CT 3D x-ray system at the Idaho National Laboratory. Figure 3 shows the setup we used to extract the radiographs for the corn stover bales. Figure 3a shows the detector and 3c shows the x-ray source. Here corn stover materials are set upon a 2.54 cm foam block in the form of a mini-round bale with a 12.7 cm diameter and 15.24 cm height (see figure 3b). In this experiment, scanning was performed with an x-ray tube voltage of 150 kV and corresponding tube current-time settings at 60 μA.

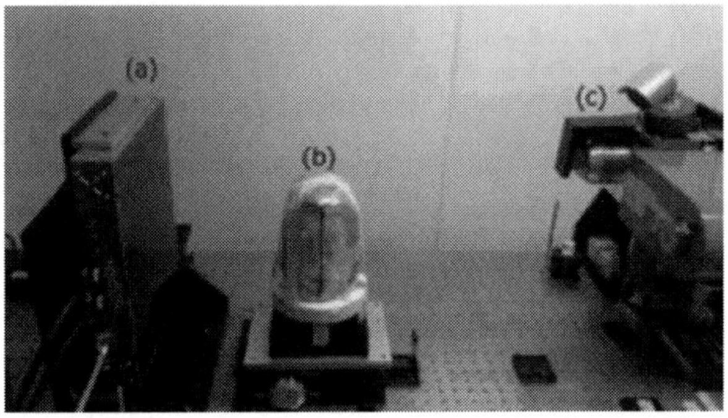

Figure 3: CT setup used to extract the radiographs of corn stover bale with (a) detector (b) mini-round corn stover bale (c) x-ray source.

3.2 Pre-processing

Once the higher density materials like rocks, and soil are removed, the next step involves detecting the corn stover material composition. The raw reconstructed images have background noise which do not contribute significantly to ethanol yield. Figure 4a shows a slice of the bale with background noise. We see in figure 4a that background and contents of corn stover bale have similar voxel intensity levels. In this work a few image preprocessing steps are incorporated to reduce the background noise and highlight the contents of the bale. In the preprocessing stage we thresholded and applied a morphological operation to remove background noise. Then the connected components are used to remove the small isolated voxels in the background noise. Connected components are extracted using the built-in function "bwlabeln" in MATLAB. The connected component labeling approach scans the given data and groups its voxels into components based on voxel connectivity. The isolated voxels are represented with small connected components that are defined to be noise and are removed by resetting their voxel values to zero. Figure 5 shows the 3D reconstructed data before and after removing the isolated voxels. The processed volume can now be analyzed for volumetric content of corn stover fractions.

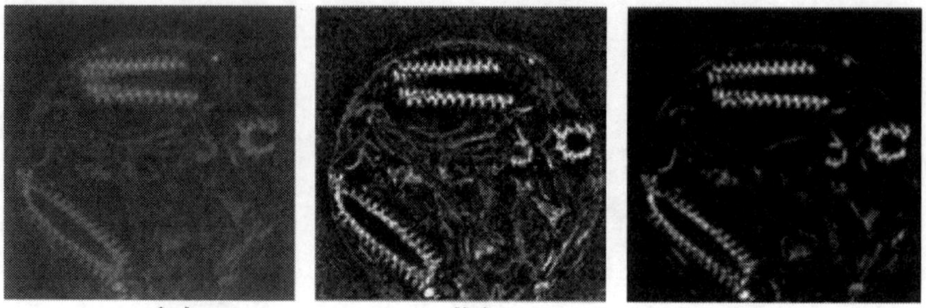

(a) (b) (c)

Figure 4: A cross-sectional slice of mini-round bale at various stages of preprocessing (a) RAW reconstructed slice (b) Thresholded (c) After Morphological Processing.

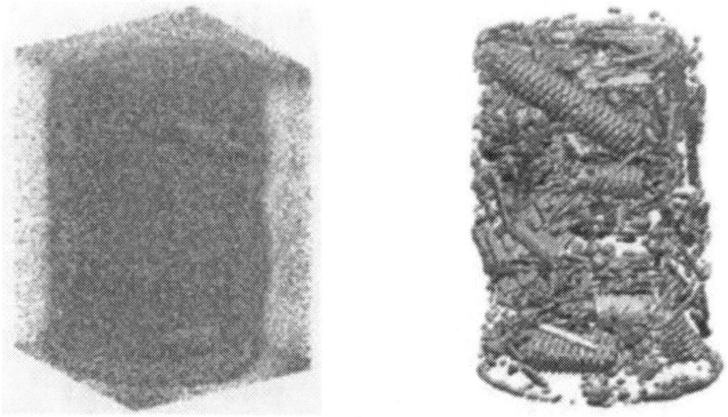

(a) (b)

Figure 5: Bale CT (a) Before pre-processing (b) After pre-processing

3.3 Automatic detection and volume estimation of corn cob fraction

We applied a combination of image processing techniques to extract corn cob volume content in a given corn stover bale. Firstly, to simplify the detection problem, skeletonization is applied to extract the shape information while preserving its topology. Subsequently, graphs are used to represent the skeletons with nodes and edge information, which provides a pairwise relationship between each node. Figure 6a, b, c shows an example corn cob, the skeleton for the corn cob, and the skeleton's graph structure. X-ray reconstruction showed the corn cobs are hollow in the center, which forms a cyclic loop in the skeleton (see figure 6b). With this, the corn cob detection problem is simplified to loop detection. The cyclic loop is detected by removing all the branches (see figure 7a).

Since fully intact corn cobs resemble a cylinder, volume information of a corn cob can be roughly estimated by measuring the radius and height. Starting with the cyclic loop, the radius is estimated by determining the center of mass for the loop and calculating the distance to the outer boundary. To start tracing the height, a regression plane is fitted to the cyclic loop. We moved the plane along the normal direction, and tracing is continued until a change in the radius is observed.

(a) (b) (c)

Figure 6: Example (a) Corn cob (b) Skeleton for the cob (C) Graph of the cob

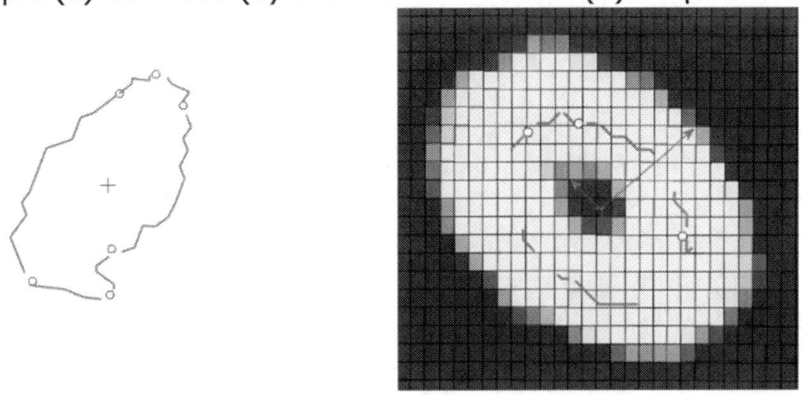

(a) (b)

Figure 7: (a) Cyclic Loop with center of mass represented with blue cross (b) 2D cross-sectional image of cyclic loop with arrows pointing the two distances which we measure

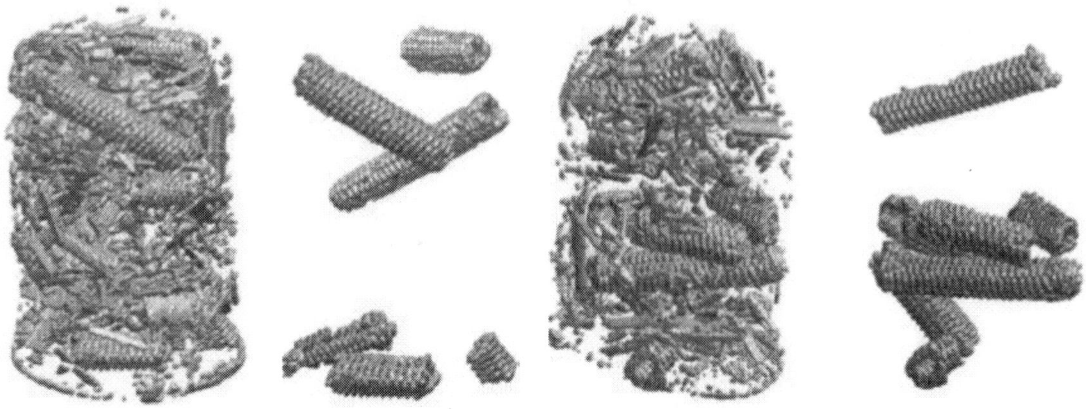

(a) (b) (c) (d)

Figure 8: (a) Baleset # 1 (b) Cob Extracted for Baleset #1 (c) Baleset # 2 (d) Cob Extracted for Baleset #2

Table 1: Volume of cob extracted from the test bale sets

	Total Volume	Max deviation from true volume
Bale set # 1	18.07 cubic inch	2.50 cubic inch
Bale set # 2	20.07 cubic inch	1.33 cubic inch

3.4 Experimental results

We have tested this approach on two mini-round bale sets. Figure 8 shows the 3D view of the two bales and the corn cobs extracted from them using this procedure. Table 1 shows the estimated volume of the corn cobs extracted from the bale set.

4 Discussion & Future Work

We are able to detect, locate, and measure corn cobs in minibales. Further research is required to devise a more efficient and optimized detection model. There are several limitations with the proposed idea as it cannot be extended to a broken corn cobs (see figure 9). Using just cyclic loop detection may not be optimal in a real-world application. Our next research goals include optimizing the model to differentiate between a loop formed with a corn cob and fully intact corn stalk. We are also working on devising a model for detecting the broken corn cobs by incorporating corn cob texture information.

Long term goals include extracting the other plant anatomical fractions like stalks, leaves, and husks. We are already in the process of using the curvature information of the corn stalks and diameter to classify between upper stalks and lower stalks. We are also working on detecting the rocks and metals in the corn stover bale picked up by the baler. These rocks and metals are higher density materials compared to corn stover fractions. Since the density difference is significant, a high contrast can be observed between the rocks and metals to corn stover fractions in the radiographs. These rocks and metals could lead to a significant downtime at the biorefinery if they damaged the grinding equipment. They can also cause sparks, which

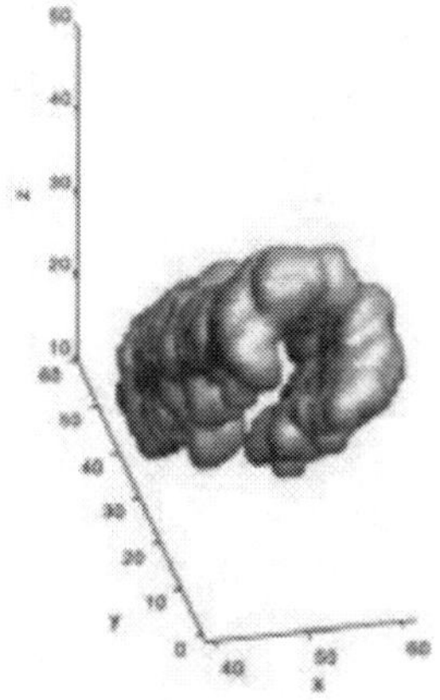

Figure 9: Example of an broken cob piece

can easily cause ignition and increase the chance for fire. It is essential to detect the shape, dimensions, and approximate location of these rocks and metals in the bale early in the screening process. Necessary measures can then be implemented to remove the contaminants of the bale before proceeding to the next steps. Our future plan includes adapting a similar analysis method to other lignocellulosic biomass materials such as forest residues and waste-derived feedstocks.

5 Acknowledgements

This work is supported by U.S. Department of Energy's Office of Energy Efficiency and Renewable Energy, Bioenergy Technologies Office, under DOE Idaho Operations Office Contract DE-AC07-05ID14517.

References

1. Update, B. T. (2016). "2016 billion-ton report: Advancing domestic resources for a thriving Bioeconomy".
2. Li, C., Aston, J. E., Lacey, J. A., Thompson, V. S., & Thompson, D. N. (2016). "Impact of feedstock quality and variation on biochemical and thermochemical conversion". Renewable and Sustainable Energy Reviews, 65, 525-536.
3. Bonner, I. J., Smith, W. A., Einerson, J. J., & Kenney, K. L. (2014). "Impact of harvest equipment on ash variability of baled corn stover biomass for bioenergy". BioEnergy research, 7(3), 845-855.
4. Kenney, K. L., Smith, W. A., Gresham, G. L., & Westover, T. L. (2013). "Understanding biomass feedstock variability". Biofuels, 4(1), 111-127.
5. Zhang, L., Xu, C. C., & Champagne, P. (2010). "Overview of recent advances in thermo-chemical conversion of biomass". Energy conversion and management, 51(5), 969-982.
6. Siddiqi, K., & Pizer, S. (Eds.). (2008). "Medial representations: mathematics, algorithms and applications (Vol. 37)". Springer Science & Business Media.
7. Blum, H. (1967). "A transformation for extracting new descriptors of shape (Vol. 4)". Cambridge: MIT press.

607943 Population Balance of Hammer Milled Loblolly Pine in 'Once-through' and 'Fractional Milling' Configurations

Abstract:

In this work we develop a predictive population balance model for hammer milling of Loblolly pine , based on a probability and breakage function to predict the milled particle size distribution as a function of variation in mill tip speed, feed rate, initial particle size, and mill chamber retention screen size. These probability and breakage parameters are dependent on material properties like the fracture resistance and minimum specific energy. Batch scale experiments are used to determine material and mill parameters within the breakage law. A breakage matrix will be developed using fracture resistance and minimum specific energy, which are found from calibration tests with iterative regression to measured product size distributions as well as inputs for feed size, flow rate, screen classification efficiency, amount of recirculated material and the number of impacts that a particle experiences. High-speed camera imaging and velocimetry will be utilized to be illustrative of the later variable. In the hammer mill experiments, the initial particle size is obtained from analytical splits of the sample along with separations with an orbital screen. Screen opening is altered by interchangeable screen sizes while the mill speed is controlled with a VFD. and the mill feed rate is controlled via a loss-in-mass feeder. These variables are associated with the impact energy, number of impacts and the size of the particles under impact. This work reports on the imparted control over resulting particle size distribution as a function of controllable operational parameters (e.g. tip speed, feed rate, initial material size and retention screen size). In this work, results for both a 'once-through' configuration in addition to a recirculating fractional milling technique will be evaluated for degree of control over the comminution process, as well as potential energy savings and optimal configuration/recirculating rate. The model will provide an estimate for daughter particle size distribution which will then be validated by experimental results, and be able to calculate the number of impacts required to achieve desired classification. Determining the lowest number of impacts to achieve a specific particle size distribution will allow optimization on energy costs and milling configuration.

609062 Assessing the Effect of Biomass Properties on Biomass to Biofuel System Performance Using Overall Operating Effectiveness

Abstract:

The development of a sustainable bioeconomy is threatened by the operational inefficiencies that are a result of the wide variability in the compositional and physical attributes of the feedstocks. Examining performance of supply systems at mean feedstock properties provides an often optimistic picture of system performance. We are proposing the use of a new metric, "overall operating effectiveness" (OOE), which uses dynamic performance information of the system to assess the composite impacts of feedstock quality on the operational efficiency and productivity of the system. For this analysis, discrete event simulation was used to conduct stochastic feedstock property-dependent throughput analysis of a conventional corn stover supply logistics-preprocessing-conversion system. Pilot scale preprocessing equipment data and experiential information were used to inform failure frequency and down time. With a 55% total carbohydrate specification for conversion, the modeled OOE values for the supply logistics, preprocessing and conversion subsystems were 56.9%, 24.4%, and 101%, respectively. The severity of throughput impacts depended primarily on moisture, while performance impacts depended on the range of corn stover compositional variability. Costs varied significantly as a function of assumed minimum acceptable carbohydrate contents from 45-59 wt% (proxy for conversion robustness), with a mean feedstock cost of $249.93/dry Mg (median of $146.39/dry Mg) for a 55% minimum carbohydrate content. As the floor was relaxed from an nth-plant specification of 59%, a higher proportion of the preprocessed feedstock was fed to conversion leading to feedstock contributions to biofuel costs ranging from $4.15-$8.73/gge (2016$). It is concluded that while conversion robustness can serve to minimize discarded feedstock, properties other than yield potential are more significant contributors to the feedstock contribution to biofuel cost.

Nth plant vs. Nth supply: Feedstock costs and supply uncertainty as function of number of biorefineries

Matthew Langholtz[1], Damon Hartley[2], Nicole Samu[1], Daniela Jones[2,3], Craig Brandt[1], Maggie Davis[1], Michael Hilliard[1], Chad Hellwinckel[4], Erin Webb[1], Chris Daly[5], Mike Halbleib[5]

[1]*Oak Ridge National Laboratory, Oak Ridge, TN, USA*
[2]*Idaho National Laboratory, Idaho Falls, ID, USA*
[3]*North Carolina State University, Raleigh, NC, USA*
[4]*University of Tennessee, Knoxville, TN, USA*
[5]*Oregon State University, Corvallis, OR, USA*

Introduction

"Pioneer plant", and "Nth-plant" are terms used to characterize the advancement of the bioenergy industry, technology, and associated cost conditions for biorefineries. Though pioneer plants are likely to be located in areas where feedstock is most abundant and least costly, they are dependent on new technology and consequently have higher investment risks. Conversely, Nth-plants are not likely to be located in the most resource-rich areas, but they would take advantage of a mature technology and lower investment risks. Thus, information on biorefinery-specific feedstock costs and cost variability can be used in conjunction with information on pioneer- or Nth-plant willingness to pay for feedstock and risk tolerance to estimate the number and location of biorefineries that may be economically accessible. To elucidate biorefinery costs that may be incurred by both pioneer and Nth-plant biorefineries, this analysis reports supply curves in terms of biorefinery-specific feedstock costs (i.e., each of *n* number of biorefineries within a fleet of biorefineries) and associated cost variability. The potential for depots to mitigate exposure to feedstock supply variability is also explored. Results shown here should be considered tentative pending peer-reviewed manuscripts in preparation. Further analysis in progress includes stochastic simulations, additional feedstocks, and alternative logistics scenarios.

Methods

To quantify n-specific feedstock supplies and prices, availability is modeled first from field to roadside, and then from roadside trucked to biorefineries. More details will be provided in a manuscript in preparation.

Roadside supplies

Potential county-level farmgate supplies are quantified at the farmgate using the Policy Analysis System (POLYSYS), a partial-equilibrium model that simulates changes in the U.S. agricultural sector[1] . Example POLYSYS results are available in the USDOE 2016 Billion-ton Report[2]. As a simplifying assumption in the present analysis, only corn stover is simulated;

analysis in preparation includes other feedstock types. Potential supplies are simulated at $5 increments ranging from $30 to $80 per ton.

Delivered supplies

Given county-level farmgate supplies, transportation cost depends on allocation of feedstocks to biorefineries. We used the Supply Characterization Model (SCM) to allocate feedstocks and calculate transportation costs. SCM solves for the least-cost biorefinery-specific weighted-average price of biomass resources. In this analysis we simulated a 50-mile grid of 1,300 potential biorefinery locations. Results provided biorefinery-specific weighted-average delivered costs.

Key Assumptions

Several key assumptions were explored in the analysis. Harvesting efficiency, i.e. the proportion of sustainably available corn stover that can be collected, is assumed 56% (low 48%, high 65%). Stover residue retention is modeled as described in the Billion-ton Report[2], and is subject to a sensitivity of +/-10%. Payment for right to harvest is assumed $10 per dry ton of biomass, +/- $5 per ton. County-level yields assume a 1:1 grain-to-residue ratio, with county-level yields derived from the USDA National Agricultural Statistics Service[3] and a sensitivity analysis of +/-2%. Stochastic results are derived from Monte Carlo simulations showing 100 executions. Deviations are derived from 30 years of yield data from the USDA National Agricultural Statistics Service [4].

Results

Results show biorefinery-specific feedstock prices within a fleet of over 100 biorefineries. Delivered corn stover prices start at about $60 per dry ton and remain nearly flat for approximately the first 50 biorefineries. In the second half of biorefineries, prices and price uncertainty increases precipitously. Prices are shown most sensitive to parameters of residue harvest efficiency, sustainability constraint, corn yield, and opportunity cost, in that order (Figure 1).

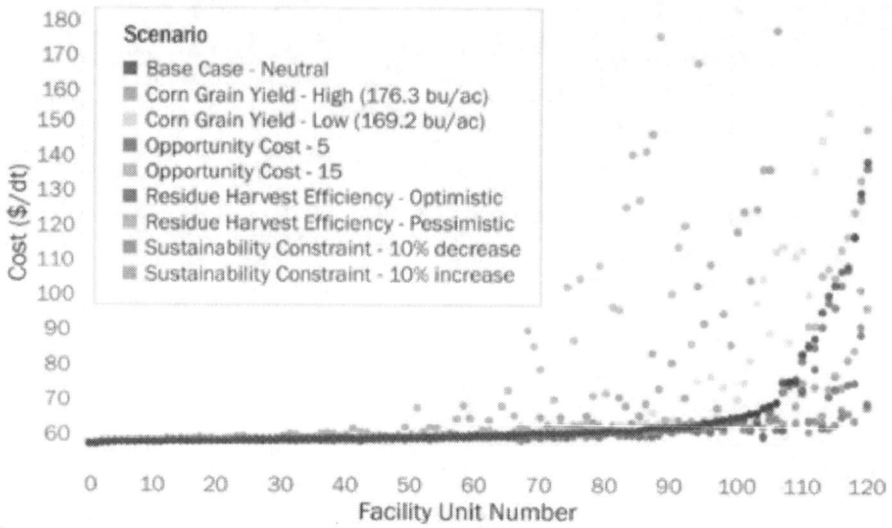

Figure 1. Delivered corn stover price per biorefinery under sensitivity to four parameters, deterministic simulations.

Stochastic results show risk exposure to variability in yields attributable to annual climate variability. Results show an upper and lower quartile of about $1 per ton, and upper and lower extremes of about $3 per ton, for the first twenty biorefineries. Variability increases with the number of biorefineries. Price variability and, in turn, feedstock price risk, increase dramatically for approximately the second half of biorefineries.

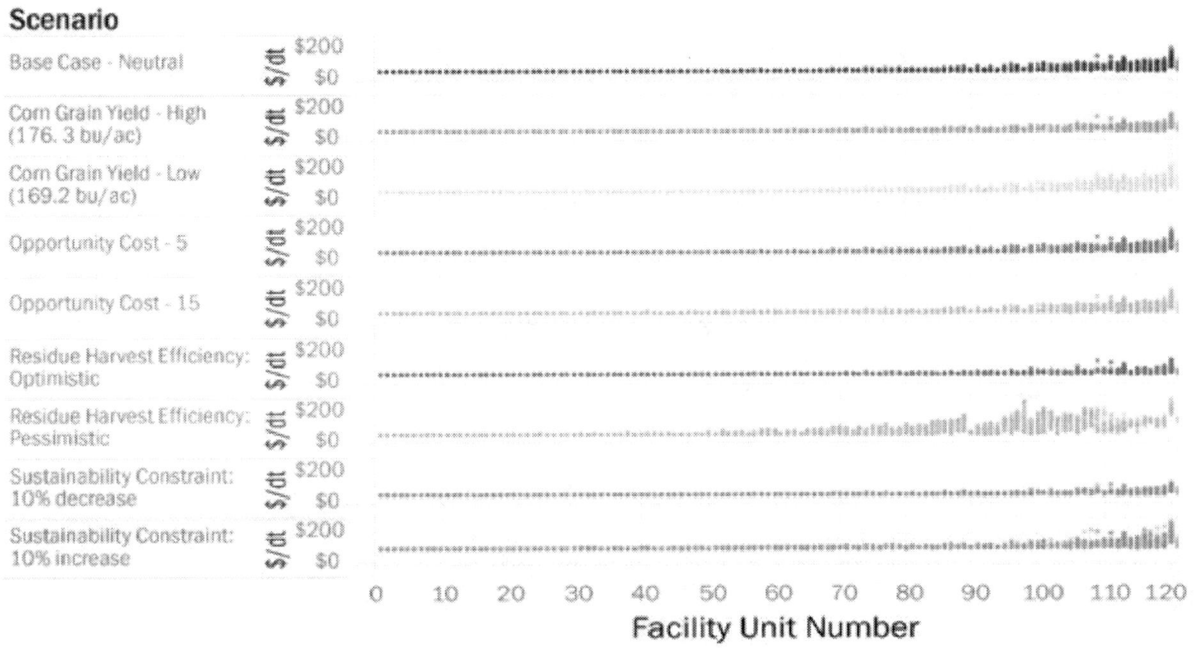

Figure 2. Delivered corn stover price per biorefinery under sensitivity to four parameters, stochastic simulations.

Summary

National biomass resource assessments that assess average feedstock prices are useful for assessing broad potential, but the economic accessibility of these feedstocks depends on factors such as spatial distribution, operational efficiency, climate variability. This analysis illustrates that while there is enough total supply of corn stover to support a fleet of over one hundred biorefineries, approximately the first half of this fleet may have access to feedstock of low enough price and price certainty. As expected, these best-candidate biorefineries for corn stover lie in the upper Plains and Midwest regions of the US. Work in progress uses this approach to explore additional feedstocks and alternative logistics scenarios that may mitigate risk exposure.

REFERENCES

1. De la Torre Ugarte, D. and D.E. Ray, *Biomass and bioenergy applications of the POLYSYS modeling framework.* Biomass & Bioenergy, 2000. **18**(4): p. 291-308.
2. USDOE, *2016 Billion-Ton Report: Advancing Domestic Resources for a Thriving Bioeconomy, Volume 1: Economic Availability of Feedstocks*, M. Langholtz, B. Stokes, and L. Eaton, Editors. 2016, Oak Ridge National Laboratory: Oak Ridge, TN. p. 448.
3. USDA-NASS, *NASS - Quick Stats. USDA National Agricultural Statistics Service*, U.S.D.o.A.N.A.S. Service, Editor. 2018: Washington, D.C.
4. USDA-NASS, *Statistics by Subject*, U.S. Department of Agriculture National Agricultural Statistics Service, Editor. 2020: Washington, D.C.

610776 Locating Nth-Plants for Blended Feestock Conversion and Preprocessing Nationwide: Biorefineries and Depots

Abstract:

Department of Energy laboratories have proposed that the nth-plant indicates a scenario with a mature conversion technology that takes advantage of economies of scale. Defining the nth-plant scenario is imperative to ultimately transform the nation's renewable biomass resources into cost-competitive, high-performance feedstocks for production of biofuels and bioproducts. To determine the nth-plant scenario for blended feedstock (switchgrass and corn stover harvested by either a 2-pass or a 3-pass method), we used a two-step process with a modified version of INL's (Idaho National Laboratory) Least-Cost Formulation model, a Mixed Integer Linear Programming (MILP) Model to deliver on-spec biomass of multiple crops that considers biomass quantity and quality at a total delivered feedstock cost that is less than $79.07/ dry ton (2016$).. First, plants were located to represent either biorefineries in the conventional supply chain system or preprocessing depots in the advanced supply chain system. In this first part of the analysis, 247 depots were supplied with 70.7 million dry tons of corn stover from 416 sources and 32 million dry tons of switchgrass from 751 sources. The second part of the MILP model will determine the location of biorefineries and depots in the conventional and advanced logistic system. Additional inputs to the model were the resource quantities presented in the BT16 by Oak Ridge National Laboratory, the design case by the National Renewable Energy Laboratory for biochemical conversion, and the targeted delivery feedstock cost to the reactor throat as defined by Idaho National Laboratory. The results from this study will be compared with outputs from the Supply Characterization Model (SCM) from Oak Ridge National Laboratory.

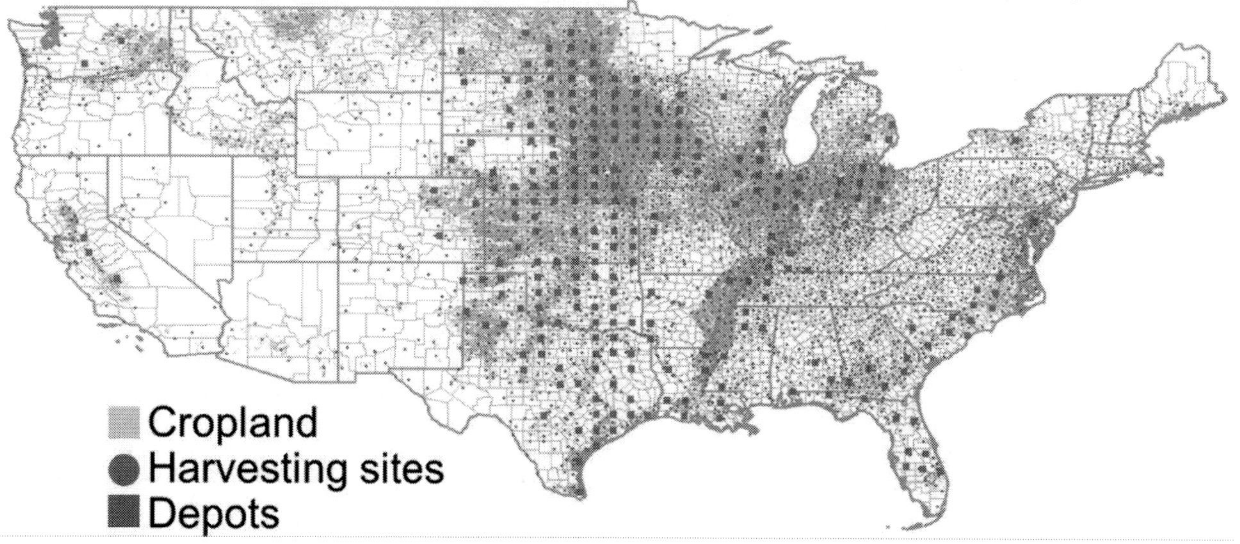

594026 Production of Rare Earth Elements and Critical Minerals from Coal-Based Resources

Abstract:

DOE-NETL's Rare Earth Element and Critical Minerals (REE-CM) Program has demonstrated the technical feasibility of recovering REEs from coal and coal-based materials, and has recently achieved production of small quantities of 80-90% pure, mixed rare earth oxide concentrates, in first-of-a-kind, domestic bench/pilot-scale extraction, separation and recovery facilities. Technology development is now focused on addressing system scale-up with process optimization and efficiency improvements, with improved economics and recovery of critical minerals, ultimately targeting production of salable rare earth oxides and metals, as well as critical minerals for incorporation into commodity and national defense intermediate and end products. Accomplishments in each of these areas will be discussed.

605866 Enabling Early Planning and Development of Eco Industrial Parks for Photovoltaic Circular Economy Using Multi Objective Optimization Techniques

Abstract:

Eco-Industrial Parks (EIPs) are widely studied as a potential solution to waste reutilization as it allows industries to work in synergy for reutilization of waste, thus avoiding landfills and other potential environmental impact. Our previous work has demonstrated the environmental benefits as a result of implementing Life Cycle Symbiosis (LCS), an extension of Industrial Symbiosis (IS) (Mathur et al., 2020). In order to implement such an LCS network, it is necessary to provide additional guiding principles for decision making, which we aim to address through two objectives in this work. The first is the expansion of an already developed hypothetical solar photovoltaic (PV) centric LCS network in a given geographic region. The second is the identification of tools that will assist in the design of such an LCS driven EIPs. Decision-making in the early stages of EIP development is challenging due to conflicting objectives of cost and environmental impact. Hence, the problem can be formulated as a multi-objective optimization (MOO) problem. In this work, we explore the development of an EIP as a combinatorial problem and formulate it as a constrained MOO problem. Given that there exist several different optimization techniques, identification of right method for solving EIP problem at an early stage is an open problem. We address this problem by comparative evaluation of a traditional scalarization optimization (the weighted sum (WS)) method with a heuristic (Genetic Algorithm (GA)) optimization technique. Further, a hybrid optimization technique (using GAs and the ε-constraint method in combination) was used to leverage the strengths of both the scalarization and heuristic optimization methods. It was found that in order to run a MOO problem using the WS method means effectively reducing the problem to a single objective optimization problem based on user preferences and results in the generation of one solution. In contrast, it was found that GAs effectively explore and exploit a given search space, thus resulting in several near-optimal solutions as opposed to a single optimal solution. One challenge in using GAs is the generation of multiple sub-optimal solutions, thus requiring additional decision-making criteria to identify the best solution for the implementation of EIP. Finally, the hybrid optimization method successfully explored the search area and based on subsequent higher-level decision-making to select a single objective would enable the identification of one global optimal solution. We present the results on comparative analysis of these methods for implementing EIP in Arizona, USA, based on emerging PV waste. To conclude, this study looks at enriching our understanding of the suitability or unsuitability of a given MOO method in the context of early design and evaluation of EIPs as opposed to simply comparing the MOOs from a mechanistic point of view for a generic MOO problem.

608718 The Life Mather Tool: Chemical Monitoring of Data and Evaluation of Alternative Materials on Consumer Products

Abstract:

In the European Union, the manufacture and use of chemical substances as well as produced and imported goods must comply with a comprehensive legislative framework (e.g. REACH, RoHS). The main focus of this legislation is the protection of human health and the environment that can be posed by chemicals. Thus, substances can be restricted or banned if they are characterized as hazardous. The manufacture of consumer products, such as home appliances, is a complicated process involving a wide range of materials and chemicals. Health and environmental impacts of such processes should be evaluated, monitored and adjusted to the lowest achievable level taking into account the legislative framework and the competitiveness of the industry.

The LIFE MATHER Tool, developed in the context of the LIFE MATHER Project, constitutes a new chemical monitoring data platform that integrates corporate data (materials, chemicals, manufacturing process) with health and environmental data in order to evaluate possible alternative materials according to existing and/or upcoming legislation. The principles of LCA (Life Cycle Analysis) are applied, thus building a rich inventory regarding consumer products (home appliances) with detailed data flows: items of the appliance (Bill of Materials, BoM), construction of materials (Full Material Disclosure, FMD) and manufacturing data (energy, electricity etc.). The inventory is enriched with open and commercial LCA databases (e.g. Ecoinvent). The platform conceptualizes health and environmental aspects using knowledge models, enables reasoning by employing semantics and ontologies, builds properties and its automated technology enables high-throughput screening and analysis.

The substitution of chemicals in existing products leads to "new products" (green products) ahead of design and new product features emerge. It is noteworthy that the approach is generalized, thus enabling the incorporation of other classes of appliances, while the challenge lies with the inclusion of processes beyond appliances. The LIFE MATHER Platform is prepared to manage large volumes of data streams as may be available from customers or external provides in general.

599846 Resilience and Risk Management for Pandemics in the Chemical Industry

Abstract:

Plants have well planned and practiced procedures for extreme weather events such as hurricanes, tornadoes, freeze and flood. However, the recent COVID-19 Pandemic proved to need some different approaches. Risk Management and policy changes will be presented to help the chemical industry to be better prepared for another pandemic along with best practice sharing from what was learnt in 2020.

609610 Development of Energy Efficient and Depolluting Ceramic Roof Tiles

Abstract:

Last years, the energy needs of roof systems are greater than before and the design of the so-called "cool roof" is systematically developed. A cool roof is defined as a roof covering system of a building that is capable of reflecting as much sunlight as possible and absorbing as little heat as possible. As a result, is a stable and cool temperature inside. Titanium dioxide (TiO2) is a chemical known for its photocatalyst advantage, its zero toxicity, and its ability to degrade organic pollutants. Due to its white color and high refractive index, titanium dioxide is the proper material. This research aims to develop a new bioclimatic product which will have photo-reflective and photocatalytic properties for use in cool roofs. Aqueous suspensions of TiO_2 nanoparticles were developed along with the addition of PEG (polyethylene glycol), as a surface-active material and sprayed in the form of a photocatalytic active suspension onto the surface of the fired clay roofing tiles. Characterization techniques were carried out, determination of the phases compositions of the samples were examined by XRD Analysis. The diffraction patterns exhibited strong diffraction peaks corresponding to anatase and rutile phases. According to the results, the particle size distribution analysis shown a particle size in average of 50 to 70 nm. Textural characteristics (BET Analysis) of the TiO2 suspensions indicated that temperature and PEG addition can change the surface area. The hydrophilicity of the surfaces of all samples was determined by contact angle measurements, monitoring the change in the glycerine contact angle with UV-irradiation time. The photocatalytic activity of all samples was obtained by monitoring the decomposition/decrease in methylene blue concentration. The results indicate that all the samples of TiO2 photocatalytic coating, exhibit good photocatalytic activity with high hydrophilicity and anti-microbiological activity.

600054 Deep Eutectic Solvent Treatment of Corn Stover Hydrolysate for the Production of High Purity Lignin

Abstract:

The production of alternative materials from sustainable biopolymer resources such as lignin has the potential to alleviate the negative environmental impact of many current industrial manufacturing practices. A major obstacle to lignin utilization is its inherent structural complexity and inconsistency, however recent advancements in deep eutectic solvent (DES) treatments have demonstrated the ability to obtain a homogeneous and highly pure lignin product.[1-3] Using a bio-ethanol hydrolysate rich in lignin, this study investigated the ability of a DES to extract and partially depolymerize lignin from corn stover. Extraction efficiencies of up to 90% were achieved, yielding lignin samples of approximately 80–96% purity. Due to the purity, structural homogeneity, and low molecular weight and polydispersity, DES extracted lignin has an increased potential to act as a feedstock for the production of a broad range of bio-based products.

As the emerging biorefinery industry develops, a large volume of lignin-containing byproducts will become available. The proper utilization of such materials is essential for both the economic feasibility of the industry, and the environmental stability of future technologies. Corn stover hydrolysate is one such material, and exists as a byproduct of the production of bio-ethanol via enzymatic hydrolysis of the cellulose portion of the biomass. Corn stover hydrolysate is comprised of residual cellulose and hemicellulose, monomeric sugars, and the majority of the original biomass's lignin. Lignin is of great interest to researchers in the field of bio-sustainability, as it is both the second most abundant material on Earth (after cellulose), and the only source of sustainable aromatic compounds. Traditionally, lignin has been burnt as a low value heating fuel for electricity generation, however, greater potential exists in using lignin to produce bio-plastic and bio-composite materials.

A DES system is the combination of two or more chemicals which act as either a hydrogen-bond acceptor (HBA), or a hydrogen-bond donor (HBD). The powerful solvation properties combined with a mild acid–base catalysis mechanism allow for the disruption of the lignocellulose matrix, resulting in the partial depolymerization and solubilization of the hemicellulose and lignin fractions. In this study, one such DES (choline chloride/lactic acid) was investigated for its potential to extract a relatively homogeneous (in terms of chemical composition and molecular weight distribution), and high purity lignin from corn stover hydrolysate.

Materials and Methods

Corn stover hydrolysate from the National Renewable Energy Laboratory (NREL) was received as a dilute solution, which was then air dried and ground into a fine powder feedstock for DES treatment. A portion of the dried hydrolysate was washed with warm water to remove the excess sugars and oligosaccharides present, in order to investigate their influence on lignin yield and purity.

The feedstock was mixed with one of three DES compositions and heated at 90-120 °C, for 1-6 h, resulting in the extraction and partial depolymerization of the lignin. After treatment, the lignin was precipitated from the mixture by the addition of water and then washed. In addition to the lignin, this work-up yielded a residual solid phase, as well as the dried DES for recycling.

The lignin and residual solid fractions obtained through the DES treatments were analyzed using the standard Klason lignin analysis process. This was performed in order to evaluate the purity of the lignin, as well as the efficiency of different treatment conditions.

Results and Discussion

The combination of lignin yield and Klason characterization allowed for the evaluation of reaction temperature, time, and DES composition, on the quantity and quality of lignin obtained from the corn stover hydrolysate. It was determined that a reaction temperature of 90 °C produced a maximum lignin yield of only 35 wt%, having a purity of 93%. Increasing the reaction time had little influence on the yield and purity of the lignin obtained, as did the DES composition used.

Increasing the treatment temperature to 120 °C improved the lignin yield without compromising purity. At this temperature, increasing the reaction time from 1 to 6 h also generally improved lignin yield without a loss of purity. Further increasing reaction time to 24 h showed no improvement in yield when compared to 6 h, indicating an idealized reaction time of ~6 h. The DES composition showed little impact on yield, however a higher ratio of choline chloride to lactic acid (5:1) produced lignin of slightly higher purity. At the relatively mild temperature of 120 °C and under atmospheric conditions,

Significance

This study has contributed to the new field of biomass treatment using DESs. A novel treatment process was developed for producing high purity lignin from a waste stream generated in the production of bioethanol. At the relatively mild temperature of 120 °C and under atmospheric conditions, lignin of ~94% purity was obtained at an extraction efficiency of ~65%. Due to its unusually high homogeneity and purity, this lignin has a greater potential for successful industrial application in the biopolymer field.

References

1. Shen, X.J., et al., *Facile fractionation of lignocelluloses by biomass-derived deep eutectic solvent (DES) pretreatment for cellulose enzymatic hydrolysis and lignin valorization.* Green Chemistry, 2019. **21**(2): p. 275-283.
2. Mamilla, J.L.K., et al., *Natural deep eutectic solvents (DES) for fractionation of waste lignocellulosic biomass and its cascade conversion to value-added bio-based chemicals.* Biomass & Bioenergy, 2019. **120**: p. 417-425.
3. Alvarez-Vasco, C., et al., *Unique low-molecular-weight lignin with high purity extracted from wood by deep eutectic solvents (DES): a source of lignin for valorization.* Green Chemistry, 2016. **18**(19): p. 5133-5141.

Figure 1. a) Corn stover hydrolysate and b) DES treated lignin

607683 Conversion of Lignin into Bioplastic By *Rhodopseudomonas Palustris CGA009*

Abstract:

Polyhydroxyalkanoates (PHAs) are biopolymers produced by bacteria under certain stress conditions that have similar properties as petroleum-based plastics. Due to their biodegradation and thermoprocessing properties, PHAs have been used in a wide array of applications from packaging to drug delivery systems. Poly(hydroxybutyrate-co-hydroxyvalerate) (PHBV) is a PHA that is more desirable due to its fraction of hydroxyvalerate that significantly enhances thermal properties compared to PHB alone. However, PHBV it is inhibited by higher production costs primarily due to the carbon source. While it has been shown that metabolically robust *Rhodopseudomonas palustris* CGA009 produces PHBV, most studies use hydroxybutyrate or valerate as the carbon source, which is costly and not ideal for real-world processes. Thus, the objective of this study was to engineer *R. palustris* to produce PHBV for the first time on lignocellulosic biomass. *R. palustris* was grown anaerobically on the lignin breakdown products p-coumarate (1mM) or coniferyl alcohol (1mM) and on Kraft lignin (3 g/L). Cells were nitrogen starved at mid-exponential growth to promote PHBV production. Samples were harvested after each day of nitrogen starvation, and the PHBV concentration was quantified using methanolysis and gas chromatography-mass spectrometry. For the lignin breakdown products, the titer ranged from 0.16 - 0.79 g/L, reaching maximum production at six days. The quantification of PHBV production on Kraft lignin supplemented with acetate is ongoing. We found that *R. palustris* does store PHBV without nutrient deprivation, but that nitrogen starvation results in significantly higher accumulations. In a comparison between phosphorous and nitrogen deprivation, nitrogen deprivation resulted in significantly higher PHBV production. To conduct a holistic analysis of the PHBV metabolism for this non-model organism, RNA samples from PHBV-producing cells grown on either p-coumarate, coniferyl alcohol, or acetate were submitted for Total RNA Sequencing. Furthermore, our ongoing efforts using RT-PCR and q-PCR are focused on illuminated the gene expression of the granule associated proteins involved in *R. palustris* PHBV production. In particular, *R. palustris* has four phasins that are merely predicted by protein homology that have shown significant variations in gene expression. Lastly, samples were submitted for Transmission Electron Microscopy to assess variations in PHBV granules. Ultimately, this study is novel in that for the first time it (i) produces PHBV from *R. palustris* on lignin breakdown products and Kraft lignin, and (ii) delivers a holistic analysis of *R. palustris'* PHBV metabolism and granule-associated proteins.

602113 Oxidative Conversion of Lignin to Value Added Products Using Oxygen and Hydrogen Peroxide

Abstract:

One of the most significant challenges in the 21st century is the development of a sustainable global economy. This calls for development of processes and technologies that allow sustainable production of materials from renewable natural resources. Kraft lignin is one such resource with annual production of 45 million metric tons/year from pulp and paper mills. Biorefinery lignin, the solid stream of biomass after enzymatic hydrolysis is another huge energy and carbon reserve. Lignin valorisation can be performed using oxidation strategy to produce value added products. The aim of this work is to deepen the fundamental understanding of factors governing the oxidative conversion processes using biorefinery and kraft lignin to further integrate into current pulp and paper industry. Vanillin with a market value of (approx.) $13,000/ton is one of the main value-added products obtained from alkaline oxidation of lignin. However, the current challenge of this process is to selectivity control and improve vanillin yield. In this study, softwood Lignoboost kraft lignin, (LBoost) and enzymatic hydrolysis (H-lignin) were used as starting materials to perform bench scale oxidations with oxygen and peroxide. Vanillin (V) was obtained as the main product along with vanillic acid (VA), and carboxylic acids such as formic and oxalic acids. In the oxidation with O_2, H-lignin provided 2.85% w/w of V, and 2.25% of VA compared to LBoost with 1.84% of V and 1.25% of VA per phenolic OH (mmol/g). The higher yield of H-lignin (5.1% w/w against LBoost - 3.09% w/w) is due to its structural characteristics close to native lignin. The concentration profiles of V and VA, the variation of O_2 partial pressure in the reactive medium and the variation of temperature with H_2O_2 oxidation were also investigated. It was also observed that oxygen partial pressure was the critical parameter determining vanillin yield reaching optimum yields at lower pressures & increased vanillin oxidation rate at higher pressures. The results obtained provide a further understanding on factors affecting phenolic aldehyde yield which paves a way to selectively improve vanillin yield. These results also show a potential to develop a lignin valorization pathway integrating with current pulp and paper manufacturing processes, thereby making a sustainable bio-refinery.

607580 Lignin-Derived Compounds for the Production of Polyurethane Plastics and Foams

Abstract:

This study investigated the production of polyurethane (PU) materials, wherein the polyol component was partially substituted for lignin. The work established a process for lignin extraction from wood sawdust using a deep eutectic solvent (DES) system, which yielded a lignin of high purity and low polydispersity. A range of lignin based polyols were then prepared, from which PU foams were produced and tested. More than 100 PU foam compositions of varying lignin content were prepared, using a range of different lignin, polyol and isocyanate combinations (see Figure 1). The structural integrity and mechanical properties of each sample were evaluated in order to determine a structure/property relationship between the components utilized and the resulting PU materials.

PUs are one of the most important classes of industrial polymers due to their incredible versatility, ease of use in manufacturing, and low cost. The world market for PU is expected to reach $79 billion by 2021, and flexible PU foams account for 31% of the PU market.[1] The flexible PU market is largely supported by the furniture and automotive industries which use PU foams as upholstery cushioning. The current industry standard for PU foam production is entirely dependent on petrochemical feedstocks for the supply of the major polyol and isocyanate components. This means that the development of a bio-based alternative polyol will not only improve the sustainability of the associated industries, but presents a highly lucrative financial opportunity.

Lignin was pursued for PU production because it is rich in the hydroxyl groups necessary to form urethane linkages, and due to the fact that it is widely available as a waste product from numerous well-established commercial industries. The incorporation of lignin may also lead to PU foams with enhanced physiochemical properties such as biodegradability and ultraviolet light stability.[2]

Materials and Methods

The lignin used in this study was extracted from pine sawdust using a DES comprised of a 2:1 molar ratio of lactic acid and choline chloride. This process is known to produce lignin of very high purity and comparably low polydispersity.[3] The DES lignin was derivatized via oxypropylation in a Parr reactor. This process was conducted in order to increase the accessibility of hydroxyl groups, thus improving the lignin's reactivity towards the urethane linkage reaction, and the flexibility of the resultant PU material.

The PU foams were produced by first mixing a small amount of additives (catalysts, surfactants and blowing agents) with the prepared lignin polyol mixture. The additives make up less than 0.5% of the mass of the resulting foam. The isocyanate component was then added, and the mixture stirred for several seconds to ensure a homogeneous foam. The reacting mixture was then poured into an open top mold, and allowed to expand upwards. After a curing time of 24 h the foams were removed from their mold. In addition to a qualitative visual inspection, each sample was analyzed in terms of its density, compressive force deflection value (CFDV), and thermal conductivity. A range of samples were also analyzed via scanning electron microscopy (SEM). The results obtained were compared to a control formulation, currently utilized within the automobile industry in upholstery cushion applications.

Results and Discussion

This project combines the societal need to effectively utilize forestry and agricultural wastes, with the commercial interest of producing bio-based materials and plastics from carbohydrates and lignin. By utilizing sawdust, a low-value byproduct of the lumber production process, the lignin feedstock was prepared in a manner which will provide a novel revenue stream to this industry. The lignin was prepared using our research group's novel, DES extraction procedure. A mild, industrially scalable process, which yields a product of high purity, and more importantly of high structural homogeneity. If lignin is to be successfully applied as a high substitution polyol alternative in semi-rigid foams, these characteristics will be of significant benefit.

PU is mainly used in the form of rigid and semi-rigid foams, and represents a $79bn/year market producing materials for the construction, transportation, furniture, and packaging industries.[1] Utilizing lignin as an alternative low-cost and sustainable resource for PU production represents a great economic opportunity.

Limited by lignin's rigid aromatic structure and poor isocyanate reactivity, lignin-based polyols have been primarily used for rigid PU foams. Such foams generally exhibit an upper threshold of ~30 wt.% polyol substitution, beyond which the properties of the resultant foams depreciate rapidly.[4] Semi-rigid lignin-based PU foam production has a more lucrative potential, due to semi-rigid foams being produced in significantly greater volume than rigid foams. This field has been previously explored, however current research suggests that the foams become too rigid upon exceeding a lignin content of ~10 wt.%. This study has explored the production of lignin based PU foams using a broad range of compositions, including numerous different lignin types, preparation methods, and degrees of polyol substitution. The work has succeeded in producing PU foams of lignin concentration that meets or exceeds the published literature, whilst maintaining key structural and mechanical characteristics such as density and compressive force deflection value (CFDV).

The results of this project to date confirm the limitation of native lignin as a polyol replacement for semi-rigid PU foam materials. Incorporating underivatized lignins at only 20 wt.% polyol substitution consistently resulted in rigid, structurally inconsistent PU materials. However, when the lignin was oxypropylated prior to its use, it was able to be included at higher degrees of substitution (30 – 40 wt.%), without significantly compromising the mechanical properties of the material.

The oxypropylation approach led to foams with a more consistent cell structure and a higher degree of incorporation of the lignin into the polymer matrix. These foams exhibited higher flexibility and lower density, resulting in materials that more closely emulated the mechanical properties of the commercial formulation.

When preparing composite materials using lignin, a certain degree of increased structural integrity will always be imparted as a result of its aromatic rigid nature. Therefore, in order to produce a PU foam of acceptable rigidity, the type, quantity and ratio of the polyol and isocyanates chosen will have to compensate for this impact. Understanding and controlling the impact of compositional choices will allow for the control of PU foam properties, which will provide opportunities across a range of industrial products. The potential also exists to explore the usage of the residual pulp obtained after DES lignin extraction of the sawdust. One possible application for this byproduct that our group is currently exploring, is as a nanofibril feedstock for the production of alternative Styrofoam packaging materials.

References

1. Gama, N.V., A. Ferreira, and A. Barros-Timmons, *Polyurethane Foams: Past, Present, and Future.* Materials, 2018. **11**(10): p. 1841.
2. Pucciariello, R., et al., *Physical properties of straw lignin-based polymer blends.* Polymer, 2004. **45**(12): p. 4159-4169.
3. Alvarez-Vasco, C., et al., *Unique low-molecular-weight lignin with high purity extracted from wood by deep eutectic solvents (DES): a source of lignin for valorization.* Green Chemistry, 2016. **18**(19): p. 5133-5141.
4. Mahmood, N., et al., *Depolymerization of lignins and their applications for the preparation of polyols and rigid polyurethane foams: A review.* Renewable Sustainable Energy Rev., 2016. **60**: p. 317-329.

Figure 1. A range of semi-rigid PU foam materials of varying lignin content

593328 Separation and Processing of Coffee Chaff By Deep Eutectic Solvents

Abstract:

Many methods have been used to process biomass unsuitable as animal feed for various uses. However, most of these processes are environmentally hazardous. We have focused on the valorization of the waste biomass coffee chaff using a solvent made of non-toxic, biodegradable components. Deep eutectic solvents (DESs) have low volatility and are low cost, non-flammable, nontoxic, and highly biodegradable. Our chosen secondary agricultural residue, coffee chaff, is generated in coffee processing. Coffee chaff or "silverskin" is the waste formed when raw coffee beans are roasted. Its low density makes it expensive to ship for valorization. We pre-treated coffee chaff with three different DESs: formic acid + choline chloride (FA:CC), lactic acid + choline chloride (LA:CC), and acetic acid + choline chloride (AA:CC). Little is available in the literature about coffee chaff upcycling because they are a niche waste. We found that DES pretreatment can separate lignin from coffee chaff, making lignin precipitated from the DESs and the biomass solid product residue both more dense and suitable for pelletization.

601783 Structural Integrity of Biomass-Derived Materials for Hydrogen Storage at High Pressure and Cryogenic Temperature

Abstract:

With the global transition towards a sustainable hydrogen (H_2) economy, the potential of H_2 is explored due to its competent aspect of being renewable and of zero-emission. However, the main challenge of its application is storing in solid-state material which often requires high pressure (around 700 bar) and low temperature (around -196 °C). Therefore, the use of H_2 energy is still struggling to thrive. Along with the adsorption parameters (e.g., high microporous surface area, large pore volume, etc.), the adsorption sites and the interaction between adsorption sites with H_2 are vital in storing H_2.

Moreover, it has been recently observed that the presence of oxygen functional groups also plays an important role in uptake H_2. However, to the best of authors' knowledge, no study has been done on the stability of storage materials as they are experienced with low temperature and high pressure. Thus, the objective of this study was to investigate the structural stability of the H_2 storage materials. Henceforth, a set of biomass-derived superactivated hydrochars were prepared from cellulose acetate by hydrothermal carbonization at 220 and 260 °C followed by KOH activation at 700 °C for 1 h. The volumetric H_2 storage experiments were carried out in High-Pressure Volumetric Analyzer (HPVA) at 100 bar and -196 °C. A total of 9 consecutive cycles of operation were conducted for each superactivated hydrochar. The BET surface area, pore volume, pore size, and volumetric H_2 uptake of the material were determined after each cycle. Additionally, surface morphology of the superactivated hydrochars were evaluated by Scanning Electron Microscope (SEM) analyzer. Analyses showed that the superactivated hydrochars underwent only a 4.5% decrease in the specific surface area and around 1.7% decrease in pore volume after 9 cycles of consecutive H_2 adsorption and desorption, signifying high stability of the material for cyclic H_2 storage.

612790 Life Cycle Modeling for Emerging Technologies: A Parametric Approach to Guide Research and Development for Additive Manufacturing

Abstract:

A quantitative understanding of the environmental and economic implications of emerging technologies in the early stages is vital to further research and development (R&D). Life Cycle Assessment (LCA) and Techno-Economic Analysis (TEA) has been used to assess the environmental impacts and economic feasibility of emerging technologies in different manufacturing sectors. However, traditional LCA do not model relationships between Life Cycle Inventory (LCI) data and key technical parameters, leading to difficulties in understanding the impacts of R&D improvement in the future and determining R&D priorities in the short-term and long-term to reduce the environmental footprints and costs of emerging technologies.

In this work[1], a parametric life-cycle modeling framework was developed to address this methodology gap. The framework integrates LCA, Life Cycle Cost (LCC) analysis and optimization. The LCI data was parameterized through process-based engineering and supply chain models, and the key R&D parameters were modeled as decision variables. The objective function is to maximize the total possibility of improving specific technical parameters through R&D in a given timeframe. Such improvements are subject to constraints based on the goals of reducing specific environmental impacts and life-cycle costs. The framework is capable of identifying the most feasible pathways to achieve economic, environmental, and energy expectations of emerging technology and identify the R&D priority areas within a reasonable timeframe (e.g., future 10 years). The framework has the potential to be used by policymakers, scientists, and engineers for accelerating the development and deployment of emerging technologies to reduce industrial-wide energy and GHG footprints.

The framework is demonstrated through a case study of the next-generation manufacturing technology - additive manufacturing (AM). The framework was applied to an AM case study for tooling and the optimization model was solved using the genetic algorithm. Four conceptual goals were set in this case study for future 20 years, including 20% reduction of life-cycle cost, 5% reduction of unit energy consumption of the parts, 20% reduction of unit costs of the parts, and no increase in the upfront costs. The model identified technical parameters that are most likely to be improved by R&D to achieve those goals simultaneously, which shed light on synergetic opportunities to improve both energy efficiency and cost-effectiveness. Such insights could be useful for future integrated design and applications of AM in the manufacturing industries. Although this case study mainly focused on energy and costs, the life cycle modeling framework and the parametric approaches presented can be applied to other indicators such as GHG emissions, as well as other emerging technologies.

1. Yao, Y.; Huang, R., A Parametric Life Cycle Modeling Framework for Identifying Research Development Priorities of Emerging Technologies: A Case Study of Additive Manufacturing. Procedia CIRP 2019, 80, 370-375.

602656 Incorporating Ecosystem Services in Life Cycle Assessment: Managed Pollinators As a Case Study

Abstract:

Ecosystem services including pollination mediated by insects are crucial to industrial communities and yet remain largely unaccounted for in life cycle assessment studies. While renewable natural resources are considered limitless, the overconsumption and degradation of ecosystem goods and services beyond the rate of replenishment threaten the sustainability of many of these services and the industries that rely upon them. Wild and managed species of insects provide several valuable ecosystem services including the pollination of the world's angiosperms which is instrumental to human nutrition and industrial activities. Dependence on pollination services has grown as demand and production of pollination-dependent crops increases. Simultaneously, the biomass and biodiversity of all insects have faced significant declines due to a number of stressors including increased pesticide use, climate change, pests, and diseases. Insect-mediated pollination accounts for over 36 billion US dollars of pollination-dependent crop production alone making the ecosystem service a highly valuable asset to the economic welfare of the agricultural sector. In addition, there are related sectors that depend upon pollination-dependent crop production (fertilizers, pesticides, equipment) as well as non-agricultural industry sectors (pharmaceutical, fuel, real estate) that share linkages with crop production and agricultural sectors, leading to intricate indirect dependence upon pollination service mediated by both commercially-managed and wild species of bees and other insects throughout the economy. Our previous work highlighted that indirect value of pollination service can be up to an additional 65 percent of the direct value of crops attributable to pollination service by insects. Given the considerable indirect impacts of pollination service on many industry sectors, it is important to include the impact of insect pollinators when considering processes or products on a life cycle basis.

We present an environmentally-extended input-output framework for quantifying the dependence on insect pollinators in life cycle assessment (LCA) through the incorporation of a new environmental vector. Reviewing available literature, we assess the pollination requirements for each of 75 pollination-dependent crops grown in the US. Using publicly available data from the USDA (United States Department of Agriculture) NASS (National Agricultural Statistics Service), the pollination requirements of each crop, and BEA (Bureau of Economic Analysis) 2012 Benchmark I-O (Input-Output) account data, we quantify the honey bee hive requirements for each agricultural sector growing pollination-dependent crops. This information is used to develop an environmental vector (***E***) for managed honey bees. This new environmental vector allows for assessment of both direct and indirect impacts of managed insect pollinators in product and process life cycle assessment. Further, it makes an important step toward valuing the impacts of ecosystem services in life cycle assessment. It also highlights the industrial sectors with the highest dependence on insect pollinators. These results provide better understanding of the role of insect pollinators in life cycles of agricultural and non-agricultural products and processes, while highlighting the need for data concerning the role of wild insect pollinators. The implications of these findings for quantifying the value of ecosystem services in product and process life cycles will be described.

610748 Exploring a Computational Framework for Spatially-Explicit Absolute Sustainability Assessment Based on a Multi-Regional Hybrid Approach

Abstract:

To give more reliable sustainability assessment results to decision-makers, life cycle assessment (LCA) methodologies have been developed in various ways. Some of the developments have focused on how to represent the complete life cycle boundary while utilizing detailed inventory data and how to take account of the spatially heterogeneous inventory data. The former one was accomplished by a multiscale modeling approach (hybrid LCA) that integrates process-based LCA (PLCA) model with environmentally-extended input-output (EEIO) model [1]. The later one has been addressed by developing multi-regional models for each of the PLCA [2] and EEIO [3] models. Multi-regional models are needed to conduct a spatially-explicit sustainability assessment since activities and interventions are different by region. For instance, activities in one region may have different commodity inputs, technological efficiencies, supply chains, and emission coefficients from the other regions. Different demographics by region also require region-specific impact characterization factors. Many existing PLCA and EEIO models such as the U.S. NREL PLCA model and the USEEIO model [4] are based on national average data, and thus, they do not take account of the above regional heterogeneity.

In this work, we explore a computational framework for spatially-explicit absolute LCA (SEA-LCA) utilizing local process models as well as regional/national/global EEIO models. For such multi-regional models, data availability is one of the primary limiting factors because it is challenging to collect region-specific inventory data. To minimize such time-consuming tasks, a hybrid modeling approach that integrates existing databases and models is necessary. The resulting hybrid model consists of multi-regional inventories at multiple spatial scales. If local process data are available at the facility or county scale, such data are included in the hybrid model as local process inventories. If any local process data are not easily available at fine scales, multi-regional EEIO data and national/global EEIO data can be employed to represent the complete life cycle boundary. U.S. EPA has published the U.S. state-level EEIO model [3] and the USEEIO model [4].

Another effort to advance the LCA methodology is to include the supply of ecosystem services in the PLCA model. Techno-ecological synergy in LCA (TES-LCA) model [5] and its regionalized model [6] have been developed to account for ecosystem services in calculating absolute sustainability indicators. Multi-regional ecosystem services can also be included in the hybrid model at multiple scales depending on their data availability. The resulting SEA-LCA model could assess the absolute sustainability of activities at various scales (e.g., local, state-level, national, and global scales). Sustainability indicators at different scales are useful in addressing various stakeholder's interests and serviceshed scales.

A case study is performed on local activities in the U.S. Midwest. We demonstrate how local process models can be connected to various scale's economic models. We discuss how sensitive to spatial scales the LCA indicators are and why a spatially-explicit assessment needs to be considered.

Through model integration, the SEA-LCA model advances sustainability assessment methodology by accounting for three aspects: multiple spatial scales, spatial heterogeneity, and ecosystem services. Additional aspects can be considered as well to advance the method further. One could be the consideration of cross-disciplinary effects of market changes and social behavioral changes on the sustainability indicators. The temporal dynamics of inventory data also need to be taken into account. A dynamic computable general equilibrium model and behavioral change model could be potentially integrated with the LCA model to address such additional aspects.

References

[1] Suh, S. (2004). Functions, commodities and environmental impacts in an ecological–economic model. Ecological Economics, 48(4), 451-467.

[2] Yang, Y., & Heijungs, R. (2017). A generalized computational structure for regional life-cycle assessment. The International Journal of Life Cycle Assessment, 22(2), 213-221.

[3] Yang, Y., Ingwersen, W. W., & Meyer, D. E. (2018). Exploring the relevance of spatial scale to life cycle inventory results using environmentally-extended input-output models of the United States. Environmental modelling & software, 99, 52-57.

[4] Yang, Y., Ingwersen, W. W., Hawkins, T. R., Srocka, M., & Meyer, D. E. (2017). USEEIO: A new and transparent United States environmentally-extended input-output model. Journal of cleaner production, 158, 308-318.

[5] Liu, X., Ziv, G., & Bakshi, B. R. (2018). Ecosystem services in life cycle assessment-Part 1: a computational framework. Journal of Cleaner Production, 197, 314-322.

[6] Liu, X., Ziv, G., & Bakshi, B. R. (2018). Ecosystem services in life cycle assessment-Part 2: Adaptations to regional and serviceshed information. Journal of Cleaner Production, 197, 772-780.

607724 Plastics: Redefining Single Use

Abstract:

There is currently over 7.3 billion tons of plastic in the Earth System, and it will grow to be 40 billion tons by 2050. Approximately 10 mega tons of plastics reach the oceans every year; 12 billion tons are projected to be in landfill by 2050. As these are released, the flux of plastics to surface waters, coastlines and oceans is amplified, with significant environmental, social and economic cost. In the global north, plastic ends up in the environment through leakages from waste collection, recycling and disposal, driven by the growth in single-use plastic, with a significant amount of egregious litter and informal disposal. The problem is now part of public consciousness and is on the political agenda nationally and internationally. The work presented here is part of the £1m Redefining Single Use project at the University of Sheffield which aims to stimulate creative thinking across disciplines and explore novel ideas and solutions with potential to deliver more circular economic approaches to plastics manufacture and utilisation.

A whole systems analysis approach, including life cycle assessment (LCA), has been conducted for proof of concept studies on agricultural plastics and reuse of plastic packaging. LCA and behaviour change results will be presented for dairy farms, where plastic is used to wrap silage and transport feed as well as to transport milk, and for horticultural farms, where plastic is used as mulches on fields. LCA and behaviour change results will also be presented for packaging of food and home goods, including consideration of light, single use packaging, alternative single use packaging materials and reusable packaging materials. The impact of bio plastics compared to fossil fuel plastics will also be discussed as will the impacts of reuse in a circular economy and the eventual end of life of the materials.

604023 Life Cycle Assessment (LCA) of Iron Oxide Nanoparticles () Synthesis By Co-Precipitation: Traditional and Novel Microfluidic Methods

Abstract:

Iron oxide nanoparticles (IONPs) have found application in a wide variety of fields including, remediation of contaminated soil and water, drug delivery, medical diagnostics, memory data storage, and targeted cancer therapies [1][2]. In addition, IONPs can be used as a support to other functional materials, and they can be easily separated by the application of magnetic fields, thereby facilitating their recovery from the reaction medium.

Shape, size distribution, and surface chemistry of the IONPs are largely dependent on the synthesis method [2]. Additionally, it also determines to a great extent the distribution and type of structural defects or impurities in the obtained materials, and ultimately all the factors might impact the magnetic response. IONPs can be produced through different processes based on physical, chemical and biological methods with differing mechanisms, inputs, yields, and reaction conditions [3][4]. In this regard, some of the simplest and most common methods to produce IONPs include sol-gel and chemical coprecipitation of iron chlorides [5]. Alternatively, less popular methods due to high associated costs and complexity include those based on sonochemistry and thermal decomposition [6][7][8]. Most recently, synthesis enabled by microfluidic devices have gained significant attention because they provide a powerful platform for preparing, functionalizing, and manipulating nano/micro materials [9], in addition to a remarkable ability to provide effective mixing and create a uniform environment for material preparation [10][11]. One main drawback of microfluidics is the need of sophisticated cleanroom manufacturing techniques to precisely control the features of the devices [12]. We have overcome this major issue by introducing a low-cost manufacturing method for the devices that relies on widely available and inexpensive laser cutting techniques [13][14].

To evaluate whether the synthesis methods induce potential environmental impacts, it is possible to apply a number of methodologies including, Environmental Impact Assessment (EIA), Strategic Environmental Assessment (SEA), and Life Cycle Assessment (LCA). This approach has been successful in estimating environmental impacts, respect to a baseline, within a geographical framework for a number of processes, including solid waste management [15], biogas production [16], and nanomaterials [17]. LCA can be conducted either by considering all involved activities during the life of a product in the so-called "cradle to grave" approach or alternatively by only looking at the unit operations of the production process itself in the so-called "cradle to gate" approach [18][19]. In the former, the analysis will include activities from extraction of raw materials to manufacturing to recycling and disposal at the end-of-life stage. In contrast, the latter will only require details of the manufacturing process. According to a recent study by Feijoo et al. [20], when assessing the outcomes from the LCA, they observed that the manufacturing of nanoproducts is mainly dominated by the environmental impacts associated with energy and chemical use. Moreover, this work highlighted the lack of comparable reports and insufficient data availability as the two main challenges currently encountered with the application of LCA to nanoproducts. These two limitations directly increase the uncertainty level associated with the estimation of the environmental impact indexes for the production of nanoparticles [20]. As a result, studies on the environmental impact of nanoproduct production are rather scarce and report contradictory results [17][20]. This is problematic as the number of processes enabled by nanotechnology has dramatically increased during the past decade.

This study was therefore aimed at evaluating the possible environmental impacts associated with two coprecipitation methods for the synthesis of IONPs, i.e., the traditional one and that enabled by microfluidic devices. Here, we explored the implementation of the "cradle to gate" descriptive approach to laboratory-scale processes. In this case, we needed to consider different aspects such as required energy reagents and the generated waste. The functional unit to conduct the analysis for both synthesis methods was defined as 0.5 g of the NPs produced per batch. However, this functional unit is only valid for this work, because the selection of a weight based functional unit makes sense when comparing production processes to produce an equivalent amount of nanomaterial. For our first approach, we defined the system boundaries starting with use of raw materials and ending with the production of the IONPs. While the water and energy consumption was taken into account, emissions and wastewater treatment assessment were disregarded.

To create the inventory report for the LCA, real-time data of both processes was collected and processed. It included consumption's data of water and energy, the exact weight of each reagent, and waste generated. Missing data such as environmental impacts associated to iron (II) chloride tetrahydrate, and tetramethylammonium hydroxide (TMAH) were estimated from reports, handbooks, and protocols. The LCA was carried out with the aid of the Ecoinvent 3.6 database of Simapro®. Additionally, characterization factors were from the International Reference Life Cycle Data System (ILCD). The eight impact categories considered in this study were human toxicity, ecotoxicity, climate change, resource depletion, acidification, and human health photochemical ozone formation.

Regarding the results, impact assessment shows that microfluidic method presented higher environmental impacts than the traditional co-precipitation method. The microfluidic method required more energy and materials and, consequently, it had the largest impacts in all categories compared to the traditional co-precipitation method. Additionally, a new impact category was considered, which contemplated the reuse of the microfluidic device to synthesize additional amounts of IONPs . However, results show that for the new category the overall importance with respect to the other analyzed categories was almost negligible. These results can be explained by our omission of the energy and materials used during the manufacture stage where significant impacts to the environment and human health might be considerable.

REFERENCES

[1] Marimón-Bolívar, W., & González, E. E. (2018). Green synthesis with enhanced magnetization and life cycle assessment of Fe3O4 nanoparticles. *Environmental Nanotechnology, Monitoring and Management*, *9*(42), 58–66.

[2] Sadhukhan, J., Joshi, N., Shemfe, M., & Lloyd, J. R. (2017). Life cycle assessment of sustainable raw material acquisition for functional magnetite bionanoparticle production. *Journal of Environmental Management*, 116-125.

[3] Willard, M.A., Kurihara, L.K., Carpenter, E.E., et al. (2004). In: Nalwa, H.S. (Ed.), *Encyclopedia of Nanoscience and Nanotechnology*, vol. 1. American Scientific Publishers, Valencia, California, p. 815.

[4] Tartaj, P., Morales, M.P., Veintemillas-Verdaguer, S., et al. (2006). Synthesis, Proper- ties and Biomedical Applications of Magnetic Nanoparticles. *Handbook of Magnetic Materials*. Elsevier, Amsterdam. The Netherlands.

[5] Laurent, S., Forge, D., Port, M. (2008). 1. Magnetic iron oxide nanoparticles: synthesis, stabilization, vectorization, physicochemical characterizations, and biological applications. *Chem. Rev.* 108 (6).

[6] Ali, M., Abdul, A., & Jameel, M. S. (2019). Mechanisms of effective gold shell on Fe3O4 core nanoparticles formation using sonochemistry method. *Ultrasonics - Sonochemistry*, November, 104865.

[7] Bagheri, S., Aghaei, H., Ghaedi, M., Asfaram, A., Monajemi, M., & Akbar, A. (2018). Synthesis of nanocomposites of iron oxide / gold (Fe3O4/Au) loaded on activated carbon and their application in water treatment by using sonochemistry: Optimization study. *Ultrasonics - Sonochemistry*, *41*(July 2017), 279–287.

[8] Xu, L., Jiao, R., Tao, X., Yi, X., & Wei, D. (2020). One-step thermal decomposition of C4H4FeO6 to Fe3O4 @carbon nano-composite for high-performance lithium-ion batteries. *Materials Chemistry and Physics*, *239*(August 2019), 122024.

[9] Luo, G., Du, L., Wang, Y., & Wang, K. (2019). Recent developments in microfluidic device-based preparation, functionalization, and manipulation of nano- and micro-materials. *Particuology*, vol. 45, 1–19.

[10] Whitesides, G. (2006). The origins and the future of microfluidics. *Nature,* vol. 442**,** 368–373.

[11] Edel, J., Fortt, R., deMello, J., deMello, A. (2002). Microfluidic routes to the controlled production of nanoparticles. *Chemical Communications*. 10. 1136-1137.

[12] Faustino, V., Catarino, S. O., Lima, R., & Minas, G. (2016). Biomedical microfluidic devices by using low-cost fabrication techniques: A review. *Journal of Biomechanics. 49*, 2280–2292.

[13] Po, Y., Vasiliy, G. (2010). Low-cost rapid prototyping of flexible microfluidic devices using a desktop digital craft cutter. *Lab Chip.* 10 (3). 384-387.

[14] Jinfang, N., Yuanzhi, L., Yun, Z., Shangwang, L., Dunnan, L., Songbai, Z. (2013). One-step patterning of hollow microstructures in paper by laser cutting to create microfluidic analytical devices. *Analyst.* 138 (2). 671-676.

[15] Khandelwal, H., Dhar, H., Kumar, A., & Kumar, S. (2019). Application of life cycle assessment in municipal solid waste management: A worldwide critical review. *Journal of Cleaner Production*, vol. *209*, 630–654.

[16] Izzah, N., Aziz, H. A., Hana, M. M., & Gheewala, S. H. (2019). A review on life cycle assessment of biogas production: Challenges and future perspectives in Malaysia. *Biomass and Bioenergy. 122*(January), 361–374.

[17] Salieri, B., Turner, D. A., Nowack, B., & Hischier, R. (2018). Life cycle assessment of manufactured nanomaterials: Where are we? *NanoImpact*, *10*(December 2017), 108–120.

[18] Korol, J., Burchart-korol, D., & Pichlak, M. (2016). Expansion of environmental impact assessment for eco-ef fi ciency evaluation of biocomposites for industrial application. *Journal of Cleaner Production*, vol. *113*, 144–152.

[19] Hesser, F. (2015). Environmental advantage by choice: Ex-ante LCA for a new Kraft pulp fi bre reinforced polypropylene composite in comparison to reference materials. *Composites Part B*, vol. *79*, 197–203.

[20] Feijoo, S., González-García, S., Moldes-Diz, Y., Vazquez-Vazquez, C., Feijoo, G., & Moreira, M. (2017). Comparative life cycle assessment of different synthesis routes of magnetic nanoparticles. *Journal of Cleaner Production*, 528-538.

612263 Efficient Design of a Metal-Oxide Mediated Red-Ox Cycle to Produce Hydrogen from Ammonia

Abstract:

Ammonia (NH_3) is proposed to be an alternative H_2 carrier, due to its lower transportation costs and it is widely produced on an industrial scale using Haber-Bosch chemistry. While the use of NH_3 to overcome transportation limitations to increase market penetration of Hydrogen based technologies is attractive, there exists a technological need to convert NH_3 to H_2. Ohio State University has developed a high-efficiency iron-oxide based redox scheme to produce H_2 from NH_3 (Process name: Ammonia to Hydrogen or ATH).

This presentation will initially talk about the design principles behind of the redox scheme for ATH such that H_2 and N_2 obtained from NH_3 decomposition are intrinsically produced in two different product streams. racking. The in-situ production of pure H_2 from NH_3 offers process intensification by combining reaction and separation in a single reactor module. The study will then detail the de-risking of thermodynamic and kinetic technology gaps for scale-up using a combination of lab-scale experiments and process simulations. The data obtained from de-risking technology gaps will be used to develop a commercially relevant ATH module design. Sensitivity analysis using surrogate modeling based black-box optimization will characterize the tradeoff between performance and cost for variables including hourly space velocities, compression costs and heat recovery. Finally, a technology to market plan based on the results of sensitivity analysis will be presented to identify short-term and long-term goals for continued development of ATH technology.

609059 Desulfurization of JP-8 By Cu-Y Zeolite at Elevated Temperatures for Remote Electricity Generation Has Two Distinct Stages: Chemisorption and Surface Reaction

Abstract:

Introduction and Prior Work

Because of their high energy density, ease of transportation, and ready availability, liquid hydrocarbon fuels are an attractive option for generating electricity in remote off-the-grid locations. This can be achieved with auxiliary power units (APUs) containing a reformer followed by a fuel cell [1]. However, the sulfur content of liquid fuels can quickly deactivate the reformer catalyst and fuel cell electrodes [2]. As a consequence, low-sulfur fuels (ideally with total sulfur content of around or even less than 1 ppm) are required for trouble-free operation of such power units.

Adsorptive desulfurization using metal exchanged zeolites has emerged as a promising technology for providing low-sulfur fuel because it does not require high volumes of solvents or reactive gases, which are not available in remote locations [3]. Several studies have addressed the problem of adsorptive desulfurization at near ambient temperatures [4], [5]. Benzo- and dibenzothiophenes are among the most difficult compounds to remove from jet fuels (JP-8) with high sulfur content. These sulfur compounds adsorb on the active metal sites of a zeolite via weak reversible complexation bonds. However, these sulfur compounds compete for the active sites with the aromatic hydrocarbons that are present in the fuel in much higher concentrations (10-25 %vol for aromatic hydrocarbons vs. at most a few thousand ppmw for sulfur compounds). As a result, metal active zeolites operating at ambient temperatures are not very effective for desulfurizing hydrocarbon fuels and specifically JP-8 [6], [7].

Using batch experiments with model fuels and JP-8, we recently demonstrated that elevated temperatures (130 – 180 °C) can help us overcome the competition between aromatic and sulfur compounds that limit the adsorption capacity of Cu exchanged Na-Y zeolites [8]. Operating at 180 °C instead of 30 °C gave a 14-fold increase in the ability of a Cu exchanged Na-Y zeolite to remove sulfur from JP-8 fuel with high sulfur concentration (2,230 ppmw). TPD experiments with spent CuNa-Y zeolites used for adsorbing sulfur compounds from model fuels convincingly showed that Cu sites retained hydrocarbons more strongly when desulfurization was carried out at temperatures above 130 °C. This was probably due to a stronger (most likely covalent) bond formed between sulfur compounds and the Cu sites in the zeolite. Desulfurization tests with JP-8 showed that both Cu sites and high temperatures (130 °C or higher) were needed for substantial sulfur removal of such JP-8 sample, with the maximum sulfur removal observed at 180 °C [8].

Adsorptive Desulfurization in a Flow System

To test the feasibility of our novel desulfurization method to produce a continuous stream of desulfurized JP-8 for remote electricity generation, we performed experiments in a continuous flow adsorber column packed with pellets of Cu-Y zeolite. The 12-inch in length by 1-inch ID column was maintained at a constant temperature of 180 °C and a pressure of 200 psi. JP-8 containing 2,230 ppmw sulfur was fed to the reactor at three flow rates (0.2, 1 and 5 mL/min) and the total outlet sulfur concentration was measured using an XRF total sulfur analyzer. We also analyzed the treated fuel with gas chromatography (GC-FID and GC-PFPD) to evaluate the overall distribution of hydrocarbon compounds and the distribution of organosulfur compounds in the treated fuel. Lastly, we characterized the spent adsorbent using SEM and studied its regeneration via thermogravimetric analysis (TGA).

Ultra-low sulfur concentrations (< 1ppmw) were reached for all flow rates. At 0.2 mL/min, ultra-low concentrations were maintained for several days and breakthrough at 1, 10 and 100 ppmw were observed after 660, 770, and 847 mL of desulfurized fuel was collected. The measured sulfur uptake at 1 ppmw breakthrough was 15.57 mg-S/g-ads, or 1.03 mol-S/mol-Cu. After 111 hours of operation at this flow rate, 1.33 L of JP-8 was processed and the sulfur uptake was 27.00 mg-S/g-ads, or 1.79 mol-S/mol-Cu, approaching the maximum values observed with our batch experiments. Ultra-clean fuel (<1 ppmw sulfur) was also obtained with the higher flow rates. As expected, however, breakthrough occurred much earlier as the JP-8 flow rate increased. While 660 mL of JP-8 with < 1 ppmw sulfur was obtained with a flow rate of 0.2 mL/min, the corresponding volumes of ultra-clean JP-8 were 162.6 mL for 1 mL/min flow rate and 96 mL for 5 mL/min.

Gas chromatography data using a flame ionization detector (GC-FID) showed a hydrocarbon matrix in the treated JP-8 similar to that of untreated JP-8, with the main peaks corresponding to paraffinic hydrocarbons ranging from C9 to C15. GC-FID chromatograms of treated samples at higher flow rates (1 and 5 mL/min) showed similar results, suggesting that the overall hydrocarbon matrix in the fuel did not change upon treatment. These results agree with our earlier report in which JP-8 was desulfurized in batch reactors using Cu-Y zeolite at the same temperature [8]. Additionally, the organosulfur compounds found during breakthrough using gas chromatography coupled with a pulsed flame photometric detector (GC-PFPD) corresponded to the alkylated benzothiophenes originally present in the untreated JP-8.

The most significant observation, however, involved a dramatic deviation from the breakthrough behavior one would expect from an adsorption system. After reaching a level that strongly depended on the flow rate, the outlet sulfur concentration did not continue to increase as expected. Instead, the outlet sulfur concentration leveled off and we detected the release of hydrogen sulfide. This is a strong indication that the mode of sulfur removal shifted from chemisorption on the partially saturated adsorbent to surface reactions of the organosulfur compounds that involved cleavage of C-S bonds [9]. These reactions gave us 60 to 70 wt% sulfur removal that was maintained for a significant period of time until the release of hydrogen sulfide forced us to end the experiments for safety reasons. In further agreement with literature data [9], the catalytic decomposition reactions of organosulfur compounds deposited significant amounts of carbon on the Cu-Y zeolite. Elemental analysis of the spent adsorbent showed that it contained 27-31%wt carbon, while the initial total carbon content in the pellets before desulfurization was 0.012%wt, as determined by ASTM E1019-18.

Regeneration tests were performed using thermogravimetric analysis (TGA). Because carbon was deposited on the surface of the adsorbents, the regeneration procedure was carried out in air to convert the carbon and sulfur depositions to carbon dioxide and sulfur dioxide. High regeneration temperatures (550 or 600 °C) were necessary to restore the adsorbent to its initial color. Regeneration under these conditions restored the adsorptive desulfurization performance of our zeolite to about 90% of its initial value.

We conclude that desulfurization of JP-8 by Cu-Y zeolite at elevated temperature proceeds in two distinct stages. Sulfur compounds are first removed by chemisorption and the utilization of the Cu active sites reaches values that may be significantly higher than 1 mol of S per mol of Cu (1.79 mol-S/mol-Cu for 0.2 mL/min). When the saturation of the adsorbent becomes very high, however, the mode of sulfur removal shifts from chemisorption to surface reaction that cleaves C-S bonds, releases hydrogen sulfide and deposits significant amount of carbon on the zeolite.

Our study also demonstrates that continuous flow adsorbers packed with Cu-Y zeolite and operating at elevated temperatures can be used for adsorptive desulfurization of JP-8 for remote electricity generation using fuel cells. We will present calculations showing how production requirements of clean fuel can be met by careful selection of column lengths, flow rates and by using parallel column bundles. Finally, the catalytic reactions we observed at the end of the chemisorption stage open new possibilities for employing adsorber/reactor cascades to meet low sulfur content and production rate specifications.

REFERENCES

[1] D.T. Tran, J.M. Palomino, and S.R.J. Oliver, "Desulfurization of JP-8 jet fuel: challenges and adsorptive materials," *RSC Adv.*, pp. 7301–7314, 2018.

[2] K. Liu, C. Song, and S. Velu, *Hydrogen and Syngas Production and Purification Technologies*. Hoboken, NJ: John Wiley & Sons, Inc., 2009.

[3] A. Stanislaus, A. Marafi, and M.S. Rana, "Recent advances in the science and technology of ultra low sulfur diesel production," *Catal. Today*, vol. 153, no. 1–2, pp. 1–68, 2010.

[4] S. Velu, C. Song, M. Namazian, S. Sethuraman, and G. Venkataraman, "Desulfurization of JP-8 by Selective Adsorption over a Ni-based Adsorbent for Micro Solid Oxide Fuel Cells," pp. 1116–1125, 2005.

[5] A.J. Hernández-Maldonado, F.H. Yang, G. Qi, and R.T. Yang, "Desulfurization of transportation fuels by pi-complexation sorbents: Cu(I)-, Ni(II)-, and Zn(II)-zeolites," *Appl. Catal. B Environ.*, vol. 56, no. 1-2 SPEC. ISS., pp. 111–126, 2005.

[6] L. Wang, B. Sun, F.H. Yang, and R.T. Yang, "Effects of aromatics on desulfurization of liquid fuel by p-complexation and carbon adsorbents," *Chem. Eng. Sci.*, vol. 73, pp. 208–217, 2012.

[7] H.T. Mayfield, "JP-8 Composition and variability.," *Tech. Rep. AL/EQ-TR-1996-0006, EQL, Armstrong Lab.*, 1996.

[8] P. Dias da Silva, S.K. Samaniego Andrade, K. Zygourakis, and M.S. Wong, "Adsorptive Desulfurization of Liquid Fuels at Elevated Temperatures Using Metal Exchanged Zeolite Y," *Ind. Eng. Chem. Res.*, 2019.

[9] L. Jaimes and H. De Lasa, "Catalytic Conversion of Thiophene under Mild Conditions over a ZSM-5 Catalyst," *Ind. Eng. Chem. Res.*, vol. 48, no. 16, pp. 7505–7516, 2009.

609085 Modular Electrochemical Treatment of Concentrated Animal Feeding Operations Waste for Simultaneous Nutrient Reduction and Recovery

Abstract:

Land application of manure is a beneficial practice which improves soil quality by adding important nutrients, with Concentrated Animal Feeding Operations (CAFOs) being the largest users of land-applied manure as a management practice for animal waste. Although CAFOs are subject to state and federal regulations requiring approved Manure Management Plans, agricultural runoff is still a widespread issue due to the complexity of controlling this distributed source of excess nutrients. This land application has led to significant pollution of water systems due to nutrient runoff in the form of phosphorus (P) and nitrogen (N) causing harmful algal blooms in water bodies such as the Great Lakes[1]. Therefore, the reduction of P and N present in manure and the recovery of these species as solid form could serve as a viable approach to reducing environmental impact of runoff, providing an additional source of revenue for farm operations and providing a less variable form of fertilizer for land application.

Ohio University (OHIO) is developing an electrochemical technology, via funding from the Ohio Water Development Authority, to enhance manure management practices for farming operations. Specifically, the technology – the e-CAFO – controls solution pH at the electrode interface and can recover phosphorous and nitrogen species as struvite ($MgNH_4PO_4$) or reduce phosphorous content through crystallization of multiple phosphate species. This offers an opportunity to operate the technology with a primary sustainability focus on wastewater remediation and a secondary economically lucrative focus on fertilizer generation. Furthermore, the modularity of the e-CAFO provides an opportunity to evaluate deployment as a distributed treatment technology depending on farm size and density in the geographical region. This distributed option would provide better control over non-point sources that lead to harmful algal blooms.

The utility of this e-CAFO has been demonstrated by OHIO [2], with maximum phosphorous recovery dependent on cell voltage, temperature, flow conditions and calcium content. This previous study showed reduced phosphorous recovery in the presence of calcium and highlighted the need for multi-factor analyses with additional variables. Furthermore, process simulations have been developed in ASPEN Plus to understand the equilibrium conditions that maximize nutrient recovery as struvite or nutrient reduction in solution. These simulations provide further insight into the operating window for the e-CAFO technology and will accompany the multi-factor analyses to be presented at the meeting.

References

[1] S. Basu, "Controlling Nitrogen and Phosphorous in Wastewater," *Chemical Engineering Progress*, vol. 116, no. 4, pp. 41–47, Apr. 2020.

[2] Z. Belarbi and J. P. Trembly, "Electrochemical Processing to Capture Phosphorus from Simulated Concentrated Animal Feeding Operations Waste," *J. Electrochem. Soc.*, vol. 165, no. 13, pp. E685–E693, Jan. 2018, doi: 10.1149/2.0891813jes.

603289 *State-By-State Comparison of the Economic, Environmental, and Energy Impacts of Manufacturing Facilities Integrating Solar Photovoltaic or Combined Heat and Power Systems*

Abstract:

Manufacturing facilities are at the nexus of significant environmental impacts, economic activity, and source energy consumption [1]. Any process modifications or facility upgrades require careful analysis through each of these lenses. With growing concerns about greenhouse gases and their long-term impacts, manufacturing facilities are under increasing scrutiny to reduce their environmental footprint. Two environmentally oriented technical upgrades to a manufacturing facility, a solar photovoltaic (PV) installations (Scenario 1) [2-4], and a natural gas combined heat and power system (Scenario 2) [5-7], are compared for their environmental, economic, and source energy impacts.

In this study, a manufacturing facility that consumes 2 MW of power and 5200 kg/h of 700 kPa saturated steam is examined. In the baseline scenario the facility receives all its electricity from the grid and produces all steam with in-house boilers. In Scenario 1, a 2 MW PV installation offsets some of the daily electrical consumption and the rest of the utilities are consumed in the same manner as the baseline. In Scenario 2, a natural gas combined heat and power system generates the required electricity and steam.

To compare Scenario 1 and 2 the facility impacts are normalized based on the capital investments, including annual operations and maintenance costs, for each technological upgrade. The environmental impact is the ratio of the capital cost to the reduction in CO_2 and NO_x emissions ($/kg CO_2 or NO_x reduced). The economic impact is the discounted payback period (DPP) with savings generated from reduced utility costs. The energy impact is the ratio of capital cost to the reduction in total source energy ($/MWh reduced). Many factors of this analysis change regionally, including electricity and natural gas costs, grid emission factors, PV electricity generation, line losses, and location cost factors. As a result, Scenario 1 and 2 are compared on a state-by-state level. Sensitivity analyses are performed with respect to solar capital costs and grid emission factors.

Although they vary state-by-state, the general trends of the technological comparison are the following: (1) CO_2 reduction costs for Scenario 1 and 2 are similar and NO_x reduction costs are always lower for Scenario 2, (2) Scenario 2 is economically viable (DPP < 25 years) in a much wider range of states, and (3) source energy reduction is less expensive with Scenario 1. The main driver for PV economic viability is grid electricity costs rather than PV electricity generation. The sensitivity analysis revealed that even with 33% cheaper PV, Scenario 1 is not economically competitive with Scenario 2. The sensitivity analysis also showed that the environmental benefits of Scenario 2 are highly dependent on grid emission factors.

[1] Gao, Qi, et al. "An economic and environmental assessment model for microchannel device manufacturing: part 1–Methodology." Journal of Cleaner Production 120 (2016): 135-145.

[2] Sun, Honghang, et al. "China's solar photovoltaic industry development: The status quo, problems and approaches." Applied Energy 118 (2014): 221-230.

[3] Taboada, Heidi, et al. "Exploring a solar photovoltaic-based energy solution for green manufacturing industry." 2012 IEEE International Conference on Automation Science and Engineering (CASE). IEEE, 2012.

[4] Abdallah, Tarek, Ali Diabat, and Jasper Rigter. "Investigating the option of installing small scale PVs on facility rooftops in a green supply chain." International Journal of Production Economics 146.2 (2013): 465-477.

[5] Henning, Moriah, Derek Machalek, and Kody M. Powell. "Integrating a microturbine into a discrete manufacturing process with combined heat and power using smart scheduling and automation." Computer Aided Chemical Engineering. Vol. 47. Elsevier, 2019. 293-298.

[6] D. Machalek, M. Henning, K. Mohammedi, and K. Powell "Economic and environmental impacts of a non-traditional combined heat and power systems for a discrete manufacturing process" Journal of Cleaner Production (Submitted)

[7] Gladysz, Bartlomiej, and Aldona Kluczek. "A framework for strategic assessment of far-reaching technologies: A case study of Combined Heat and Power technology." Journal of Cleaner Production 167 (2017): 242-252.

603043 Modular Plate Cold DBD Plasma Reactors for Dry Methane Reforming

Abstract:

Conversion of stranded natural gas resources into a hydrocarbon stream is of critical importance for distributed manufacturing. Conventionally, syngas is synthesized from methane via high temperature processes which require high capital investment. In this work, a modular dielectric discharge barrier plasma reactor was devised and tested to perform methane dry reforming with CO2 to yield syngas at atmospheric pressure and ambient temperature. Deposition methods were devised to obtain a thin film catalyst to achieve more than 95 % single pass methane conversion and better than 50% CO2 conversion at circumroom temperatures. Series of catalysts, including Ni-Al2O3 as well as high dielectric constant non-conventional inorganic compounds were explored and optimal conversion conditions were identified. CO2 generated plasma was also shown to achieve stable catalyst performance via Boudouard reaction.

599837 Overview of Potential Carbon Storage Complexes in the Offshore United States Gulf of Mexico and Atlantic

Abstract:

Estimates of regional geologic storage resource are key to advancing the widespread deployment of carbon capture and storage (CCS). This paper consolidates the results of several recent studies to estimate the storage resource on the continental shelf off the Atlantic and Gulf coasts of the United States. Much of the recent work has been carried out through a series of Offshore Resource Assessment Projects supported by $13.8M from the U.S. DOE FE Carbon Storage Program. Results of five projects are discussed, which used a variety of methodologies and covered different geologic settings offshore of the Gulf of Mexico and the Atlantic coasts. Sediments beneath the Gulf of Mexico represent a vast storage resource in oil and gas reservoirs and saline formations, and resource estimates benefit from extensive oil and gas exploration and development data. Using production data, it is estimated that there is 20GT of storage in just the oil and gas reservoirs. Including saline formations, analysis of seismic data and well data indicates many tens of gigatonnes, or more, storage resource is available in the Gulf. Storage resource estimates are less well defined off the Atlantic coast due to lack of data. Nonetheless, improved stratigraphic models, guided by biostratigraphy, and combined with available seismic data indicate many tens of gigatonnes, or more, of storage resource is also available throughout a large area studied from New York to northern Florida.

596798 Laboratory Investigations of CO_2-Enhanced Oil Recovery and Associated Storage in Reservoir and Source Rocks from the Bakken Petroleum System

Abstract:

Laboratory studies are being performed in support of field-based projects to both increase oil recovery and store CO_2 in unconventional reservoirs, with a focus on the Bakken Petroleum System in North Dakota (USA). As is common in unconventional reservoirs, oil recoveries in the Bakken are low, typically < 10% of known oil in place,[1,2] which has led to increasing interest for injecting CO_2 both for increasing oil recoveries as well as achieving associated CO_2 storage.[3-5] Although CO_2-based enhanced oil recovery (EOR) in conventional reservoirs has been practiced for decades, the tight-fractured nature of unconventional systems like the Bakken are likely to require different approaches for successful field operations, and will require deeper mechanistic understandings of the parameters controlling oil recoveries based on laboratory studies designed to mimic processes in tight fractured systems.[6]

Methods have been developed at EERC to investigate CO_2 EOR and CO_2 storage in the Bakken system at reservoir temperature (230 °F) and include: (1) Measuring minimum miscibility pressures (MMPs) using a capillary-rise vanishing interfacial tension (VIT) technique for crude oils produced from the target Middle Bakken and Three Forks formations,[7] (2) determining the ability of CO_2 to dissolve light and heavy hydrocarbons via vaporization gas drive from crude oil at different pressures, (3) determining the ability of CO_2 at different pressures to recover crude oil hydrocarbons from both Middle Bakken and Three Forks reservoir rocks and Upper and Lower source shales,[6] and (4) measure sorption isotherms of CO_2 with Bakken source and reservoir rocks using a magnetic suspension balance.

Results from these four experimental approaches can be summarized as: (1) Measurements of MMP for crude oils produced from the target Middle Bakken and Three Forks formations show that MMP can be reached at pressures about ½ or lower than the original reservoir pressure of about 5000 to 7000 psi. (2) Hydrocarbon compositions dissolved via vaporization gas drive from crude oil into the CO_2-dominated "miscible" phase show dramatic increases with higher pressures in both the total oil hydrocarbons dissolved and in the ability of CO_2 to dissolve heavier hydrocarbons. (3) Oil recoveries from rock core samples exposed to a "bath" of CO_2 (rather than a typical core-flooding experiment) under conditions designed to mimic fracture flow of injected CO_2 also show that higher pressures yield better total oil recoveries as well as better recoveries of heavier hydrocarbons. Mechanistic studies demonstrate that rock surface area, CO_2 pressure, and CO_2/rock contact time are the major factors controlling oil recovery from Middle Bakken and Three Forks rocks, as well as from the Upper and Lower shales. (4) Finally, measured isotherms show low to moderate storage capacities of the Middle Bakken and Three Forks rocks (ca. 0.5 to 2 mg CO_2 per gram rock), but very high capacities in the Upper and Lower Bakken shales (ca. 12 to 16 mg CO_2 per gram rock), indicating that the Bakken shales represent an opportunity for substantial storage of injected CO_2, as well as acting as effective seal caps to retain CO_2 in the Middle Bakken and Three Forks formationsMcNally,

M.S.;

1. Brandt, A.R. The Productivity and Potential Future Recovery of the Bakken Formation of North Dakota. *Unconventional Oil Gas Res.* **2015**, *11*, 11-18.
2. Hoffman, B.T.; Evans, J.G. In *Improved Oil Recovery IOR Pilot Projects in the Bakken Formation.* SPE Low Perm Symposium; Denver, CO, 2016. DOI: 10.2118/180270-MS.
3. Shoaib, S.; Hoffman, B.T. In *CO_2 Flooding the Elm Coulee Field.* SPE Rocky Mountain Petroleum Technology Conference; Denver, CO, 2009. DOI: 10.2118/123176-MS.
4. Yu, W.; Lashgari, H.R.; Wu, K.; Sepehrnoori, K. CO_2 Injection for Enhanced Oil Recovery in Bakken Tight Oil Reservoirs. *Fuel* **2015**, *159*, 354-363.
5. Jin, L.; Sorensen, J.A.; Hawthorne, S.B.; Smith, S.A.; Pekot, L.J.; Bosshart, N.W.; Burton-Kelly, M.E.; Miller, D.J.; Grabanski, C.B.; Gorecki, C.D.; Steadman, E.N. Improving Oil Recovery by Use of Carbon Dioxide in the Bakken Unconventional System: A Laboratory Investigation. *SPE Reservoir Evaluation & Engineering* **2017,** *20,* 602-612.
6. Hawthorne, S.B.; Grabanski, C.B.; Miller, D.J.; Kurz, B.A. and Sorensen, J.A. Hydrocarbon Recovery from Williston Basin Shale and Mudrock Cores with Supercritical CO_2: Part 2. Mechanisms That Control Oil Recovery Rates and CO_2 *Energy and Fuels* **2019**, *33*, 6867-6877.
7. Hawthorne, S.B.; Miller, D.J.; Jin, L.; Gorecki, C.D. Rapid and Simple Capillary-Rise/Vanishing Interfacial Tension Method to Determine Crude Oil Minimum Miscibility Pressure: Pure and Mixed CO_2, Methane, and Ethane. *Energy Fuels* **2016,** *30*, 6365-6372.

608454 Optimization of EOR in Horizontal Fracturing through Injection of Recovered CO2 from Amine Sweetening Process

Abstract:

Rajib Mukherjee, Sepehr Arbabi*

Department of Chemical Engineering, University of Texas Permian Basin, TX 79762

**corresponding:* sepehr.arbabi@utpb.edu

Abstract

Injection of CO_2 into a suitable reservoir has long been practiced for co-optimization of enhanced oil recovery (EOR) and CO_2 sequestration [1]. Recently, economic optimization of EOR through injection of gas, which is otherwise flared, to stimulate oil production as well as sequestration of the injected gas have been studied [2]. The injection process is generally carried out in multiple huff and puff stages. Even though EOR through gas injection may result in capture of the gas in-situ, a substantial amount of the injected gas flows back with production of the oil and gas from the wells. It has been found that when CO_2 is used for EOR, the amount of CO_2 increases significantly in the produced gas following EOR process. The CO_2 can be recovered during natural gas processing through amine sweetening process. In this presentation, EOR through injection of acid gas recovered during gas processing have been studied to understand the economic and environmental implications of recovered acid gas injection in shale oil developments, which is otherwise vented. A surrogate model-based optimization is proposed to integrate the development of a shale oil and gas reservoir along with EOR projects. The framework optimizes decisions such as drilling scheduling, workover of depleted production wells, natural gas processing unit, location of injection sites, injection rates, and duration of EOR operations. The EOR is implemented in the Bakken formation through a reservoir with 10 wellpads and a modular amine sweetening processing unit. EOR in the subsurface is modeled through compositional reservoir simulation with its production functions scaled and utilized as input to the simulation for gas processing on the surface. It is found that and as much as 50 % of the injected gas can be reused in subsequent stages of EOR. Results also show the expected enhancement in oil production and recovery factor. Sensitivity analysis was also performed to find the importance of different operational variables on the EOR efficiency.

Keywords: carbon management, process design and development, carbon capture & storage, environmental sustainability

[1] Wang, X., van't Veld, K., Marcy, P., Huzurbazar, S., & Alvarado, V. (2018). Economic co-optimization of oil recovery and CO2 sequestration. Applied Energy, 222, 132-147.

[2] Calderón, A. J., & Pekney, N. J. (2020). Optimization of enhanced oil recovery operations in unconventional reservoirs. Applied Energy, 258, 114072.

604196 SMART: A New Initiative to Transform Subsurface Visualization and Prediction through Machine Learning

Abstract:

The U.S. Department of Energy's (DOE) Office of Fossil Energy aims to advance the science and technology of the fossil energy industry to enable the reliable, efficient, affordable, and environmentally sound use of fossil fuels. Carbon capture, utilization, and storage (CCUS) is a prominent technology being developed by DOE to reduce carbon dioxide (CO_2) emissions from anthropogenic sources (e.g. fossil fuel power generation and industrial facilities). This technology depends on safe, secure, and cost-effective storage of large volumes of CO_2 in the subsurface. The Carbon Storage Program, managed by the Office of Fossil Energy through the National Energy Technology Laboratory(NETL), has partnered with the Oil and Natural Gas Program to embark on a new initiative that focuses on further elucidating the subsurface through innovative artificial intelligence and machine learning (ML) techniques.

The SMART Initiative (**S**cience-informed **M**achine Learning to **A**ccelerate **R**eal **T**ime Decisions in Subsurface Applications) is an NETL-led multi-organizational effort which leverages the expertise of 15 different research organizations—national laboratories, industry partners, universities and Carbon Storage Regional Initiative Partnerships. The initiative has three primary goals: 1) real-time visualization of key subsurface features and processes, 2) virtual learning for rapid investigation of reservoir behavior, and 3) real-time forecasting of select attributes at active storage and production sites for optimization and risk mitigation. The SMART Initiative capitalizes on recent advances in geophysics and data collection methods, as well as high fidelity and rapid predictive modeling. The SMART team was built from long-term collaboration through previous initiatives and partnerships and benefits from over 15 years of data collection and analysis sponsored by DOE. This uniquely positions the team to have the resources and understanding of relevant field data to develop, train, and test ML algorithms. The initiative, currently in its feasibility phase, is expected to benefit subsurface operators, regulators, and stakeholders by providing transformative solutions to visualizing and managing the subsurface in near real-time, thus reducing risk and increasing efficiency. This presentation will focus on the latest progress from the carbon storage tasks of the SMART Initiative.

603163 Improving CO_2 Enhanced Oil Recovery Performance through Data Analytics and Next-Generation Controllable Completions

Abstract:

The Energy & Environmental Research Center (EERC) is conducting a 5-year project to field-test an advanced machine learning approach integrating controllable completions (interval control valves [ICVs]) to enable active (smart) well control during carbon dioxide (CO_2) enhanced oil recovery (EOR). The concepts to be tested are that 1) ICVs can be used to direct fluids into selected zones of horizontal wells, improve conformance, and maximize enhanced recovery potential; and 2) real-time monitoring data can be integrated in a machine learning approach to develop a "human-in-the-loop" (semiautonomous) active control system for both injection and production wells. The concept will be validated through design and execution of a pilot CO_2 EOR test in the carbonates of the Ordovician Red River Formation of southwestern North Dakota.

The pilot test focuses on a horizontal injection well paired with a horizontal production well, each completed with ICVs (up to ten ICVs per well). Rock properties are being assessed through laboratory investigations using core samples. Fast-flow pathways in the reservoir will be identified using a combination of downhole monitoring technologies. An active control system will be developed that uses machine learning to analyze real-time data sets and manage ICV operation (select which stages to open/close and their timing).

Geomodel construction and numerical simulation were conducted to predict key EOR performance metrics (e.g., incremental production and CO_2 utilization) and inform pilot test operation. Sensitivity studies were conducted to investigate potential performance improvements in a range of scenarios and generate information applicable to other reservoir types. These studies included variations of permeability, heterogeneity, injection/production well ICV combinations, and operational schedule. These scenarios and their modeled impact on EOR performance will be discussed.

Successful completion of this field test will help reduce the uncertainty in CO_2 EOR performance and improve overall EOR project economics. These improvements will help accelerate the adoption of the approach across existing and future CO_2 EOR operations and will support the expansion of CO_2 EOR into locations where current marginal economic outlook deters investment.

608699 Thermodynamic Modelling of $FeCO_3$ Solubility Using the Extended Uniquac Model

Abstract:

CO_2 corrosion is receiving more attention these days, as we need to reduce the emission of greenhouse gases. CO_2 is present in many processes including biogas, carbon capture and storage, the cement-, and in the oil and gas industries, and their presence result in corrosion of the equipment and pipelines, which over time will lead to significant production losses and costly shutdowns.

One way to reduce corrosion is to engineer/manipulate/tailor the precipitation of the protective layer formed by $FeCO_3$. Detailed knowledge of the $FeCO_3$ solubility is important in the prediction on CO_2 corrosion as studies suggest that the protective layer is not forming below 60 °C[1]–[5].

This study focus on the thermodynamic modelling of $FeCO_3$ solubility in water, salt solutions, and under the influence of CO_2 partial pressure. The purpose is to create a thermodynamic model, which describes these systems. It is important that the model is able to accurately predict phase equilibria and physical/chemical properties of this system. The Extended UNIQUAC model is applied as it has previously been shown to be able to model these kind of systems [6]. Parameters in the thermodynamic model are determined based on the obtained experimental data.

The interaction parameters, u^0 and u^T, the surface, and volume parameters, r and q used in the Extended UNIQUAC model are fitted using experimental SLE data. The parameters in the model are optimized by minimizing the sum of squared residuals of each data point.

Through comparing the obtained experimental SLE data with the model, the validity/ability of the Extended UNIQUAC is corroborated.

This work is relevant for future understanding, and process simulation of systems dealing with CO_2 corrosion.

References

[1] S. Kakooei, M. C. Ismail, B. Raja, H. Mohebbi, S. S. Emamian, and M. Moayedfar, "Formation of nano-scale FeCO3 protective corrosion product in carbon dioxide-saturated 3% sodium chloride solution," *Key Eng. Mater.*, vol. 740 KEM, no. June, pp. 3–8, 2017.

[2] J. Hernandez, A. Munoz, and J. Genesca, "Formation of iron-carbonate scale-layer and corrosion mechanism of API X70 pipeline steel in carbon dioxide-saturated 3% sodium chloride," *AFINIDAD LXIX*, vol. 560, pp. 251–258, 2012.

[3] W. Farida, T. Hemmingsen, T. Berntsen, and P. Rabindran, "Effect of Precorrosion and Temperature on the Formation Rate of Iron Carbonate Film," *Pipeline Technol. Conf.*, 2012.

[4] A. Kahyarian, M. Achour, and S. Nesic, *CO2 corrosion of mild steel*. Elsevier Ltd, 2017.

[5] G. Schmitt and Michaela Hörstmeier, "Fundamental aspects of CO2 metal loss corrosion - Part II: Influence of different parameters on CO2 corrosion mechanisms," *Corrosion*, vol. 06112, pp. 1–26, 2006.

[6] K. Thomsen, "Aqueous electrolytes : model parameters and process simulation," Ph.D thesis, DTU, 1997. https://doi.org/10.11581/dtu:00000074

601792 Catalyst Screening for the Hydrothermal Liquefaction of Food Waste

Abstract:

Hydrothermal Liquefaction (HTL) is a promising route for conversion of food waste into biocrude oil.[1–3] However, the obtained crude oil has a high heteroatom content, similar to that of the initial food waste. This negatively influences the quality and heating value of the biocrude oil. Hence, removing the N, S and O from the biocrude oil is required to produce a finished fuel. Deoxygenation, denitrogenation and desulfurization can be promoted by using proper catalysts in the HTL process.[4,5] Therefore, in this study, we focused on catalytic HTL (350 °C, 40 min) of food waste by testing several heterogeneous supported metal catalysts (Ni/C, Pt/C, Ru/C, Pd/C, Ni/SiO_2-Al_2O_3, Pt/Al_2O_3 and Ru/Al_2O_3) in the presence and absence of high-pressure hydrogen, metal oxides (CaO, Al_2O_3, CeO_2, La_2O_3 and SiO_2), and a series of acid and base additives. Supported metal catalysts and additives did not increase biocrude yield, but three of the metal oxides led to higher yields with the following order $SiO_2 > La_2O_3 > CeO_2$. The elemental compositions and heating values of the biocrudes were sensitive to the catalysts used, especially with the presence of high pressure hydrogen. The supported Ni catalysts regardless of the support types, with or without H_2 gas, resulted in sulfur content below the detection limit (500 ppm). Supported metal catalysts all caused high H content in the crude oil with the highest amount of 8.57 wt%, achieved by Ni/C under high pressure hydrogen. The biocrude oil from HTL with Ni/C under high pressure hydrogen also had the heating value of 33.6 MJ/kg, which was higher than 29.7 MJ/kg for the bio-oil obtained from the HTL with no catalyst. The chemical compositions of the biocrudes were examined by conducting Gas Chromatography- Mass Spectrometry (GC-MS) analysis. The results showed that the amount of fatty acids, the most abundant compounds in biocrude oil from HTL of this particular simulated food waste, is sensitive to the type of catalysts used. Finally, since the biocrude oil yields were affected the most by metal oxide catalysts, these biocrudes were analyzed by thermogravimetric analysis to see how the catalysts affected the relative amounts of heavy and light compounds. Ultimately, the results of this study promotes our understanding of the role of different type of catalysts in HTL of food waste and can guide the selection of catalysts or additives to improve the quality of the biocrude oil.

References:

1. Maag, Alex R., Alex D. Paulsen, Ted J. Amundsen, Paul E. Yelvington, Geoffrey A. Tompsett, and Michael T. Timko. "Catalytic hydrothermal liquefaction of food waste using CeZrOx." *Energies*11, no. 3 (2018): 564.
2. Aierzhati, Aersi, Michael J. Stablein, Niki E. Wu, Chih-Ting Kuo, Buchun Si, Xu Kang, and Yuanhui Zhang. "Experimental and model enhancement of food waste hydrothermal liquefaction with combined effects of biochemical composition and reaction conditions." *Bioresource technology*284 (2019): 139-147.
3. Zastrow, Dustin Jay, and Paul A. Jennings. "Hydrothermal liquefaction of food waste and model food waste compounds." PhD diss., Florida Institute of Technology, 2013.
4. Duan, Peigao, and Phillip E. Savage. "Hydrothermal liquefaction of a microalga with heterogeneous catalysts." *Industrial & Engineering Chemistry Research*50, no. 1 (2011): 52-61.
5. Sánchez-Bayo, Alejandra, Rosalía Rodríguez, Victoria Morales, Nima Nasirian, Luis Fernando Bautista, and Gemma Vicente. "Hydrothermal Liquefaction of Microalga Using Metal Oxide Catalyst." *Processes*8, no. 1 (2020): 15.

595870 Bifunctional Acid/Base Catalysis for Improved Biocrude Recovery from Food Waste Hydrothermal Liquefaction

Abstract:

Millions of tons of food waste are thrown into landfills annually and the resultant degradation products lead to a variety of detrimental wastes. Hydrothermal liquefaction (HTL) is an emerging technology in the waste-to-energy field that utilizes high temperature and pressure to convert this waste into bio-crude. With appropriate upgrading, such as hydrodeoxygenation, this oil is suitable for the transportation and heating industries. HTL's benefit is using complex, high water-content feedstocks without inefficient drying steps.

Homogeneous base catalysts are popular in HTL systems, however, pose process complexity and expense with separation and neutralization. Recently, solid catalysts, with bi-functional characteristics have been used successfully with in-situ upgrading of food waste to bio-oil. Hydroxyapatite (HAp) is a crystalline mineral found in bone and is an attractive catalyst due to its tuneable acid-base sites and potentially renewable sources. Six different synthesized and commercial HAp samples were screened for the highest oil yield. The commercial HAp-200 sample (200 nm particle size) resulted in an oil yield over 3 times greater than that of non-catalytic HTL. Commercial samples outperformed synthesized samples in all experiments. We have also tested and shown the crystalline structure of HAp to be stable for at least 100 hours, as verified through XRD and catalyst re-use.

In addition to improved oil yield, catalyst stability and economic feasibility are important factors in the success and future commercialization of food waste HTL. Through a technoeconomic analysis we produce a bio-crude with selling price < \$3.00 (GGE).

Hydroxyapatite serves as a beneficial HTL catalyst by tripling oil yield and quadrupling energy recovery compared to non-catalytic experiments. This research serves to provide an economically promising process for the conversion of waste to a useful energy product by producing a bio-crude with competitively high yield and energy recovery.

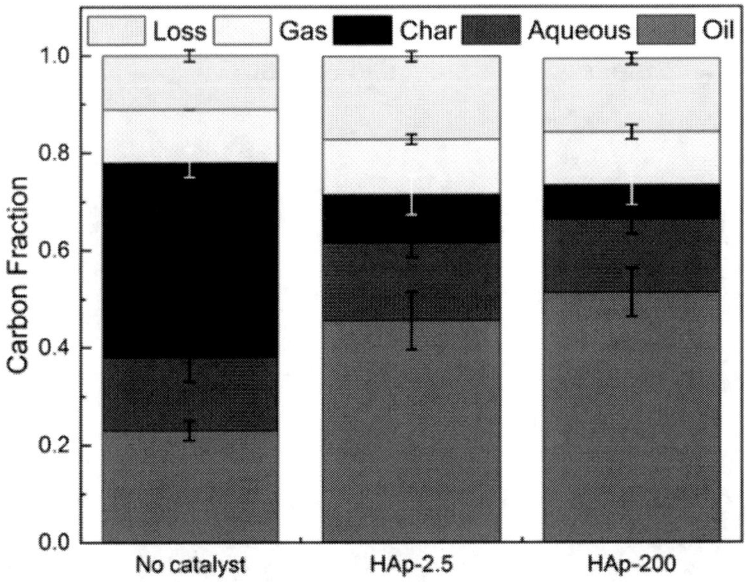

Figure 1. Carbon yields for each product fraction for thermal HTL and catalytic HTL of the two best HAp catalysts. Names signify 2.5μm and 200nm particle sizes.

607573 Simultaneous Biogas Upgrading and Hydrogen Sulfide Removal through Enhanced Electrochemically Assisted Anaerobic Digestion of Food Waste

Abstract:

Currently, one of the most sustainable methods to dispose of food waste is anaerobic digestion, which exhibits the highest primary energy production and leaves the smallest environmental footprint. The biogas generated through anaerobic digestion of food waste typically contains 50-70% methane, 30-50% carbon dioxide, and small amounts of impurities such as hydrogen sulfide. The high carbon dioxide reduces the heating value of biogas, whilst the hydrogen sulfide is extremely toxic to personnel, corrosive to facilities, and inhibitive to methanogenesis. Therefore, measures to reduce carbon dioxide and hydrogen sulfide in the biogas are essential to increase its further applications. Current methods typically require separate units to reduce the carbon dioxide and hydrogen sulfide in the biogas stream after anaerobic digestion, which are very costly and cannot offset the inhibitory effects.

To tackle these issues, an innovative system (Fig. 1) incorporating multifunctional bioelectrochemical treatment into the anaerobic digestion of food waste was established for in-situ biogas upgrading and hydrogen sulfide removal. Food waste was hydrolyzed in the upper layer of reactor, and volatile fatty acids were produced and washed down through the percolation layer into the leachate in the lower chamber. A cyclic flow between the lower and upper chambers was created to soak the solids and increase the mass transfer using a peristaltic pump. A pair of electrodes were submersed in the leachate and connected with a direct current power supply. A voltage of 0.3-1.0 V below the requirement of water electrolysis (~1.23 V) was applied. The low carbon steel on the anode was oxidized to release ferrous ions into the leachate, which further combined with aqueous sulfide to generate ferrous sulfide precipitate, thus significantly reducing the gaseous hydrogen sulfide evolution. Meanwhile, on the cathode, some carbon dioxide was reduced to methane through both hydrogenotrophic methanogenesis and homoacetogenesis/acetoclastic methanogenesis pathways. Different combinations of electrode materials (carbon rod, low carbon steel, etc.) and conductive materials (biochar, activated carbon, etc.) were tested to enhance the experimental performances, achieving a high hydrogen sulfide removal efficiency while maintaining minimal material sacrificing rate and electrical charge consumption. Consequently, a hydrogen sulfide removal efficiency over 75% was secured, and the methane content in the biogas was improved to over 80%.

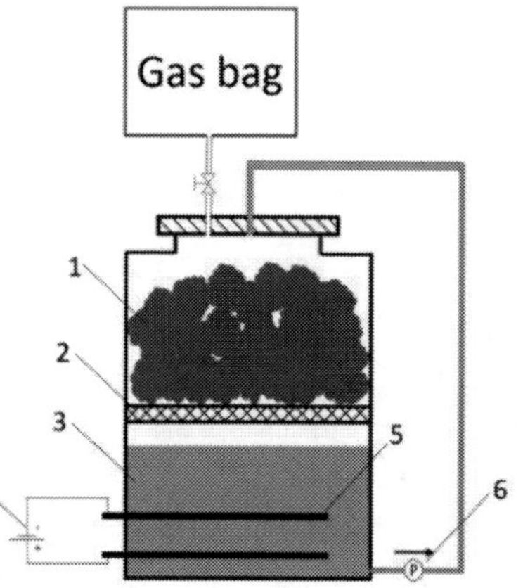

Fig. 1 Schematic of the system integrating anaerobic digestion of food waste and bioelectrochemical treatment of liquid leachate (1. Mixed food waste and seed inoculum; 2. Percolation layer; 3. Liquid leachate; 4. DC power supply; 5. Electrode pair; 6. Peristaltic pump)

Application of Black Soldier Fly Larvae to Convert Municipal Organic Waste to Surfactants and Other Value-Added Chemicals

Eleanor O'Doherty and Dr. James McGregor

Current waste management practices are struggling to cope with economic growth and an ever-increasing global population. UN estimates indicate that 1.3 billion tonnes of food waste are generated globally per annum. Municipal organic waste particularly represents a challenge in urban environments. The most relevant large-scale approach to recycling of this waste is the time-consuming process of compositing. Simultaneously, declining reserves of fossil resources and environmental concerns around the use of coconut and palm kernel oil present a need to identify new sources of the raw materials required for the production of fuels and chemicals. In this work, an approach to tackle both of these grand challenges is proposed using black solider fly larvae (BSFL). BSFL are highly effective in their consumption of waste, and the biological transformation of this into higher value organic oils with a variety of chemical applications. Therefore, BSFL have the potential to be a safe sustainable means of synthesising functionalised chemical feedstocks from municipal waste.

In this work, waste biomass is converted into proteins and lipids, the latter representing roughly 30% of the total larval body mass. Lipids extracted using soxhlet extraction with hexane have been converted into chemical products such as bio-fuels, solvents, and surfactants. This process takes advantage of the capability of BSFL to consume municipal waste streams to accumulate body fat, behaving as living "bio-catalysts", thereby avoiding many of the health and environmental issues associated with conventional methods of processing municipal waste. GC-MS analysis shows this to have a very similar composition to coconut and palm oils, i.e. free long chain carboxylic acids and in particular palmitic acid and lauric acid. Thus, BSFL can provide a renewable source of fats, and can be grown anywhere within the world, including developed countries where there are no local sources of free fatty acids. It can also be applied globally as BSFL can be artificially reared in countries where weather is unsuitable for their growth.

This innovation provides a novel and sustainable route to consumer products via an alternative oil source. It can also exert a significant environmental benefit as food waste is a vastly overlooked driver of climate change, with the FAO UN 2011 assessment estimating the total carbon footprint of food wastage, including land use change, at around 4.4 Gt CO_2 per year. Our proposal therefore provides a sustainable resolution in a sector res
total global greenhouse gas emissions (International Energy Agency 2014

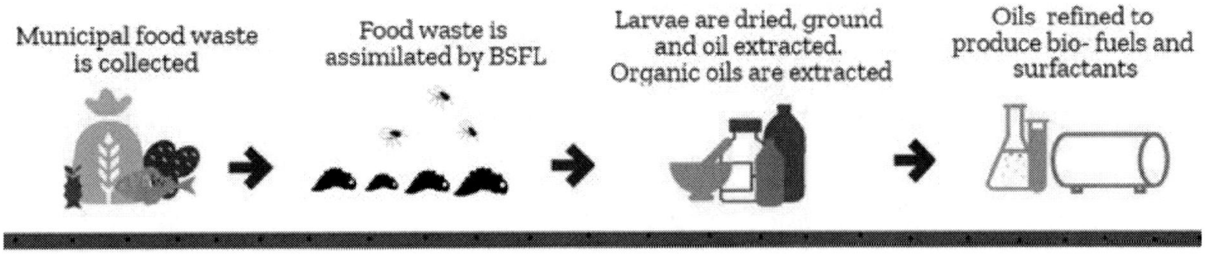

611156 Algae Cultivation in a Space Station-Based Biorefinery

Abstract:

In a biorefinery for isolated environments such as a space station, human activity-derived waste can be treated using biological systems to produce recovered potable water, usable returned cabin air, and edible protein cake. This biorefinery concept in confined spaces is similar to a municipal wastewater treatment where there are processes to treat solids by anaerobic digestion, to reduce chemical oxygen demand in the water through aerobic biotreatment, to reduce nitrogen and phosphorus by algae cultivation and to remove the metals, chlorophyll in algae and other organic substances by adsorption. In this study, algae cultivation systems are designed to produce algae biomass for protein, carbohydrates, and lipids, to reduce nitrogen and phosphorus in the water, and to convert carbon dioxide from the space station cabin air to oxygen. Visual observations and gravimetric measurements of the cell concentration is preferred over spectrophotometry because the algae cells aggregate. Contamination of other microorganisms and maintaining a viable algae culture were challenges in algae cultivation studies. Experiments to measure the biomass yield for cultivation of the algae were conducted. The *Chlorella vulgaris* UTEX 2714 strain was grown in suspended growth photobioreactors using three media for 18 days of cultivation. Experimental results showed that the media containing sulfur has the higher percent utilized phosphate and nitrogen for algae growth, which is explained by higher algae biomass concentrations than the biomass yield in another media without sulfur. Another experiment was performed to formulate a maintenance media for acclimation of the algae for 18 days to prevent any nutrient deficiency prior to feeding the algae with effluent from the aerobic biotreatment or anaerobic digestion processes. Nutrients are also fed intermittently to prevent any substrate inhibition. A review of studies reported in the literature on the effect of sodium chloride, sodium bicarbonate, ammonium salt or nitrate salts on the growth of Chlorella vulgaris suggested a range of concentrations as nutrient requirements for growth but inhibitory effects are specific to the algae strain, which were experimentally determined under the cultivation conditions for wastewater treatment. A factorial design of experiments approach on two levels of 3 nutrient sources (carbon, nitrogen and phosphorus) were performed without contamination. A regression model from statistical analysis from the experimental data serves as an effects model on the levels of carbon, nitrogen, and phosphate with the interaction parameters, which are valuable for future algae cell viability studies. Hence, a maintenance media is formulated for effluent from aerobic biotreatment and can replace the commercially available maintenance media. The known levels of carbon, nitrogen and phosphorus for maintaining a viable algae culture would be the basis for intermittent feeding rates of the actual wastewater effluent for algae cultivation. Moreover, an analysis of the carbohydrates, lipid, and protein content of the harvested biomass show the relationship of the substrates in the media to the biochemical pathways for production of metabolites in the algae. These metabolites provide nourishment for astronauts in a space station. Furthermore, a viable algae culture maintained in a photobioreactor for wastewater treatment in a space station would serve as a platform for studies on measuring conversion of carbon dioxide to oxygen, thereby producing recovered usable oxygen-enriched cabin air. As far as a literature review on carbon dioxide sequestration and photosynthesis efficiency of algae has revealed that studies on oxygen production are limited. Experiments to show oxygen production are also performed to know the feasibility of enrichment of the cabin air with oxygen. Ultimately, the goal of this project is to have data on material and energy balances in a closed loop system where the human-derived wastes are utilized by biorefinery systems to replenish food, water and oxygen for astronauts in a space station.

611470 Keynote: Solar Hydrocarbons from Air Captured Carbon Dioxide and Water

Abstract:

The value of liquid hydrocarbons is hard to overestimate. If they did not already exist, they would need to be invented. Today, mining fossil carbon provides access to hydrocarbons for fuels and chemicals. However, there are other ways to produce liquid hydrocarbons. For steel, aluminum and many other metal resources, societies have long accepted the energy cost of reducing metal oxides or sulfides to metal. The energy requirements are often large. The oxygen in aluminum oxide is far more tightly bound than the oxygen in carbon dioxide. Electricity – often from renewable hydropower – is used to reduce aluminum oxide to aluminum. Renewable energy could also be used to reduce H_2O and CO_2. The resulting syngas would serve as the starting point for hydrocarbon synthesis. The cost of solar energy is now so low that raw photovoltaic energy is comparable in price to natural gas. Water is cheap, and the promise of direct air capture technologies is that the cost of CO_2 from air is not dominating the cost of fuel. Once it is affordable to capture CO_2 from the air for sequestration, it is equally affordable to capture it for new fuels. The biggest remaining challenge is intermittency. The high capital cost of energy conversion systems makes it impossible to use them only for short times in the day. Overcoming this problem will require a different approach to deploying capital. Small mass-produced systems come down learning curves with the promise of dramatic cost reductions. Solar photovoltaic energy is not the only example, but it is the posterchild for the energy transition. As air capture and electrolyzers or other energy conversion equipment comes down the learning curve, it becomes possible to operate a fuel synthesis infrastructure on low-cost power that is only available for hours in the day. On a power basis, the capital cost of mass-produced car engines is two orders of magnitude lower than that of a power plant. As a result, car owners won't fret because the car stands idle most of the day.

607388 Invited Talk: Plasma-Assisted CO_2recycling: A Tool for Storing Solar Electricity

Abstract:

Solar electricity generation inherently suffers from intermittency and unequally distributed irradiation conditions. The conversion of solar energy into chemical energy carriers, i.e. solar fuels - which can be long term stored and transported - can tackle some of the drawbacks linked to the utilization of this renewable resource [1]. Water and/or CO_2 splitting, either using concentrated solar power or though photo-chemical and photo-catalytic processes, have been intensively studied in the last decades. New materials and innovative solar reactors have been proposed, highly contributing to the development of solar fuels technologies [2,3].

Carbon dioxide is indeed a very stable molecule. Any process envisaging its transformation into fuels or chemicals requires important amounts of energy and generally implies stringent conditions in terms of temperature and/or pressure. Methane production from CO_2 is not an exception. Carbon dioxide methanation, through the well-known Sabatier's reaction (Eq. 1), implies its hydrogenation/reduction, yielding methane and water. When coupled to water electrolysis, the production of synthetic methane from CO_2 is seen as a promising tool for storing off-peak renewable electricity production (power-to-gas concept).

$CO_{2(g)} + 4H_{2(g)} = CH_{4(g)} + 2H_2O_{(g)}$ $\Delta H°_{25°C} = -165.3$ kJ/mol (Eq. 1)

The Sabatier's reaction is exothermic and thermodynamically favorable at ambient temperature and pressure. However, it is strongly hindered by its sluggish reaction kinetics. Even in the presence of a catalyst, moderate temperatures - normally over 350°C - are needed to overcome this kinetic barrier. Increasing the operating pressure up to 20-30 MPa helps controlling the formation of by products, such as CO, and catalyst deactivation.

(Fig. 1. Plasma-catalytic redox cycle in the presence of a Ni/CeZrO$_x$catalyst.)

Lately, we have demonstrated that the plasma-intensification of CO_2methanation can lead to high methane yield under pseudo-adiabatic conditions (no external heating source) and atmospheric pressure [4-6]. A Dielectric Barrier Discharge (DBD) cold plasma was generated within two electrodes, leading to the ionization of the reactant gas mixture. In the presence of a Ni-containing CeZrO$_x$-supported catalyst, the reaction was 100% selective to methane and no sign of deactivation was observed over 100 h operating time. The information gathered through a detailed physicochemical characterization of the spent catalyst led us to conclude that the use of a CeZrO$_x$support, a non-stoichiometric oxide played a very important role in a redox cycle that is furthermore sustained by the transfer of electrons in the gas plasma [7]. In the presence of a DBD plasma, our catalyst works as a PN junction: electrons are transferred from the Ni-active sites to the Ce-cations in the support. Oxygen transfer follows, leading to continuous oxidation and reduction of both the active phase and the non-stoichiometric oxide, as depicted in Fig. 1. In the absence of the DBD plasma, the random-Brownian movement of gas molecules impedes an efficient electron transfer, the redox cycle is not completed, oxygen accumulates on the surface and the catalyst deactivates with time-on-stream.

Electron transfer is therefore a key issue in CO_2 reduction or splitting reactions. Both the dielectric properties of the catalyst and its physicochemical features can therefore be tailored in order to increase the yield in the targeted products under plasma-catalytic operation. Our latest results open up the door for further research on plasma catalytic processes for CO_2 recycling, offering new solutions for renewable energy storage and the production of synthetic fuels from CO_2.

[1] A. Steinfeld, A. Meier, Encyclopedia of Energy, Ed. Elsevier, 5 (2004) 623-637.

[2] W.C. Chueh, C. Falter, M. Abbott, D. Scipio, P. Furler. S. M. Haile, A. Steinfeld, Science 330 (2010) 1797-1801.

[3] A. Corma, H. Garcia. J. Catal. 308 (2013) 168-175.

[4] M. Nizio, A. Albarazi, S. Cavadias, J. Amouroux, M. E. Galvez and P. Da Costa, Int. J. Hydrog. Energy 41 (2016) 11584–11592.

[5] M. Nizio, R. Benrabbah, M. Krzak, R. Debek, M. Motak, S. Cavadias, M. E. Gálvez and P. Da Costa, Catal. Commun. 83 (2016) 14–17.

[6] M. Mikhail, B. Wang, R. Jalain, S. Cavadias, M. Tatoulian, S. Ognier, M. E. Gálvez and P. Da Costa, React. Kinet. Mech. Catal. 126 (2019) 629–643.

[7] M. Mikhail, P. Da Costa, J. Amouroux, S. Cavadias, M. Tatoulian, S. Ognier, M.E. Gálvez. Cat. Sci. Technol. *In press*(2020). https://doi.org/10.1039/D0CY00312C.

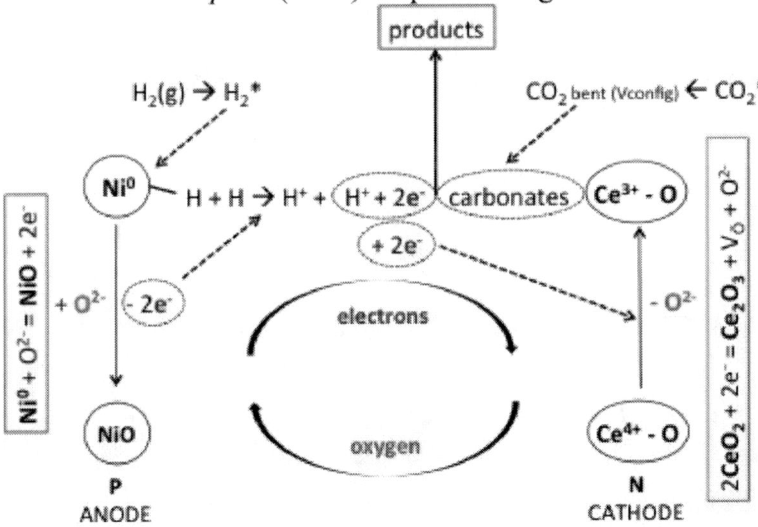

610961 Solar-Thermal Ammonia Production: A Renewable, Carbon-Neutral Route to Ammonia

Abstract:

Ammonia (NH_3) is an energy-dense chemical and a vital component of fertilizer. In addition, it is a carbon-neutral liquid fuel and a potential candidate for thermochemical energy storage for high-temperature concentrating solar power (CSP). NH_3 is currently synthesized via the Haber-Bosch process, which requires pressures of 15- 25 MPa and temperatures of 400-500 °C. Nitrogen (N_2) and hydrogen (H_2) are essential feedstocks for producing NH_3: H_2 is generally derived by steam reforming of methane, and N_2 is sourced from air, after O_2 removal, via combustion of methane. Burning hydrocarbons produces the heat and mechanical energy required to drive the NH_3 reaction. All three processes (sourcing the energy, producing H_2 and N_2) produce CO_2 emissions. The development of a renewable pathway to NH_3 synthesis that utilizes concentrated solar irradiation for the process heat instead of hydrocarbon combustion and operates under relatively low pressure, will result in both a decrease (or elimination) in greenhouse gas emissions and avoid the cost, complexity, and safety issues inherent in high-pressure processes.

The aim of the Solar-Thermal Ammonia Production (STAP) project is to develop a solar thermochemical looping technology to produce and shuttle N_2 from air for the subsequent production of NH_3 in an advanced two-stage process. The endothermic thermal reduction of redox-active metal oxide particles is driven by concentrated solar irradiation; subsequent exposure to air re-oxidizes the particles, removing O_2 and producing relatively pure N_2 gas. The N_2 serves as a feedstock for a renewable NH_3 production process in an advanced two-step, low pressure, looping process. In the first step, reduced metal (or nitrogen deficient metal nitride) particles react with N_2 to form a nitride. In the second step, H_2 reacts with the resulting nitride to produce NH_3. The net result is NH_3 produced from sunlight, air, and (green) H_2, while the metal oxide and nitride particles are recycled.

The STAP project consists of four thrusts: (1) Systematic synthesis, characterization, and thermodynamic analysis of oxides for N_2 recovery from air; (2) materials investigation and development (synthesis, characterization, and thermodynamic analysis) of nitrides for NH_3 production and re-nitridation; (3) process and reactor design, modeling, fabrication, and small-scale demonstration, and (4) full system and techno-economic analyses. The system concept will be summarized in this presentation with a brief summary of the results to date.

602596 Sustainable Thermochemical High-Purity Nitrogen Production Using Redox Oxides

Abstract:

In 2019, the global production of ammonia was 150 million metric tons of contained nitrogen, whereby 88 % of the ammonia was used for fertilizer production [1]. Therefore, the production of ammonia is fundamental to ensure the nutrition of the world population within the existing agriculture and diet. Due to the growing world population a further increase in demand for ammonia is expected.

Ammonia is synthesized via the Haber-Bosch process. The process itself has high energy consumption (32 MJ kg^{-1}) [2] and the required nitrogen is predominantly produced via energy intensive cryogenic air separation. An alternative method utilizes thermochemical cycles where a redox material is reduced at high temperature accompanied by a release of oxygen. At lower temperatures the material is oxidized and thus takes the oxygen from the incoming gas (see Fig. 1). An advantage of this method is the opportunity to use concentrated solar radiation (CSR) for the heat supply and therefore, providing a sustainable nitrogen production.

Perovskites are often used as redox material. They are described with the general formula $AMO_{3-\delta}$, where A are alkali, alkaline earth or rare earth metal cations and the M site is occupied in most cases by transition metal cations. δ ($\delta = 0 - 0.5$) indicates an oxygen non-stoichiometry. The partial reduction, which is described by Eq. 1, depends on the partial oxygen pressure p_{O2} and Temperature T. During the reduction the perovskites maintain their crystal structure, only smaller distortions are observed. This enables fast redox kinetics (reactions are completed in less than one min) and highly reversible redox cycles with long lifetime expectancy [3].

$AMO_3 \leftrightarrow AMO_{3-\delta} + \delta/2\ O_2$ (1)

For the Haber-Bosch process oxygen impurities of 3 ppm in N_2 can be tolerated [2]. Therefore, it is essential to produce high-purity N_2. A recent publication has proven that this is possible with thermochemical cycling using pre-purified air (1% O_2) and $SrFeO_{3-\delta}$ as redox material [5]. For the pre-purification of the air a pressure swing absorption devise (PSA) can be used. Economic analysis reveals that a combined solar-electrical driven PSA with a thermochemical system requires a lower thermal energy input than both solely solar-electrical driven PSA and solar-electrical cryogenic distillation (see Fig 2.).

Based on this analysis, the focus of the current work is the coupling of a PSA devise to a thermochemical reactor for the production of high-purity nitrogen. We present the concept design for this completely new approach, as well as an optimization of the process parameters based on thermochemical process simulations. Another field of research for improving the efficiency of thermochemical air separation is material science. The used perovskites determine the required temperatures and the achievable oxygen partial pressures. Doping of $SrFeO_{3-\delta}$ and $CaMnO_{3-\delta}$ leads to improved redox performance by either stabilizing the crystal structure and preventing phase changes, increasing the oxygen affinity or reducing the reduction and oxidation temperatures. We show the results of this material development and its positive effects on nitrogen production.

References

1. https://pubs.usgs.gov/periodicals/mcs2020/mcs2020.pdf, 7th April 2020
2. J.R. Jennings, *Catalytic ammonia synthesis: fundamentals and practice*, Springer Science & Business Media, 2013
3. Vieten, J.; Bulfin, B.; Call, F.; Lange, M.; Schmücker, M.; Francke, A.; Roeb, M.; Sattler, C.; J. Mater. Chem. A 2016, 4 (35), 13652-13659
4. J. Vieten, D. Guban, B. Lachmann, B. Bulfin, D. Krauzner, 2019, **patent no** DE 10 2019 126 114.7
5. Bulfin, B.; Lapp, J.; Richter, S.; Gubàn, D.; Vieten, J.; Brendelberger, S.; Roeb, M.; Sattler, C.; Chem. Engineering Science, 2019, 203, 68-75

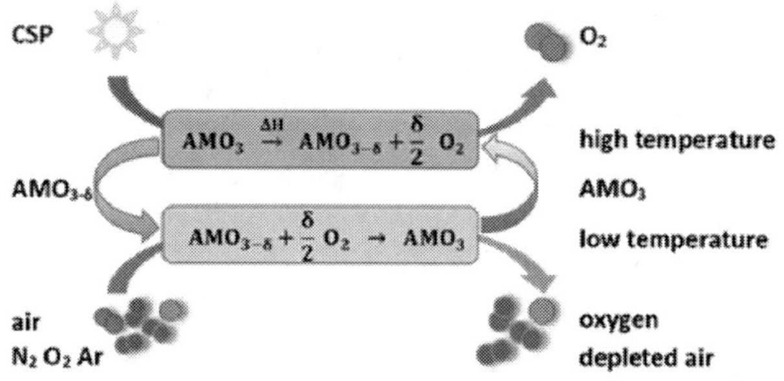

Fig. 1 Thermochemical redox cycle of a perovskite $AMO_{3\text{-}\delta}$ for air separation. The perovskite is reduced at high temperatures provided by CSR which leads to oxygen vacancies δ within the perovskites and a release of pure oxygen. The following re-oxidation occurs at lower temperatures and leads to an oxygen depletion of the gas stream [3].

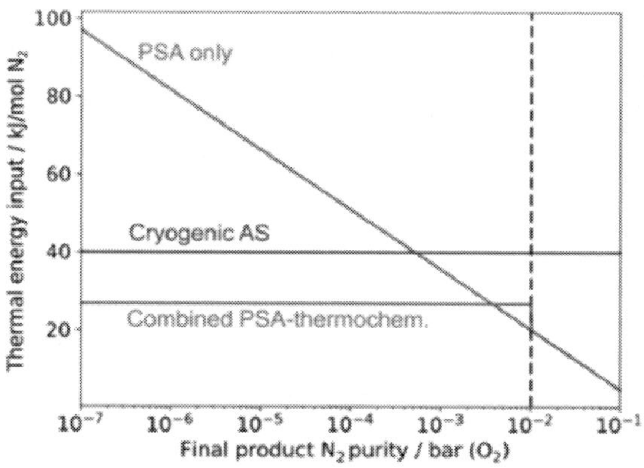

Fig. 2 Calculated thermal energy input for air separation. The thermal energy input is estimated as solar heat for all scenarios. The dotted line displays the oxygen concentration at the PSA outlet [4].

601864 Invited Talk: From Carbon Dioxide and Water to Carbon-Neutral Fuels and Chemicals

Abstract:

The contractual objective to limit the global temperature rise well below 2°C represents a milestone in international climate policy and requires the timely development of strategies that provide solutions to how effective energy-saving pathways can be realized. The defossilisation of the energy system plays a key role here. But the transport sector must also be reflected in strategies to become carbon-neutral because of its high share of GHG emissions. Hydrogen and synthetic hydrocarbon fuels are foreseen to play an essential role as energy carrier for the transport sector. Hence the development of renewable pathways to produce those fuels is very important.

Promising approaches to utilize CO_2 and water as a feedstock are solar heated thermochemical processes. Our group investigates such processes amongst which are the solar thermal dissociation of water and carbon dioxide to generate a mixture of CO and H_2 (= synthesis gas), which is then converted to a liquid fuels like methanol, gasoline or kerosene. This is at the same time a method to chemically store renewable energy using a two-step process based on multi-valent metal oxides capable to abstract and reversibly incorporate oxygen from water or CO_2.

R&D on such processes is carried out on European level. The technology has been developed from laboratory scale experiments to demonstration plants for hydrogen production at the Plataforma Solar de Almería (PSA) operated by the Spanish organisation CIEMAT and for kerosene production at a solar tower in Mostoles, Spain operated by IMDEA Energia.

In an accompanying German national project the complete production chain of solar fuels is being investigated. The aim is to obtain the most advantageous process design possible by carrying out a techno-economic analysis and optimisation, with which the target fuels can be produced at competitive prices. This chain not only considers the source of feedstock and the production route but also the use of the fuel inside a motor or turbine, as drop-in fuel or as newly designed fuel. The thermochemical pathways under consideration for this comparative analysis comprise routes based on porous reactive ceramics as solar absorbers as well as others based on ceramic particle flows. Heat recovery plays an essential role to make those processes feasible since they are carried out in a temperature swing type of operation. Heating the redox material accounts for about 60 % of the total heat demand. With a thermal storage heat that would otherwise be lost in the cooling phase is saved and used to preheat the material before the next cycle. The time of irradiation is thereby decreased. Calculations show that up to 40 % of sensible heat can be saved.

Acknowledgement: This work received funding from received funding from the European Union's Horizon 2020 research and innovation programme under grant agreement No. 654408 (Sun2Liquid) and from Fuel Cell and Hydrogen Joint Undertaking (Hydrosol_beyond) under grant agreement No. 826379 and from German BMWi under grant agreement No. 03EIV221 (SolareKraftstoffe).

605149 Invited Talk: Screening and Characterization of Reduction/Oxidation-Active Reactions of Mixed Ionic-Electronic Conducting Materials for Air Separation

Abstract:

An investigation was performed on the impact of A- and B-site substitution on $SrFeO_{3-\delta}$ reduction/oxidation (redox) extents. $SrFeO_{3-\delta}$ is a promising oxygen carrier for thermochemical air separation due to considerable oxygen capacity and chemical stability across a range of temperature and oxygen partial pressures. A- and B-substitution may increase reducibility while preserving the phase stability of $SrFeO_{3-\delta}$ [1,2]. Singly- and doubly-substituted $Sr_{1-x}A'_x Fe_{1-y} B'_yO_{3-\delta}$ (A' = La, Ba, Ca; B' = Co, Cr, Cu Mn), where x = 0 to 0.2 and y = 0 to 0.2, were synthesized using the sol-gel method [5] and screened based on their redox activity.

Redox activity screenings were performed via thermogravimetric analysis (TGA, Netzsch STA 449 F3 Jupiter) for a temperature range of 300 to 1200 °C, in 0 and 20% O_2/Ar. Changes in oxygen non-stoichiometry were quantified from measured mass losses to determine air separation capacity. Larger non-stoichiometry changes were seen in the A-sited doped samples. Of B-site substituents, only Co enhanced $SrFeO_{3-\delta}$ reducibility. Most samples reduced continuously as temperature increased. The largest reduction extents were observed for A'= La and Ba. Samples with A'= Ca generally showed faster reduction kinetics under pure Ar; however, there was indication of a potentially reversible phase change around 850 °C.

Powder X-ray diffraction was performed on samples after synthesis and after TGA to verify phase stability. Scanning electron microscopy (SEM) and energy-dispersive X-ray spectroscopy (EDS) were performed on $La_{0.1}Sr_{0.9}FeO_{3-\delta}$ and $Ba_{0.1}Sr_{0.9}FeO_{3-\delta}$ to investigate particle morphology and elemental distribution before and after thermal reduction and oxidation cycling experiments. SEM showed some smoothing and rounding of particle surfaces after cycling, without significant changes in bulk characteristics and particle sizes. SEM revealed no particle sintering, while EDS showed no evidence of phase segregation in the samples, suggesting promising chemical stability after several thermal cycles.

References:

[1] H. J. M. Bouwmeester, A. J. Burggraaf, Dense ceramic membranes for oxygen separation, in Membr. Sci.Technol., ed. A. J. Burggraaf and L. Cot., Elsevier, 1996, vol.4, pp. 435–528.

[2] C. C. Chang and H. S. Weng, Ind. Eng. Chem. Res., 1993, 32(11), 2930–2933.

[3] H.E. Bush, KP. Schlichting, R.J. Gill, S.M. Jeter, P.G. Loutzenhiser, Design and Characterization of a Novel Upward Flow Reactor for the Study of High-Temperature Thermal Reduction for Solar-Driven Processes, J. Sol. Energy Eng, 2017.

[4] R. Gill, H.E. Bush, P. Haueter, P.G. Loutzenhiser, "Characterization of a 6 kW high-flux solar simulator with an array of xenon arc lamps capable of concentrations of nearly 5000 suns." 2015, *86* (12), 8.

[5] J. Vieten, B. Bulfin, M. Roed, C. Sattler, Citric acid auto-combustion synthesis of Ti-containing perovskites via aqueous precursors. Solid State Ionics, 2018. 315: p. 92-97.

609280 Perspectives of Solar Thermochemical Fuels for Aviation

Abstract:

One of the most important challenges of the future is to mitigate climate change. Solar energy as a vast renewable energy source is a major pillar of the energy transition. However, efficient long-term storage is required for its widespread every-day use. The storage of solar energy in chemical compounds in the form of liquid fuels represents one important solution with high specific energy, which allows to decarbonise those parts of the transportation sector, where electrical storage is no option. The European Green Deal aims at a 90% reduction in transport emissions to achieve climate neutrality by 2050 [1]. Long-haul shipping and aviation are most dependent on renewable energy-dense liquid fuels. For aviation, being the fastest growing transport mode, solar thermochemical conversion of water and carbon dioxide to synthesis gas and further to "drop-in"-capable synthetic paraffinic kerosene (SPK) can fill this need and take advantage of the existing infrastructure and aircraft fuel systems for rapid deployment in the global fleet.

Setting the scene for solar fuels: decarbonizing aviation

Two targets are particularly prominent on the global agenda. The aviation industry has set a target of "50% reduction of CO_2 emission in 2050 compared to 2005" (ATAG target) [2]. To translate the magnitude of the challenge into the metrics of the renewable fuel domain, we show that in 2050 a challenging 90% emission reduction and a production volume of renewable SPK >550 Mt/a is required to meet the ATAG target. The ratio of CO_2 emissions in 2050 compared to 2005 can be represented as the product of (1) the growth in revenue passenger kilometers (RPK) according to estimates of the industry (red, approximately doubling every 15 years, or 4.7%/a growth rate), (2) the efficiency gains of aircraft and operations (green, assuming 1.5-2%/a efficiency gain) and (3) the fraction of substitution of fossil energy by "zero-carbon energy" (blue) required to meet the ATAG target (Figure 1, taken from ref. 3). Obviously, the path towards deep emission reduction requires the use of alternative jet fuels. The five to six-fold increase in fuel burn (green) in 2050 at 50% of emissions relative to 2005 requires then to remove approximately 90% of the emissions with the use of alternative jet fuel of several hundred megatons per year globally and several tens of megatons per year in Europe [4].

The second target, set out by the Paris Agreement, concerns the limitation of global warming [5]. According to the IPCC 2018 Special Report (on the impacts of global warming of 1.5 °C above pre-industrial levels and related global greenhouse gas emission pathways), "all pathways that limit global warming to 1.5°C [...] project the use of carbon dioxide removal (CDR) [...] to compensate for residual emissions and, in most cases, achieve net negative emissions to return global warming to 1.5°C following a peak" [6]. The feasibility and sustainability of CDR via direct air capture (DAC) of CO_2 and storage is one option in a wider portfolio of technologies.

Technology pull for CO2-based solar fuel and direct air capture

On the one hand, the "technology pull" of the two targets promotes the development of cost-effective DAC as a necessary resource of renewable CO_2 at a scale of gigatonnes p.a. for the global demand of carbon-neutral renewable SPK for aviation to fulfil the ATAG target. On the other hand it promotes DAC as a technology option for the negative emission regime from 2050 on to limit global warming to 1.5 °C according to the IPCC report. The technology pull for DAC-CO_2-based fuels is also motivated by the fact that current biofuel technology does not meet the sustainability and availability requirements at the scale of future global fuel demand [7]. A more efficient technology for storing solar energy with smaller CO_2 footprint is the production of solar thermochemical fuels from water and non-fossil CO_2.

The solar thermochemical fuel pathway

CO_2 and H_2O are converted into a mixture of CO and H_2 (syngas) through a redox cycle of a metal oxide operated in a temperature-pressure swing using concentrated solar energy as the source of heat [8,9]. The produced syngas is then converted into liquid hydrocarbons in a Fischer-Tropsch synthesis (FT).

The energy conversion efficiency of the solar thermochemical step is the most relevant performance indicator and the key to economic competitiveness. This is because – analogous to concentrated solar power (CSP) plants, where the solar concentrating subsystem represents about half of the total investment cost – higher efficiency implies a smaller solar concentrating system for the same fuel yield, which results in lower specific fuel cost. The long-term goal is 20% to ensure commercial viability [10].

Research on the solar thermochemical fuel pathway has progressed significantly over the last years, e.g. in the European projects SOLAR-JET and SUN-to-LIQUID, progressing the energy conversion efficiency from <1% to ca. 1.7% [8,11] and further to 5.6%±1% [12], respectively. In the future, the novel ordered redox materials [13] (enabling homogeneous reactor temperatures and higher mass loadings, increasing the yield of the process) and the recuperation of heat from the thermal redox cycle [14,15] represent an opportunity to significantly enhance the efficiency.

Life-cycle emission reduction potential

For a baseline case with CO_2 from DAC and a thermochemical reactor efficiency of 19%, the greenhouse gas emissions of the solar thermochemical fuel pathway are shown in Figure 2 (taken from ref. 16). As can be seen, the favorable environmental performance is achieved through counterbalancing the positive emissions by negative emissions through air capture of CO_2.

Geographical techno-economic potential

Through the exclusion of unsuitable areas, the net available land for solar fuel production is identified. The following areas are excluded: areas with existing ground structures, water bodies, shifting sands, slopes ≥5%, protected areas, as well as areas covered by forest, closed shrubland, woody savannas, wetland, cropland, urban settlements, or snow/ice. In the United States, 0.6×10^6 km^2 or about 8% of the contiguous country area is available, mostly in the Southwest. Here, the USA has highly irradiated areas and very favorable costs of capital, enabling low-cost production of solar fuels (Figure 3, left; green line indicates US jet fuel demand and the red line (in both graphs) the global demand. Figure taken from ref. 17). In the baseline case with an assumed reactor efficiency of 19%, production costs of under 1.75 €/L can be achieved, whereas the national demand for jet fuel could be covered at costs between 1.74 and 1.82 €/L, and at average costs of 1.79 €/L (global demand: 1.74-1.88 €/L; average 1.84 €/L). For a reactor efficiency of 25%, production costs of under 1.6 €/L are possible at the best locations.

Assuming a global perspective, the production costs can be further decreased in very sunny locations with favorable financial conditions such as Chile. The right diagram in Figure 3 shows the cost-supply curves for the baseline case with varying nominal weighted average cost of capital (WACC) between countries, and constant nominal WACC in each country of 4% and 10%, respectively. The colors of the pie charts indicate the distribution of total production volumes over the regions for each case in the different regions of Europe, North Africa, and the Middle East (MED, red), the USA (USA, blue), and South America (SAM, green) (Figure taken from ref. 17).

Roadmap for development of the thermochemical reactor efficiency

The main technology gap and the main risks for the further development of the fuel path indisputably relate to the thermochemical solar reactor subsystem. The main drivers of efficiency improvement are seen to be the adoption of ordered structures for the redox material that will enable a homogeneous temperature distribution in the reactor and heat recuperation. Proposed maturity gates for the solar thermochemical reactor efficiency on the way to commercialization are 10-15% in 2020-2025, 15-20% in 2025-2030, and >20% from 2030 on.

[1] https://eur-lex.europa.eu/legal-content/EN/TXT/HTML/?uri=CELEX:52019DC0640&from=EN

[2] www.atag.org

[3] A. Sizmann, 4th Bauhaus Luftfahrt Symposium, 2019

[4] Transport and Environment, *Roadmap to decarbonising European aviation*, 2018, https://www.transportenvironment.org/sites/te/files/publications/2018_10_Aviation_decarbonisation_paper_final.pdf

[5] 21st Conference of Parties of the UNFCCC to adopt the Paris Agreement, Decision 1/CP.21, par. 21, 2016

[6] IPCC: Summary for Policymakers. In: Global warming of 1.5°C. An IPCC Special Report [V. Masson-Delmotte, et al. (eds.)]. World Meteorological Organization, Geneva, Switzerland, 2018

[7] A. Mohr, S. Raman, *Energy Policy*, **63**, 2013, 114; S. Y. Searle, C.J. Malins, Proceedings of the 20th European Biomass Conference and Exhibition, Milan, **53**, 2012, and references therein.

[8] W.C. Chueh, C. Falter, et al., *Science*, **330**, 2010, 1797.

[9] M. Romero, A. Steinfeld, *Energy Environ. Sci.* **5**, 2012, 9234.

[10] E.B. Stechel, J.E. Miller, *J. of CO2 Utilization*, **1**, 2013, 28.

[11] D. Marxer, P. Furler, J. Scheffe, et al., *Energy Fuels*, **5**, 2015, 3241

[12] S. Zoller, Diss. ETH No. 26451, 2020

[13] M. Hoes, S. Ackermann, D. Theiler, et al., *Energy Technology*, 2019, 1900484

[14] J. Scheffe, A. Steinfeld, *Energy & Fuels*, **26**, 2012, 1928

[15] C. Falter, R. Pitz-Paal, *Solar Energy*, 2018, **176**, 320

[16] C. Falter, A. Valente, A. Habersetzer, et al., SUN-to-LIQUID (654408) Public Report D1.7, 2019.

[17] C. Falter, N. Scharfenberg, A. Habersetzer, *Energies*, **13**, 2020, 802

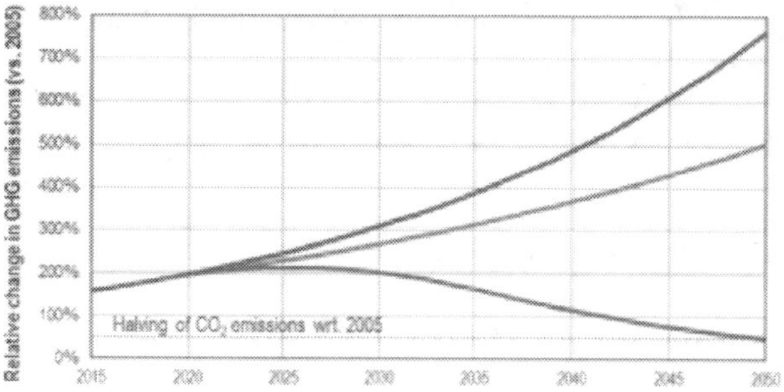

$$\dot{M}_{CO_2,eq}(2050) = \dot{M}_{CO_2,eq}(2005) \times \begin{bmatrix} \text{GROWTH} \\ \text{in} \\ \text{RPK} \end{bmatrix} \times \begin{bmatrix} \text{GAIN} \\ \text{in} \\ \text{EFFICIENCY} \end{bmatrix}^{-1} \times \left(1 - \begin{bmatrix} \text{FRACTION of 2050's} \\ \text{"ZERO-CARBON"} \\ \text{ENERGY} \end{bmatrix}\right)$$

Figure 1 : Development of emissions of the aviation sector until 2050.

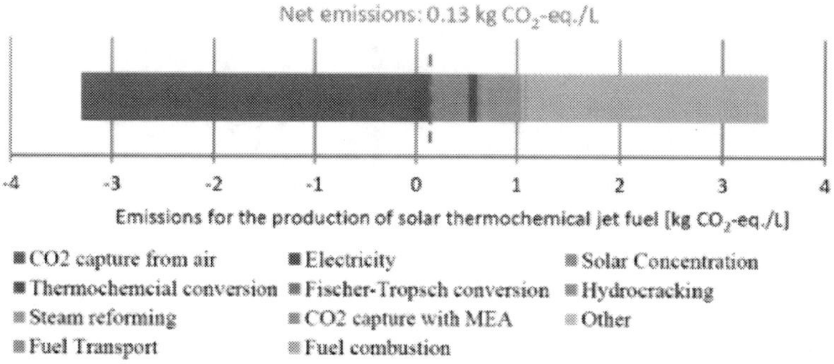

Figure 2 Greenhouse gas emissions from baseline case of solar thermochemical fuel production.

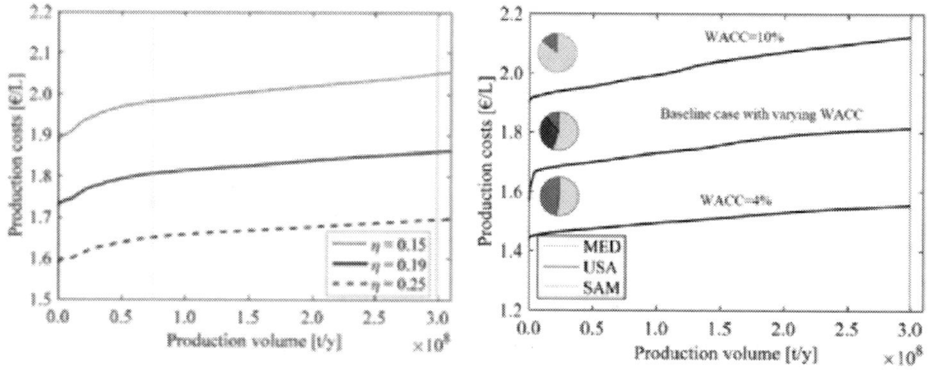

Figure 3 Left: Cost-supply curves of solar thermochemical fuels in the USA (left) and globally (right).

596980 Methanol from Sunlight and Air Using a Modular Solar Dish Reactor System

Abstract:

We report on the experimental pilot demonstration of a modular solar dish-reactor system for the production of methanol from ambient air using concentrated solar energy. We describe the complete process chain and its key components, and present representative on-sun runs with fully-automated consecutive CO_2-splitting and H_2O-splitting redox cycles.

The system integrates three thermochemical units operated in series, namely: 1) the capture of CO_2 and H_2O directly from ambient air via an adsorption-desorption cyclic process; 2) the co-splitting of CO_2 and H_2O to generate a specific mixture of CO and H_2 (syngas) via a reduction-oxidation cyclic process using concentrated solar energy; and, 3) the gas-to-liquid synthesis of methanol. The experimental validation of the entire process chain to solar methanol under realistic field conditions was accomplished by moving from a laboratory setup (Marxer et al. 2017) to a modular solar system, thereby advancing the technological readiness and its industrial implementation.

The 2nd process in the aforementioned process chain is the 2-step thermochemical splitting of CO_2 and H_2O via ceria redox reactions, represented by:

1st step) Reduction: $CeO_2 \rightarrow CeO_{2-\delta} + \delta/2\ O_2$ (1)

2nd step)

Oxidation with H_2O: $CeO_{2-\delta} + \delta\ H_2O \rightarrow CeO_2 + \delta\ H_2$ (2a)

Oxidation with CO_2: $CeO_{2-\delta} + \delta\ CO_2 \rightarrow CeO_2 + \delta\ CO$ (2b)

where the non-stoichiometry δ denotes the ceria reduction extent. Ceria is selected as the redox material because of its fast redox kinetics and high crystallographic stability. In the first, high-temperature solar-driven endothermic reduction step (eq. 1), ceria is partially reduced to a non-stoichiometric state. In the subsequent lower-temperature non-solar exothermic oxidation step, the reduced ceria is reacted with H_2O and/or CO_2 to generate H_2 and/or CO (eqs. 2a and 2b). The solar reactor for effecting this redox cycle consists of a cavity-receiver containing a reticulated porous ceramic (RPC) foam structure made of pure CeO_2. The RPC is directly exposed to the high-flux solar intensity and provides efficient radiant absorption and combined heat/mass transport within the reaction site. Two identical solar reactors are mounted side-by-side on a solar concentrator subsystem (Dähler et al. 2018). It encompasses a sun-tracking parabolic primary reflector and a planar secondary reflector. The later allows to rotate the focal point among 4 positions: the two solar reactors, a water calorimeter for solar radiative power input measurements, and a Lambertian target for solar flux distribution measurements. This optical arrangement enables the operation of the two solar reactors for performing both redox steps simultaneously by alternating the solar radiative input between them while making continuous and uninterrupted use of the incoming concentrated sunlight. Thus, while one solar reactor is performing the endothermic reduction step on sun, the second solar reactor is performing the exothermic oxidation step off sun. During the reduction step (eq. 1), the solar reactor is heated with concentrated sunlight up to 1500 °C and the total pressure is lowered to 4 mbar by the vacuum pump to evolve a pure stream of O_2 from CeO_2. During the oxidation step (eqs. 2a and 2b), CO_2 and H_2O are injected into the reactor's cavity, react with the reduced ceria, and are transformed into syngas - a specific mixture of CO and H_2. Downstream of the two solar reactors, the syngas is compressed up to 200 bar and subsequently processed to methanol in a catalytic gas-to-liquid packed-bed reactor. The complete solar system is fully automatized to perform consecutive redox cycles.

Full day on-sun consecutive cycles demonstrate the stability and robustness of the system. Solar runs with varying H_2O/CO_2 mixtures further show the flexibility of the system to produce syngas in the desired quality and stoichiometry suitable for the downstream gas-to-liquid process, e.g. Fischer-Tropsch or methanol synthesis.

Acknowledgements: This work was funded in part by the Swiss Federal Office of Energy (Grant No. SI/501213-01), the Swiss National Science Foundation (Grant No. 200021-162435), and the European Research Council under the European Union's ERC Advanced Grant (SUNFUELS – Grant No. 320541). We thank Patrick Basler, Thomas Cooper, Yago Gracia, Philipp Good, Andrea Pedretti, David Rast, Max Schmitz, Nikolas Tzouganatos, Alex Muroyama, and Michael Wild for their valuable contributions to the technology development.

References:

Marxer, D., Furler, P., Takacs, M., Steinfeld, A., 2017. Solar thermochemical splitting of CO2 into separate streams of CO and O2 with high selectivity, stability, conversion, and efficiency. Energy Environ. Sci. 10, 1142–1149.

Dähler, F., Wild, M., Schäppi, R., Haueter, P., Cooper, T., Good, P., Larrea, C., Schmitz, M., Furler, P., Steinfeld, A., 2018. Optical design and experimental characterization of a solar concentrating dish system for fuel production via thermochemical redox cycles. Solar Energy 170, 568–575.

611759 Integrated Reactor/Heat-Exchanger for Direct Synthesis of Dimethyl-Ether

Abstract:

Highly efficient, microscale-based technologies have provided a route to modularized solar-thermochemical energy storage, as is the case with the Solar Thermal Advanced Reactor System (STARS) [1]. As a platform to provide thermocatalytic conversion of solar syngas into value added carbon carrying species, an integrated reactor/heat-exchanger for the direct synthesis of dimethyl ether (DME) is presented here. Three goals were addressed in the design of this process intensified unit operation: 1) facilitation of a differential reaction temperature profile in order to maximize use of exergy, 2) recovery of heat of reaction for use in steam generation on board the STARS system, and 3) implementation of a lamina-plate architecture that provides flexible capacity and serviceable catalyst carrying structures via an openable design with commercially available gasketing. Multiphysics modeling of the reactor is presented. Plate fabrication, fluidic characterization, and catalyst coating development are discussed.

[1] R. Zheng *et al.*, "Integrated Solar Thermochemical Reaction System for Steam Methane Reforming," in *Energy Procedia*, 2015, pp. 1192–1200.

AUTHOR INDEX

Abad, Keemia 121
Abdelrahman, Omar 6, 109, 187, 189
Abedsoltan, Hossein 391
Aboagye, Emmanuel A. 152
Abolhasani, Milad 167
Ackermann, Simon 278
Adewuyi, Yusuf G. 112
Adhikari, Sushil 385
Agblevor, Foster 252, 420
Agrawal, Rakesh 35
Aguda, Remil 528
Aguilar-Dominguez, Donovan 53
Aguirre-Villegas, Horacio 156
Ahmad, Ayyaz 115
Ahmed, Razan 138
Ahn, Hyeunguk(Henry) 261
Al-Mohannadi, Dhabia 38, 138
Al-Salik, Yahya 94
Alamani, Bryan G. 160
Alaqeeli, Anas 224
Alba-Rubio, Ana 2, 391
Albarracín-Suazo, Sandra 12
Albrecht, Kevin 532
Alenzi, Adel 328
Alhashim, Mohammed 258
Aliabadi, Majid Moradi 36
Allen, David 42, 331
Alsmail, Jamal 258
Alturki, Abdulaziz 119
Altuzar-Coello, Patricia E. 285
Alvin, Mary Anne 487
Alzahrani, Hashim A. 193
Amador-Noguez, Daniel 127
Amar, Vinod S. 202, 205
Amato, Kelly 141
Ambrosetti, Gianluca 278, 279
Ambrosini, Andrea 532, 536
Ammal, Salai C. 10
Andersen, Suzanne 199
Anderson, Eric 8
Anderson, Ryther 194, 409
Anderson, Shelby 165
Androshchuk, Iryna 94
Ankathi, Sharath 213
Arancibia-Bulnes, Camilo A. 285
Arbabi, Sepehr 520
Argot, Jonathan 257
Arjomandi, Maziar 179
Asadi, Mohammad 29
Asafu-Adjaye, Osei 385
Ashman, Peter J. 179
Asif, Usman 300
Atiyeh, Hasan 131
Auad, Maria 385
Auyeung, Nick 271, 543
Avendano, Marco 446
Avraamidou, Styliani 78
Ayala-Cortés, Alejandro 285
Azzolina, Nicholas 522
Babauta, Jerome 274
Bai, Peng 32
Bajaj, Ishan 1, 269
Baker, Sarah 220
Bakshi, Bhavik 44, 132, 185, 242, 244, 503
Baldea, Michael 42, 436
Baltrusaitis, Jonas 62, 423, 516
Bamunuarachchi, Vimantha 2
Bansode, Archana 385
Bao, Ying 426
Baratsas, Stefanos G. 78
Barbato, Maurizio C. 279
Barbee, David 14
Barde, Mehul 385
Bardow, André 74, 212
Barrett, William 236
Bassas, Hugo 75
Basso, Bruno 40
Batchu, Sai Praneet 188
Bathena, Tanmayi 108
Batteiger, Valentin 537
Bayon, Alicia 94
Beauchemin, Suzanne 81
Beckham, Gregg T. 8
Beis, Sedat H. 252, 420
Bejarano, Lorena 65
Belden, Elizabeth 421
Bell, Alexis T. 105, 107, 191, 192
Bellamy, Tim 141
Benson, Tracy 16, 25
Berquist, Zachary 359
Bettinson, Lance A. 107
Bhattacharjee, Tiasha 294, 480
Bhattacharyya, Debangsu 17, 50, 139, 434
Bhosale, Rahul 146, 412
Bielicki, Jeffrey M. 503
Bilec, Melissa M. 135

Birnbaum, Amanda ... 78
Black, Brenna ... 207
Black, Joseph M. ... 20
Bluck, David ... 22
Bo, Alessandro ... 180
Bokis, Constantinos P. ... 23
Bollas, George M. ... 407
Boman, Roelof ... 55
Bommarius, Andreas ... 165
Bommarius, Bettina ... 165
Bond, Jesse ... 4, 6, 148
Boorla, Veda Sheersh ... 127
Bose, Abhishek ... 46
Bosshart, Nicholas ... 522
Boualavong, Jonathan ... 430
Bradley, William ... 251
Brandt, Craig ... 482, 486
Bravo-Suárez, Juan J. ... 193
Brendelberger, Stefan ... 535
Brezicki, Gordon ... 190
Brody, Leo ... 413
Bromhal, Grant ... 521
Brown, Brandi ... 494
Brown, Robert C. ... 72, 208
Brown, Solomon F. ... 48, 53
Brunser, Sebastian Sas ... 206
Bukas, Vanessa J. ... 199
Bukhovko, Maxim P. ... 143
Burli, Pralhad ... 486
Burrows, Laron ... 407
Bush, H. Evan ... 532, 536
Bustillos, Steven ... 119
Byun, Jaewon ... 302
Bécsy-Jakab, Villo E. ... 384
Béguin, François ... 285
Cabezas, Heriberto ... 55, 184
Caicedo, Andrés ... 65
Campanella, Osvaldo H. ... 304
Capece, Sofia H. ... 524
Caratzoulas, Stavros ... 109, 188
Cardona-Martínez, Nelson ... 147
Carey, J. W. ... 296
Carpenter, John R. ... 352
Carreon, Moises A. ... 409
Carter, Emily A. ... 102
Casalins, Miguel A. ... 506
Castañeda, Verónica ... 65
Castaño, Sara Vallejo ... 232
Castellano, Elena ... 90
Cavagnaro, Robert ... 201
Cavalcante Jr., Célio L. ... 293
Cen, Jiajie ... 264
Ceron, Maira R. ... 220
Chakraborty, Debasish ... 199
Chandran, Rohini Bala ... 357
Chang, Chaoyi ... 110
Chang, Cheng ... 333
Chang, Hochan ... 1
Chapman, Bryan R. ... 442
Chapman, Walter G. ... 330
Charles, Michael ... 44
Chartier, Ryan ... 14
Chatterton, Lindsey ... 141
Chatzidimitriou, Theodore ... 136
Chatzidmitriou, Anargyros ... 4
Chea, John D. ... 152
Chen, Chien-Hua ... 516
Chen, Dan ... 116
Chen, Jingguang G. ... 10
Chen, Peipei ... 356
Chen, Qining ... 331
Chen, Siwei ... 284
Chen, Tso-Hsuan ... 109
Chen, Yonru ... 155
Chen, Zheng ... 163
Chen, Zihao ... 28
Chen, Zijie ... 357
Cheng, Feng ... 525
Cheng, Juiyen ... 396
Cheng, Yan ... 126
Chi-Hwa, Wang ... 263, 265, 289, 379
Chiang, Yadong ... 75
Chikan, Viktor ... 196
Chinnici, Alfonso ... 179
Choi, Changhyeok ... 200
Choi, Hoon ... 207
Choi, Wonjae ... 176
Chorkendorff, Ib ... 199
Chowdhury, Ratul ... 127
Chu, Libing ... 116
Chueh, William ... 102
Chun, Soomin ... 503
Cisneros-López, Erick O. ... 388
Clewett, Catherine F. M. ... 104
Coblyn, Matthew ... 543
Cody, Lucas ... 14
Coleman, Dr. Maria R. ... 2, 391
Conway, Jonathan M. ... 419
Cooney, Gregory ... 225
Cooper, Thomas ... 99, 358
Coplien, Jason ... 130, 292
Corazza, Marcos ... 137
Corbin, David R. ... 164, 303, 424
Cordeiro Jr., Jose C. ... 257

Cordiner, Joan 490
Coronella, Charles 422
Crandall, Dustin 328
Crawford, James M. 409
Cremaschi, Selen 249
Criado, Arantza 137
Crocker, Mark 222
Cronin, Dylan 492, 496
Crossley, Steven 5
Crunkleton, Daniel W. 257
Cruz, Juan C 506
Cruz, Mutya A. 28
Cuentas-Gallegos, Ana K. 285
Cussler, Edward L. 262
Cárcamo, Jonnathan 267
D. Srinivasan, Priya 193
Da Silva, Priscilla Dias 510
Dai, Xiaoyu 174
Dai, Yanjun 379
Dalal, Ashit 159
Dalamagas, Theodore 136
Dale, Evelyn 225
Dally, Bassam B. 179
Daly, Chris 482
Damiani, Darin 447, 517, 521
Dangl, Jeffery L. 419
Danish, Muhammad 115
Daoutidis, Prodromos 203
Daramola, Damilola 125, 513
Das, Srashtasrita 181
Das, Tapas K. 209
Dasgupta, Neil P 359
Dash, Satyakam 127
Dauenhauer, Paul 187, 262
Dauson, Erin 296
Davis, Maggie 486
Davis, Robert J. 190
De La Calle, Alberto 286, 287, 532
Deak, Csaba 184
Deng, Shuguang 175
Dent, Peter 304
Derespino, Sophie 161
Didas, Stephanie 434
Dien, Bruce S. 129
Ding, Lingkan 526
Diwekar, Urmila 55
Doan, Linh 25
Doliente, Stephen S. 76
Donahue, Carly M. 296
Donohue, Timothy J. 130, 292
Dotan, Hen 255, 376
Du, Chunxiao 335
Dudek, Ryan B. 413
Dugar, Deepak 253
Dumesic, James A. 1, 3, 104
Dunbar, Alan 53
Dunn, Jennifer B. 79, 211, 331
Dähler, Fabian 541
D'Angelo, Sebastiano C. 241
Echtermeyer, Alexander 74
Edgar, Thomas F. 42, 436
Ehrenstein, Michael 54
Ejeh, Jude O. 48, 53
El-Halwagi, Mahmoud 446
Ela, Raisa Carmen Andeme 387
Elder, Thomas 385, 495
Ermanoski, Ivan 98, 175, 286, 287, 532
Escobar, Jennifer 356
Esmaeilirad, Mohammadreza 29
Estrada, Claudio 277, 290
Ezeobinwune, Chinyere 434
Falter, Christoph 537
Falzone, Gabriel 118
Fan, L. S. 509
Fang, Tianqi 151
Farag, Ramsis 385
Farooq, Usman 115
Farr, Tyler 532, 536
Fathima, Navaira 268
Faturay, Futu 45
Fend, Thomas 535
Fichthorn, Kristen 28
Finkel, Omri M. 419
Flores-Tlacuahuac, Antonio 41, 441
Flórez, Juan Sebastián Rodríguez 299
Fosbøl, Philip L. 523
Foster, Charles 127
Fox, Brian G. 104
Frash, Luke 296
Freeman, Charles J. 211
Freiberg, Lucas 543
Fuentes, Olga P. 506
Fuentes-Cortes, Luis Fabian 41
Fujimoto, Sho 88
Fujimura, Yasushi 88
Fukushima, Yasuhiro 354
Furler, Philipp 278, 541
Gaffney, Anne M. 392
Galán-Martín, Ángel 301
Gamil, Ahmed 360
Ganguly, Arna 72
Gani, Rafiqul 96, 186
Gao, Xiang 286, 287, 532
Gao, Xin 6

Gao, Yunfei ... 144, 410
Gao, Zhennan ... 342
Garrido, Javier ... 65
Gastelum-Arellan, Argel ... 295
Gautam, Gopalakrishnan Sai ... 102
Gayle, Andrew J ... 359
Geissbühler, Lukas ... 278
Geng, Zhigang ... 342
Gentimis, Thanassis ... 136
Gençer, Emre ... 46
Ghasemi, Shokoofeh ... 214
Ghoniem, Ahmed F. ... 176
Ghosh, Arpa ... 208
Giannikopoulos, Ioannis ... 42
Gigantino, Marco ... 206
Gilcher, Elise B. ... 104
Giles, Neil ... 20
Giridhar, Nishant V. ... 139
Giron, Bruna ... 291
Glenn, Gregory M. ... 381
Glezakou, Vassiliki-Alexandra ... 201
Gogar, Ravikumar ... 2
Goldstein, Hannah M. ... 220
Gomez, Laura A. ... 5
Gonzalez, Michael A. ... 236
Good, Philipp ... 278, 279
Gopalakrishnan, Saratram ... 127
Gorski, Christopher ... 430
Gorte, Raymond J. ... 188, 189
Goulas, Konstantinos ... 108
Gounaris, Chrysanthos E. ... 50, 139
Grabanski, Carol ... 518
Grader, Gideon S. ... 255, 376
Griffel, L. Michael ... 481, 486
Grozinger, Christina ... 502
Gu, Geun Ho ... 200
Gualdron, Diego Gomez ... 194, 409
Gualle, Arleth ... 65
Guban, Dorottya ... 533
Guillén-Gosálbez, Gonzalo ... 54, 241, 301
Gunukula, Sampath ... 210
Guo, Ping ... 306
Gupta, Geet ... 11
Gupta, Vijay ... 14, 141
Gutierrez, Jennifer A. ... 5
Guzman-Urbina, Alexander ... 354
Guzmán, José Luis ... 415
Gálvez, Maria E. ... 530
Gándara, Javier ... 65
Haile, Sossina M. ... 92, 95
Hajer, Ahmad Abu ... 125
Halabi, Rawan ... 376
Halbleib, Mike ... 482
Hamling, John A. ... 522
Han, Jeehoon ... 302
Han, Suyong ... 167
Hanumante, Neeraj ... 55
Hao, Yong ... 95, 101
Harhara, Ahmed ... 248
Harris, Kylee ... 71
Harrison, Andrew ... 166
Hart-Cooper, William ... 381
Hartley, Damon ... 481, 482, 486
Hasan, M M Faruque ... 248
Hashimoto, Jayni ... 181
Hatzell, Marta ... 195
Hau, Liwei ... 296
Haueter, Philipp ... 541
Haugen, Stefan ... 207
Haus, Moritz ... 212
Haussener, Sophia ... 174, 375, 377
Hawkins, Troy R. ... 79
Hawthorne, Steve ... 518
He, Guannan ... 46
He, Q. Peter ... 418
Head-Gordon, Martin ... 107
Hellwinckel, Chad ... 482
Henni, Amr ... 19
Heras, Jorge ... 246
Hermans, Ive ... 3
Hernandez, Rafael ... 528
Hernández-Romero, Ilse María ... 41
Heuberger, Clara F. ... 46
Heyden, Andreas ... 6, 10
Hietsoi, Oleksandr ... 252, 420
Hilliard, Matthew ... 418
Hilliard, Michael ... 482
Hinkle, John ... 385
Ho, Y ... 528
Hodge, David ... 384
Holmes, William ... 528
Holmes-Gentle, Isaac ... 377
Holsen, Thomas ... 113
Hong, Bo ... 436
Hong, Xiaodong ... 442
Hopkinson, David ... 432
Hortal-Sánchez, Isabel ... 147
Hoshino, Yuki ... 88
Hossain, Tasmin ... 486
Hostert, Jacob ... 414
Houck, Joseph ... 202
Howe, Gary ... 141
Hu, Bo ... 526
Hu, Yicheng ... 47, 156

Huang, Baogang 342
Huang, Chaoran 196
Huang, Haoyong 333
Huang, Nan 111
Huang, Runze 501
Huang, Wenyu 150
Huang, Yinlun 36
Huber, George W. 1, 3
Hughes, Ryan 17, 434
Huhnke, Raymond L. 131
Huo, Jiajie 251
Hwang, Gyeong S. 30, 444
Häusler, Johannes 240
Hülsey, Max J. 13
Idriss, Hicham 94
Immethun, Cheryl 417, 494
Ince, Nevin Gerek 22
Ingenhoven, Philip 179
Innocenti, Giada 110
Ioannou, Iasonas 241
Ishii, Hiroki 89
Isobe, Hiroyuki 88
Isowamwen, Faith 113
Ito, Takamasa 89
Iv, Thomas A. Knotts 20
Iyer, Jaisree K. 438
Jacobson, Tyler B. 127
Jafarian, Mehdi 179, 272
Jahromi, Hossein 420
Jain, Varsha 26
Jaiswal, Yash 428
Jameel, Kashif 300
Jeong, Joonsoo 305
Ji, Yichen 188, 189
Jiang, Zhen 29
Jiang, Zhihua 249, 495
Jiao, Feng 120
Jin, Baitang 142
Jin, Wencheng 294
Jin, Xin 7, 149, 151
Johnson, Paul 296
Jokiel, Michael 169
Jones, Christopher 143
Jones, Daniela 482, 486
Jordan, Alex 502
Joseph, Babu 415
Joshi, Chandni 158
Jovanovic, Goran 543
Ju, Bin 306
Jucar, John Paul Renzo 160
Judbarovski, David 84
Jung, Howoun 266
Jung, In-Ho 102
Jung, Yousung 200
Kabatnik, Christoph 74
Kai, Mototaka 88
Kakekhani, Arvin 29
Kakosimos, Konstantinos E. 268
Kale, Matthew 262
Kalidindi, Subhash 2
Kalyva, Agni E. 268
Kamlet, Olivia 414
Kancharla, Rahul Reddy 472
Kang, Dongseong 302
Kanhere, Sagar 384
Kao, Chingti 396
Karacasulu, Kaan 74
Karakitsios, Spyros 491
Karimi, Iftekhar A. 442
Karimi, Maryam 276
Karlen, Steven D. 130, 292
Karp, Eric M. 207
Kasick, Andrew 125
Kastner, James 239
Kathe, Mandar 509
Kauffman, Douglas R. 122
Kawamata, Miho 88
Kazantzis, Nikolaos 421
Keller, Michael 257
Kelley, Stephen S. 215
Kelly, Nolan 356
Kermenidou, Marianthi 491
Ketter, Jacob 274
Khajeh, Afsaneh 395
Khanal, Sami 242
Khanna, Vikas 59, 135, 502
Kibsgaard, Jakob 199
Kim, Dokyoum 425
Kim, Junhwan 425
Kim, Myung Jun 28
Kim, Semie 231
Kinchin, Christopher 71
King, Keith 180
Kiziltas, Alper 380
Klaas, Lena 533
Klamczynski, Artur P. 381
Klausner, James F. 180
Klinger, Jordan 294, 472, 480
Knipe, Jennifer M. 220
Koelling, Kurt W. 304
Kokossis, Antonis C. 136, 250, 489
Komarneni, Mallikharjuna R. 5
Kondori, Alireza 29
Kong, Jaemin 256

Konstandopoulos, Athanasios G. 268
Kontur, Wayne 130, 292
Koronaios, Peter 328
Kotamreddy, Goutham 17, 434
Kouroutzi, Maria 491
Kraus, George 251
Krempl, Kevin 199
Krishna, Siddarth 3
Kubic Jr., William L. 211
Kuch, Nathaniel 104
Kuech, Thomas F. 3
Kuhl, Kendra 124
Kulkarni, Soham 22
Kumar, Apurv 101, 182, 288
Kumar, Arun 263, 265
Kumar, Gaurav 187
Kumar, Manish 161
Kumar, Narendra 499
Kuo, Li-Jung 201
Kurup, Parthiv 86
Kury, Matthew W. 532
Kurz, Bethany 518
Kuznetsov, Anatoliy 187
Kwon, Oseok 302
La Plante, Erika Callagon 119, 232
Lachmann, Bruno 533
Lackner, Klaus S. 529
Lagos, Andrés S. 65, 65
Lagos, Tomas 59
Lail, Marty 14, 141
Lan, Kai 60, 215
Landman, Avigail 255, 376
Landázuri, Andrea C. 65
Lane, Todd W. 220
Langholtz, Matthew 482, 486
Lapkin, Alexei A. 82
Larson, Rebecca 156
Lauterbach, Jochen 12
Lawal, Ajibola 189
Leblanc, Brandon 528
Leclerc, Heather O. 525
Lee, Doyeon 231
Lee, Hyungwoo 425
Lee, Jacob 14
Lee, Jay H. 266
Lee, Joo-Youp 431
Lee, Kyuha 97, 242, 503
Lee, Nagyeong 305
Lee, Taek Soon 126
Lehr, Austin L. 152
Lemaire III, Peter 328
Lenert, Andrej 359
Lenzen, Manfred 45
Leonhardt, Branden 105
Li, Bingjia 357
Li, Fanxing 144, 410, 413
Li, Lifeng 182, 283
Li, Peiwen 360
Li, Rui 113
Li, Sha 101, 187, 288
Li, Shiguang 142
Li, Xian 289, 379
Li, Xinyi 284
Li, Xin 395
Li, Zhenghong 128
Liang, Chloe 79
Liang, Xinhua 142
Liao, Mochen 217
Lim, Kipil 102
Lim, Young-Il 231
Lin, Bosong 87
Lin, Hsi-Hsin 126
Lin, Yingqian 486
Lin, Yvonne 263
Lin, Zhexi 10
Lindsay, Michael 264
Linke, Patrick 138
Linke, Steffen 169
Lipinski, Wojciech 101, 182, 283, 288, 289, 379
Lipton, Jason 256
Liu, Bin 196
Liu, Bowei 4, 148
Liu, Jian 201
Liu, Jiefeng 149
Liu, Jinlu 330
Liu, Junchen 144, 410
Liu, Kunlei 121, 222
Liu, Mengyuan 149, 151
Liu, Weilin 342
Liu, Yong 426
Lively, Ryan P. 75
Liò, Pietro 82
Lobato-Peralta, Diego R. 285
Lombardini, Leonardo 78
Long, Jeffrey R. 434
Lou, Jiahui 95
Lougou, Bachirou Guene 172
Loutzenhiser, Peter 532, 536
Lu, Yingxi 128
Ludwig, Thomas 198
Luna, F. Murilo T. 293
Lykokanellos, Filopoimin 489
Lynam, Joan G. 499
Lynd, Lee R. 127

López-Bautista, Luis Angel 441
Løge, Isaac A. 523
Ma, Hanyu 197
Ma, Ting 284
Ma, Yao 431
Macaggi, Martina 155
Macedo, Beatriz L. B. C. 355
Machalek, Derek 514
Maclean, Stephen A. 256
Macqueen, Blake 12
Maddipudi, Bharath 202, 205
Madugula, Adhish 16
Maffia, Gennaro 392
Mahajan, Kanwal 447, 517, 521
Mahal, Jushwin 358
Majumdar, Arun 102
Mallapragada, Dharik 46
Malmali, Mahdi 87
Malviya, Shubham 32
Maranas, Costas 127
Maravelias, Christos T. 1, 40, 73, 269
Margolis, Robert 86
Marshall, Bennett D. 23
Martin, Irene Mas 72
Martín, Antonio J. 241
Martín, Mariano 47, 90, 137, 246, 297
Martín, Ángel Galán 54
Martín-Hernández, Edgar 47, 297
Marín, Danitza 267
Masanet, Eric 86
Mathias, Paul M. 21
Mathur, Nehika 488
Matsumura, Yukihiko 254
Matuszewski, Michael S. 17, 434
McBride, Kevin 169
McCaffrey, Zachariah 381
McCaskill, Andrew 274
McCormick, Alon V. 262
McCormick, Lisa 276
McEvoy, Joanne Gamage 81
McGregor, James 527
McMillan, Colin 86
Medford, Andrew 9, 110
Medlin, James 110
Meereboer, Kjeld 388
Mehdipour, Iman 118
Mei, Haiyan 306
Mele, Fernando Daniel 301
Mendez-Perez, Daniel 126
Meyer, David E. 236
Miao, Yu 271
Mielewski, Debbie F. 380
Migliore, Calogero 258
Mikhail, Maria 530
Miller, David 518
Miller, William 261
Millican, Samantha L. 94
Misra, Manju 388
Mitch, William 111
Mitsos, Alexander 74
Miyake, Naomi 190
Mlayah, Bouchra Benjelloun 250
Mobley, Paul 14
Mohammad, Roni 481
Mohan, Babu Cadiam 263, 265
Mohanty, Amar K. 383, 388
Moni, Sheikh 225
Monnerie, Nathalie 535
Monteiro, Ana 434
Montorfano, Davide 279
Moore, Thomas 438
Mora, Tomas 447
Morais, Ana R. C. 8
Morales, Emilia 65
More, Ajinkya 495
Moreno, Daniel 121
Mortensen, Peter M. 228
Morton, Frank C. 145
Motavaf, Bita 524
Mounfield III, William P. 8
Mountraki, Aikaterini D. 250
Muhich, Christopher L. 101, 181
Mukherjee, Rajib 520
Muleja, Adolph A. 114
Murcia, Diana Carolina Figueroa 523
Murdoch, Lawrence 296
Musgrave, Charles B. 94
Mutchek, Michele 225
Myer, Larry 517
Nair, Sankar 75
Najmi, Sean 9, 110
Nakagawa, Seiji 296
Nam, Joonhyun 102
Narayan, Ramani 382
Narita, Keisuke 88
Nathan, G. J. 179, 272
Navarrete, Danny 65
Nazari, Rouzbeh 276
Neal, Luke 413
Neerup, Randi 523
Neurock, Matthew 103
Ngo, Duong T. 106
Nguyen, Nhu "ty" 532, 536
Nguyen, Quang 481

Nguyen, Riley 528
Nguyen-Phan, Thuy-Duong 122
Nikolic, Heather 222
Nisal, Apoorva 55
Noguera, Daniel R. 130, 292
Noh, Juhwan 200
Norskov, Jens 198, 199
Nápoles-Rivera, Fabricio 41
O'Doherty, Eleanor 527
O'Neill, Eric 40
Ogale, Amod 384
Ogasawara, Kazuto 226
Ognier, Stéphanie 530
Oh, Seokyoung 425
Ohno, Hajime 354
Ojha, Deepak 262
Olsen, Nora L. 129
Olson, Daniel 127
Omell, Benjamin P. 17, 434
Omodolor, Ibeh S. 2
Omosebi, Ayokunle 121
Orejuela, Lourdes 65
Orgeron, Cory 528
Orjuela, Alvaro 299
Orlov, Alexander 264
Orts, William J. 381
Oshinowo, Lanre 258
Osma, Johann F. 506
Ostroski, Anais 59
Otanicar, Todd 257
Ou, Longwen 215
Ouchi, Kakeru 354
Overdeep, Kyle 264
O'Dowd, William 517
Pacheco-Catalán, Daniella E. 285
Padgett, Kevin 135
Pagan-Torres, Yomaira J. 12
Pal, Akhilesh K. 388
Pal, Sunderlal 428
Palkovits, Regina 212
Palys, Matthew 203
Parambathu, Arjun Valiya 330
Park, Ah-Hyung Alissa 80
Park, Hoyoung 302
Park, Sang Jae 143
Park, Sunkyu 215
Parker, Surya T. 434
Parkhi, Amod 249
Parvathikar, Sameer 352
Pasel, Joachim 240
Patankar, Aniket 176
Patch, Harland 502
Patil, Santosh 522
Patil, Vivek 385
Paulson, Joel 244
Paz, Renata 65
Pearlman, Howard 516
Pedersen, Jakob B. 199
Peng, Peng 80
Peng, Xinyue 269
Penman, Jennifer 234, 259
Pennington, Ashley M. 237
Perez, Miguel 130, 292
Peters, Ralf 240
Peters, Robert 112, 276
Petersen, Stephan 533
Petrasch, Joerg 180
Petrecca, Michael 509
Pfromm, Peter H. 196
Philip, Firuz 19
Phung, Truc 108
Pillai, Hemanth S. 37
Pillai, Shylesh 253
Pilon, Laurent 232
Pistikopoulos, Efstratios N. 78
Podolak, Kyle 251
Pontes, Ricardo F. F. 355
Pope, Roger 14
Pottas, Johannes J. 182
Pourhashem, Ghasideh 214
Powell, Kody M. 514
Prajapati, Aditya 443
Prasher, Ravi 34
Preisner, Nicole 533
Prentice, Dale 119
Prokopyev, Oleg A. 59
Psarras, Peter C. 194
Psycha, Melina 489
Pye, John 283
Pérez, José A. 267
Pérez-Ramírez, Javier 241
Qaramaleki, Saeed Vahed 422
Qi, Liang 191
Qian, Xin 95
Quiroz-Ramírez Sr., Juan José 295
Raghuvanshi, Keshav 167
Rahmatian, Nima 180
Rai, Neeraj 26
Rajput, Harsheen 181
Ralph, John 130, 292
Ramasamy, Karthikeyan K. 253
Ramesh, Narayan 233
Ramsurn, Hema 257
Ramírez-Cárdenas, Lucía 65

Randhir, Kelvin 180
Rapp, Vi 261
Rappe, Andrew M. 29
Rashid, Suliman 146, 412
Rasmussen, Mathew 110
Rautenbach, Marené 228
Rayer, Aravind V. 141
Redd, Mark E. 20
Redinbo, Matthew R. 419
Ren, Chris 296
Ren, Limin 187
Renner, Julie 414
Resasco, Daniel E. 106
Resnik, Kevin P. 432
Restrepo-Florez, Juan Manuel 73
Revellame, Emmanuel 528
Reza, Toufiq 500
Ridge, Claron 264
Rivera-Goyco, Christian G. 147
Roberts, Diarmid 48
Roberts, Nathan 418
Roeb, Martin 533, 535
Rojas, Jimmy 102
Roling, Luke 11, 251
Román-Leshkov, Yuriy 8
Root, Thatcher W. 104
Rosenstiel, Andreas 535
Rosenthal, Alex 211
Rothman, Rachael H. 505
Rothschild, Avner 255, 376
Roy, Nitin 38
Ruiz-Mercado, Gerardo 47, 178, 236
Rutz, David 278, 541
Ryan, Bradley J. 251
Röhr, Jason A. 256
Saboe, Patrick 207
Saccoccio, Mattia 199
Saha, Nepu 500
Saha, Rajib 417, 494
Salas-González, Isai 419
Samineni, Laxmicharan 161
Samsatli, Sheila 76, 234, 259
Samu, Nicole 482
Sanchez-Ramirez, Eduardo 295
Sankaranarayanapilla, Shylesh 107
Sant, Gaurav 118, 119, 232
Santos, Ronaldo 291
Santosa, Daniel 252
Sarigiannis, Dimosthenis 491
Sattler, Christian 533, 535
Sautet, Philippe 13
Savage, Phillip E. 389, 393, 524
Savelski, Mariano 152
Saw, Woei L. 179
Sayama, Shogo 226
Scheffczyk, Jan D. 74
Schneider, William F. 197
Schoeneberger, Carrie 86
Schoepke, Andrew R. 129
Schäfer, Florian 182
Schäppi, Remo 541
Seaton, Glenn 20
Seay, Jeffrey 158
Segovia-Hernández, Juan Gabriel 295
Seiichi, Kunitomi 226
Sener, Canan 130, 130, 292
Seo, Kyeongjun 436
Seo, Myung Won 231
Seo, Yongdeuk 425
Serena, Teo Lay Ming 265
Seyfaee, Ahmad 179, 272
Shah, Utkarsh 244
Shahbazi, Abloghasem 395
Shamsuzzaman, Farooq 442, 442
Shan, Nannan 196
Shang, Zeyu 142
Shanks, Brent H. 126, 251
Shao, Yuyan 201
Shapel, Sarah G. 199
Sharma, Pradeepkumar 352
Shastri, Yogendra 55
Shea-Andersh, Maureen A. 129
Sheaffer, Paul 261
Sheikhi, Amir 386
Shen, Chenghsien 396
Shen, Jian-Ping 141
Shende, Anuradha 202, 205
Shende, Rajesh 202, 205
Shi, Wei 432
Shiflett, Mark B. 164, 303, 424
Shimoda, Sho 232
Shin, Dongil Peter 305
Shonnard, David 213
Shuai, Yong 172
Sibi, Mukund 214
Sievers, Carsten 9, 110
Silva, Manoj 62
Simoneaux, Sarah 528
Simonetti, Dante 119, 224
Singh, Shweta 45, 134, 155, 219, 260, 488
Sizmann, Andreas 537
Skone, Timothy 225
Skouteris, Alkiviadis 42
Slater, Stewart 152

Slininger, Patricia J. 129
Smith, Dr. Elisa Barney 472
Smith, Eli 135
Smith, Raymond 178, 236
Smith, Ryan G. 208
Smith, Steve 518
Smith, William A. 472
Song, Yi 333
Song, Yurou 7
Sorensen, James A. 518, 522
Stadtherr, Mark A. 42, 436
Stallings, Justin 499
Stanford, John P. 208
Statt, Michael J. 199
Stavitski, Eli 110
Stechel, Ellen B. 98, 175, 286, 287, 532
Steinfeld, Aldo 206, 541
Stelly, Cody 528
Stengel, Jake P. 152
Stevens, Miriam 260
Steward, Darlene 71
Stolaroff, Joshuah K. 438
Stolten, Detlef 240
Stone, Kevin 102
Stone, Michael 8
Stratidakis, Antonis 491
Su, Yanchun 342
Su, Zihang 414
Subramanya, Seshasayee Mahadevan 389, 393
Suda, Toshiyuki 89
Suh, Chungho 50
Sullivan, Mary 447
Sultana, Al Ibtida 500
Sun, Geng 13
Sun, Qiming 289, 379
Sundmacher, Kai 169
Suter, Clemens 377
Sutherland, John 488
Svadlenak, Scott Ryan 108
Szczuka, Aleksandra 111
Sánchez, Antonio 90
Taifouris, Manuel 137
Tan, Eric 71, 178, 211
Tan, Heping 172
Tang, Christina 166
Tanner, Ralph S. 131
Tanthana, Jak 14
Taylor, André D. 256
Teixeira, Andrew R 525
Tembhurne, Saurabh 377
Ternes, Mary Ellen 183, 275
Thagard, Selma Mededovic 113
Thakker, Vyom 132
Thaore, Vaishali B. 442
Thibault, Yves 81
Thies, Mark 384
Thompson, David N. 481, 486
Thompson, Jesse G. 121
Thompson, Vicki S. 480
Thomsen, Kaj 523
Thornburg, Jonathan 14
Thunuguntla, Rahul 131
Tian, Yuan 413
Tian, Zhao 179
Tian, Zhenyu 95
Timko, Michael T. 421, 525
Tindall, Graham W. 384
Tompsett, Geoffrey 525
Toney, Michael 102
Tong, Xiao 264
Tongwen, Jiang 334
Toriello, V. A. Suárez 295
Torres, Lennard 381
Tran, Dat T. 210
Tran, Justin T. 93, 94
Tran-Gyamfi, Mary Bao 220
Trembly, Jason 125, 513
Trottier, Ryan 94
Trávez, Alina L. 65
Tsapatsis, Michael 187
Tsay, Calvin 436
Tshangana, Charmaine S. 114
Tsilomelekis, George 128
Tulus, Victor 54
Umana, German 130, 292
Underwood, M. Kylee 521
Usadi, Adam K. 442
Valentino, Lauren 211
Valera-Medina, Agustin 52
Vasefi, Seyedehsan 239
Vekaria, Ashil 128
Velegol, Stephanie B. 161
Vella, Joseph R. 23
Velraj, Samgopiraj 125
Venkataraman, Mahesh B. 101
Verma, Ankit 164, 303, 424
Vernaza, María Gabriela 65
Vesborg, Peter C. K. 199
Via, Brian 385
Viamajala, Sridhar 2
Viell, Jörn 74
Villafán-Vidales, Heidi I. 285
Villarreal, Juan S. 65
Vlachos, Dionisios G. 109, 188

Vodovotz, Yael 304
Von Jouanne, Annette 271
Vunnava, Venkata Sai Gargeya 45, 134, 155, 219
Walker, Theodore 104
Walton, William G. 419
Walz, Sarah A. 2
Wan, Houpeng 396
Wan, Weiming 10
Wang, Bin 106, 150
Wang, Bo 182, 283
Wang, Haotian 123
Wang, Huamin 252
Wang, Jianlong 116
Wang, Jinyao 7, 149, 151
Wang, Jin 418
Wang, Lijun 395
Wang, Lin-Wang 27
Wang, Peiyu 284
Wang, Qiuwang 284
Wang, Siwen 4, 148
Wang, Xiaonan 128, 442
Wang, Xijun 410
Wang, Xi 126
Wang, Ziwei 103
Warren, Kent J. 93
Warren, Quinta Nwanosike 235
Watkins, David W. 213
Webb, Erin 482
Weber, Jana M. 82
Webster, Dean 214
Wei, Xie 334
Weimer, Alan W. 93, 94
West, Alan 80
Westmoreland, Phillip R. 413
Wheeler Sr., Jonathan 301
Wheeler, Vincent M. 101, 182, 288
Wiesner, Theodore 87
Wilcox, Jennifer 194
Wilding, W. Vincent 20
Wiley, Benjamin 28
Wilkins, Mark 494
Wilson, Woodrow 26
Winter, Benedikt 212
Woltmann, Friederike 240
Wong, Michael S. 510
Wood, Delilah F. 381
Woodley, John M. 165
Worthington, Matthew A. 220
Wright, Mark Mba 72
Wu, Jianfa 333
Wu, Xiaoyu 176
Xi, William 86
Xia, Qi 149
Xia, Yidong 294
Xiao, Hai 296
Xie, Chuang 426
Xin, Hongliang 37
Xu, Jinhai 284
Xu, Mai 175
Xueming, Dong 207
Yamamoto, Seiji 226
Yan, Ning 13, 238, 289
Yan, Yu 106, 150
Yancey, Neal 480
Yang, Chaohe 7, 149, 151
Yang, Haotian 264
Yang, Jun 335, 342
Yang, Song 283
Yao, Yuan 60, 215, 217, 501
Yelvington, Paul 33
Yenkie, Kirti M. 152
Yokochi, Alexandre 271
Yokoyama, Hiroki 254
Yong, Wang 334
Yoon, Bohak 30, 444
You, Kyung-Eun 10
Yu, Xiao 149
Zappi, Mark 528
Zappitelli, Joseph 135
Zavala, Victor M. 41, 47, 156
Zavattoni, Simone A. 279
Zeevi, Michael 166
Zeng, Fanhua 306, 335, 342
Zhai, Shang 102
Zhang, Bin 284
Zhang, Bo 395
Zhang, Guangyu 151
Zhang, Haoran 128
Zhang, Jingyi 86
Zhang, Juwei 89
Zhang, Lifeng 3
Zhang, Maolin 306
Zhang, Quanxing 7
Zhang, Wenxiang 7, 149
Zhang, Xiao 492, 496
Zhang, Yanfei 192
Zhang, Yizhong 306
Zhang, Yuxuan 284
Zhang, Zhong 111
Zhao, Jiupeng 172
Zhao, Mingyu 7
Zhao, Shengxian 333
Zhao, Xiaoying 304
Zheng, Fan 27

Zheng, Qinghe 141
Zhu, Hongda 193
Zhu, Jiaqi 356
Zhu, Ming 39
Zhu, Ran 4, 148
Zhu, Yan 426
Zhu, Yunting...356
Zuber, Adam 128
Zygourakis, Kyriacos...510